PATH INTEGRALS AND HAMILTONIANS

Providing a pedagogical introduction to the essential principles of path integrals and Hamiltonians, this book describes cutting-edge quantum mathematical techniques applicable to a vast range of fields, from quantum mechanics, solid state physics, statistical mechanics, quantum field theory, and superstring theory to financial modeling, polymers, biology, chemistry, and quantum finance.

Eschewing use of the Schrödinger equation, the powerful and flexible combination of Hamiltonian operators and path integrals is used to study a range of different quantum and classical random systems, succinctly demonstrating the interplay between a system's path integral, state space, and Hamiltonian. With a practical emphasis on the methodological and mathematical aspects of each derivation, this is a perfect introduction to these versatile mathematical methods, suitable for researchers and graduate students in physics, mathematical finance, and engineering.

Belal E. Baaquie is a Professor of Physics at the National University of Singapore, specializing in quantum field theory, quantum mathematics, and quantum finance. He is the author of *Quantum Finance* (2004), *Interest Rates and Coupon Bonds in Quantum Finance* (2009), and *The Theoretical Foundations of Quantum Mechanics* (2013), and co-author of *Exploring Integrated Science* (2010).

PATH INTEGRALS AND HAMILTONIANS

Principles and Methods

BELAL E. BAAQUIE
National University of Singapore

CAMBRIDGE
UNIVERSITY PRESS

CAMBRIDGE
UNIVERSITY PRESS

University Printing House, Cambridge CB2 8BS, United Kingdom

Cambridge University Press is part of the University of Cambridge.

It furthers the University's mission by disseminating knowledge in the pursuit of education, learning and research at the highest international levels of excellence.

www.cambridge.org
Information on this title: www.cambridge.org/9781107009790

First published 2014

A catalogue record for this publication is available from the British Library

Library of Congress Cataloguing in Publication data

Baaquie, B. E.
Path integrals and Hamiltonians : principles and methods / Belal E. Baaquie, National University of Singapore.
pages cm
ISBN 978-1-107-00979-0 (hardback)
1. Differential operators. 2. Differential equations. 3. Hamiltonian operator. 4. Path integrals. I. Title.
QC174.12.B24 2014
530.1201′51539–dc23 2013044323

ISBN 978-1-107-00979-0 Hardback

This book is dedicated to the memory of
Kenneth Geddes Wilson (1936-2013).
Intellectual giant, visionary scientist,
exceptional educator, altruistic spirit.

Contents

Preface

Quantum mechanics is undoubtedly one of the most accurate and important scientific theories in the history of science. The theoretical foundations of quantum mechanics have been discussed in depth in Baaquie (2013e), where the main focus is on the interpretation of the mathematical symbols of quantum mechanics and on its enigmatic superstructure. In contrast, the main focus of this book is on the mathematics of path integral quantum mechanics.

The traditional approach to quantum mechanics has been to study the Schrödinger equation, one of the cornerstones of quantum mechanics, and which is a special case of partial differential equations. Needless to say, the study of the Schrödinger equation continues to be a central task of quantum mechanics, yielding a steady stream of new and valuable results.

Interestingly enough, there are two other formulations of quantum mechanics, namely the operator approach of Heisenberg and the path integral approach of Dirac–Feynman, that provide a mathematical framework which is independent of the Schrödinger equation. In this book, the Schrödinger equation is never directly solved; instead the Hamiltonian operator is analyzed and path integrals for different quantum and classical random systems are studied to gain an understanding of quantum mathematics.

I became aware of path integrals when I was a graduate student, and what intrigued me most was the novelty, flexibility and versatility of their theoretical and mathematical framework. I have spent most of my research years in exploring and employing this framework.

Path integration is a natural generalization of integral calculus and is essentially the integral calculus of infinitely many variables, also called functional integration. There is, however, a fundamental feature of path integration that sets it apart from functional integration, namely the role played by the Hamiltonian in the formalism. All the path integrals discussed in this book have an underlying linear structure that is encoded in the Hamiltonian operator and its linear vector state space. It is this combination of the path integral and its underlying Hamiltonian that provides a powerful and flexible mathematical machinery that can address a vast variety and range of diverse problems. Path integration can also address systems that do not have a Hamiltonian and these systems are not studied. Instead, topics have been chosen that can demonstrate the interplay of the system's path integral, state space, and Hamiltonian.

The Hamiltonian operator and the mathematical formalism of path integration make them eminently suitable for describing quantum indeterminacy as well as classical randomness. In two chapters of the book, namely Chapter 7 on stochastic processes and Chapter 17 on compact degrees of freedom, path integrals are applied to classical stochastic and random systems. The rest of the chapters analyze systems that have quantum indeterminacy.

The range and depth of subjects that come under the sway of path integrals are unified by a common thread, which is the mathematics of path integrals. Problems seemingly unrelated to indeterminacy such as the classification of knots and links or the mathematical properties of manifolds have been solved using path integration. The applications of path integrals are almost as vast as calculus, ranging from finance, polymers, biology, and chemistry to quantum mechanics, solid state physics, statistical mechanics, quantum field theory, superstring theory, and all the way to pure mathematics. The concepts and theoretical underpinnings of quantum mechanics lead to a whole set of new mathematical ideas and have given rise to the subject of *quantum mathematics*.

The ground-breaking and pioneering book by Feynman and Hibbs (1965) laid the foundation for the study of path integrals in quantum mechanics and is always worth reading. More recent books such as those by Kleinert (1990) and Zinn-Justin (2005) discuss many important aspects of path integration and cover a wide range of applications. Given the complex theoretical and mathematical nature of the subject, no single book can conceivably cover the gamut of worthwhile topics that appear in the study of path integration and there is always a need for books that break new ground. The topics chosen in this book have a minimal overlap with other books on path integrals.

A major field of theoretical physics that is based on path integrals is quantum field theory, which includes the Standard Model of particles and forces. The study of quantum field theory leads to the concept of nonlinear gauge fields and to the concept of renormalization, both of which are beyond the scope this book.

The purpose of the book is to provide a pedagogical introduction to the essential principles of path integrals and of Hamiltonians; for this reason many examples have been worked out in full detail so as to elucidate some of the varied methods and techniques that have proven useful in actually performing path integrations. The emphasis in all the derivations is on the methodological and mathematical aspect of the problem – with matters of interpretation being discussed only in passing. Starting from the simplest examples, the various chapters lay the ground work for analyzing more advanced topics. The book provides an introduction to the foundations of path integral quantum mechanics and is a primer to the techniques and methods employed in the study of quantum finance, as formulated by Baaquie (2004) and Baaquie (2010).

Acknowledgements

I would like to acknowledge and express my thanks to many outstanding teachers, scholars, and researchers whose work motivated me to study path integral quantum mechanics and to grapple with its mathematical formalism.

I had the singular privilege of doing my Ph.D. thesis under the guidance of Nobel Laureate Kenneth G. Wilson; his visionary conception of quantum mechanics and of quantum field theory – rooted in the path integral – greatly enlightened and inspired me, and continues to do so today. As an undergraduate I had the honor of meeting and conversing a number of times with Richard P. Feynman, the legendary discoverer of the path integral, and this left a permanent impression on me.

I thank Frederick H. Willeboordse for his consistent support and Wang Qinghai, Andreas Keil, Tang Pan, Duxin, and Cao Yang for many helpful discussions.

I thank my wife Najma for being a wonderful companion and for her uplifting approach to family and professional life. I thank my precious family members Arzish, Farah, and Tazkiah for their delightful company and warm encouragement. They have made this book possible.

I am deeply indebted to my late father Muhammad Abdul Baaquie for being a life long source of encouragement and whose virtuous qualities continue to be a beacon of inspiration.

1

Synopsis

This book studies the mathematical aspect of path integrals and Hamiltonians – which emerge from the formulation of quantum mechanics. The theoretical framework of quantum mechanics provides the mathematical tools for studying both *quantum indeterminacy* and *classical randomness*. Many problems arising in quantum mechanics as well as in vastly different fields such as finance and economics can be addressed by the mathematics of quantum mechanics, or *quantum mathematics* in short. All the topics and subjects in the various chapters have been specifically chosen to illustrate the structure of quantum mathematics, and are not tied to any specific discipline, be it quantum mechanics or stochastic systems.

The book is divided into the following six parts, in accordance with the *Chapter dependency flowchart* given below.

- Part one addresses the *Fundamental principles* of path integrals and (Hamiltonian) operators and consists of five chapters. Chapter 2 is on the *Mathematical structure of quantum mechanics* and introduces the mathematical framework that emerges from the quantum principle. Chapters 3 to 6 discuss the mathematical pillars of quantum mathematics, starting from the Feynman path integral, summarizing Hamiltonian mechanics and introducing path integral quantization.
- Part two is on *Stochastic processes*. Stochastic systems are dissipative and are shown to be effectively modeled by the path integral. Chapter 7 is focused on the application of quantum mathematics to classical random systems and to stochastic processes.
- Part three discusses *Discrete degrees of freedom*. Chapters 8 and 9 discuss the simplest quantum mechanical degree of freedom, namely the double valued Ising spin. The Ising model is discussed in some detail as this model contains all the essential ideas that unfold later for more complex degrees of freedom. The general properties of path integrals and Hamiltonians are discussed in the context of the Ising spin. Chapter 10 on *Fermions* introduces a degree of freedom

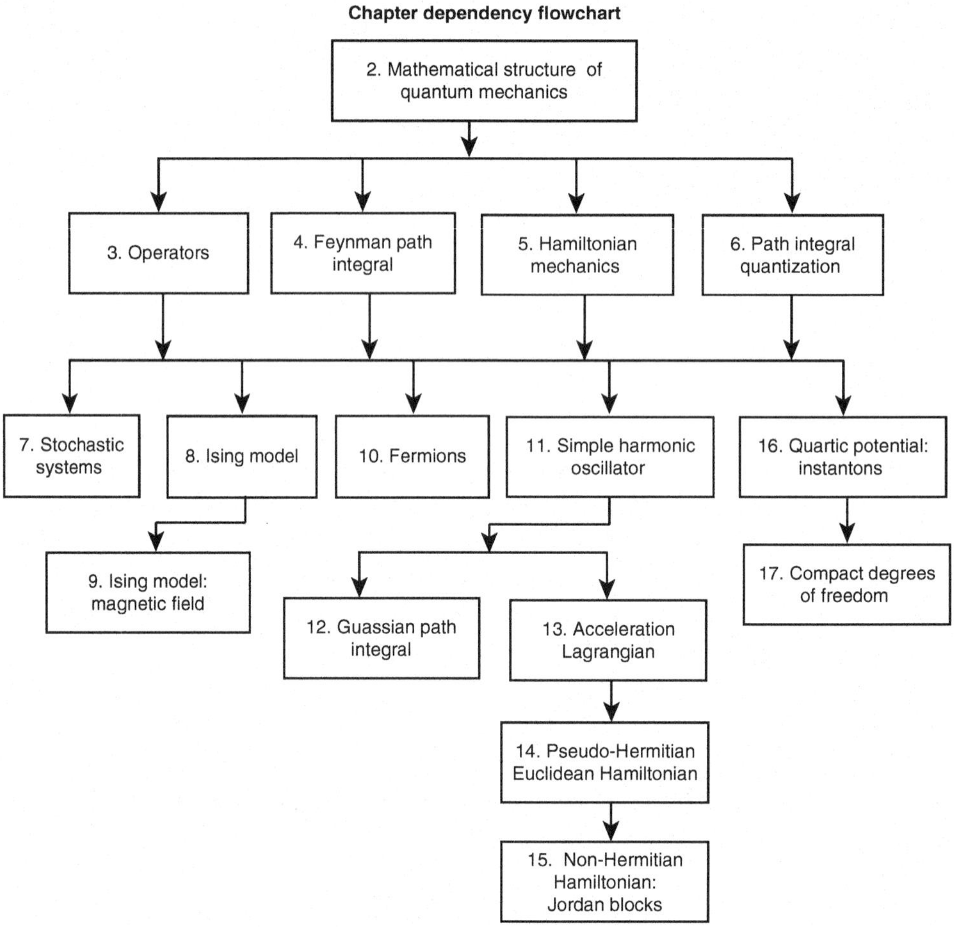

that is essentially discrete – but is represented by fermionic variables that are distinct from real variables. The calculus of fermions, including the key structures of quantum mathematics such as the Hamiltonian, state space, and path integrals are discussed in some detail.

- Part four covers *Quadratic path integrals*. Chapter 11 is on the simple harmonic oscillator – one of the prime exemplars of quantum mechanics – and it is studied using both the Hamiltonian and path integral approach. In Chapter 12 different types of Gaussian path integrals are evaluated using techniques that are useful for analyzing and solving path integrals.
- Part five is on the *Acceleration action*. An action with an acceleration term is defined for Euclidean time and is shown to have a novel structure not present in usual quantum mechanics. In Chapter 13, the Lagrangian and path integral are

analyzed and shown to be equivalent to a constrained system. The Hamiltonian is obtained using the Dirac constraint method. In Chapter 14, the acceleration Hamiltonian is shown to be pseudo-Hermitian and its state space and propagator are derived. Chapter 15 examines a critical point of the acceleration action and the Hamiltonian is shown to be essentially non-Hermitian, being block diagonal and with each block being a Jordan block.

- Part six is on *Nonlinear path integrals*. Chapter 16 studies the nonlinear quartic Lagrangian to illustrate the qualitatively new features that nonlinear path integrals exhibit. The double well potential is studied in some detail as an exemplar of nonlinear path integrals that can be analyzed using the semi-classical expansion. And lastly, in Chapter 17 degrees of freedom are analyzed that take values in a compact manifold; these systems have a nonlinearity that arises from the nature of the degree of freedom itself – rather than from a nonlinear piece in the Lagrangian. Semi-classical expansions of the path integral about multiple classical solutions, classified by a winding number and path integrals on curved manifolds, are briefly touched upon.

Part one

Fundamental principles

2

The mathematical structure of quantum mechanics

An examination of the postulates of quantum mechanics reveals a number of fundamental mathematical constructs that form its theoretical underpinnings. Many of the results that are summarized in this Chapter will only become clear after reading the rest of the book and a re-reading may be in order.

The dynamical variables of classical mechanics are superseded by the quantum degree of freedom. An exhaustive and complete description of the indeterminate degree of freedom is given by its state function, which is an element of a state space that, in general, is an infinite-dimensional linear vector space. The properties of the indeterminate degree of freedom are extracted from its state vector by the linear action of operators representing experimentally observable quantities. Repeated applications of the operators on the state function yield the average value of the operator for the state [Baaquie (2013e)].

The conceptual framework of quantum mechanics is discussed in Section 2.1. The concepts of degree of freedom, state space and operators are briefly reviewed in Sections 2.3–2.5. Three distinct formulations of quantum mechanics emerge from the superstructure of quantum mechanics and these are briefly summarized in Sections 2.7–2.9.

2.1 The Copenhagen quantum postulate

The Copenhagen interpretation of quantum mechanics, pioneered by Niels Bohr and Werner Heisenberg, provides a conceptual framework for the interpretation of the mathematical constructs of quantum mechanics and is the standard interpretation that is followed by the majority of practicing physicists [Stapp (1963), Dirac (1999)].

The Copenhagen interpretation is not universally accepted by the physics community, with many alternative explanations being proposed for understanding quantum mechanics [Baaquie (2013e)]. Instead of entering this debate, this book is based on the Copenhagen interpretation, which can be summarized by the following postulates:

- The quantum entity consists of its degree of freedom $\mathcal{F}$ and its state vector $\psi(t, \mathcal{F})$, where t is a real number parameterizing time. The foundation of the quantum entity is its degree of freedom, which takes a range of values and constitutes a space $\mathcal{F}$. The quantum degree of freedom is completely described by the quantum state $\psi(t, \mathcal{F})$, a complex valued function of the degree of freedom that is an element of state space $\mathcal{V}(\mathcal{F})$.
- The quantum entity is an *inseparable pair*, namely, the degree of freedom and its state vector.
- All physically observable quantities are obtained by applying Hermitian operators $\mathcal{O}(\mathcal{F})$ on the state $\psi(t, \mathcal{F})$.
- Experimental observations collapse the quantum state and repeated observations yield $E_\psi[\mathcal{O}(\mathcal{F})]$, which is the expectation value of the operator $\mathcal{O}(\mathcal{F})$ for the state $\psi(t, \mathcal{F})$.
- The Schrödinger equation determines the time dependence of the state vector, namely of $\psi(t, \mathcal{F})$, but does not determine the process of measurement.

It needs to be emphasized that the state vector $\psi(t, \mathcal{F})$ provides only statistical information about the quantum entity; the result of any particular experiment is impossible to predict.[1]

The organization of the theoretical superstructure of quantum mechanics is shown in Figure 2.1.

The quantum state $\psi(t, \mathcal{F})$ is a complex number that describes the degree of freedom and is more fundamental than the observed probabilities, which are always real positive numbers. The scheme of assigning expectation values to operators, such as $E_\psi[\mathcal{O}(\mathcal{F})]$, leads to a generalization of classical probability to quantum probability and is discussed in detail in Baaquie (2013e).

To give a concrete realization of the Copenhagen quantum postulate, consider a quantum particle moving in one dimension; the degree of freedom is the real line, namely $\mathcal{F} = \Re = \{x | x \in (-\infty, +\infty)\}$ with state $\psi(t, \Re)$. Consider the position operator $\mathcal{O}(x)$;[2] a measurement projects the state to a point $x \in \Re$ and collapses the quantum state to yield, after repeated measurements

$$P(t, x) \equiv E_\psi[\mathcal{O}(x)] = |\psi(t, x)|^2, \quad P(t, x) > 0, \quad \int_{-\infty}^{+\infty} dx P(t, x) = 1. \quad (2.1)$$

[1] There are special quantum states called eigenstates for which one can exactly predict the outcome of some experiments. But for even this special case the degree of freedom is indeterminate and can never be directly observed.

[2] The position projection operator $\mathcal{O}(x) = |x\rangle\langle x|$; see Chapter 3.

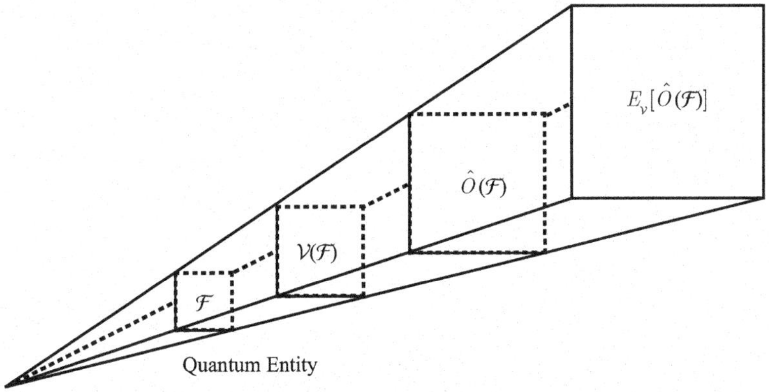

Figure 2.1 The theoretical superstructure of quantum mechanics; the quantum entity is constituted by the degree of freedom $\mathcal{F}$ and its state vector, which is an element of state space $\mathcal{V}(\mathcal{F})$; operators $\mathcal{O}(\mathcal{F})$ act on the state vector to extract information about the degree of freedom and lead to the final result $E_{\mathcal{V}}[\mathcal{O}(\mathcal{F})]$; only the final result, which is furthest from the quantum entity, is empirically observed.

Note from Eq. 2.1 that $P(t, x)$ obeys all the requirements to be interpreted as a probability distribution. A complete description of a quantum system requires specifying the probability $P(t, x)$ for all the possible projection operators of the quantum system. For a quantum particle in space, these are labeled by the different positions $x \in [-\infty, +\infty]$.

The position of the quantum particle is indeterminate and $P(t, x) = |\psi(t, x)|^2$ is the probability of the state vector collapsing at time t and at $\mathcal{O}(x)$ – the projection operator at position x. The moment that the state $\psi(t, \mathfrak{R})$ is observed at *specific* projection operator $\mathcal{O}(x)$, the state $\psi(t, \mathfrak{R})$ instantaneously becomes zero everywhere else. The transition from $\psi(t, \mathfrak{R})$ to $|\psi(t, x)|^2$ is an expression of the collapse of the quantum state. It needs to be emphasized that no classical wave undergoes a collapse on being observed; the collapse of the state $\psi(t, \mathfrak{R})$ is a purely quantum phenomenon.

The pioneers of quantum mechanics termed it as "wave mechanics" since the Newtonian description of the particle by its trajectory $x(t)$ was replaced by the state $\psi(t, \mathfrak{R})$ that looked like a classical wave that is spread over (all of) space $\mathfrak{R}$. Hence the term "wave function" is used by many physicists for denoting $\psi(t, \mathfrak{R})$.

The state $\psi(t, \mathcal{F})$ of a quantum particle is *not* a classical wave; rather, the only thing it has in common with a classical wave is that it is sometimes spread over space. However, there are quantum states that are *not spread over space*. For example, the up and down spin states of a quantum particle exist at a single point; such quantum states are described by a state that has no dependence on space and hence is *not* spread over space.

In the text, the terms *state, quantum state, state function,* or *state vector* are henceforth used for $\psi(t, \mathcal{F})$, as these are more precise terms than the term wave function.

The result given in Eq. 2.1 is an expression of the great discovery of quantum theory, namely, that *behind what is directly observed* – the outcome of experiments from which one can compute the probabilities $P(t, x) = |\psi(t, x)|^2$ – there lies an *unobservable world of the probability amplitude* that is fully described by the quantum state $\psi(t, \mathcal{F})$.

2.2 The superstructure of quantum mechanics

The description and dynamics of a quantum entity require an elaborate theoretical framework. The quantum entity is the foundation of the mathematical superstructure that consists of five main constructs:

- The quantum degree of freedom space $\mathcal{F}$.
- The quantum state vector $\psi(t, \mathcal{F})$, which is an element of the linear vector state space $\mathcal{V}(\mathcal{F})$.
- The time evolution of $\psi(t, \mathcal{F})$, given by the Schrödinger equation.
- Operators $\mathcal{O}(\mathcal{F})$ that act on the state space $\mathcal{V}(\mathcal{F})$.
- The process of measurement, with repeated observations yielding the expectation value of the operators, namely $E_\psi[\mathcal{O}(\mathcal{F})]$.

The five mathematical pillars of quantum mechanics are shown in Figure 2.2.

2.3 Degree of freedom space $\mathcal{F}$

Recall that in classical mechanics a system is described by dynamical variables, and its time dependence is given by Newton's equations of motion. In quantum mechanics, the description of a quantum entity starts with the generalization of the classical dynamical variables and, following Dirac (1999), is called the quantum *degree of freedom.*

Degree of freedom	State space	Dynamics	Operators	Observation
$\mathcal{F}$	$\mathcal{V}(\mathcal{F})$	$\frac{\partial \psi(t,\mathcal{F})}{\partial t}$	$O(\mathcal{F})$	$E_\psi[O(\mathcal{F})]$

Figure 2.2 The five mathematical pillars of quantum mechanics.

The degree of freedom is the root and ground on which the quantum entity is anchored. The degree of freedom embodies the *qualities* and *properties* of a quantum entity. A single quantum entity, for example the electron, can simultaneously have many degrees of freedom, such as spin, position, angular momentum and so on that all, taken together, describe the quantum entity. The symbol $\mathcal{F}$ is taken to represent all the degrees of freedom of a quantum entity.

A remarkable conclusion of quantum mechanics – validated by experiments – is that a quantum degree of freedom *does not* have any precise value before it is observed; the degree of freedom is *inherently indeterminate* and does not have a determinate objective existence before it is observed. One interpretation of the degree of freedom being intrinsically indeterminate is that it *simultaneously* has a *range of possible values*; the collection of all possible values of the degree of freedom constitutes a *space* that is denoted by $\mathcal{F}$; the space $\mathcal{F}$ is *time independent*.

The entire edifice of quantum mechanics is built on the degree of freedom and, in particular, on the space $\mathcal{F}$.

2.4 State space $\mathcal{V}(\mathcal{F})$

In the quantum mechanical framework, a quantum degree of freedom is inherently indeterminate and, metaphorically speaking, simultaneously has a range of possible values that constitutes the space $\mathcal{F}$.

Consider an experimental device designed to examine and study the properties of a degree of freedom. For a quantum entity that has spin ℓ, the degree of freedom consists of $2\ell + 1$ discrete points. A device built for observing a spin ℓ system needs to have $2\ell + 1$ possible distinct outcomes, one for each of the possible values of the degree of freedom.

The experiment needs to be repeated many times due to the indeterminacy of the quantum degree of freedom. The outcome of each particular experiment is *completely uncertain* and *indeterminate*, with the degree of freedom inducing the device to take any *one* of its (the device's) many possible values.[3] However, the cumulative result of repeated experiments shows a pattern – for example, with the device pointer having some positions being more likely to be observed than others.

How does one describe the statistical regularities of the indeterminate and uncertain outcomes of an experiment carried out on a degree of freedom? As mentioned in Section 2.3, the subject of quantum probability arose from the need to describe quantum indeterminacy. A complex valued state vector, also called the state function and denoted by ψ, is introduced to describe the observable properties of the

[3] It is always assumed, unless stated otherwise, that a quantum state is not an eigenstate.

degree of freedom. The quantum state ψ maps the degree of freedom space $\mathcal{F}$ to the complex numbers $\mathbb{C}$, namely

$$\psi : \mathcal{F} \rightarrow \mathbb{C}.$$

In particular, for the special case of coordinate degree of freedom $x \in \Re = \mathcal{F}$ the state vector ψ is a complex function of x and hence

$$x \in \Re \Rightarrow \psi(x) \in \mathbb{C}.$$

Noteworthy 2.1 **Dirac's formulation of the quantum state.**

- The foundation of the quantum entity is the degree of freedom $\mathcal{F}$; the *quantum state* (state, state vector, and state function) provides an exhaustive description of the quantum entity.
- The term state or *state vector* refers to the quantum state considered as a vector in state space $\mathcal{V}(\mathcal{F})$, usually denoted by $\psi(t, \mathcal{F})$.
- In Dirac's bracket notation, a state vector is denoted by $|\psi(t, \mathcal{F})\rangle$ or $|\psi\rangle$ in short, and is called a *ket* vector.
- The dual to the ket vector is denoted by $\langle\psi(t, \mathcal{F})|$ or $\langle\psi|$ in brief and is called a *bra* vector.
- The scalar product of two state vectors $|\chi\rangle, |\psi\rangle$ is a complex number $\in \mathbb{C}$ and is denoted by the full bracket, namely $\langle\chi|\psi\rangle$.
- The term *state function* refers to the *components* of the state vector and is denoted by $\langle x|\psi(t, \mathcal{F})\rangle \equiv \langle x|\psi_t\rangle \equiv \psi(t, x)$, where $x \in \mathcal{F}$, namely x is a representation of the degree of freedom $\mathcal{F}$.

For degrees of freedom taking discrete values, Dirac's bra and ket vectors are nothing except the row and column vectors of a finite-dimensional linear vector space, with the bracket of two state vectors being the usual scalar product of two vectors.

When the degree of freedom becomes continuous, Dirac's notation carries over into functional analysis and allows for studying questions of the convergence of infinite sequences of state vectors that go beyond linear algebra.

One of the most remarkable properties of the quantum state vector $|\psi\rangle$ is that it is an element of a state space $\mathcal{V}$ that is a *linear vector space*. The precise structure of the linear vector space $\mathcal{V}$ depends on the nature of the quantum degree of freedom $\mathcal{F}$. From the simplest quantum system consisting of two possible states, to a system having N degrees of freedom in four-dimensional spacetime, to quantum fields having an infinite number of degrees of freedom, there is a linear vector space $\mathcal{V}$ and a state vector defined for these degrees of freedom.

Euclidean space $\Re^N$ is a finite-dimensional linear vector space; the linear vector spaces $\mathcal{V}$ that occur in quantum mechanics and quantum field theory are usually state spaces that are an infinite-dimensional generalization of $\Re^N$. Infinite-dimensional linear vector spaces arise in many applications in science and engineering, including the study of partial differential equations and dynamical systems and many of their properties, such as the addition of vectors and so on, are the generalizations of the properties of finite-dimensional vector spaces.

The state vector is an element of a time independent normed linear vector space, namely

$$|\psi\rangle \in \mathcal{V}(\mathcal{F}).$$

The following are some of the main properties of a vector space $\mathcal{V}$:

1. Since they are elements of a linear vector space, a state vector can be added to other state vectors. In particular, *ket vectors* $|\psi\rangle$ and $|\chi\rangle$ are *complex valued vectors* of $\mathcal{V}$ and can be added as follows

$$|\eta\rangle = a|\psi\rangle + b|\chi\rangle, \tag{2.2}$$

   where a, b are complex numbers $\in \mathbb{C}$, and yield another element $|\eta\rangle$ of $\mathcal{V}$. Vector addition is commutative and associative.
2. For every ket vector $|\psi\rangle \in \mathcal{V}$, there is a dual (bra) vector $\langle\psi|$ that is an element of the *dual linear vector space* $\mathcal{V}_D$. The dual vector space is also linear and yields the following

$$\langle\eta| = a^*\langle\psi| + b^*\langle\chi|.$$

   The collection of all (dual) bra vectors forms the dual space $\mathcal{V}_D$.
3. More formally, $\mathcal{V}_D$ is the collection of *all* linear mappings that take elements of $\mathcal{V}$ to C by the *scalar product*. In mathematical notation

$$\mathcal{V}_D : \mathcal{V} \to \mathbb{C}.$$

   The vector space and its dual are not necessarily isomorphic.[4]
4. For any two ket $|\psi\rangle$ and bra $\langle\eta|$ vectors belonging to $\mathcal{V}$ and $\mathcal{V}_D$, respectively, the *scalar product*, namely $\langle\eta|\psi\rangle$, yields a complex number and has the following property:

$$\langle\eta|\psi\rangle = \langle\psi|\eta\rangle^*,$$

[4] Two spaces are isomorphic if there is an invertible mapping that maps each element of one space to a (unique) element of the other space.

where * stands for complex conjugation. The scalar product is linear and yields

$$\langle\eta|\zeta\rangle = a^*\langle\psi|\zeta\rangle + b^*\langle\chi|\zeta\rangle.$$

In particular, $\langle\psi|\psi\rangle \equiv |\psi|^2$ is a real number – a fact of far reaching consequence in quantum mechanics.

5. One of the fundamental properties of quantum states is that *two states are distinct* if they are *linearly independent*. In particular, two states $|\psi\rangle$ and $|\chi\rangle$ are completely distinct if and only if they are *orthogonal*, namely

$$\langle\chi|\psi\rangle = 0 \quad : \quad \text{orthogonal}. \tag{2.3}$$

2.4.1 Hilbert space

Starting in the 1900s, Hilbert space was studied by David Hilbert, Erhard Schmidt, and Frigyes Riesz as belonging to the class of infinite-dimensional function space. The main feature that arises in a Hilbert space is the issue of *convergence* of an infinite sequence of elements of Hilbert space, something that is absent in a finite-dimensional vector space.

To allow for the probabilistic interpretation of the state vector $|\psi\rangle$, all state vectors that represent physical systems must have unit norm, that is

$$\langle\psi|\psi\rangle \equiv |\psi|^2 = 1 \quad : \quad \text{unit norm}.$$

The restriction of the linear vector space $\mathcal{V}$ to be a *normed vector space* defines a *Hilbert space*. For a Hilbert space, the dual state space is isomorphic to the Hilbert space, namely $\mathcal{V} \simeq \mathcal{V}_D$, shown in Figure 2.3.

The state space of quantum entities is a Hilbert space. However, there are classical random systems, for example that occur in finance and for quantum dissipative processes, where the state space is not a Hilbert space and in particular leads to a dual state space: $\mathcal{V}_D$ is not isomorphic to the state space $\mathcal{V}$ [Baaquie (2004)].

For the continuous degree of freedom $\mathcal{F} = \Re$, an element of $|\psi\rangle$ of Hilbert space has unit norm and hence yields

$$\langle\psi|\psi\rangle \equiv |\psi|^2 = \int_{-\infty}^{+\infty} dx|\psi(x)|^2 = 1 \quad : \quad \text{unit norm}.$$

2.5 Operators $\mathcal{O}(\mathcal{F})$

The connection of the quantum degree of freedom with its observable and measurable properties is indirect and is always, of necessity, *mediated* by the process of

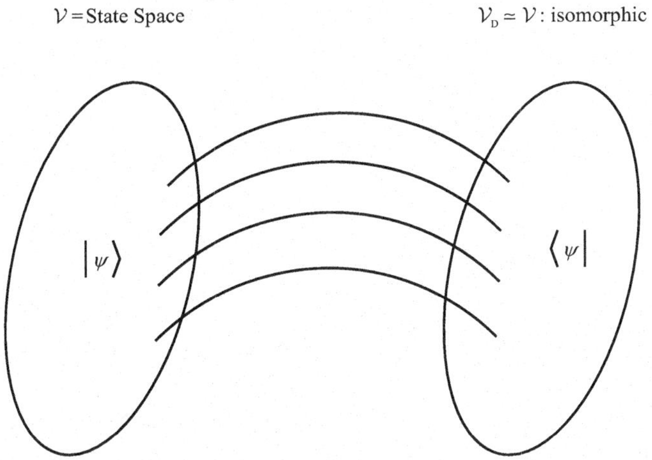

Figure 2.3 Hilbert space is a unit norm state space with $\mathcal{V} \simeq \mathcal{V}_D$.

measurement. A consistent interpretation of quantum mechanics requires that the measurement process plays a central role in the theoretical framework of quantum mechanics.

In classical mechanics, observation and measurement of the physical properties plays no role in the definition of the classical system. For instance, a classical particle is fully specified by its position and velocity at time t and denoted by $x(t)$, $v(t)$; it is immaterial whether a measurement is performed to ascertain the position and velocity of the classical particle; in other words, the position and velocity of the classical particle $x(t)$, $v(t)$ exist *objectively*, regardless of whether its position or velocity is measured or not.

In contrast to classical mechanics, in quantum mechanics the degree of freedom $\mathcal{F}$, *in principle*, can *never* be directly observed. All the observable physical properties of a degree of freedom are the result of a process of measurement carried out on the state vector ψ. Operators, discussed in Chapter 3, are mathematical objects that represent physical properties of the degree of freedom $\mathcal{F}$ and act on the state vector; the action of operators on the state vector is a mathematical representation of the process measuring the physical properties of the quantum entity.

The degree of freedom $\mathcal{F}$ and its measurable properties – represented by the operators $\mathcal{O}_i$ – are *separated* by the quantum state vector $\psi(t, \mathcal{F})$ [Baaquie (2013e)]. An experiment can only measure the *effects* of the degree of freedom – via the state vector $\psi(t, \mathcal{F})$ – *on the operators* $\mathcal{O}_i$. Furthermore, each experimental device is designed and tailor-made to measure a specific physical property of the degree of freedom, represented by an operator $\mathcal{O}_i$.

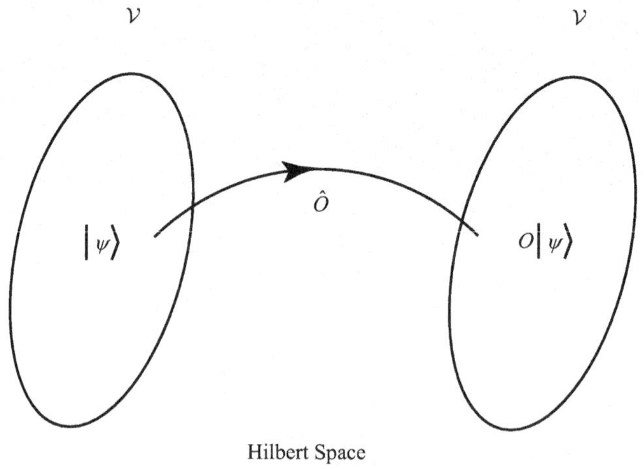

Figure 2.4 An operator $\mathcal{O}$ acting on element $|\psi\rangle$ of the state space $\mathcal{V}$ and mapping it to $\mathcal{O}|\psi\rangle$.

Every degree of freedom $\mathcal{F}$ defines a state space $\mathcal{V}$ and operators $\mathcal{O}$ that act on that state space. All operators $\mathcal{O}$ are mathematically defined to be *linear mappings* of the state space $\mathcal{V}$ into itself, shown in Figure 2.4, and yield, for constant a, b

$$\mathcal{O}: \ |\psi\rangle \to \mathcal{O}|\psi\rangle \ \Rightarrow \ \mathcal{O}: \ \mathcal{V} \to \mathcal{V}$$

$$\mathcal{O}\Big(a|\psi_1\rangle + b|\psi_2\rangle\Big) = a\mathcal{O}|\psi_1\rangle + b\mathcal{O}|\psi_2\rangle \ : \ \text{linear.}$$

Operators are the generalization of matrices; an arbitrary element of an operator $\mathcal{O}$ is given by

$$\langle\chi|\hat{O}|\psi\rangle \ \text{ with } |\psi\rangle \in \mathcal{V}, \ \langle\chi| \in \mathcal{V}_D.$$

The diagonal matrix element of an operator is given by

$$\langle\psi|\hat{O}|\psi\rangle \ \text{ with } |\psi\rangle \in \mathcal{V}, \ \langle\psi| \in \mathcal{V}_D.$$

Important physical quantities associated with a particle such as its position, momentum, energy, angular momentum, and so on are physical observables that are represented by Hermitian operators, discussed in Section 3.3. Physical quantities are indeterminate; the best that we can do in quantum mechanics is to measure the average value of a physical quantity, termed as its *expectation value*.

For example, a quantum particle, in general, has no fixed value for its observable properties, but only has an average value. For example, the expectation value (average value) of the particle's position $\hat{x}$ is given by

$$E[\hat{x}] = \int dx x P(x) = \int dx \, x \, |\psi_t(x)|^2 . \tag{2.4}$$

The quantum particle's average value of the position is interpreted as the diagonal values of the **position operator** $\hat{x}$ since Eqs. 2.4 and 3.29 yield the following

$$\begin{aligned} E[\hat{x}] &= \int dx \langle \psi_t | x \rangle x \langle x | \psi_t \rangle \\ &= \langle \psi_t | \hat{x} | \psi_t \rangle. \end{aligned}$$

Let $\mathcal{O}$ be an observable, which could be the position operator $\hat{x}$, or momentum operator $\hat{p}$, or the Hamiltonian operator H and so on. Generically, for an operator $\hat{O}$ we have

$$\hat{O} : \mathcal{V} \to \mathcal{V}.$$

Hence, an operator is an element of the space formed by the outer product of $\mathcal{V}$ with its dual $\mathcal{V}_D$, that is

$$\hat{O} \in \mathcal{V} \otimes \mathcal{V}_D. \tag{2.5}$$

A *fundamental postulate* of quantum mechanics that follows from Eq. 2.1 is the following: on repeatedly measuring the value of the observable $\mathcal{O}$ in some state $|\chi\rangle$, the expectation value (average value) of the observable is given by

$$E[\mathcal{O}] \equiv \langle \chi | \mathcal{O} | \chi \rangle. \tag{2.6}$$

In other words, the expectation value of the observable is the diagonal value of the operator $\mathcal{O}$ for the given state $|\chi\rangle$. The expected value of a physical quantity is always a real quantity.

Consider some physical quantity, such as a particle's position, and let it be represented by a Hermitian operator $\hat{O}$ with eigenvalues λ_i and eigenstates ψ_i defined by

$$\hat{O}|\chi_i\rangle = \lambda_i |\chi_i\rangle, \quad \langle \chi_i | \chi_j \rangle = \delta_{i-j}. \tag{2.7}$$

For Hermitian operators the eigenvalues λ_i are all real. Any physical state can always be expressed as a superposition of the eigenstates of a Hermitian operator with amplitude c_i and can hence be written as

$$|\psi\rangle = \sum_i c_i |\chi_i\rangle.$$

The result of measuring the physical quantity $\hat{O}$ for the state $\psi(x)$ *always* results in the state function $\psi(x)$ "collapsing" (being projected), with probability $|c_i|^2$, to one of the eigenstates of the operator $\hat{O}$, say $|\chi_i\rangle$ – whose eigenvalue λ_i is then physically observed.

After repeated measurements on the system – each made in an identical manner and hence represented by $|\psi\rangle$ – the average value of $\hat{O}$ is given by

$$E_\psi[O] = \langle\psi|\hat{O}|\psi\rangle = \sum_i |c_i|^2 \langle\chi_i|\hat{O}|\chi_i\rangle = \sum_i |c_i|^2 \lambda_i. \tag{2.8}$$

The measured values of the position, energy, momentum, and so on of a quantum particle are always real numbers. Hence, all physical quantities such as the average position, momentum, energy, and so on must correspond to operators that have *only real eigenvalues*, namely, for which all λ_i are real; this is the reason why all physical quantities are represented by Hermitian operators.

2.6 The process of measurement

Ignore for the moment details of what constitutes an experimental device. What is clear from numerous experiments is that the experimental *readings* obtained by observing a quantum entity by the experimental device cannot be explained by deterministic classical physics and, in fact, require quantum mechanics for an appropriate explanation.

Consider a degree of freedom $\mathcal{F}$; the existence of a range of possible values of the degree of freedom is encoded in its state vector $\psi(\mathcal{F})$. Let physical operators $\mathcal{O}(\mathcal{F})$ represent the observables of the quantum degree of freedom. Recall the degree of freedom cannot be directly observed; instead, what can be measured is the effect of the degree of freedom on the operators mediated by the state vector $\psi(\mathcal{F})$.

The *preparation* of a quantum state yields the quantum state $\psi(\mathcal{F})$, which is then subjected to repeated measurements.

Operators $\mathcal{O}(\mathcal{F})$ are the mathematical basis of measurement theory. The experimental device is *designed* to measure the properties of the operator $\mathcal{O}(\mathcal{F})$. Measurement theory requires knowledge of special quantum states, namely the eigenstates χ_n of the operator $\mathcal{O}(\mathcal{F})$, which are defined in Eq. 2.7.

The process of measurement ascertains the properties of the degree of freedom by subjecting it to the experimental device. The measurement is mathematically represented by applying the operator $\mathcal{O}(\mathcal{F})$ on the state of the system $\psi(\mathcal{F})$ and projecting it to one of the eigenstates of $\mathcal{O}(\mathcal{F})$, namely

$$|\psi(\mathcal{F})\rangle \;\rightarrow\; \text{measurement} = \mathcal{O}(\mathcal{F})|\psi(\mathcal{F})\rangle \;\rightarrow\; \chi_n \quad : \text{ collapse of state } \psi(\mathcal{F}).$$

Applying $\mathcal{O}(\mathcal{F})$ on the state vector causes it to collapse to one of $\mathcal{O}(\mathcal{F})$'s eigenstates. The projection of the state vector ψ to one of the eigenstates χ_n of the operator $\mathcal{O}(\mathcal{F})$ is *discontinuous* and *instantaneous*; it is termed as the *collapse of the state vector* ψ. The result of a measurement has to be *postulated* to lead to the collapse of the state vector and is a feature of quantum mechanics that *is not governed* by the Schrödinger equation.

Unlike classical mechanics, where the same initial condition yields the same final outcome, in quantum mechanics the same initial condition leads to a wide range of possible final states. The result of identical quantum experiments is inherently uncertain.[5] For example, radioactive atoms, even though identically prepared, decay randomly in time precisely according to the probabilistic predictions of quantum mechanics.

After many repeated observations performed on state $\psi(\mathcal{F})$, all of which in principle are identical to each other, the experiment yields the average value of the physical operator $\mathcal{O}(\mathcal{F})$, namely

$$\mathcal{O} \rightarrow \text{measurements on } \psi(\mathcal{F}) \rightarrow E_\psi[\mathcal{O}(\mathcal{F})].$$

The process of measurement *cannot* be modeled by the Schrödinger equation, and this has long been a point of contention among physicists. Many theorists hold that the fundamental equations of quantum mechanics should determine both the evolution of the quantum state as well as the collapse of the state caused by the process of measurement. As of now, there has been no resolution of this conundrum.

2.7 The Schrödinger differential equation

The discussion so far has been kinematical, in other words, focused on the framework for *describing* a quantum system. One of the fundamental goals of physics is to obtain the dynamical equations that *predict* the future state of a system. This requirement in quantum mechanics is met by the Schrödinger partial differential equation that determines the *future time evolution* of the state function $\psi(t, \mathcal{F})$, where t parameterizes time. The Schrödinger equation is time reversible.

To exist, all physical entities must have energy; hence, it is reasonable that the *Hamiltonian* operator H should enter the Schrödinger equation. The Hamiltonian operator H represents the energy of a quantum entity; H determines the form and numerical range of the possible allowed energies of a given quantum entity. Furthermore, energy is the quantity that is conjugate to time, similar to position being conjugate to momentum and one would consequently expect that H should play a central role in the state vector's time evolution. However, in the final analysis, there is no derivation of the Schrödinger equation from any underlying principle and one has to simply postulate it to be true.

The Schrödinger equation is expressed in the language of state space and operators and determines the time evolution of the state function $|\psi(t)\rangle$, with t being

[5] Except, as mentioned earlier, for eigenstates.

the time parameter. One needs to specify the degrees of freedom of the system in question, that in turn specifies the nature of the state space $\mathcal{V}$; one also needs to specify the Hamiltonian H.

The celebrated Schrödinger equation is given by

$$-\frac{\hbar}{i}\frac{\partial|\psi(t)\rangle}{\partial t} = H|\psi(t)\rangle. \tag{2.9}$$

For the case of the degree of freedom being all the possible positions of a quantum particle, $\mathcal{F} = \Re$, in the position basis $|x\rangle$, the state vector is

$$\langle x|\psi(t)\rangle = \psi(t, x)$$

and the Schrödinger equation given in Eq. 2.9, yields the following

$$\begin{aligned} & -\frac{\hbar}{i}\langle x|\frac{\partial}{\partial t}|\psi(t)\rangle = \langle x|H|\psi(t)\rangle \\ \Rightarrow \quad & -\frac{\hbar}{i}\frac{\partial\psi(t, x)}{\partial t} = H(x, \frac{\partial}{\partial x})\psi(t, x), \end{aligned} \tag{2.10}$$

where note that the Hamiltonian operator acts on the dual basis.

For a quantum particle with mass m moving in one dimension in a potential $V(x)$, the Hamiltonian is given by

$$H = -\frac{\hbar}{2m}\frac{\partial^2}{\partial x^2} + V(x) \tag{2.11}$$

and yields Schrödinger's partial differential equation

$$-\frac{\hbar}{i}\frac{\partial\psi(t, x)}{\partial t} = -\frac{\hbar}{2m}\frac{\partial^2\psi(t, x)}{\partial x^2} + V(x)\psi(t, x).$$

A variety of techniques has been developed for solving the Schrödinger equation for a wide class of potentials as well as for multi-particle quantum systems [Gottfried and Yan (2003)].

Let $|\psi\rangle$ be the initial value of the state vector at $t = 0$ with $\langle\psi|\psi\rangle = 1$. Equation 2.9 can be integrated to yield the following formal solution

$$|\psi(t)\rangle = e^{-itH/\hbar}|\psi\rangle = U(t)|\psi\rangle. \tag{2.12}$$

Similar to the momentum operator translating the state vector in space, as in Eq. 3.39, the Hamiltonian H is an operator that translates the initial state vector in time, as in Eq. 2.12. The evolution operator $U(t)$ is defined by

$$U(t) = e^{-itH/\hbar}, \quad U^\dagger(t) = e^{itH/\hbar}$$

and is unitary since H is Hermitian; more precisely

$$U(t)U^\dagger(t) = \mathbb{I}.$$

The unitarity of $U(t)$, and by implication the Hermiticity of H, is crucial for the conservation of probability. The total probability of the quantum system is conserved over time since unitarity of $U(t)$ ensures that the normalization of the state function is time-independent; more precisely

$$\langle\psi(t)|\psi(t)\rangle = \langle\psi|U^\dagger(t)U(t)|\psi\rangle = \langle\psi|\psi\rangle = 1.$$

The operator $U(t)$ is the exponential of the Hamiltonian H that in many cases, as is the case given in Eq. 2.11, is a differential operator. The Feynman path integral is a mathematical tool for analyzing $U(t)$ and is discussed in Chapter 4.

The Schrödinger equation given in Eq. 2.9 is a *linear equation* for the state function $|\psi(t)\rangle$. Consider two solutions $|\psi_1(t)\rangle$ and $|\psi_2(t)\rangle$ of the Schrödinger equation; then their linear combination yields yet another solution of the Schrödinger equation given by

$$|\psi(t)\rangle = \alpha|\psi_1(t)\rangle + \beta|\psi_2(t)\rangle, \tag{2.13}$$

where α, β are complex numbers. The *quantum superposition* of state vectors given in Eq. 2.13 is of far reaching significance and in particular leads to the Dirac–Feynman formulation of quantum mechanics discussed in Section 2.9.

The mathematical reason that state vector $|\psi(t)\rangle$ is an element of a normed *linear vector space* is due to the *linearity* of the Schrödinger equation and yields the result that all state vectors $|\psi(t)\rangle$ are elements of a linear vector space $\mathcal{V}$.

The fact that $|\psi(t)\rangle$ is an element of a linear vector space leads to many nonclassical and unexpected phenomena such as the existence of entangled states and the quantum superposition principle [Baaquie (2013e)].

The Schrödinger equation has the following remarkable features:

- It is a first order differential equation in time, in contrast to Newton's equation of motion that is a second order differential equation in time. At $t = 0$, the Schrödinger equation requires that the initial state function be specified for *all values* of the degree of freedom, namely $|\psi(\Re)\rangle$, whereas in Newton's law, only the position and velocity *at the starting point* of the particle are required.
- At each instant, Schrödinger's equation specifies the state function for all values of the *indeterminate* degree of freedom. In contrast, Newton's law of motion specifies only the *determinate* position and velocity of a particle.
- The state vector $|\psi(t)\rangle$ is complex valued. In fact, the Schrödinger equation is the first equation in natural science in which complex numbers are essential and not just a convenient mathematical tool for representing real quantities.

Quantum mechanics introduces a great complication in the description of Nature by replacing the dynamical variables x, p of classical mechanics, which consist of only six real numbers for every instant of time, by an entire space $\mathcal{F}$ of the indeterminate degree of freedom; a description of the quantum entity requires, in addition, a state vector ψ that is a *function* of the space $\mathcal{F}$. According to Dirac (1999), the *great complication* introduced by quantum indeterminacy is "offset" by the *great simplification* due to the linearity of the Schrödinger equation.

2.8 Heisenberg operator approach

Every physical property of a degree of freedom is mathematically realized by a Hermitian operator $\mathcal{O}$. Generalizing Eq. 2.8 to time dependent state vectors and from Eq. 2.12, the expectation value of an operator at time t, namely $\mathcal{O}(t)$, is given by

$$\begin{aligned} E_\psi[\mathcal{O}(t)] &= \langle\psi(t)|\mathcal{O}|\psi(t)\rangle = \langle\psi|e^{itH/\hbar}\mathcal{O}e^{-itH/\hbar}|\psi\rangle \\ &= tr\big(\mathcal{O}(t)\rho\big) \ : \ \rho = |\psi\rangle\langle\psi|. \end{aligned} \tag{2.14}$$

The density matrix ρ is a time-independent operator that encodes the initial quantum state of the degree of freedom.

From Eq. 2.14, the time-dependent expectation value has two possible interpretations: the state vector is evolving in time, namely, the state vector is $|\psi(t)\rangle$ and the operator $\mathcal{O}$ is constant, or equivalently, the state vector is fixed, namely $|\psi\rangle$ and instead, the operator is evolving in time and is given by $\mathcal{O}(t)$. The *time-dependent* Heisenberg operator $\mathcal{O}(t)$ is given by

$$\begin{aligned} O(t) &= e^{itH/\hbar}\mathcal{O}e^{-itH/\hbar} \\ i\hbar\frac{\partial O(t)}{\partial t} &= [\mathcal{O}(t), H]. \end{aligned} \tag{2.15}$$

In the Heisenberg formulation of quantum mechanics, quantum indeterminacy is completely described by the algebra of Hermitian operators.

All physical observables of a quantum degree of freedom are elements of the Heisenberg operator algebra, and so are the density matrices that encode the initial quantum state of the degree of freedom. Quantum indeterminacy is reflected in the spectral decomposition of the operators in terms of its eigenvalues and projection operators (eigenvectors), as given in Eq. 3.21. For example, the single value of energy for a classical entity is replaced by a whole range of eigenenergies of the Hamiltonian operator for a quantum degree of freedom, with the eigenfunctions encoding the inherent indeterminacy of the degree of freedom.

The time dependence of the state vector given by the Schrödinger equation is replaced by the time dependence of the operators given in Eq. 2.15. All expectation values are obtained by performing a trace over this operator algebra, namely by $tr\big(\rho\mathcal{O}(t)\big)$ as given in Eq. 2.14.

From the point of view of quantum probability, Heisenberg's operator formulation goes far beyond just providing a mathematical framework for the mechanics of the quantum, but instead, also lays the foundation of the quantum theory of probability [Baaquie (2013e)].

2.9 Dirac–Feynman path integral formulation

The time evolution of physical entities is fundamental to our understanding of Nature. For a classical entity evolving in time, its *trajectory* exists objectively, regardless of whether it is observed or not, with both its position $x(t)$ and velocity $v(t)$ having exact values for each instant of time t.

We need to determine the mode of existence of quantum indeterminacy for the case of the time evolution of a quantum degree of freedom.

Consider a quantum particle with degree of freedom $x \in \Re = \mathcal{F}$. Suppose that the particle is observed at time t_i, with the position operator finding the particle at point x_i and a second observation is at time t_f, with the position operator finding the particle at point x_f. To simplify the discussion, suppose there are N slits between the initial and final positions, located at positions $x_1, x_2, \ldots, x_N$, as shown in Figure 2.5.

There are two cases for the quantum particle making a transition from x_i, t_i to x_f, t_f, namely when the path taken at an intermediate time t is *observed* and when it is *not observed.* For the case when the path taken at an intermediate time t is observed, one simply obtains the classical result discussed in Section 4.3.

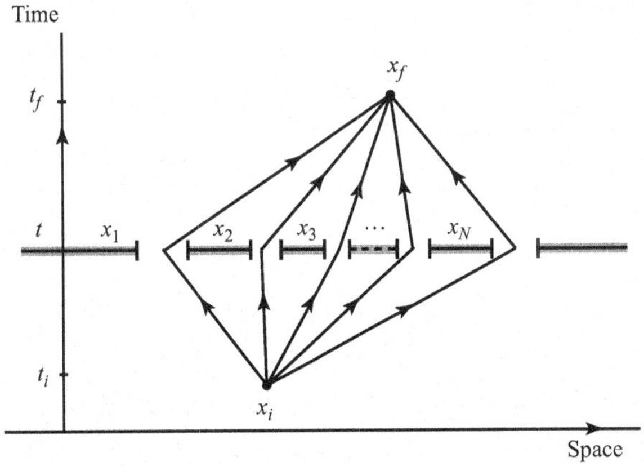

Figure 2.5 A quantum particle is observed at first at initial position x_i at time t_i and a second time at final position x_f at time t_f. The quantum particle's path being indeterminate means that the single particle simultaneously *exists in all the allowed paths.*

What is the description of the quantum particle making a transition from x_i, t_i to x_f, t_f when it is *not observed* at an intermediate time t? The following is a summary of the conclusions:

- The quantum indeterminacy of the degree of freedom, together with the linearity of the Schrödinger equation, leads to the conclusion that the *path* of the quantum particle is *indeterminate.*
- The indeterminacy of the path is realized by the quantum particle by existing in all possible paths simultaneously; or metaphorically speaking, the single quantum particle simultaneously "takes" all possible paths.

The concept of the *probability amplitude*, which is a complex number, is used for describing the indeterminate paths of a quantum system.

To start with, a *probability amplitude* is assigned to each *determinate path.* In the case of no observation being made to determine which path was taken, all the paths are indistinguishable and hence the particle's path is indeterminate, with the particle *simultaneously existing in all the N-paths*, as shown in Figure 2.5. The probability amplitude for the quantum particle having an *indeterminate* path is obtained by combining the probability amplitudes for the different *determinate* paths using the quantum superposition principle.

Let probability amplitude ϕ_n be assigned to the determinate path going through a slit at x_n with $n = 1, 2, \ldots N$, as shown in Figure 2.5, and let $\phi(x_f, t_f|x_i, t_i)$ be the net probability amplitude for a particle that is observed at position x_i at time t_i and then observed at position x_f at later time t_f. The probability amplitude $\phi(x_f, t_f|x_i, t_i)$ for the transition is obtained by *superposing* the probability amplitudes for all *indistinguishable determinate paths* and yields

$$\phi(x_f, t_f|x_i, t_i) = \sum_{n=1}^{N} \phi_n : \text{indistinguishable paths.} \tag{2.16}$$

Once the probability amplitude is determined, its modulus squared, namely $|\phi|^2$ yields the *probability* for the process in question. For the N-slit case

$$|\phi(x_f, t_f|x_i, t_i)|^2 = P(x_f, t_f|x_i, t_i), \quad \int dx_f P(x_f, t_f|x_i, t_i) = 1,$$

where $P(x_f, t_f|x_i, t_i)$ is the *conditional probability* that a particle, observed at position x_i at time t_i, will be observed at position x_f at later time t_f.

Quantum mechanics can be formulated entirely in terms of indeterminate paths, a formulation that is independent of the framework of the state vector and the Schrödinger equation; this approach, known as the Dirac–Feynman formulation, is discussed in Chapter 4.

2.10 Three formulations of quantum mechanics

In summary, quantum mechanics has the following three independent, but equivalent, mathematical formulations for describing quantum indeterminacy:

- *The Schrödinger equation* for the state vector postulates that the quantum state vector encodes all the information that can be extracted from a quantum degree of freedom. The degree of freedom always remains indeterminate since all measurements only encounter the quantum state vector, causing it to collapse to an observed manifestation.
- *The Heisenberg operator formalism.* The state vector is completely dispensed with and instead a density matrix, which is an operator, represents the quantum entity. All observations consist of detecting the collapse of the density matrix, which makes a transition from the pure to a mixed density matrix, discussed in Section 3.9; the detection of the mixed density matrix by projection operators results in the experimental determination of the probability of the various projection operators detecting the quantum entity.

  Quantum probability assigns probabilities to projection operators. The indeterminate nature of the degree of freedom is reflected in that it is never detected by any of the operators. The violation of the Bell-inequality shows that the quantum indeterminacy cannot be explained by classical probability theory; in particular, the degree of freedom has no determinate value before an observation – and hence no objective existence – showing its indeterminate nature.[6]
- *The Dirac–Feynman path integral formulation.* The path integral is the sum over all the indeterminate (indistinguishable) paths, from the initial to the final state, and reflects quantum indeterminacy which is at the foundation of quantum mechanics. The state vector appears as initial and final conditions for the indeterminate paths that are being summed over.

  In the path integral approach, the quantum degrees of freedom appear as integration variables and provide the clearest representation of the indeterminate nature of the degree of freedom. An integration variable has no fixed value but, rather, takes values over its entire range; for the degree of freedom this means that the entire degree of freedom space $\mathcal{F}$ is integrated over. The freedom to change variables for path integration is equivalent to changing the representation chosen for the degree of freedom, and is similar to the freedom in choosing basis states for Hilbert space.

[6] Quantum probability is fundamentally different from classical probability. The difference was crystallized by the ground-breaking work of Bell (2004) and is discussed in detail by Baaquie (2013e).

Each framework has its own advantages, throwing light on different aspects of quantum mechanics that would otherwise be difficult to express. For example, the Schrödinger equation is most suitable for studying the bound sates of quantum entities such as atoms and molecules; the Heisenberg formulation is most suitable for studying the measurement process; and the Feynman path integral is most appropriate for studying the indeterminate quantum paths.

2.11 Quantum entity

In light of the mathematical superstructure of quantum mechanics, what is a quantum entity? A careful study of what is an entity, a thing, an object leads to the remarkable conclusion that the quantum entity is intrinsically indeterminate and its description requires a framework that departs from our classical conception of Nature.

The quantum entity's foundation is its degree of freedom $\mathcal{F}$ and quantum indeterminacy is due to to the intrinsic indeterminacy of the degree of freedom. A landmark step, taken by Max Born, was to postulate that quantum indeterminacy can be described by a state vector $\psi(\mathcal{F})$ that obeys the laws of quantum probability. The state vector is inseparable from the degree of freedom and encodes all the information that can be obtained from the indeterminate degree of freedom, and is illustrated in Figure 2.6.

The state vector $\psi(\mathcal{F})$ encompasses the degree of freedom, but does not do so in physical space; rather, Figure 2.6 illustrates the fact that all observations carried out on the degree of freedom always encounter the state vector and no observation can ever come into direct "contact" with the degree of freedom itself. All "contact" of the measuring device with the degree of freedom is mediated by the state vector.

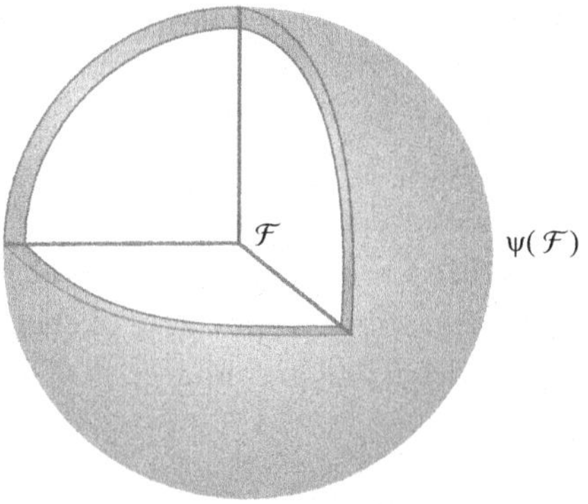

Figure 2.6 A quantum entity is constituted by its degree of freedom $\mathcal{F}$ and the state vector $\psi(\mathcal{F})$ that permanently encompasses and envelops its degrees of freedom.

In brief, quantum mechanics provides the following as a definition of the quantum entity: A quantum entity is constituted by a pair, namely the degree(s) of freedom $\mathcal{F}$ and the state vector $\psi(\mathcal{F})$ that encodes all of its properties. This inseparable pair, namely the degree of freedom and the state vector, embodies the condition in which the quantum entity exists.

2.12 Summary

Classical physics is based on explaining the behavior of Nature based on attributing mathematical properties directly to the observed phenomenon; for example, a tangible force acts on a particle and changes its position. The logic of quantum mechanics is quite unlike classical physics. An elaborate mathematical superstructure connects the experimentally observed behavior of the particle's degree of freedom – enigmatically enough the degree of freedom can never in principle ever be empirically observed – with its mathematical description [Baaquie (2013e)].

All our understanding of a quantum entity is based on theoretical and mathematical concepts that, in turn, have to explain a plethora of experimental data. In the case of quantum mechanics, the mathematical construction has led us to infer the existence of the quantum degree of freedom. The theoretical constructions of quantum mechanics are far from being arbitrary and ambiguous; on the contrary, given the maze of links from the quantum entity to its empirical properties, it is highly unlikely that there are any major gaps or redundancies in the theoretical superstructure of quantum mechanics.

Quantum mechanics and quantum field theory – bedrocks of theoretical physics and of modern technology – synthesize a vast range of mathematical disciplines that constitutes its mathematical foundations and has given rise to the discipline of quantum mathematics. *Quantum mathematics* includes such diverse mathematical fields as calculus, linear algebra, functional analysis and functional integration, probability and information theory, dynamical systems, logic, combinatorics and graph theory, Lie groups and representation theory, differential and algebraic geometry, topology, knot theory, and number theory, to name a few.

The relation of quantum mathematics to quantum mechanics is analogous to the connection of calculus to Newtonian mechanics: although calculus was discovered by Newton for explaining classical mechanics, calculus as a discipline goes far beyond Newtonian mechanics – having applications in almost every branch of science. Similarly, it is worth noting that quantum mathematics is a discipline that is far greater than quantum mechanics – with possible applications in all fields of science as well as the social sciences that are based on uncertainty and randomness.

Quantum mathematics describes random, uncertain and indeterminate systems using the concept of the degree of freedom, which in turn defines a linear vector state space; the dynamics of the degrees of freedom is determined by the analog of the Hamiltonian or the Lagrangian, which are defined on the state space. The expectation values of random quantities – which are functions of the degrees of freedom – can be obtained by using either the techniques of operators and state space or by employing the Feynman path integral (functional integration) that entails summing over all possible configurations of the degrees of freedom.

A leading example of quantum mathematics is the explanation of critical phenomena. Classical random systems undergoing phase transitions – such as a piece of iron becoming a magnet when it is cooled – are examples of critical phenomena and are described by classical statistical mechanics. Wilson (1983) solved the problem of classical phase transitions by describing it as a system that has infinitely many degrees of freedom and which is mathematically identical to a (Euclidean) quantum field theory. Experiments later validated the explanation of critical classical systems by quantum mathematics, and in particular by the mathematics of quantum field theory.

In fact, based on the common ground of quantum mathematics, there is a two way relation between classical random systems and quantum mechanics. For example, the work of Wilson (1983) showed that all renormalizable quantum field theories, in turn, are mathematically equivalent to classical systems that undergo second order phase transitions.

Phase transitions are mathematically described by quantum field theories in Euclidean time. If one restricts quantum mathematics to quantum mechanics, then one may ask questions such as "is probability conserved in phase transitions?" – questions that are clearly meaningless since systems undergoing phase transitions are in equilibrium and hence there is no concept of time evolution in phase transitions. Instead, using quantum mathematics, Wilson (1983) computed classical quantities such as critical exponents that characterize phase transitions, exponents that can be experimentally measured [Papon *et al.* (2002)].

Quantum mathematics has been applied to finance in Baaquie (2004) and Baaquie (2010). The classical random behavior of financial instruments is interpreted in terms of the symbols of quantum mathematics; the framework for connecting the observed market behavior of these instruments with their mathematical description turns out to be quite different from that of quantum mechanics.

A major area of application of quantum mathematics is the field of pure and applied mathematics. S. Donaldson received the Fields medal in 1986 for a result on four dimensional manifolds that he obtained by studying the classical solutions of Yang–Mills gauge theory in four dimensions. Witten (1989) applied the techniques of quantum field theory to derive the Jones polynomial, which classifies knots and links in three space dimensions – and received the Fields medal in 1989.

More recently, superstring theory has led to a plethora of results in pure mathematics using quantum mathematics.

Needless to say, neither mathematics nor mathematical finance is a quantum system. The interpretations of quantum mathematics that are valid for quantum systems are not directly applicable to these problems; instead one has to reason from first principles to interpret the structures of quantum mathematics. In particular, the results obtained in mathematics and finance using quantum mathematics entail a radically different interpretation of its symbols.

From the example of phase transitions and mathematics it can be seen that the symbols of quantum mathematics, when applied to other fields such as finance [Baaquie (2004), Baaquie (2010)], the human psyche [Baaquie and Martin (2005)], the social sciences [Haven and Khrennikov (2013)] and so on, have interpretations that are quite different from quantum mechanics The interpretations of quantum mathematics in these diverse fields have no fixed prescription but, instead, have to be arrived at from first principles [Baaquie (2013a)].

The main thrust of the remaining chapters is on the mathematics of quantum mechanics, leaving aside questions of how these mathematical results are applied to physics, finance, and other disciplines. Various models are analyzed to develop the myriad and multi-faceted principles and methods of quantum mathematics.

3
Operators

Operators represent physically observable quantities, as discussed in Section 2.5. The structure and property of operators depend on the nature of the degree of freedom; operators act on the state space and in particular on the state vector of a given degree of freedom. The significance of operators in the interpretation of quantum mechanics has been discussed in Baaquie (2013e).

The operators discussed in this chapter are mostly based on the continuous degree of freedom, which is analyzed in Section 3.1. Hermitian operators represent physically observable properties of a degree of freedom and their mathematical properties are defined in Section 3.3. The coordinate and momentum operators are the leading exemplar of a pair of noncommuting Hermitian operators and these are studied in some detail in Section 3.4. The Weyl operators yield, as in Section 3.5, a finite-dimensional example of the shift and scaling operators; Section 3.8 provides a unitary representation of the coordinate and momentum operators.

The term self-adjoint operator is used for Hermitian operators when there is a need to emphasize the importance of the *domain* of the Hilbert space on which the operators act – a topic not usually discussed in most books on quantum mechanics. Sections 3.10 and 3.11 discuss the concept of self-adjoint operators, in particular the crucial role played by the domain for realizing the property of self-adjointness. It is shown in Section 3.12 how the requirement of self-adjointness yields a nontrivial extension of Hamiltonians that include singular interactions.

3.1 Continuous degree of freedom

Continuous and discrete degrees of freedom occur widely in quantum mechanics. An in-depth analysis of a discrete degree of freedom is presented in Chapter 8. In this chapter, the focus is on analysis of a continuous degree of freedom and its state space and operators. The structure of the continuous degree of freedom is seen to

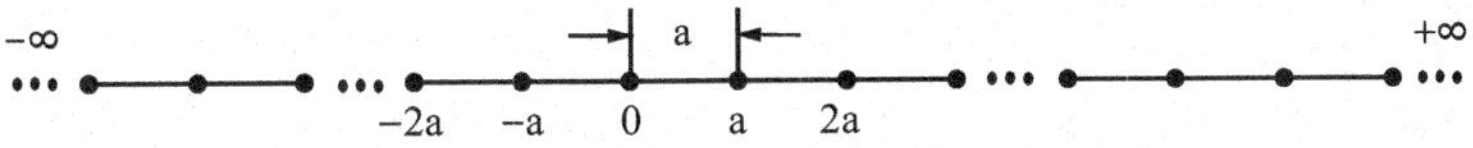

Figure 3.1 Discretization of a continuous degree of freedom space $\mathcal{F} = \Re$.

emerge naturally by taking the continuum limit of an underlying system consisting of a discrete degree of freedom.

Consider a quantum particle that can be detected by the position projection operators at any point of space; to simplify the discussion suppose the particle can move in only one dimension and hence can be found at any point $x \in [-\infty, +\infty] = \Re$. Hence, the degree of freedom is $\mathcal{F} = \Re$ and the specific values of the degree of freedom x constitute a real *continuous variable*. Since there are infinitely many points on the real line, the quantum particle's degree of freedom has *infinitely* many possible outcomes.

As shown in Figure 3.1, let the continuous degree of freedom x, $-\infty \leq x \leq +\infty$, take only discrete values at points $x = na$ with lattice spacing a and with $n = 0, \pm1, \pm2, \ldots$; in other words, the lattice is embedded in the continuous line $\Re$ and the lattice point n identified with the point na in $\Re$. To obtain the continuous position degree of freedom $\mathcal{F}$, let $a \to 0$ and the allowed values of the particle's position x can take any real value, that is, $x \in \Re$, and hence $\mathcal{F} \to \Re$.

The discrete basis vectors of the quantum particle's state space $\mathcal{V}$ are represented by infinite column vectors with the only nonzero entry being unity in the nth position. Hence

$$|n\rangle \ : \ n = 0, \pm1, \pm2, \ldots \pm \infty,$$

where, more explicitly

$$|n\rangle = \begin{bmatrix} \cdots \\ 0 \\ 1 \\ 0 \\ \cdots \end{bmatrix} \ : \ n\text{th position.}$$

The basis vectors for the dual state space $\mathcal{V}_D$ are given by

$$\langle n| = \begin{bmatrix} \cdots 0 & 1 & 0 & \cdots \end{bmatrix}$$
$$\Rightarrow \langle n|m\rangle = \delta_{n-m}. \tag{3.1}$$

The completeness of the basis states yields the following:

$$\sum_{n=-\infty}^{+\infty} |n\rangle\langle n| = \text{diagonal}(\ldots, 1, 1, \ldots) = \mathbb{I} \ : \ \text{completeness equation,}$$

where $\mathbb{I}$ above is the infinite-dimensional unit matrix.

The limit of $a \to 0$ needs to be taken to obtain a continuous x; in terms of the underlying lattice, the continuous point x is related to the discrete lattice point n by

$$-\infty \leq x \leq +\infty \ : \ x = \lim_{a\to 0}[na], \quad n = 0, \pm 1, \pm \ldots . \pm \infty.$$

The state vector for the particle is given by the ket vector $|x\rangle$, with its dual vector given by the bra vector $\langle x|$. The basis state $|n\rangle$ is dimensionless; the ket vector $|x\rangle$ has a dimension of $1/\sqrt{a}$ since, from Eq. 3.14, the Dirac delta function has dimension of $1/a$. Hence, due to dimensional consistency

$$|x\rangle = \lim_{a\to 0} \frac{1}{\sqrt{a}}|n\rangle, \quad \langle x| = \lim_{a\to 0} \frac{1}{\sqrt{a}}\langle n|. \tag{3.2}$$

The position projection operator is given by the outer product of the position ket vector with the bra vector and hence

$$|x\rangle\langle x| = \lim_{a\to 0} \frac{1}{a}|n\rangle\langle n|. \tag{3.3}$$

The scalar product, for $x = na$ and $x' = ma$, in the limit of $a \to 0$, is given, from Eqs. 3.1, 3.2, and 3.14, by the Dirac delta function

$$\langle x|x'\rangle = \lim_{a\to 0} \frac{1}{a}\delta_{m-n} \Rightarrow \langle x|x'\rangle = \delta(x - x'). \tag{3.4}$$

The completeness equation above has the following continuum limit:

$$\mathbb{I} = \sum_{n=-\infty}^{+\infty} |n\rangle\langle n| = \lim_{a\to 0} a \sum_{n=-\infty}^{+\infty} |x\rangle\langle x| \tag{3.5}$$

$$\Rightarrow \int_{-\infty}^{\infty} dx|x\rangle\langle x| = \mathbb{I} \ : \ \text{completeness equation.} \tag{3.6}$$

Equation 3.5 shows that the projection operators given in Eq. 3.3 are complete and span the entire state space $\mathcal{V}$.

$\mathbb{I}$ is the *identity operator* on state space $\mathcal{V}$; namely for any state vector $|\psi\rangle \in \mathcal{V}$, it follows from the completeness equation that

$$\mathbb{I}|\psi\rangle = |\psi\rangle.$$

The completeness equation given by Eq. 3.6 is a key equation that is central to the analysis of state space, and yields

$$\langle x|\mathbb{I}|x'\rangle = \int_{-\infty}^{\infty} dz\langle x|z\rangle\langle z|x'\rangle = \int_{-\infty}^{\infty} dz\delta(x - z)\delta(z - x') = \delta(x - x'),$$

that follows from the definition of the Dirac delta function $\delta(x - x')$. The above equation shows that $\delta(x - x')$ is the matrix element of the identity operator $\mathbb{I}$ for the continuous degree of freedom $\mathcal{F} = \Re$ in the x basis.

The state space $\mathcal{V}(\mathcal{F})$ of a continuous degree of freedom $\mathcal{F}$ is a *function space* and it is for this reason that the subject of functional analysis studies the mathematical properties of quantum mechanics.

For the case of $\mathcal{F} = \Re$, the state vector $|f\rangle$ is an element of $\mathcal{V}(\Re)$ and yields a state function $f(x)$ given by $f(x) = \langle x|f\rangle$; hence functions of x, namely $f(x)$, can be thought of as elements of a state space $\mathcal{V}(\Re)$. Being an element of a state space endows the function $f(x)$ with the additional property of linearity that needs to be consistent with all the other properties of $f(x)$. It should be noted that not all functions are elements of a (quantum mechanical) state space.

Noteworthy 3.1 **Dirac delta function**

The Dirac delta function is useful in the study of continuum spaces, and some of its essential properties are reviewed. Dirac delta functions are not ordinary Lebesgue measureable functions since they have a support set with measure zero; rather they are generalized functions also called distributions. In essence, the Dirac delta function is the continuum generalization of the discrete Kronecker delta function.

Consider a continuous line labelled by coordinate x such that $-\infty \leq x \leq +\infty$, and let $f(x)$ be an infinitely differentiable function. The Dirac delta function, denoted by $\delta(x-a)$, is defined by the following:

$$\delta(x-a) = \delta(a-x) \ : \ \text{even function},$$

$$\delta(c(x-a)) = \frac{1}{|c|}\delta(x-a),$$

$$\int_{-\infty}^{+\infty} dx f(x)\delta(x-a) = f(a), \tag{3.7}$$

$$\int_{-\infty}^{+\infty} dx f(x)\frac{d^n}{dx^n}\delta(x-a) = (-1)^n \frac{d^n}{dx^n} f(x)|_{x=a}. \tag{3.8}$$

The Heaviside step function $\Theta(t)$ is defined by

$$\Theta(t) = \begin{cases} 1 & t > 0 \\ \frac{1}{2} & t = 0 \\ 0 & t < 0 \end{cases}. \tag{3.9}$$

From its definition $\Theta(t) + \Theta(-t) = 1$. The following is a representation of the δ-function:

$$\int_{-\infty}^{b} \delta(x-a) = \Theta(b-a), \tag{3.10}$$

$$\Rightarrow \int_{-\infty}^{a} \delta(x-a) = \Theta(0) = \frac{1}{2}, \tag{3.11}$$

where the last equation is due to the Dirac delta function being an even function. From Eq. 3.10

$$\frac{d}{db}\Theta(b-a)=\delta(b-a).$$

A representation of the delta function based on the Gaussian distribution is

$$\delta(x-a)=\lim_{\sigma\to 0}\frac{1}{\sqrt{2\pi\sigma^2}}\exp\{-\frac{1}{2\sigma^2}(x-a)^2\}. \tag{3.12}$$

Moreover

$$\delta(x-a)=\lim_{\mu\to\infty}\frac{1}{2}\mu\exp\{-\mu|x-a|\}.$$

The definition of Fourier transform yields a representation of the Dirac delta function. It can be shown that

$$\delta(x-a)=\int_{-\infty}^{+\infty}\frac{dp}{2\pi}e^{ip(x-a)}. \tag{3.13}$$

A proof of Eq. 3.13 is found in the book on quantum mechanics by Landau and Lifshitz (2003). One can perform the following consistency check of Eq. 3.13. Integrate both sides of Eq. 3.13 over x as follows:

$$\begin{aligned}
\text{L.H.S} &= \int_{-\infty}^{+\infty} dxe^{-ikx}\delta(x-a)=e^{-ika},\\
\text{R.H.S} &= \int_{-\infty}^{+\infty} dxe^{-ikx}\int_{-\infty}^{+\infty}\frac{dp}{2\pi}e^{ip(x-a)}\\
&= \int_{-\infty}^{+\infty}\frac{dp}{2\pi}e^{-ipa}2\pi\delta(p-k)=e^{-ika},
\end{aligned}$$

where Eq. 3.13 was used in performing the x integration for the right hand side. Hence, one can see that Eq. 3.13 is self-consistent.

To make the connection between the Dirac delta function and the discrete Kronecker delta function consider the discretization of the continuous line into a discrete lattice with spacing a. As shown in Figure 3.1, the continuous degree of freedom x, $-\infty\le x\le+\infty$, takes only discrete values at points $x=na$ with $n=0,\pm1,\pm2,\ldots$ The discretization of Eq. 3.7, for $x=na$ and $y=ma$, yields

$$\delta_{n-m}=\begin{cases}0 & n\neq m\\ 1 & n=m\end{cases}.$$

Discretize continuous variable x into a lattice of discrete points $x = n\epsilon$, and let $a = m\epsilon$; then $f(x) \to f_n$. Discretizing Eq. 3.7 gives

$$\int_{-\infty}^{+\infty} dx f(x)\delta(x-a) \to \epsilon \sum_{n=-\infty}^{+\infty} f_n \delta(x_n - a_m) = f_m = \sum_{n=-\infty}^{+\infty} \delta_{n-m} f_m$$

$$\Rightarrow \delta(x-a) \to \frac{1}{\epsilon}\delta_{n-m}.$$

Taking the limit of $\epsilon \to 0$ in the equation above yields

$$\delta(x-a) = \lim_{\epsilon \to 0} \frac{1}{\epsilon}\delta_{n-m} = \begin{cases} 0 & x \neq a \\ \infty & x = a \end{cases}. \tag{3.14}$$

3.2 Basis states for state space

The bra and ket vectors $\langle x|$ and $|x\rangle$ are the basis vectors of $\mathcal{V}_\text{D}$ and $\mathcal{V}$ respectively. For the infinite-dimensional state space, a complete basis set of vectors must satisfy the completeness equation, which for the co-ordinate basis $|x\rangle$ is given by Eqs. 3.6 and 3.4, namely

$$\int_{-\infty}^{\infty} dx |x\rangle\langle x| = \mathbb{I}, \quad \langle x|x'\rangle = \delta(x-x').$$

In general, state vectors $|\psi_n\rangle$ – with components given by $\psi_n(x) = \langle x|\psi_n\rangle$ – form a complete basis if

$$\sum_{n=-\infty}^{+\infty} |\psi_n\rangle\langle\psi_n| = \mathbb{I} \;\Rightarrow\; \sum_{n=-\infty}^{+\infty} \psi_n(x)\,\psi_n^*(x') = \delta(x-x').$$

The completeness equation is also referred to as the *resolution of the identity* since only a complete set of basis states can yield the identity operator on state space.

An element of the state space $\mathcal{V}$ is a ket vector $|\psi\rangle$, and can be thought of as an infinite-dimensional vector with *components* given by $\psi(x) = \langle x|\psi\rangle$. The vector $|\psi\rangle$ has the following representation in the $|x\rangle$ basis:

$$|\psi\rangle = \int_{-\infty}^{\infty} dx |x\rangle\langle x|\psi\rangle = \int_{-\infty}^{\infty} dx \psi(x)|x\rangle, \quad \psi(x) = \langle x|\psi\rangle. \tag{3.15}$$

The vector $|\psi\rangle$ can be mapped to a unique *dual* vector denoted by $\langle\psi| \in \mathcal{V}_\text{D}$; in components $\psi^*(x) = \langle\psi|x\rangle$ and

$$\langle\psi| = \int_{-\infty}^{\infty} dx \langle\psi|x\rangle\langle x| = \int_{-\infty}^{\infty} dx \psi^*(x)\langle x|, \quad \psi^*(x) = \langle\psi|x\rangle.$$

Note that the state vector and its dual are related by complex conjugation, namely

$$\langle\chi|\psi\rangle = \langle\psi|\chi\rangle^* \quad\Rightarrow\quad \langle x|\psi\rangle = \langle\psi|x\rangle^*. \tag{3.16}$$

The scalar product of two state vectors is given by[1]

$$\langle\chi|\psi\rangle \equiv \int dx\,\chi^*(x)\psi(x).$$

The vector $|\psi\rangle$ and its dual $\langle\psi|$ have the important property that they define the "length" $\langle\psi|\psi\rangle$ of the vector. The completeness equation, Eq. 3.6, yields

$$\langle\psi|\psi\rangle = \int_{-\infty}^{\infty} dx\,\psi(x)^*\psi(x) \geq 0.$$

3.3 Hermitian operators

An operator acting on state space defines a *linear mapping* of the state space $\mathcal{V}$ onto itself, and for a Hilbert space is an element of the tensor product space $\mathcal{V} \otimes \mathcal{V}_{\mathrm{D}}$.

For a two-state system, the state space is a two-dimensional Euclidean space and operators are 2×2 complex valued Hermitian matrices. Operators on linear vector state space are infinite-dimensional generalizations of $N \times N$ matrices, with $N \to \infty$ and have new properties that are absent in finite matrices.

The Hermitian conjugate of a square matrix M is defined by $M^\dagger_{ij} \equiv M^*_{ji}$. Similar to a matrix, the Hermitian conjugate of an arbitrary operator $\mathcal{O}$, denoted by $\mathcal{O}^\dagger$, is defined by

$$\langle\psi|\mathcal{O}^\dagger|\chi\rangle \equiv \langle\chi|\mathcal{O}|\psi\rangle^* \quad : \quad \text{Hermitian conjugation.}$$

An operator is Hermitian if the Hermitian conjugate operator is equal to the operator itself, that is, if

$$\mathcal{O}^\dagger = \mathcal{O} \quad\Rightarrow\quad \langle\psi|\mathcal{O}|\chi\rangle \equiv \langle\chi|\mathcal{O}|\psi\rangle^*. \tag{3.17}$$

One of the reasons for studying the Hermitian conjugate operator is because one can ascertain the state space on which its conjugate acts. It is not enough that the form of a Hermitian operator be invariant under conjugation, as in Eq. 3.17. For self-adjoint (Hermitian) operators, it is also necessary that the *domains* of the operator and its conjugate be isomorphic, and this is discussed in Section 3.10.

[1] A more direct derivation of the completeness equation is the following:

$$\langle\chi|\psi\rangle = \langle\chi|\{\int_{-\infty}^{\infty} dx|x\rangle\langle x|\}|\psi\rangle \Rightarrow \mathbb{I} = \int_{-\infty}^{\infty} dx|x\rangle\langle x|.$$

However, for non-Hermitian Hamiltonians, and these are the ones that occur in the acceleration (higher derivative) Lagrangian discussed in Chapter 14, as well as for non-Hermitian Hamiltonians describing classical random systems, the difference between the domains of the operator and its conjugate is non-trivial.

Note that all the *diagonal elements* of a Hermitian operator $\mathcal{O}$ are real, since for any arbitrary state vector $|\psi\rangle$, the diagonal element is

$$\langle\psi|\mathcal{O}|\psi\rangle = \langle\psi|\mathcal{O}^\dagger|\psi\rangle = \langle\psi|\mathcal{O}|\psi\rangle^* \ : \ \text{real}.$$

Furthermore, similar to matrices, the Hermitian adjoint of a sum and products of operators is given by

$$(A + B + \ldots)^\dagger = A^\dagger + B^\dagger \ldots, \quad (AB\ldots)^\dagger = \ldots B^\dagger A^\dagger.$$

The trace operation for an operator $\mathcal{O}$, similar to matrices, is defined as a sum of all its "diagonal elements." To make this statement more precise, one needs a resolution of the identity operator on state space $\mathcal{V}$. Consider for concreteness, the continuous degree of freedom with the completeness equation given by Eq. 3.6 as follows:

$$\mathbb{I} = \int_{-\infty}^{\infty} dx|x\rangle\langle x|.$$

Trace is a *linear* operation on $\mathcal{O}$ and is defined by

$$tr(\mathcal{O}) = tr(\mathcal{O}\mathbb{I}) = \int_{-\infty}^{\infty} dx\, tr\Big(\mathcal{O}|x\rangle\langle x|\Big) = \int_{-\infty}^{\infty} dx\langle x|\mathcal{O}|x\rangle. \tag{3.18}$$

Trace operation is defined in Eq. 3.18, and is summarized below.

$$\begin{aligned} tr(A + B + \ldots) &= tr(A) + tr(B)\ldots \\ tr(A^\dagger) &= tr^*(A) \\ tr(ABC) &= tr(CBA) \ : \ \text{cyclic}. \end{aligned}$$

A unitary operator, the generalization of the exponential function $\exp i\phi$, is given in terms of a Hermitian operator $\mathcal{O}$ by

$$\begin{aligned} U &= e^{i\phi\mathcal{O}} \ \Rightarrow \ UU^\dagger = \mathbb{I} \\ V &= \frac{1 - ia\mathcal{O}}{1 + ia\mathcal{O}} \ \Rightarrow \ VV^\dagger = \mathbb{I}. \end{aligned}$$

3.3.1 Eigenfunctions; completeness

Consider a Hermitian operator $\mathcal{O}$. All Hermitian operators have the following important properties:

- The eigenfunctions are a very special set of state functions that are only *rescaled* when the operator acts on them and are given, as in Eq. 2.7, by

$$\mathcal{O}|\psi_n\rangle = \lambda_n|\psi_n\rangle,$$

where λ_n are eigenvalues of the operator $\mathcal{O}$.

- The eigenfunctions are orthonormal and complete, namely

$$\langle\psi_n|\psi_n\rangle = \delta_{n-m}, \quad \sum_n |\psi_n\rangle\langle\psi_n| = \mathbb{I}. \tag{3.19}$$

Every state vector $|\psi\rangle$ has an expansion in terms of the basis states given by

$$|\psi\rangle = \mathbb{I}|\psi\rangle = \sum_n \langle\psi_n|\psi\rangle|\psi_n\rangle. \tag{3.20}$$

- The Hilbert space and its dual are isomorphic and hence $\mathcal{O} \in \mathcal{V} \otimes \mathcal{V}_D \equiv \mathcal{V} \otimes \mathcal{V}$; the *spectral resolution* of the Hermitian operator yields

$$\mathcal{O} = \sum_n \lambda_n |\psi_n\rangle\langle\psi_n|, \tag{3.21}$$

$$\mathcal{O}^\dagger = \sum_n \lambda_n^* \left(|\psi_n\rangle\langle\psi_n|\right)^\dagger = \mathcal{O}.$$

In other words, an operator is completely equivalent to the set of all of its eigenfunctions and eigenvalues.

- Hermitian operator $\mathcal{O}$ is represented by an experimental device. Experimental observations carried out on the state $|\psi\rangle$ function are represented by the operator acting on the quantum state, namely $\mathcal{O}|\psi\rangle$. Repeated observations cause the state $|\psi\rangle$ to collapse to one of the eigenfunctions $|\psi_n\rangle$ of $\mathcal{O}$ with probability $|\langle\psi_n|\psi\rangle|^2$. This procedure is the mathematical basis for making measurements on a quantum entity, as discussed in Section 2.6.

Functions of Hermitian operators are fundamental to a quantum system; for an arbitrary operator valued function $f(\mathcal{O})$, the spectral resolution given in Eq. 3.21 yields

$$f(\mathcal{O}) = \sum_n f(\lambda_n)|\psi_n\rangle\langle\psi_n|, \tag{3.22}$$

where $f(\lambda_n)$ is an ordinary numerical valued function of the eigenenergies λ_n. If the function $f(\lambda) = f^*(\lambda)$: real, then from its definition, for Hermitian operator $\mathcal{O}$

$$f^\dagger(\mathcal{O}) = f(\mathcal{O}) \quad : \text{ Hermitian.}$$

3.3.2 Hamiltonian for a periodic degree of freedom

The eigenfunctions of every Hermitian operator yield a resolution of the identity as given in Eq. 3.19. This result is derived for the Hamiltonian operator of a particle freely moving on a circle S^1 with radius L.

The free particle Hamiltonian operator H, with the potential $V(x) = 0$, is given, from Eq. 2.11, by

$$H = -\frac{\hbar}{2m}\frac{\partial^2}{\partial x^2}. \tag{3.23}$$

Since $x \in S^1$ is defined on a *periodic domain* $[0, 2\pi L]$, the degree of freedom has the periodicity

$$x = x + 2\pi L.$$

Consider the normalized eigenfunctions of H given by

$$H\psi_n(x) = E_n\psi_E(x), \quad \int_0^{2\pi L} dx|\psi_n(x)|^2 = 1.$$

The ground state (having lowest energy) is nondegenerate and is given by

$$\psi_0(x) = \frac{1}{\sqrt{2\pi L}}, \quad E_0 = 0.$$

All the other energy eigenfunctions are two fold-degenerate and given by

$$H\psi_{\pm n}(x) = E_n\psi_{\pm n}(x), \ E_n = \frac{n^2}{2mL^2},$$
$$\psi_{\pm n}(x) = \frac{1}{\sqrt{2\pi L}}e^{\pm i\frac{nx}{L}}, \quad n = 1, 2, \ldots + \infty.$$

From the spectral decomposition of Hermitian operators given in Eq. 3.21, the Hamiltonian has the representation

$$H = \sum_n E_n|\psi_n\rangle\langle\psi_n| = \frac{1}{4\pi mL^3}\sum_{n=1}^{+\infty} n^2\Big(|\psi_{+n}\rangle\langle\psi_{+n}| + |\psi_{-n}\rangle\langle\psi_{-n}|\Big). \tag{3.24}$$

The general result given in Eq. 3.19 yields that the eigenfunctions of H are complete and hence provide the completeness equation

$$\mathbb{I} = |\psi_0\rangle\langle\psi_0| + \sum_{n=1}^{+\infty}(|\psi_{+n}\rangle\langle\psi_{+n}| + |\psi_{-n}\rangle\langle\psi_{-n}|). \tag{3.25}$$

Note that the completeness equation requires that *all* the eigenfunctions, including all the degenerate eigenfunctions, of the Hermitian operator, be included in the resolution of the identity operator. To prove the completeness equation, consider the following expression:

$$\begin{aligned}
&\langle x'|\{|\psi_0\rangle\langle\psi_0| + \sum_{n=1}^{+\infty}(|\psi_{+n}\rangle\langle\psi_{+n}| + |\psi_{-n}\rangle\langle\psi_{-n}|)\}|x\rangle \\
&= \frac{1}{2\pi L}(1 + \sum_{n=1}^{+\infty}(e^{i\frac{n}{L}(x'-x)} + e^{-i\frac{n}{L}(x'-x)})) \\
&= \frac{1}{2\pi L}\sum_{n=-\infty}^{+\infty} e^{i\frac{n}{L}(x'-x)} = \sum_{n=-\infty}^{+\infty} \delta\left(x - x' + 2\pi n L\right) \\
&= \delta\left(x - x'\right) \quad \text{since } x, x' \in [0, 2\pi L].
\end{aligned}$$

To obtain the final expression requires Poisson's summation formula

$$\frac{1}{2\pi L}\sum_{n=-\infty}^{+\infty} e^{i\frac{n}{L}(x'-x)} = \sum_{n=-\infty}^{+\infty} \delta\left(x - x' + 2\pi n L\right). \tag{3.26}$$

Taking the limit of $L \to \infty$ yields the result for $x \in [-\infty, +\infty]$, namely that the completeness equation given in Eq. 3.25 for a circle converges to the unit operator on the real line $\Re$ given in Eq. 3.6,

$$\lim_{L\to\infty}\frac{1}{2\pi L}\sum_{n=-\infty}^{+\infty} e^{i\frac{n}{L}(x'-x)} = \frac{1}{2\pi}\int_{-\infty}^{+\infty} dp e^{ip(x'-x)} = \delta(x' - x). \tag{3.27}$$

The limit of $L \to \infty$ shows that the eigenfunctions of the Hamiltonian H for the degree of freedom being a real line $\Re$ have a divergent normalization. To make the normalization finite for $\Re$, one can use various methods, including making the degree of freedom periodic to obtain a finite normalization for the eigenfunctions; one then needs to take the limit of the circle having an infinite radius to obtain the real line $\Re$.

3.4 Position and momentum operators $\hat{x}$ and $\hat{p}$

A quantum particle has a continuous (real) degree of freedom $x \in \Re$. The state space consists of functions of the single variable x, namely $\mathcal{V} = \{\psi(x)|x \in \Re\}$, where $\langle x|\psi\rangle = \psi(x)$.

One of the most important observables is the Hermitian coordinate operator $\hat{x}$ that represents the coordinate degree of freedom on the state space of the quantum particle. The operator notation $\hat{x}$ is often simplified to x if there is no ambiguity in its meaning.

The action of operator $\hat{x}$ is defined by the multiplication of the state function $\psi(x) \in \mathcal{V}$ by x, that is

$$\hat{x}\psi(x) \equiv x\psi(x).$$

The operator $\hat{x}$ has a *continuous spectrum* of eigenvalues and eigenstates.

Similarly to an $N \times N$ matrix M that is fully specified by its matrix elements $M_{ij}, i, j = 1, \ldots, N$, an operator is also specified by its matrix elements. In the notation of Dirac

$$\hat{x}\psi(x) \equiv \langle x|\hat{x}|\psi\rangle = x\langle x|\psi\rangle = x\psi(x).$$

In other words, the matrix element $\langle x|\hat{x}|\psi\rangle$ of the operator $\hat{x}$ is given by $x\psi(x)$. Let us choose the function $|\psi\rangle = |x'\rangle$ that yields

$$\begin{aligned} &\langle x|\hat{x}|x'\rangle = x\langle x'|x'\rangle = x\delta(x - x') \\ &\Rightarrow \hat{x}|x'\rangle = x'|x'\rangle. \end{aligned} \tag{3.28}$$

From the above it follows that the observable $\hat{x}$ has eigenfunctions $|x\rangle$ with eigenvalues $x \in \Re$; hence the spectral resolution of observable $\hat{x}$ and its *completeness equation* are given by

$$\hat{x} = \int_{-\infty}^{\infty} dx|x\rangle x\langle x|, \quad \int_{-\infty}^{\infty} dx|x\rangle\langle x| = \mathbb{I}. \tag{3.29}$$

The second equation above is the completeness equation given earlier in Eq. 3.6.

For N particles in three-dimensions, one has the following straightforward generalization of the coordinate operator $\hat{x}$ and completeness equation:

$$\hat{\mathbf{x}} = \left[\hat{x}_1 \otimes \hat{y}_1 \otimes \hat{z}_1\right] \otimes \left[\hat{x}_2 \otimes \hat{y}_2 \otimes \hat{z}_2\right] \ldots \otimes \left[\hat{x}_N \otimes \hat{y}_N \otimes \hat{z}_N\right] \tag{3.30}$$

$$\begin{aligned} \mathcal{I}_N = &\int_{-\infty}^{\infty} dx_1 dy_1 dz_1 \ldots dx_N dy_N dz_N \\ &\times |x_1, y_1, z_1\rangle \otimes \langle x_1, y_1, z_1| \ldots |x_N, y_N, z_N\rangle \otimes \langle x_N, y_N, z_N|, \end{aligned}$$

where $|x_1, y_1, z_1\rangle = |x_1\rangle|y_1\rangle|z_1\rangle$, $\langle x_1, y_1, z_1| = \langle x_1|\langle y_1|\langle z_1|$ and so on.

3.4.1 Momentum operator $\hat{p}$

Momentum is a central concept in classical physics and important classical quantities such as energy and angular momentum depend on momentum. Since the quantum particle's state function $\psi(x)$ depends only on x, how do we define the quantum generalization of classical momentum $p = mdx/dt$, where a particle's mass is m?

In Section 6.7, it is shown that in the path integral framework, the momentum of a particle continues to be $p = mdx/dt$, but with the caveat that p is an indeterminate (uncertain) quantity. The path integral is one way of defining a quantum system; another equivalent method is to *define* momentum to be an operator $\hat{p}$ on state space $\mathcal{V}$.

There are many ways of motivating the definition of the momentum operator but in the final analysis one has to *postulate* the definition, as there is no way of "deriving" this result from classical physics. Of course, the final test of whether the postulate is correct is experiment, and the definition adopted has been rigorously tested experimentally.

The momentum operator $\hat{p}$ is *postulated* to be[2]

$$\hat{p} = -i\hbar\frac{\partial}{\partial x}. \tag{3.31}$$

Note Planck's constant $\hbar$ enters due to dimensional consistency, but its actual empirical value is fixed by Nature and has to be obtained by doing an appropriate experiment.

Consider a particle moving in one dimension. The differential operator $\partial/\partial x$ maps $\psi(x) \in \mathcal{V}$ to its derivative $\partial\psi(x)/\partial x \in \mathcal{V}$. The momentum operator $\hat{p} = -i\hbar\partial/\partial x$, in Dirac's notation, is given by

$$\langle x|\hat{p}|\psi\rangle = -i\hbar\langle x|\frac{\partial}{\partial x}|\psi\rangle = -i\hbar\frac{\partial\psi(x)}{\partial x}. \tag{3.32}$$

From Eq. 3.17, a Hermitian operator satisfies the following

$$\hat{p}^\dagger = \hat{p} \quad \Rightarrow \quad \langle\psi|\hat{p}^\dagger|\chi\rangle \equiv \langle\chi|\hat{p}|\psi\rangle^* = \langle\psi|\hat{p}|\chi\rangle. \tag{3.33}$$

Doing an integration by parts yields the following proof that $\hat{p}$ is Hermitian:

$$\begin{aligned}\langle\chi|\hat{p}|\psi\rangle^* &= \left[-\int_{-\infty}^{+\infty} dx\chi^*(x)i\hbar\frac{\partial\psi(x)}{\partial x}\right]^* \\ &= -\int_{-\infty}^{+\infty} dx\psi^*(x)i\hbar\frac{\partial\chi(x)}{\partial x} = \langle\psi|\hat{p}|\chi\rangle.\end{aligned}$$

The eigenfunctions of $\hat{p}$, in the notation of Eq. 3.31, are given by

$$\hat{p}|p\rangle = p|p\rangle$$

with completeness equation

$$\int_{-\infty}^{+\infty}\frac{dp}{2\pi\hbar}|p\rangle\langle p| = \mathbb{I}. \tag{3.34}$$

[2] From Eq. 3.30, since the coordinate operator for the $3N$ degrees of freedom is a tensor product of the single degree of freedom, it is sufficient to define the momentum operator for one dimension and build up the momentum for the $3N$ degrees of freedom by an appropriate tensor product.

The operator $\hat{p}$, from the completeness equation of Hermitian operators given in Eq. 3.21, has the representation

$$\hat{p} = \int_{-\infty}^{\infty} \frac{dp}{2\pi\hbar} |p\rangle p \langle p|. \tag{3.35}$$

The eigenfunctions of $\hat{p}$, in the notation of Eq. 3.31, are given by

$$\langle x|p\rangle = e^{ipx/\hbar} \quad \Rightarrow \quad \langle x|\hat{p}|p\rangle = pe^{ipx/\hbar}. \tag{3.36}$$

The completeness equation for degree of freedom x, using Eq. 3.36, yields the following transformation between the momentum and coordinate basis states:

$$|p\rangle = \int_{-\infty}^{\infty} dx|x\rangle\langle x|p\rangle = \int_{-\infty}^{\infty} dx e^{ipx/\hbar}|x\rangle. \tag{3.37}$$

The momentum operator, acting on the state function, shifts it in space. More precisely, using Eq. 3.31, for a constant a consider the following *shift operator*:

$$T(a) = e^{i\frac{a}{\hbar}\hat{p}} = e^{a\frac{\partial}{\partial x}}, \quad T(a)T(b) = T(a+b), \tag{3.38}$$

$$\langle x|T(a)|\psi\rangle = e^{a\frac{\partial}{\partial x}}\psi(x) = \psi(x+a) = \langle x+a|\psi\rangle, \tag{3.39}$$

$$\Rightarrow T(a)|x\rangle = |x-a\rangle, \quad \langle x|T(a) = \langle x+a|. \tag{3.40}$$

One can *define* the momentum operator $\hat{p}$, from first principles, as a translation operator using Eqs. 3.39 and 3.40.

Three Hermitian operators that play a central role in quantum mechanics are the coordinate operator x, the momentum operator p and the Hamiltonian operator H; the Hamiltonian, as given in Eq. 2.11, has the form

$$H = \frac{1}{2m}p^2 + V(x) = -\frac{1}{2m}\frac{\partial^2}{\partial x^2} + V(x) = H^\dagger.$$

3.5 Weyl operators

Both the coordinate and momentum operators, $\hat{x}$ and $\hat{p}$ respectively, are infinite-dimensional Hermitian operators. Since $tr\hat{x} = \infty = tr\hat{p}$, these operators cannot have any finite-dimensional matrix representation. In contrast, when acting on a periodic lattice, both

- the exponential of the momentum operator, namely, the shift operator $T(a)$ given in Eq. 3.38
- and the exponential of the coordinate operator

have a finite-dimensional matrix representation.

The finite-dimensional realizations of the exponential of the coordinate and momentum operators are called the Weyl operators.

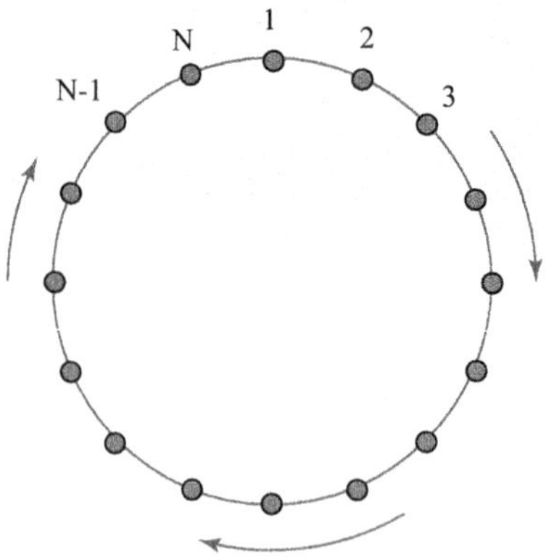

Figure 3.2 A finite periodic lattice, with N lattice sites.

Consider the periodic chain being embedded in real space and let its length be L. Let the lattice points be located at $x = na$ with $Na = L$. The periodic lattice – with sites $n = 1, 2, \ldots .N$ – is shown in Figure 3.2 and has position eigenstates $|n\rangle$ that are periodic, namely

$$|n + N\rangle = |n\rangle.$$

The *shift operator* V, the finite-dimensional analog of the momentum shift operator $T(a)$ given in Eq. 3.38, is defined by

$$V|n\rangle = |n + 1\rangle.$$

The coordinate operator U, the finite-dimensional analog of the exponential of the coordinate operator $\hat{x}$, is a multiplication operator given by

$$U|n\rangle = \lambda^n|n\rangle.$$

For the periodic chain the following N-dimensional orthonormal vectors form a complete coordinate basis:

$$|1\rangle = \begin{pmatrix} 1 \\ 0 \\ \vdots \\ 0 \end{pmatrix}, \ |2\rangle = \begin{pmatrix} 0 \\ 1 \\ 0 \\ \vdots \end{pmatrix}, \ldots \ |N-1\rangle = \begin{pmatrix} 0 \\ \vdots \\ 1 \\ 0 \end{pmatrix}, \ |N\rangle = \begin{pmatrix} 0 \\ \vdots \\ 0 \\ 1 \end{pmatrix}.$$

The shift operator is given as $V|n\rangle = |n+1\rangle$; for the state $|N\rangle$, due to periodicity, one has $V|N\rangle = |N + 1\rangle = |1\rangle$; hence

$$V|N\rangle = |N + 1\rangle = |1\rangle \ \Rightarrow \ V \begin{pmatrix} 0 \\ \vdots \\ 1 \end{pmatrix} = \begin{pmatrix} 1 \\ \vdots \\ 0 \end{pmatrix}.$$

A matrix representation of V is, consequently, given by

$$V = \begin{pmatrix} 0 & \dots & \dots & 0 & 1 \\ 1 & 0 & \dots & & \\ 0 & 1 & 0 & \dots & \\ \ddots & & \ddots & & \\ 0 & \ddots & & 1 & 0 \end{pmatrix}. \tag{3.41}$$

It can be verified that the matrix representation of V given in Eq. 3.41 yields, as expected,

$$V^N|n\rangle = |N+n\rangle = |n\rangle \quad \Rightarrow V^N = \mathbb{I}.$$

For the coordinate representation, note U is a diagonal matrix with the constraint that

$$\begin{aligned} U|N+1\rangle &= \lambda^{N+1}|N+1\rangle = U|1\rangle = \lambda|1\rangle \\ &\Rightarrow \lambda^N = 1 \quad \Rightarrow \lambda = e^{\frac{2\pi i}{N}}. \end{aligned}$$

Hence, one obtains the following matrix representation of the U and V operators:

$$U = \begin{pmatrix} \lambda & 0 & \dots & & \\ 0 & \lambda^2 & 0 & \dots & \\ 0 & 0 & \lambda^3 & 0 & \\ 0 & \dots & & \ddots & 0 \\ 0 & \dots & & & 1 \end{pmatrix}, \quad V = \begin{pmatrix} 0 & \dots & \dots & 0 & 1 \\ 1 & 0 & \dots & & \\ 0 & 1 & 0 & \dots & \\ \ddots & & \ddots & & \\ 0 & \ddots & & 1 & 0 \end{pmatrix}.$$

Note, as required by a consistent representation, one has the following

$$U^N = \begin{pmatrix} \lambda^N & 0 & \dots & & \\ 0 & \lambda^{2N} & 0 & \dots & \\ 0 & 0 & \lambda^{3N} & 0 & \\ 0 & \dots & & \ddots & 0 \\ 0 & \dots & & \lambda^{(N-1)N} & 0 \\ 0 & \dots & & 0 & 1 \end{pmatrix} = \mathbb{I}.$$

The finite Fourier transform of the basis states $|n\rangle$ yields the following:

$$\begin{aligned} |k\rangle &= \sum_{n=1}^{N} e^{-\frac{2\pi ikn}{N}}|n\rangle \\ \Rightarrow V|k\rangle &= \sum_{n=1}^{N} e^{-\frac{2\pi ikn}{N}}|n+1\rangle = e^{\frac{2\pi i}{N}k}\sum_{n=1}^{N} e^{-\frac{2\pi ikn}{N}}|n\rangle = \lambda^k|k\rangle \end{aligned}$$

$$U|k\rangle = \sum_{n=1}^{N} e^{-\frac{2\pi ikn}{N}} U|n\rangle = \sum_{n=1}^{N} e^{-\frac{2\pi ikn}{N}} e^{\frac{2\pi in}{N}} |n\rangle = |k-1\rangle.$$

Hence the ket vectors $|k\rangle$, $k = 1, 2, \ldots N$ are eigenstates that are dual to $|n\rangle$, being (momentum) eigenstates of the operator V and on which U is a shift operator.

The finite version of the commutator of U and V is given as

$$\begin{aligned} UV|n\rangle &= U|n+1\rangle = \lambda^{n+1}|n+1\rangle \\ VU|n\rangle &= \lambda^{n} V|n\rangle = \lambda^{n}|n+1\rangle \\ &\Rightarrow UV = \lambda VU. \end{aligned} \tag{3.42}$$

In Section 3.8, the matrix operators U and V on the periodic lattice are represented by the coordinate and momentum operator $\hat{x}$ and $\hat{p}$ and yield an exponential (unitary) version of the Heisenberg commutation.

3.6 Quantum numbers; commuting operator

A classical system has conserved quantities such as energy, momentum, angular momentum and so on; in fact, one usually characterizes a classical system by its conserved quantities, called constants of motion. There is a quantum mechanical generalization of classically conserved quantities.

The Hamiltonian H is the most important operator for a quantum system since it evolves the state function in time. The commutator of any two operators A and B is defined by

$$[A, B] = AB - BA.$$

Consider a collection of *commuting operators* $\mathcal{O}_i$; $i = 1, 2, \ldots N$ that also commute with H, namely

$$\begin{aligned} [\mathcal{O}_i, H] &= 0, \quad i = 1, 2, \ldots N \\ \left[\mathcal{O}_i, \mathcal{O}_j\right] &= 0, \quad i, j = 1, 2, \ldots N. \end{aligned}$$

Hermitian operators have an important property that all *commuting operators* can be simultaneously diagonalized with eigenfunctions $|\psi_{n,n_1,n_2,\ldots n_N}\rangle$ that obey

$$\begin{aligned} H|\psi_{n,n_1,n_2,\ldots n_N}\rangle &= E_n|\psi_{n,n_1,n_2,\ldots n_N}\rangle, \quad n = 1, 2, \ldots \\ \mathcal{O}_i|\psi_{n,n_1,n_2,\ldots n_N}\rangle &= \lambda^i_{n_i}|\psi_{n,n_1,n_2,\ldots n_N}\rangle, \quad i = 1, 2, \ldots N, \quad n_i = 0, \pm 1, \pm 2, \ldots \end{aligned}$$

Suppose $|\psi_{n_1,n_2,\ldots n_N}\rangle$ is the initial state function and is an eigenfunction of all the operators $\mathcal{O}_i$; $i = 1, 2, \ldots N$ – but *not* necessarily an eigenfunction of H; then, from Eq. 2.12 and due to the commutativity of $\mathcal{O}_i$ with H

$$\begin{aligned}
|\psi_{t;n_1,n_2,\ldots n_N}\rangle &= e^{-itH/\hbar}|\psi_{n_1,n_2,\ldots n_N}\rangle \\
\Rightarrow \mathcal{O}_i|\psi_{t;n_1,n_2,\ldots n_N}\rangle &= e^{-itH/\hbar}\mathcal{O}_i|\psi_{n,n_1,\ldots n_i,\ldots n_N}\rangle \\
&= \lambda^i_{n_i} e^{-itH/\hbar}|\psi_{n_1,\ldots n_i,\ldots n_N}\rangle \\
&= \lambda^i_{n_i}|\psi_{t;n_1,\ldots n_i,\ldots n_N}\rangle.
\end{aligned}$$

The result above shows that the set of observables which commute with the Hamiltonian and with each other provides eigenvalues that are *conserved* in time.

The integers n_i, $i = 1, 2, \ldots N$ are called *quantum numbers* – and replace the classical constants of motion. The *time independent* eigenvalues $\lambda^i_{n_i}$ are *quantum constants of motion* and are the generalization of classically conserved quantities.[3]

Fixing the values of the various n_is fully specifies a particular eigenstate of the observables $\mathcal{O}_i$; $i = 1, 2, \ldots N$. An arbitrary state function for such a system can be expressed by the eigenfunction expansion

$$|\psi_t\rangle = \sum_{n_1,n_2,\ldots n_N} c_{n_1,n_2,\ldots n_N}(t)|\psi_{n_1,n_2,\ldots n_N}\rangle.$$

3.7 Heisenberg commutation equation

Operators that do not commute with the Hamiltonian and with each other occur widely in quantum mechanics; that is the reason that operator algebras that occur in quantum mechanics are nontrivial. One of the most important cases of noncommuting observables is that of position and momentum, for which

$$[\hat{x}, \hat{p}] = i\hbar\mathbb{I} \neq 0. \tag{3.43}$$

Equation 3.43 defines the famous Heisenberg commutation equation, also called the Heisenberg algebra.

To explore the concept of *noncommuting operators*, Eq. 3.43 is derived from first principles. Consider the following representation of the coordinate and momentum observables in one dimension given by Eqs. 3.29 and 3.35:

$$\hat{x} = \int_{-\infty}^{\infty} dx|x\rangle x\langle x|, \quad \hat{p} = \int_{-\infty}^{\infty} \frac{dp}{2\pi\hbar}|p\rangle p\langle p|.$$

[3] There are quantum systems for which observables like momentum and position have continuous eigenvalues; the discussion for integer quantum numbers can be generalized for these systems.

The commutator of the coordinate and momentum of a quantum particle, from the above equations, is given by[4]

$$\left[\hat{x}, \hat{p}\right] \equiv \hat{x}\hat{p} - \hat{p}\hat{x} = \int_{-\infty}^{\infty} dx \frac{dp}{2\pi\hbar} xp \left[|x\rangle\langle x|p\rangle\langle p| - |p\rangle\langle p|x\rangle\langle x|\right]. \tag{3.44}$$

Expressing the commutator entirely in the coordinate basis by transforming the momentum basis using Eq. 3.37 yields

$$\left[\hat{x}, \hat{p}\right] = \int_{-\infty}^{\infty} dx dx' \frac{dp}{2\pi\hbar} xp\left[e^{ipx/\hbar}e^{-ipx'/\hbar}|x\rangle\langle x'| - e^{ipx'/\hbar}e^{-ipx/\hbar}|x'\rangle\langle x|\right]$$

$$= \frac{\hbar}{i}\int_{-\infty}^{\infty} dx dx' x\left[|x\rangle\langle x'|\frac{\partial}{\partial x}\delta(x-x') - |x'\rangle\langle x|\frac{\partial}{\partial x'}\delta(x'-x)\right], \tag{3.45}$$

$$= \frac{\hbar}{i}\int_{-\infty}^{\infty} dx dx'\left[(x-x')\frac{\partial}{\partial x}\delta(x-x')\right]|x\rangle\langle x'|$$

$$= i\hbar\int_{-\infty}^{\infty} dx|x\rangle\langle x| = i\hbar\mathbb{I}, \tag{3.46}$$

where Eq. 3.45 and the last equation follow from the identity[5]

$$(x-x')\frac{\partial}{\partial x}\delta(x-x') = -\delta(x-x').$$

Hence

$$\left[\hat{x}, \hat{p}\right] = i\hbar\mathbb{I} \ : \ \text{Heisenberg commutation equation.} \tag{3.47}$$

For N particles moving in three space dimensions, the degrees of freedom are $x_{ai}, \ p_{ai}$ with $a = 1, 2, \ldots N$ and $i = 1, 2.3$. The Heisenberg commutation equation is given by

$$\left[\hat{x}_{ai}, \hat{p}_{bj}\right] = i\hbar\delta_{a-b}\delta_{i-j}\mathbb{I}$$

$$\left[\hat{x}_{ai}, \hat{x}_{bj}\right] = 0, \quad \left[\hat{p}_{ai}, \hat{p}_{bj}\right] = 0.$$

3.8 Unitary representation of Heisenberg algebra

The Weyl operators discussed in Section 3.5 are re-examined in light of the Heisenberg algebra given in Eq. 3.47.

The periodic lattice given in Figure 3.2 is defined as being embedded in continuous space, and with the radius of the periodic lattice being R. In other words the

[4] There is an elementary derivation of $[\hat{x}, \hat{p}]$ using the chain rule of differentiation; the derivation given examines the operator structure of the momentum and coordinate operators.

[5] The identity follows from the equation

$$\frac{\partial}{\partial x}\left[(x-x')\delta(x-x')\right] = 0.$$

lattice sites of the periodic lattice correspond to a discrete set of points in continuous space. The length of the lattice is $2\pi R = L$ and lattice spacing a is given by $Na = L$.

The position eigenstates are

$$|na\rangle = |x\rangle, \quad n = 1, 2, \ldots N$$

$$\langle na|ma\rangle = \delta(an - am) = \frac{1}{a}\delta_{n-m} \quad : \text{continuum normalization.}$$

Let us define the continuous realization of the Weyl operators by (infinite-dimensional) unitary operators U, V given for the continuous coordinate basis as

$$V = e^{-\frac{i}{\hbar}a\hat{p}} = e^{-a\frac{\partial}{\partial x}}, \quad U = e^{\frac{2\pi i\hat{x}}{L}}.$$

The continuous degree of freedom yields, from Eq. 3.40

$$V|x\rangle = |x + a\rangle \;\Rightarrow\; V|na\rangle = |na + a\rangle$$

and

$$U|x\rangle = e^{\frac{2\pi i\hat{x}}{L}}|na\rangle = e^{\frac{2\pi ina}{L}}|na\rangle = \lambda^n|na\rangle$$
$$\Rightarrow \lambda = e^{2\pi i(\frac{a}{L})} = e^{\frac{2\pi i}{N}}.$$

Let $U = e^A$ and $V = e^B$. Since

$$[x, \partial_x] = x\partial_x - \partial_x x = -1,$$

the Campbell–Hausdorff formula yields

$$e^A e^B = e^{A+B+\frac{1}{2}[A,B]}.$$

Hence

$$UV = e^{\frac{2\pi ix}{L}}e^{a\frac{\partial}{\partial x}} = e^{\frac{2\pi ix}{L}+a\partial_x+\frac{2\pi ia}{2L}[x,\partial_x]} = \lambda^{-1/2}e^{A+B}$$
$$VU = \lambda^{1/2}e^{A+B},$$

and yields

$$UV = e^{\frac{2\pi i}{N}}VU \; : \text{Heisenberg commutation equation.} \tag{3.48}$$

Note Eq. 3.48 is the continuous version of the commutation equation derived for Weyl operators in Eq. 3.42.

In summary, the unitary operators

$$V = e^{\frac{ia}{\hbar}\hat{p}}, \quad U = e^{\frac{2\pi i\hat{x}}{L}} \tag{3.49}$$

provide a unitary representation of Heisenberg algebra.

The Weyl matrices discussed in Section 3.5 provide a finite-dimensional realization of the unitary representation of the algebra given in Eq. 3.49.

Noteworthy 3.2 **Position and momentum incompatible**

Position and momentum operators do not commute since $[\hat{x}, \hat{p}] = i\hbar\mathbb{I}$ and hence there is no state function that is a simultaneous eigenfunction of both $\hat{x}$ and $\hat{p}$. The noncommutativity of $\hat{x}$ and $\hat{p}$ is an operator expression of the fact that if the position of the quantum particle is known at instant t, its momentum is not known. This property of a quantum particle reflects the fact that the quantum particle *does not* have a unique trajectory – leading to the lack of a unique derivative of the position and consequently to an indeterminate momentum. The Heisenberg commutation equation is an operator expression of the fact that the path of the quantum particle is indeterminate.

3.9 Density matrix: pure and mixed states

The measurement of the expectation value of observable $\mathcal{O}$ can be expressed in terms of the density matrix. From Eq. 2.6, the expectation value of an operator $\mathcal{O}$ can be expressed as

$$E[\mathcal{O}] \equiv \langle\chi|\mathcal{O}|\chi\rangle = tr(\mathcal{O}|\chi\rangle\langle\chi|) = tr(\mathcal{O}\rho_P), \tag{3.50}$$

where the density matrix of a pure state ρ_P is given by

$$\Rightarrow \rho_P = |\chi\rangle\langle\chi| \;:\; \text{pure density matrix.} \tag{3.51}$$

Equation 2.8 for the expectation value of an operator $\mathcal{O}$ with eigenvectors $\mathcal{O}|\psi_i\rangle = \lambda_i|\psi_i\rangle$ can be re-written in terms of the mixed density matrix ρ_M as

$$E[\mathcal{O}] = \langle\psi|\hat{O}|\psi\rangle = tr(\sum_i |c_i|^2\mathcal{O}|\psi_i\rangle\langle\psi_i|) = tr(\mathcal{O}\rho_M)$$

$$\Rightarrow \rho_M = \sum_i p_i|\psi_i\rangle\langle\psi_i|, \quad p_i = |c_i|^2 \;:\; \text{mixed density matrix.}$$

The mixed density matrix ρ_M can be used for evaluating the expectation value of any function of the operator $\mathcal{O}$. However, there are uncontrollable errors if one uses ρ_M for evaluating the expectation value of another operator $\mathcal{Q}$ that does not commute with $\mathcal{O}$, namely $[\mathcal{O}, \mathcal{Q}] \neq 0$, and this is discussed in Baaquie (2013e).

Consider a quantum mechanical system, specified by a Hamiltonian H, in thermal equilibrium with a heat bath at temperature T. The system now has a quantum mechanical uncertainty as well as classical uncertainty due to thermal randomness.

Let H be the quantum mechanical Hamiltonian with the following resolution in terms of the energy eigenfunctions $|\psi_i\rangle$ and eigenvalues E_i:

$$H = \sum_i E_i |\psi_i\rangle\langle\psi_i|.$$

The canonical ensemble yields a probability distribution of energy eigenstates for the quantum system given by the Boltzmann distribution

$$\frac{1}{Z} e^{-H/k_B T} = \frac{1}{Z} \sum_i e^{-E_i/k_B T} |\psi_i\rangle\langle\psi_i|, \tag{3.52}$$

$$\text{where} \quad Z = tre^{-H/k_B T}, \tag{3.53}$$

where k_B is the Boltzmann constant.

Hence the mixed state density matrix ρ_M for the canonical ensemble is given by

$$\rho_M = \sum_i p_i |\psi_i\rangle\langle\psi_i|, \tag{3.54}$$

$$p_i = \frac{1}{Z} e^{-E_i/k_B T}, \quad \sum_i p_i = 1. \tag{3.55}$$

Consider the expectation value of the operator $\mathcal{O}$ for which $[\mathcal{O}, H] \neq 0$; then

$$\begin{aligned} E[\mathcal{O}] &= tr(\mathcal{O}\rho_M) \\ &= \sum_i p_i \alpha_i, \quad \alpha_i = \langle\psi_i|\mathcal{O}|\psi_i\rangle. \end{aligned}$$

Note ρ_M encodes both thermal and quantum uncertainty, with p_i and α_i reflecting the thermal and quantum uncertainties, respectively.

3.10 Self-adjoint operators

In this section the expression $\langle\psi_t|\hat{O}|\psi_t\rangle$ is carefully analyzed to determine precisely how self-adjoint operators are defined.[6] In particular, a point usually ignored in physics textbooks is the question of the *domain* of the self-adjoint operator. In the following sections, the importance of the domain is discussed as an independent ingredient in the definition of self-adjoint operators.

A self-adjoint operator is the generalization of the concept of a Hermitian *matrix* and takes into account the aspect of the domain of the Hermitian operator.

[6] The term self-adjoint is synonymous with Hermitian and is used when the more mathematical aspect of the operator is being discussed. Since in physics infinite-dimensional self-adjoint operators on Hilbert space are called Hermitian, the two terms will be used interchangeably.

A Hermitian matrix is only defined for complex square matrices $N \times N$ with M_{ij} with $i, j = 1, 2, \ldots N$. A matrix is Hermitian if M_{ij} satisfies

$$M^\dagger = M \Rightarrow M^*_{ji} = M_{ij}, \quad i, j = 1, 2, \ldots N. \tag{3.56}$$

Note the crucial point that the equality in Eq. 3.56 can hold for all i, j only if i, j have the same range. For finite square matrices there is no issue about the range of the indices i, j since they always have the same range.

The range over which the finite index i takes values has a generalization for state functions in Hilbert space $\mathcal{V}$. The domain of the operator $\mathcal{O}$, denoted by $D(\mathcal{O}) \subset \mathcal{V}$ is defined by all elements $|\psi\rangle$ in $\mathcal{V}$ such that $\mathcal{O}|\psi\rangle \in \mathcal{V}$. Similarly, the vector $|\chi\rangle$ is in $D(\mathcal{O}^\dagger)$, the domain of $\mathcal{O}^\dagger$, if $\mathcal{O}^\dagger|\chi\rangle \in \mathcal{V}$.

For an operator $\mathcal{O}$ acting on Hilbert space, its *adjoint* $\mathcal{O}^\dagger$ is defined, for an arbitrary state vector $|\psi\rangle$ and a dual vector $\langle\chi|$, as follows:

$$\langle\psi|\mathcal{O}^\dagger|\chi\rangle = \langle\chi|\mathcal{O}|\psi\rangle^* \text{ for all } |\psi\rangle \in D(\mathcal{O}),\ |\chi\rangle \in D(\mathcal{O}^\dagger). \tag{3.57}$$

Equation 3.57 is the definition of the adjoint operator $\mathcal{O}^\dagger$.

For operators on Hilbert space, the adjoint (Hermitian conjugate) operator can be defined for only those operators $\mathcal{O}$ for which the domain of the operator and its adjoint are isomorphic, or in other words, $D(\mathcal{O}) = D(\mathcal{O}^\dagger)$.

Once the domains of the operator and its conjugate are isomorphic, the form of the operator has to be invariant under conjugation, that is, $\mathcal{O} = \mathcal{O}^\dagger$, for the operator to be self-adjoint.

3.10.1 Momentum operator on finite interval

The example of the momentum operator $\hat{p} = -i\hbar d/dx$ is analyzed to demonstrate the subtleties of self-adjoint operators; Planck's constant $\hbar$ is set to 1 for notational convenience.

Consider the momentum operator being defined on a finite interval $[a, b]$; naively one would expect $\hat{p}$ to be self-adjoint using the rule for conjugation, namely that $(d/dx)^\dagger \rightarrow -d/dx$ and hence under conjugation

$$\hat{p} = -i\frac{d}{dx} \rightarrow \hat{p}^\dagger = -i\frac{d}{dx} : \qquad \text{Is this self-conjugate?}$$

The fact that the form of the operator is invariant under conjugation is not the whole story since we need to show that the *domains* $\mathcal{D}(\hat{p})$ and $\mathcal{D}(\hat{p}^\dagger)$ are equal; only then can it be concluded that $\hat{p}$ is self-adjoint. Since $\hat{p}$ is a first order differential operator, its domain is fixed by specifying only one boundary condition. The domain of $\hat{p}$ is a linear vector space; the boundary condition is chosen as a linear function of the state vectors to preserve the linearity of the domain; the most general boundary condition is given by

$$f(a) + \alpha f(b) = 0, \ f(x) \in \mathcal{D}(\hat{p}), \ x \in [a, b]. \tag{3.58}$$

Integration by parts yields

$$\begin{aligned}\langle f|\hat{p}^{\dagger}|g\rangle \equiv \langle g|\hat{p}|f\rangle^* &= \Big[-i \int_a^b dx g^*(x) \frac{df}{dx} \Big]^* \\ &= \langle f|\hat{p}|g\rangle + i\left[g^*(b)f(b) - g^*(a)f(a)\right] \\ \Rightarrow \langle f|\hat{p}^{\dagger}|g\rangle - \langle f|\hat{p}|g\rangle &= i\left[g^*(b)f(b) - g^*(a)f(a)\right].\end{aligned}$$

Hence, from the above $\hat{p}^{\dagger} = \hat{p}$ only if the boundary term is zero. Applying the boundary condition given in Eq. 3.58 to both $g(a)$, $f(a)$ yields

$$g^*(b)f(b) - g^*(a)f(a) = g^*(b)f(b)\left[1 - |\alpha|^2\right] = 0.$$

Hence, the momentum operator is self-adjoint if and only if

$$\alpha = e^{i\theta}.$$

In other words, for *all* functions $f(x)$ in the domains $\mathcal{D}(\hat{p})$ and $\mathcal{D}(\hat{p}^{\dagger})$, the condition of self-adjointness imposes an additional condition on the boundary values of $f(x)$, namely

$$f(a) + e^{i\theta} f(b) = 0, \ f(x) \in \mathcal{D}(\hat{p}); \ f(x) \in \mathcal{D}(\hat{p}^{\dagger}) \ \Rightarrow \mathcal{D}(\hat{p}) = \mathcal{D}(\hat{p}^{\dagger}). \tag{3.59}$$

Hence $\hat{p} = -id/dx$ is a self-adjoint on the interval $[a, b]$ since

$$\hat{p} = \hat{p}^{\dagger}$$

and at the same time the requirement of $\mathcal{D}_{\theta}(\hat{p}) = \mathcal{D}_{\theta}(\hat{p}^{\dagger})$ is fulfilled. The equality of the two domains results in the following boundary condition for the functions in the domain of $\hat{p}$:

$$f(a) = e^{i\theta} f(b).$$

In conclusion, on finite open interval $[a, b]$, there is a whole *class of momentum operators* $\hat{p}$ that are Hermitian on domain $\mathcal{D}_{\theta}$ indexed by an angular label $\theta \in [0, 2\pi)$.

Note that for the special case of a particle in an infinitely deep potential well the boundary condition is $f(a) = 0 = f(b)$, and hence the momentum operator $\hat{p}$ is self-adjoint for all values of θ.

3.11 Self-adjoint domain

From the discussion in the previous sections, an operator $\hat{O}$ is self-adjoint if and only if *both* the *form* of the operator and of its adjoint are equal, as well as the *domain* of the operator and of its adjoint being equal. The domains being equal is the analogy of only square matrices being Hermitian. Hence

$$\hat{O}^\dagger = \hat{O}, \quad D(\mathcal{O}) = D(\mathcal{O}^\dagger): \ \text{self-adjoint.} \tag{3.60}$$

A (self-adjoint) Hermitian operator $\hat{O}$ satisfies

$$\langle\chi|\hat{O}|\psi\rangle^* = \langle\psi|\hat{O}^\dagger|\chi\rangle = \langle\psi|\hat{O}|\chi\rangle$$
$$|\psi\rangle \in D(\mathcal{O}), \quad |\chi\rangle \in D(\mathcal{O}^\dagger).$$

In particular, $|\chi\rangle = |\psi\rangle$ yields

$$\langle\psi|\hat{O}|\psi\rangle^* = \langle\psi|\hat{O}|\psi\rangle \ : \ \text{real.} \tag{3.61}$$

In other words, the diagonal elements of a self-adjoint operator are always real. Since the expected values of physical quantities are the diagonal elements of the self-adjoint operator that represents a physical quantity, the self-adjoint property of the operator ensures that the result of measurements is always a real number.

3.11.1 Real eigenvalues

The reality of the eigenvalues of an operator critically hinges on the operator being self-adjoint. The condition that the eigenvalues of a Hermitian operator are real is carefully re-examined below from the point of view of the domain of the operator.

Suppose a self-adjoint operator H has an eigenfunction ψ that satisfies $H|\psi\rangle = \lambda|\psi\rangle$. Then, since $\langle\psi|\psi\rangle = 1$, it follows that

$$\begin{aligned}
\lambda = \lambda\langle\psi|\psi\rangle &= \langle\psi|H|\psi\rangle \ : \text{eigenvalue condition} \\
\Rightarrow \lambda^* &= \langle\psi|H|\psi\rangle^* \ : \text{definition} \\
&= \langle\psi|H^\dagger|\psi\rangle \ : \text{since domain is self-adjoint} \\
&= \langle\psi|H|\psi\rangle \ : \text{since } H \text{ is self-adjoint} \\
&= \lambda\langle\psi|\psi\rangle \ : \text{eigenvalue condition} \\
\Rightarrow \lambda^* &= \lambda \ : \ \text{Real.}
\end{aligned}$$

Note that it is essential that *all* the eigenvectors $|\psi\rangle$ belong in the domain of *both* H and $H^\dagger$. There is a more general class of pseudo-Hermitian operators, studied in Chapter 14, that obey $H^\dagger = SHS^T$, with SS^T=1. Many of the results that hold for Hermitian operators carry over to the case of pseudo-Hermitian operators.

3.12 Hamiltonian's self-adjoint extension

The requirement that an operator be self-adjoint is a powerful constraint. A *self-adjoint extension* of an operator refers to *fixing the boundary condition* on the allowed state functions so that the operator is self-adjoint.

The concept of self-adjoint extension allows one to exactly solve a whole class of one-dimensional problems that have point-like interactions, and which are expressed by potentials that are combinations of a Dirac delta function and its derivatives [Vanderbilt (1990)].

Consider the case of a Hamiltonian H that has a point-like interaction only at the origin, of strength represented by $V(x) = g\delta(x)$. Let the kinetic operator be denoted by T. The time independent Schrödinger equation is given by[7]

$$H\psi(x) = [T + V]\psi(x) = E\psi(x), \quad T = -\frac{d^2}{dx^2}, \quad x \in \mathbb{R}.$$

We need to ascertain $\mathcal{D}(H)$: the domain for which the Hamiltonian operator H is self-adjoint. Since the potential V is self-adjoint by inspection, it is the domain of the kinetic operator T that, in effect, determines the boundary conditions that need to be imposed on ψ to make H self-adjoint. We need to show that: a) $\mathcal{D}(H) = \mathcal{D}(H^\dagger)$; and b) $\langle\psi|H^\dagger|\chi\rangle^* = \langle\chi|H|\psi\rangle$ for all $|\psi\rangle \in \mathcal{D}(H)$ and $|\chi\rangle \in \mathcal{D}(H^\dagger)$.

All integrations in the neighborhood of $x = 0$ need to be studied carefully due to the presence of the delta function potential $V(x) = g\delta(x)$ located at $x = 0$. Define, for $\epsilon \to 0$, the following

$$\psi(\epsilon) \equiv \psi_+, \quad \psi(-\epsilon) \equiv \psi_-$$
$$\frac{d\psi(\epsilon)}{dx} \equiv \psi'_+, \quad \frac{d\psi(-\epsilon)}{dx} \equiv \psi'_-.$$

Consider the matrix element of T; in the limit of $\epsilon \to 0$

$$\begin{aligned}\langle\psi|T|\chi\rangle &= -\int_{-\infty}^{+\infty} dx\psi^*(x)\frac{d^2\chi(x)}{dx^2} \\ &= -\int_{-\infty}^{-\epsilon} dx\psi^*(x)\frac{d^2\chi(x)}{dx^2} - \int_{\epsilon}^{+\infty} dx\psi^*(x)\frac{d^2\chi(x)}{dx^2}.\end{aligned} \tag{3.62}$$

Using the identity ($d\psi/dx = \psi'$)

$$\psi^*(x)\frac{d^2\chi(x)}{dx^2} = \frac{d^2\psi^*(x)}{dx^2}\chi(x) + \frac{d}{dx}(\psi^*\chi') - \frac{d}{dx}(\psi'^*\chi) \tag{3.63}$$

and that all functions are zero at $x = \pm\infty$ yields, after an integration by parts,

$$\int_{\epsilon}^{+\infty} dx\psi^*(x)\frac{d^2\chi(x)}{dx^2} = \int_{\epsilon}^{+\infty} dx\frac{d^2\psi^*(x)}{dx^2}\chi(x) - \psi_+^*\chi'_+ + \psi'^*_+\chi_+. \tag{3.64}$$

[7] Choose m so that $2m = 1$.

Hence, from Eqs.3.62 and 3.64

$$\begin{aligned}\langle\psi|T|\chi\rangle &= -\int_{-\infty}^{+\infty} dx\psi^*(x)\frac{d^2\chi(x)}{dx^2}\\ &= -\int_{-\infty}^{+\infty} dx\frac{d^2\psi^*(x)}{dx^2}\chi(x) + R,\\ R &= \psi_+^*\chi_+' - \psi_+'^*\chi_+ - \psi_-^*\chi_-' + \psi_-'^*\chi_-. \end{aligned}\tag{3.65}$$

Hence, the definition of the adjoint of an operator yields

$$\begin{aligned}\langle\chi|T^\dagger|\psi\rangle - \langle\chi|T|\psi\rangle &= \langle\psi|T|\chi\rangle^* - \langle\chi|T|\psi\rangle\\ &= R^*,\end{aligned}\tag{3.66}$$

where the remainder R^*, in matrix notation, is given by Eq. 3.65 as

$$R^* = \begin{pmatrix}\chi_+^* & \chi_+'^*\end{pmatrix}\begin{pmatrix}0 & -1\\ 1 & 0\end{pmatrix}\begin{pmatrix}\psi_+\\ \psi_+'\end{pmatrix} - \begin{pmatrix}\chi_-^* & \chi_-'^*\end{pmatrix}\begin{pmatrix}0 & -1\\ 1 & 0\end{pmatrix}\begin{pmatrix}\psi_-\\ \psi_-'\end{pmatrix}. \tag{3.67}$$

Let us choose the (linear) boundary conditions to be the same for the domain of *both* the kinetic operator T and its adjoint $T^\dagger$, as is required for the operator to be self-adjoint. The boundary matrix $\mathcal{U}$ is hence postulated to be

$$\begin{pmatrix}\psi_+\\ \psi_+'\end{pmatrix} = \mathcal{U}\begin{pmatrix}\psi_-\\ \psi_-'\end{pmatrix}, \quad \begin{pmatrix}\chi_+^* & \chi_+'^*\end{pmatrix} = \begin{pmatrix}\chi_-^* & \chi_-'^*\end{pmatrix}\mathcal{U}^\dagger, \tag{3.68}$$

where

$$\mathcal{U} = e^{i\theta}\begin{pmatrix}\gamma & \beta\\ \alpha & \delta\end{pmatrix} \to \begin{pmatrix}\gamma & \beta\\ \alpha & \delta\end{pmatrix}, \tag{3.69}$$

since it can be shown that $\theta = 0$ for all stationary problems, such as scattering from a point-like potential located at $x = 0$. From Eqs. 3.67, 3.68, and 3.69 it follows that

$$R^* = 0$$

yields

$$\mathcal{U}^\dagger\begin{pmatrix}0 & -1\\ 1 & 0\end{pmatrix}\mathcal{U} = \begin{pmatrix}0 & -1\\ 1 & 0\end{pmatrix}. \tag{3.70}$$

Taking the determinant of both sides of Eq. 3.70 shows that $\det^2\mathcal{U} = 1$ and hence

$$\gamma\delta - \alpha\beta = 1. \tag{3.71}$$

Consequently, only three independent real constants fully determine the boundary condition matrix $\mathcal{U}$.

3.12.1 Delta function potential

Consider the special case of $\det\mathcal{U} = 1$ such that

$$\mathcal{U}^\dagger = \begin{pmatrix} 1 & \alpha \\ 0 & 1 \end{pmatrix}, \quad \mathcal{U} = \begin{pmatrix} 1 & 0 \\ \alpha & 1 \end{pmatrix}. \tag{3.72}$$

It is straightforward to show that $\mathcal{U}$ satisfies Eq. 3.70 for all α. The boundary condition matrix yields the boundary conditions

$$\begin{pmatrix} \psi_+ \\ \psi'_+ \end{pmatrix} = \begin{pmatrix} 1 & 0 \\ \alpha & 1 \end{pmatrix} \begin{pmatrix} \psi_- \\ \psi'_- \end{pmatrix} \quad \Rightarrow \psi_+ = \psi_-, \ \psi'_+ = \psi'_- + \alpha\psi_- \tag{3.73}$$

and similarly for the dual space vectors $\chi^*_\pm$, $\chi'^*_\pm$.

Note that the state function $\psi(x)$ is normalizable with $\int dx\, |\psi(x)|^2 = 1$. The boundary conditions of the state function $\psi(x)$ given in Eq. 3.73 imply that it is continuous at $x = 0$ and its derivative $\psi'(x)$ has a jump (discontinuity) at $x = 0$, as shown in Figure 3.3(a). The discontinuity at $x = 0$ leads to the second derivative of $\psi(x)$ having a delta function singularity at $x = 0$.

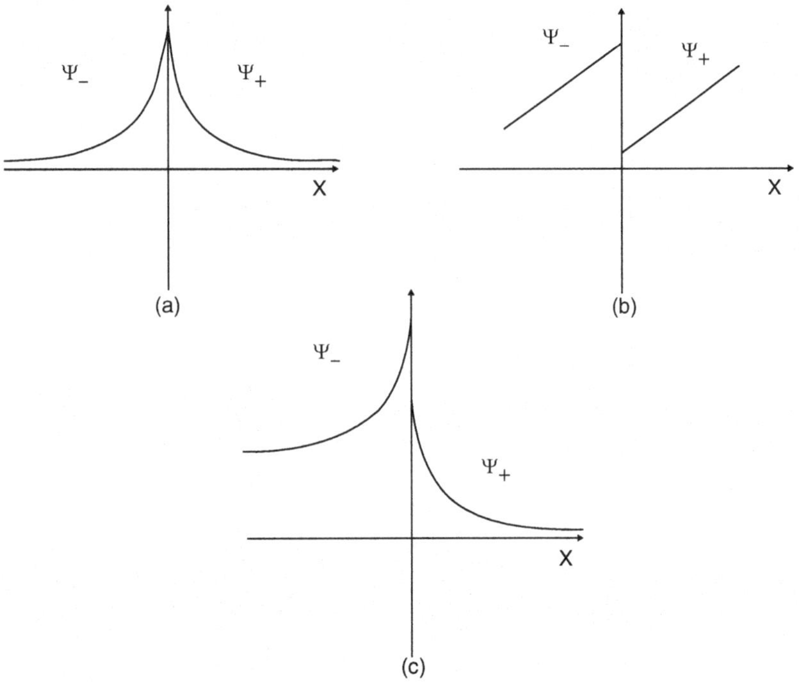

Figure 3.3 Three distinct classes of discontinuities for the state function $\psi(x)$ at the origin $x = 0$. In (a) the state function is continuous but its derivative is discontinuous; in (b) the state function is discontinuous but its derivative is continuous; and in (c) both the value of the state function and its derivative are discontinuous.

What is the potential in Schrödinger's equation that yields a state function with the discontinuity given in Eq. 3.73? The answer lies in $\psi''(0)$ being singular at $x = 0$ and this indicates the presence of a delta function potential.

The simplest case of a point interaction is a delta function potential at the origin of strength g, represented by $V(x) = g\delta(x)$. This yields a time independent Schrödinger equation given by (recall $2m = 1$)

$$\begin{aligned} E\psi(x) &= H\psi(x) = [T + V]\psi(x) \\ &= -\frac{d^2\psi(x)}{dx^2} + g\delta(x)\psi(x) \end{aligned} \tag{3.74}$$

and for which $\psi''(0)$ has a delta function singularity.

To make the connection between the potential $g\delta(x)$ and the boundary condition matrix $\mathcal{U}$, we integrate the Schrödinger equation from $-\epsilon$ to $+\epsilon$. Since $\delta(x)$ is an even function, one has $\int_0^L dx\delta(x) = 1/2 = \int_{-L}^0 dx\delta(x)$ for any $L > 0$; this yields

$$\begin{aligned} \int_{-\epsilon}^{+\epsilon} dx\delta(x)\psi(x) &= \int_{-\epsilon}^{0} dx\delta(x)\psi(x) + \int_{0}^{+\epsilon} dx\delta(x)\psi(x) \\ &= \frac{1}{2}(\psi_- + \psi_+). \end{aligned} \tag{3.75}$$

Using boundary conditions Eq. 3.73 to simplify the last two lines, Eq. 3.74 and the above equation yields

$$\begin{aligned} -\int_{-\epsilon}^{+\epsilon} dx\frac{d^2\psi(x)}{dx^2} + g\int_{-\epsilon}^{+\epsilon} dx\delta(x)\psi(x) &= E\int_{-\epsilon}^{+\epsilon} dx\psi(x) = O(\epsilon) \\ -(\psi'_+ - \psi'_-) + \frac{g}{2}(\psi_- + \psi_+) &= 0 \\ -\alpha\psi_- + g\psi_- = 0 \ &\Rightarrow \alpha = g. \end{aligned} \tag{3.76}$$

From Eq. 3.76 it follows that the discontinuity of the state function at $x = 0$ is a consequence of the delta function potential at the origin.

There are two more simple cases for U that satisfy the self-adjointness condition given in Eq. 3.70, namely

$$\mathrm{U}_I = \begin{pmatrix} 1 & \beta \\ 0 & 1 \end{pmatrix}, \quad \mathrm{U}_{II} = \begin{pmatrix} \gamma & 0 \\ 0 & \frac{1}{\gamma} \end{pmatrix}.$$

It is rather surprising that even these simple boundary matrices do not have any simple explanation in terms of an underlying potential $V(x)$ that is a combination of delta functions. The three cases discussed so far are all special limits of a very general potential called the Fermi pseudo-potential.

3.13 Fermi pseudo-potential

The time-independent Schrödinger equation for the Fermi pseudo-potential is written in a nonlocal form as

$$-\psi''(x) + \int_{-\infty}^{+\infty} dx' V(x, x')\psi(x') = E\psi(x), \tag{3.77}$$

where the real Fermi pseudo-potential, as defined by Wu and Yu (2002), is given by

$$\begin{aligned} V(x, x') &= g_1\delta(x)\delta(x') + g_2\left[\delta'_P(x)\delta(x') + \delta'(x)\delta_P(x')\right] + g_3\delta'_P(x)\delta'_P(x') \\ &= V(x', x). \end{aligned}$$

Note that $V(x, x')$ is symmetric, as required by the conservation of probability. The modified delta prime function is defined by

$$\delta'_P(x)\psi(x) \equiv \delta'(x)\tilde{\psi}(x)$$

and the definition of $\tilde{\psi}(x)$ is given by

$$\tilde{\psi}(x) \equiv \begin{cases} \psi(x) - \frac{1}{2}(\psi_+ - \psi_-) & x > 0 \\ \psi(x) + \frac{1}{2}(\psi_+ - \psi_-) & x < 0 \end{cases},$$

where we recall that $\psi_+ = \psi(\epsilon)$ and $\psi_- = \psi(-\epsilon)$. Note that $\tilde{\psi}(x)$ is continuous at $x = 0$ and in fact $\tilde{\psi}(0) = (\psi_+ + \psi_-)/2$; furthermore, even though $\delta'(x)\psi(x)$ is undefined for $\psi(x)$ that is discontinuous at $x = 0$, the combination $\delta'_P(x)\psi(x)$ is well defined at $x = 0$ even when $\psi(0)$ is discontinuous. In fact this is the main motivation for defining the (generalized) function $\delta'_P(x)$.

For the Fermi pseudo-potential, the boundary condition matrix can be shown to be given by [Wu and Yu (2002)]

$$\begin{aligned} \mathcal{U}_F &= \frac{1}{\Delta}\begin{pmatrix} (2 - g_2)^2 - g_1 g_3 & -4g_3 \\ 4g_1 & (2 + g_2)^2 - g_1 g_3 \end{pmatrix} \\ \Delta &= (2 + g_2)(2 - g_2) - g_1 g_3. \end{aligned} \tag{3.78}$$

It can be directly verified that the matrix $\mathcal{U}$ given above satisfies the self-adjointness condition given in Eq. 3.70.

The special case of the delta function potential is treated by taking $g_1 = \alpha$ and setting $g_2 = g_3 = 0$; this yields the potential $V(x, x') = g_1\delta(x)\delta(x')$.

To obtain $\mathcal{U}_I$ one sets $g_3 = -\beta$ and $g_1 = g_2 = 0$, and finally to obtain $\mathcal{U}_{II}$ one sets $g_1 = g_3 = 0$ and $\gamma = (2 - g_2)/(2 + g_2)$.

Since $\mathcal{U}_F$ has three arbitrary parameters g_1, g_2, g_3 it can be shown that $\mathcal{U}_F$ is the most general boundary matrix. Hence for a single degree of freedom the Fermi pseudo-potential is the most general point-like interaction.

3.14 Summary

Operators have a central role to play in quantum mathematics. In physics, operators represent observable quantities whereas in other applications of quantum mathematics such as geometry, finance, and statistical mechanics, operators have a different interpretation.

The structure of a Hermitian operator is realized by its spectral decomposition, in terms of all of its eigenfunctions and eigenvalues, and these also yield a representation of the completeness relation of the underlying state space. The position and momentum operator have been discussed at length as these are the leading exemplars of Hermitian operators as well as being among the most important physical observables. The Weyl operators were shown to provide a finite-dimensional unitary representation of the position and momentum operators as well as that of the Heisenberg algebra.

Physical quantities are represented in quantum mechanics by Hermitian operators. The value of a physical quantity is obtained by the operator acting on the underlying linear state space, elements of which represent the physical state of a quantum system. The set of all mutually commuting operators – which all commute with the Hamiltonian also – provide an exhaustive description of all the conserved quantities of a quantum system and are the quantum generalization of the constants of motion of classical mechanics. The Heisenberg commutation equation is a reflection of the non-commuting nature of observables and results from the underlying dynamics of the quantum system.

The concept of the *domain* of the adjoint (Hermitian conjugate) of an operator was discussed; it was shown that the domain is an independent ingredient in the definition of a self-adjoint (Hermitian) operator. The momentum operator on a finite interval exemplified the role of the domain. Self-adjoint extensions of an operator were defined for generalized delta function potentials by choosing the appropriate domain for the Hamiltonian operator. The Fermi pseudo-potential was shown to be the most general form of a generalized delta function potential that satisfies the extension of domain required for self-adjointness.

4

The Feynman path integral

A *path*, in general, is defined by a determinate trajectory in time, from an initial to a final point. The classical trajectory is only one of the possible trajectories, and in quantum mechanics *all the possible paths* between the initial and final point come into play. Recall that the probability amplitude is a complex number that is assigned to each *determinate* path. *Indeterminate* paths are defined as a collection of determinate paths that are experimentally *indistinguishable*. In the Dirac–Feynman approach, the inherent indeterminacy of the quantum entity is realized by the degree of freedom – in undergoing time evolution – "taking" indeterminate paths [Baaquie (2013e)].

For a quantum degree of freedom evolving from an *observed* initial state to the *observed* final state – and with no other observations made – the Feynman path integral is a mathematical construction that computes the probability amplitudes by summing over *all* the allowed determinate paths of the degree of freedom – discussed in Feynman and Hibbs (1965), Zinn-Justin (1993), Zinn-Justin (2005) and Baaquie (2013e).

4.1 Probability amplitude and time evolution

Recall that the description of a quantum system, at *a particular instant*, is given by its state vector, namely $|\psi\rangle$. To avoid confusion with the concept of a state vector, the term *probability amplitude* is used for describing a quantum entity undergoing transitions in time.

Consider a quantum system making a transition from an arbitrary initial state function $|\psi\rangle$ at time $t_i = 0$ to an arbitrary final state function $|\eta\rangle$ at final time $t_f = t$. Note that in quantum theory both the initial state $|\psi\rangle$ and the final state $|\eta\rangle$ can be *independently* specified.

The state vector $|\psi\rangle$ must be evolved for a duration of time t to reach the time at which the final state vector $|\eta\rangle$ is located, as shown in Figure 4.1. The initial state

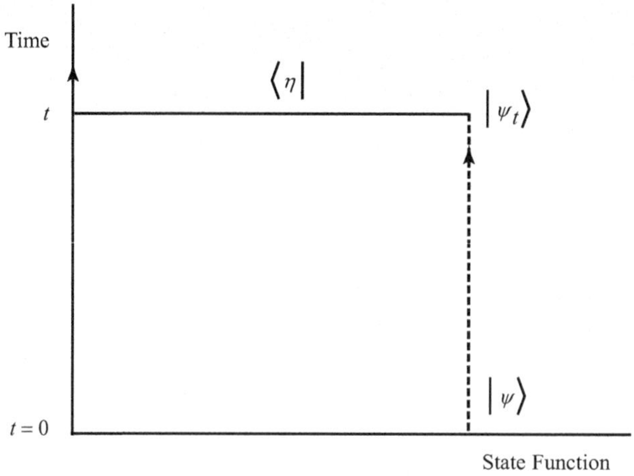

Figure 4.1 Evolving state vector $|\psi\rangle$ through time t to state vector $|\psi_t\rangle$ to find the probability amplitude $\langle\eta|\psi_t\rangle$.

vector $|\psi\rangle$ is evolved by applying the evolution operator to it, and Eq. 2.12 yields

$$|\psi_t\rangle = e^{-itH/\hbar}|\psi\rangle = U(t)|\psi\rangle. \tag{4.1}$$

The probability amplitude for the transition is written as $\langle\eta|\psi_t\rangle$. The probability amplitude and probability for the transition are given by the following:

Initial state function : $|\psi\rangle$ at time = 0, Final state function : $|\eta\rangle$ at time = t, Probability amplitude : $\langle\eta|\psi_t\rangle$, Probability of transition : $|\langle\eta|\psi_t\rangle|^2$.

Hence, the probability amplitude to go, in time t, from an arbitrary initial state $|\psi\rangle$ to another arbitrary final state $|\chi\rangle$ is given by

$$\langle\eta|\psi_t\rangle = \langle\chi|e^{-itH/\hbar}|\psi\rangle : \text{ Probability amplitude } (\psi \to \chi; t), \tag{4.2}$$

and the probability is given by

$$|\langle\eta|\psi_t\rangle|^2 = |\langle\chi|e^{-itH/\hbar}|\psi\rangle|^2 : \text{ Probability}(\psi \to \chi; t).$$

Consider superposed states given by

$$|\psi\rangle = \sum_i c_i|\psi_i\rangle, \quad |\chi\rangle = \sum_i b_i|\chi_i\rangle.$$

From Eq. 4.2, the probability amplitude is given by

$$\langle\eta|e^{-itH/\hbar}|\psi\rangle = \sum_{i,j} b_i^* c_j \langle\chi_i|e^{-itH/\hbar}|\psi_j\rangle.$$

Hence, the fundamental expression that needs to be evaluated is the general matrix element

$$\langle \chi_i | e^{-itH/\hbar} | \psi_j \rangle.$$

All probability amplitudes can be evaluated from the general matrix element of $e^{-itH/\hbar}$ that is given above.

Consider the important special case of a quantum particle with the degree of freedom given by the coordinate x. From Eq. 4.2, the probability amplitude, using the completeness equation Eq. 3.6, is given by

$$\langle \eta | e^{-itH/\hbar} | \psi \rangle = \int dx_f dx_i \eta^*(x_f) \langle x_f | e^{-itH/\hbar} | x_i \rangle \psi(x_i).$$

The *conditional probability* for the particle that starts from initial state vector $|\psi\rangle$ and, after evolving for time $t = t_f - t_i$, ends up at the coordinate $|x_f\rangle$ is given – since $\langle \psi | \psi \rangle = 1$ – by

$$\mathrm{P}(x_f | \psi; t) = \frac{|\langle x_f | e^{-itH/\hbar} | \psi \rangle|^2}{\int dx |\langle x | e^{-itH/\hbar} | \psi \rangle|^2} = |\langle x_f | e^{-itH/\hbar} | \psi \rangle|^2. \tag{4.3}$$

The normalization is necessary since one is comparing the likelihood of the particle's degree of freedom making a transition from its initial state $|\psi_0\rangle$ to x_f with the particle ending up at any other position. In particular,

$$\int dx_f \mathrm{P}(x_f | \psi; t) = 1.$$

To obtain finite results, the initial state has to be normalizable.

4.2 Evolution kernel

The time evolution of the state vector is determined by the operator $e^{-itH/\hbar}$, with its matrix elements being given by

$$\begin{aligned} K(x_f, t_f; x_i, t_i) &= \langle x_f | U(t) | x_i \rangle = \langle x_f, t_f | x_i, t_i \rangle; \quad t = t_f - t_i \\ &= \langle x_f | e^{-i(t_f - t_i)H/\hbar} | x_i \rangle. \end{aligned} \tag{4.4}$$

Simplifying the notation, the evolution kernel is written as

$$K(x, x'; t) \equiv \langle x | e^{-itH/\hbar} | x' \rangle. \tag{4.5}$$

Using the time evolution of a state vector, from Eq. 2.12, and the completeness equation given in Eq. 3.6, we obtain

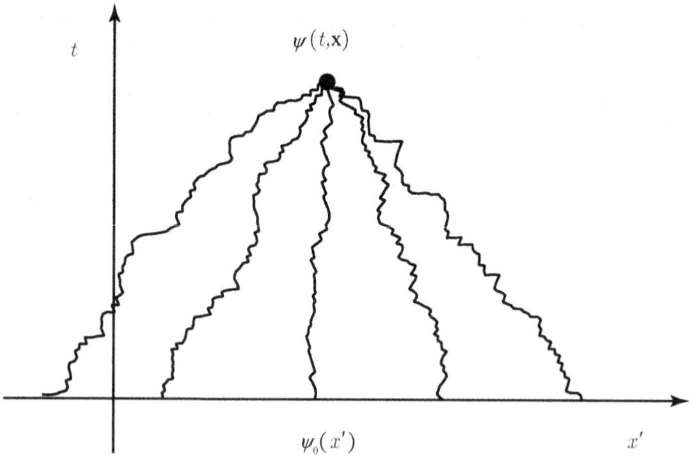

Figure 4.2 The evolution kernel $K(x, x'; t)$ propagates the initial state vector $\psi_0(x)$ through time t to state vector $\psi(t, x)$.

$$\psi(t, x) = \langle x|e^{-itH/\hbar}|\psi_0\rangle = \int dx' \langle x|e^{-itH/\hbar}|x'\rangle\langle x'|\psi_0\rangle$$
$$= \int dx' K(x, x'; t)\psi_0(x'). \tag{4.6}$$

Equation 4.6 is illustrated in Figure 4.2 and graphically shows how the values of the initial state function $\psi_0(x)$ propagate in time to determine the value of $\psi(t, x)$.

The solution of Schrödinger's equation given in Eqs. 2.12 and 4.1 is formal since one needs to evaluate the matrix elements of the evolution operator $U(t)$, which in turn requires solving for all the eigenenergies and eigenfunctions of the Hamiltonian H. Let the eigenfunctions of the Hamiltonian be given by

$$H|\psi_n\rangle = E_n|\psi_n\rangle. \tag{4.7}$$

The completeness equation, from Eq. 3.21, is then given by

$$\mathbb{I} = \sum_n |\psi_n\rangle\langle\psi_n|. \tag{4.8}$$

Using the completeness equation given in Eq. 4.8 yields the evolution kernel

$$K(x, x'; t) = \langle x|e^{-itH/\hbar}|x'\rangle = \sum_n e^{-itE_n/\hbar}\langle x|\psi_n\rangle\langle\psi_n|x'\rangle. \tag{4.9}$$

Even though the expression for the transition amplitude $K(x_f, x_i; t)$ has become greatly simplified, the sum over all eigenstates is still quite nontrivial to evaluate – even for the case of the harmonic oscillator.

All the eigenfunctions and eigenvalues of H are seldom known, and hence Eq. 4.9 is in most cases only a formal expression for the transition amplitude. One would like to have other avenues for (approximately) computing $K(x_f, x_i; t)$ and the path integral has been developed primarily to address this problem.

> Noteworthy 4.1 **Boundary conditions in classical and quantum mechanics**
>
> In classical mechanics, a classical particle is fully described by its position and velocity. Since Newton's equations of motion are based on acceleration that requires the second time derivative, one needs to specify *two* boundary conditions to uniquely specify a classical trajectory. Once the boundary conditions are specified Newton's equations of motion yield a *determinate* and unique trajectory. In particular, if the position and velocity of a particle are specified at some instant, its future trajectory is fully determined.
>
> In quantum mechanics the situation is quite different. The quantum degree of freedom is described by a state vector that yields the likelihood for experimentally observing the expectation value of a particular projection operator. The Schrödinger equation involves only the first time derivative of the state vector; hence, one can specify *either* the initial *or* the final state vector. Quantum mechanics, unlike classical mechanics, is a theory of probabilities. The initial state vector $|\psi\rangle$ evolves into a state $|\psi_t\rangle$ that has nonzero probability amplitude $\langle\chi|\psi_t\rangle$ with *many different* state vectors $\langle\chi|$. Hence, the time evolution of the quantum particle (degree of freedom) is *indeterminate*, with a likelihood of evolving from its initial state $|\psi\rangle$ to many different possible final states $\langle\chi|$.

4.3 Superposition: indeterminate paths

The probability amplitude of making a quantum transition from an initial state – that can go through many intermediate determinate paths – to a final state has two very different and distinct cases:

- The intermediate paths are *distinguishable* and the path taken is determinate and experimentally known.
- The intermediate paths are *indistinguishable*, namely the information on which path has been taken by the quantum particle is *not* experimentally determined.

When the path taken is *not known* the intermediate state of the quantum system is *indeterminate*, while it is *determinate* when the path is *known*.

The *nonclassical* content of quantum mechanics comes out in a remarkable manner for the case of the degree of freedom making a transition from an initial to a final quantum state via many *indistinguishable* intermediate paths. Hence the path taken by the degree of freedom is indeterminate.

Consider the case of an initial state vector $|x_i, t_i\rangle$ making a transition to a final state vector $|x_f, t_f\rangle$ via N intermediate slits specified by states $|x_n\rangle$, $n = 1, 2 \ldots N$

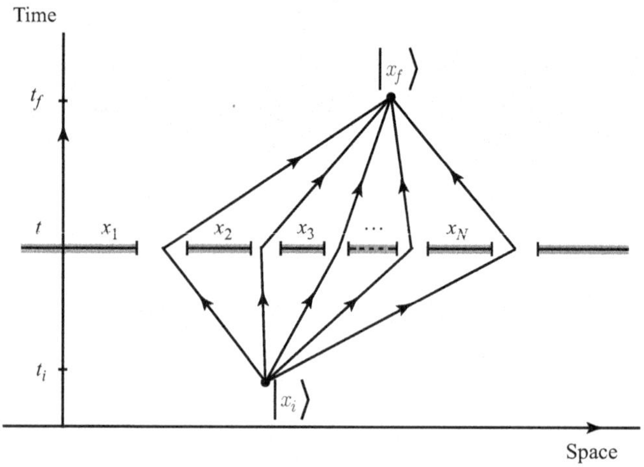

Figure 4.3 Probability amplitudes for transition from initial state vector $|x_i\rangle$ at time t_i to final state vector $|x_f\rangle$ at time t_f for N different possible intermediate paths.

and shown in Figure 4.3. In going from $|x_i, t_i\rangle$ to $|x_f, t_f\rangle$, the particle can go through any of the N-slits. The probability amplitude, in the notation of Eq. 4.4, is given by $K(x_f, t_f; x_i, t_i) = \langle x_f, t_f|x_i, t_i\rangle$.

The probability P of the transition from an initial state $|x_i, t_i\rangle$ to a final state $|x_f, t_f\rangle$, as shown in Figure 4.3, has the following two different expressions:

- The path taken for the transition is known, and hence determinate, due to a measurement being made at time t that ascertains which intermediate position x_i is taken by the particle; for this case, the *probabilities* for the different paths *are added* and yield the probability of transition P_D given by

$$P_D = \sum_{n=1}^{N} |\langle x_f, t_f|x_n, t\rangle\langle x_n, t|x_i, t_i\rangle|^2 = \sum_{n=1}^{N} P_{x_f,x_n} P_{x_n,x_i}, \tag{4.10}$$

where

$$P_{x_f,x_n} = |\langle x_f, t_f|x_n, t\rangle|^2, \quad P_{x_n,x_i} = |\langle x_n, t|x_i, t_i\rangle|^2.$$

The result for P_D follows from the classical composition of conditional probabilities, with the intermediate states being the allowed intermediate states. In particular there is no interference between the distinct paths as there is no cross-term for the different paths taken.

- For the case when the intermediate paths are indistinguishable , the *probability amplitudes* for the different determinate paths *are added* to yield the transition probability amplitude $\langle x_f, t_f|x_i, t_i\rangle$ given by

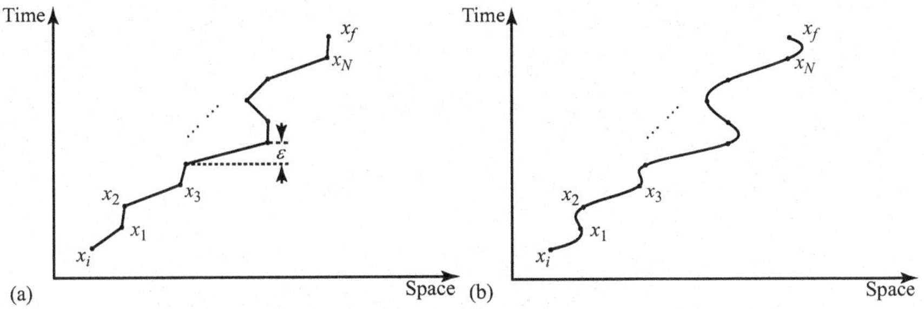

Figure 4.4 (a) A single determinate path, discretized by time steps ϵ, from initial to final position. (b) The $\epsilon \to 0$ continuum limit of the discretized path.

$$\langle x_f, t_f | x_i, t_i \rangle = \sum_{n=1}^{N} \langle x_f, t_f | x_n, t \rangle \langle x_n, , t | x_i, t_i \rangle. \tag{4.11}$$

The probability amplitude yields the following observable transition probability P_I

$$\begin{aligned} P_I &= \Big| \langle x_f, t_f | x_i, t_i \rangle \Big|^2 = \Big| \sum_{n=1}^{N} \langle x_f, t_f | x_n, t \rangle \langle x_n, , t | x_i, t_i \rangle \Big|^2 \\ &= \sum_{n=1}^{N} P_{x_f, x_n} P_{x_n, x_i} \\ &\quad + \sum_{n \neq m}^{N} \langle x_f, t_f | x_n, t \rangle \langle x_n, , t | x_i, t_i \rangle \langle x_f, t_f | x_m, t \rangle^* \langle x_m, , t | x_i, t_i \rangle^*. \end{aligned} \tag{4.12}$$

All the determinate paths interfere, as can be seen from their cross-terms given in Eq. 4.12; the interference is a purely quantum effect and there is no analog of this result in classical probability theory, which is discussed in Baaquie (2013e).

4.4 The Dirac–Feynman formula

Consider the case of a determinate and *discrete path*, with infinitesimal steps ϵ, as shown in Figure 4.4(a), that goes from x_i at t_f to final position x_f at time t_f. Let the points in the path, at intermediate times t_n, be denoted by the following:

$$x_i = x_0,\ x_1,\ x_2,\ x_3,\ x_n, \ldots,\ x_{N-1},\ x_N = x_f,\ t_n = t_i + \epsilon n,\ t_N = t_f.$$

The path is *determinate* since all the intermediate points x_n are known (by hypothetical experimental observations). Hence the principle of quantum superposition for successive steps tells us that the net amplitude ϕ[path] for the determinate path is equal to the product of the probability amplitude for each infinitesimal step [Feynman and Hibbs (1965)], and yields

$$\phi[\text{path}] = \langle x_N; t_N | x_{N-1}; t_{N-1}\rangle \ldots \langle x_{n+1}; t_{n+1} | x_n; t_n\rangle \ldots \langle x_1; t_1 | x_0; t_0\rangle.$$

Writing the probability amplitude in product notation yields

$$\phi[\text{path}] = \prod_{n=0}^{N-1} \langle x_{n+1}; t_{n+1} | x_n; t_n\rangle. \tag{4.13}$$

For each infinitesimal step $\epsilon = t_{n+1} - t_n$, the probability amplitude is given by the Dirac–Feynman formula

$$\langle x_{n+1}; t_{n+1} | x_n; t_n\rangle = \mathcal{N}(\epsilon) \exp\left\{\epsilon \frac{i}{\hbar} \mathcal{L}(x_n, x_{n+1}; \epsilon)\right\}, \tag{4.14}$$

where $\mathcal{N}(\epsilon)$ is a normalization constant and $\mathcal{L}$ is the *Lagrangian* of the particle, to be defined more precisely in Section 4.5. The fact that for an infinitesimal step the probability amplitude is an exponential that is proportional to the Lagrangian is a deep insight of Dirac that was further developed by Feynman.

The quantum particle makes a "quantum jump" from x_n at time t_n to x_{n+1} at time t_{n+1}. Unlike classical mechanics, for which the position of the particle is known for every instant t, the quantum particle can be said to be in a "trans-empirical" form of existence from time t_n to time t_{n+1}; this aspect of quantum mechanics has been discussed in Baaquie (2013e).

Using the Dirac–Feynman formula, given in Eq. 4.14, the probability amplitude for the discretized determinate path given in Eq. 4.13 has the form

$$\begin{aligned}\phi[\text{path}] &= \prod_{n=0}^{N-1} \langle x_{n+1}; t_{n+1} | x_n; t_n\rangle \\ &= \mathcal{N} \exp\{\epsilon \frac{i}{\hbar} \sum_{n=0}^{N-1} \mathcal{L}(x_n, x_{n+1}; \epsilon)\} \qquad (4.15) \\ &= \mathcal{N} \exp\{\frac{i}{\hbar} S[\text{path}]\}, \qquad (4.16)\end{aligned}$$

where the discrete and determinate path that appears in $S[\text{path}]$ is shown in Figure 4.4(a). $\mathcal{N}$ is a path-independent normalization.

The term S is the *action functional* for the discrete paths, and is given from Eqs. 4.16 and 4.15 as follows:

$$S[\text{path}] = \epsilon \sum_{n=0}^{N-1} \mathcal{L}(x_n, x_{n+1}; \epsilon). \tag{4.17}$$

4.5 The Lagrangian

Let the total time interval $t_N - t_0 = t_f - t_i$ be kept fixed, with $N = (t_f - t_i)/\epsilon$. In the continuum limit $\epsilon \to 0$ the paths become continuous. The discretized path shown in Figure 4.4(a) converges to the continuous path shown in Figure 4.4(b).

The continuum limit $\epsilon \to 0$ ($N \to \infty$) yields

$$\frac{x_{n+1} - x_n}{\epsilon} \to \frac{dx}{dt}, \quad t = n\epsilon$$

$$\mathcal{L}(x_n, x_{n+1}; \epsilon) \to \mathcal{L}(x, dx/dt)$$

$$S[\text{path}] \to S[x] = \int_{t_i}^{t_f} dt \mathcal{L}(x, dx/dt). \tag{4.18}$$

The quantum particle's Lagrangian for continuous time is $\mathcal{L}(x, dx/dt)$ and the action functional is $S[x]$; the notation used for the action is to indicate that the action depends on the *entire path* $x(t)$ with $t \in [t_i, t_f]$.

The probability amplitude for the determinate continuous path $x(t)$ – going from x_i at time t_i to final position x_f at time t_f – is given by the continuum limit of Eq. 4.16, and the continuum action $S[x]$ replaces the discretized action, namely $S[\text{path}]$. Hence

$$\phi[x] = \mathcal{N} \exp\{\frac{i}{\hbar} S[x]\}. \tag{4.19}$$

The action $S[x]$ has the dimensions of $\hbar$ and dividing it by $\hbar$ is required since only the dimensionless quantity $S/\hbar$ can be exponentiated; it is an empirical result that $\hbar$ is given by Planck's constant.

Equation 4.19 gives the probability amplitude for the quantum particle making a transition from the initial to its final position via *a specific possible path.* In other words, the path $x(t)$ is *one possible determinate path* from the initial to the final position, and not necessarily the classical path determined by classical mechanics.

The Hamiltonian given in Eq. 2.11 yields the Lagrangian

$$\mathcal{L}(x, dx/dt) = \frac{1}{2} m \left(\frac{dx(t)}{dt}\right)^2 - V(x(t)). \tag{4.20}$$

Although the action and Lagrangian given in Eqs. 4.18 and 4.20 look like classical expressions they are vastly different from the classical case. The reason is that in classical mechanics, $x(t)$ in the Lagrangian and action is *restricted to one path,* namely the *classical path* $x_c(t)$ that obeys Newton's equation of motion, and for which the particle's path is a numerical function of time t; in contrast, for quantum mechanics, the symbol $x(t)$ that appears in the Lagrangian and action can be *any possible path* from the initial to the final position.

4.5.1 *Infinite divisibility of quantum paths*

The Dirac–Feynman formula given in Eq. 4.14 is the reason that the continuum limit of $\epsilon \to 0$ exists for the probability amplitude. As one makes the step size ϵ smaller, the Dirac–Feynman formula for each infinitesimal transition, say from x_n to x_{n+1}, yields the correspondingly smaller expression in the exponential, namely $\exp\{i\epsilon\mathcal{L}(x_n, x_{n+1}; \epsilon)/\hbar\}$; this property of the paths leads, for $N \to \infty$, to a well defined limit of the infinite product for probability amplitude given in Eq. 4.13, namely $\phi[\text{path}] \to \phi[x]$ given in Eq. 4.19.

One can turn the above discussion around and argue that, for quantum mechanics to exist for continuous time, the probability amplitude for an infinitesimal step in time, of necessity, needs to be an exponential of an infinitesimal – of the form given by the Dirac–Feynman formula. This is because any determinate path is infinitely divisible in continuous time and hence requires a concomitant convergent probability amplitude.

The requirement for a convergent probability amplitude for continuous paths answers a fundamental question as to why the action S that appears in classical physics needs to be exponentiated in quantum mechanics, as in $\exp\{iS[x]/\hbar\}$ given in Eq. 4.19. The classical to quantum transition is schematically given by

$$S[x_c] \to \exp\{\frac{i}{\hbar}S[x]\}.$$

One explanation provided by the probability amplitude is that the requirement of quantum processes taking place in continuous time necessitates the exponentiation of the action. One may even state that the exponentiation of the action in quantum mechanics is also the reason why quantum mechanics is qualitatively different from, and "exponentially" more complex, than classical mechanics.

4.6 The Feynman path integral

The result of the previous section provides an expression for the probability amplitude for the quantum entity to take a specific and *determinate path* in going from its initial to its final position. What is the probability amplitude if the quantum particle is only observed at its initial and final position? Due to the quantum indeterminacy of a quantum entity the paths of the entity's degree of freedom are indeterminate and hence it "takes" all possible *determinate paths* simultaneously.

How many determinate paths are there between the initial and final positions? Clearly, there are many paths, and to develop a sense of these paths, consider putting barriers between the initial and final position to *limit* the number of possible paths, as shown in Figure 4.5, so that we can enumerate the determinate paths. Once

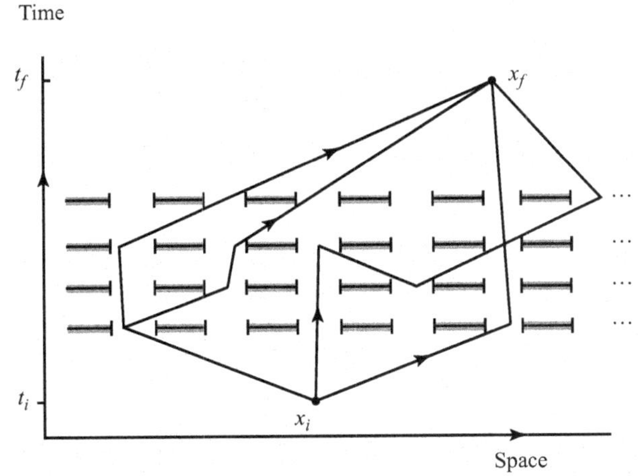

Figure 4.5 Probability amplitudes for transition from initial state vector $|x_i, t_i\rangle$ to final state vector $|x_f, t_f\rangle$ for many successive slits with *indistinguishable* trans-empirical paths.

the procedure for enumerating the determinate paths becomes clear, the barriers will be removed and all the determinate paths will then be included in our analysis.

Figure 4.5 shows a quantum particle going from initial state x_i at time t_i to final position x_f at time t_f, through barriers that restrict the number of paths available. Let the *entire continuous path* – going from initial state x_i to final state x_f through the successive slits as shown in Figure 4.5 – be denoted by path(n), with the probability amplitude denoted by ϕ[path(n)]. One can take path(n) to be straight lines from x_i, t_i to the successive slit positions and another straight line from the last slit to x_f, t_f, as shown in Figure 4.5.

Let there be N total number of different paths going from x_i to x_f. All the N determinate paths from x_i, t_i to x_f, t_f are indistinguishable. From the superposition principle given in Eq. 4.11, the total probability amplitude is given by *adding* the probability amplitudes for all the indistinguishable determinate paths, and yields

$$\langle x_f, t_f | x_i, t_i \rangle = \sum_{n=1}^{N} \phi[\text{path}(n)]. \tag{4.21}$$

The probability amplitude ϕ[path(n)] for *each determinate path* is given by Eq. 4.19 and yields

$$\phi[\text{path}(n)] = \mathcal{N} \exp\{iS_n/\hbar\}, \tag{4.22}$$

$$S_n = S[\text{path}(n)], \qquad n = 1, 2, \ldots N, \tag{4.23}$$

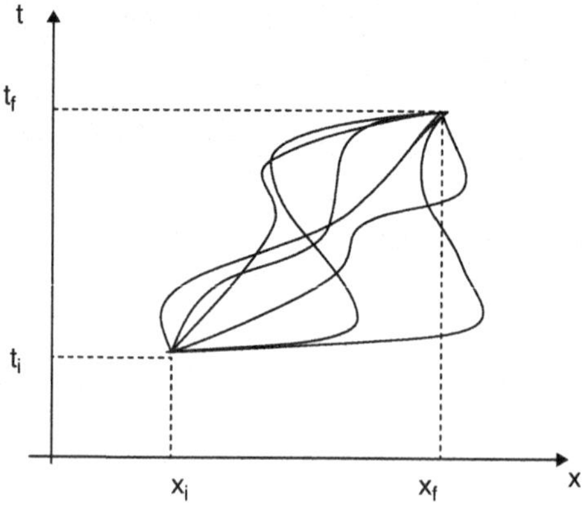

Figure 4.6 All possible trans-empirical paths from the initial to the final state vector.

where $S[\text{path}(n)]$ is the action for the continuous path(n) and $\mathcal{N}$ is a path independent normalization.

Hence, from Eqs. 4.21 and 4.22, the total spacetime probability amplitude that the initial state vector $|x_i, t_i\rangle$ makes a transition to the final state vector $|x_f, t_f\rangle$ – via *indistinguishable paths* – is given by superposing the amplitude for all the indistinguishable paths and yields

$$\begin{aligned}\langle x_f, t_f | x_i, t_i\rangle &= \mathcal{N} \sum_{n=1}^{N} e^{iS[\text{path}(n)]/\hbar} = \mathcal{N} \sum_{n=1}^{N} e^{iS_n/\hbar} \\ &= \mathcal{N}\left\{e^{iS_1/\hbar} + e^{iS_2/\hbar} + \cdots\right\}.\end{aligned}$$

From Eq. 4.4, the evolution kernel has the representation

$$K(x_f, t_f; x_i, t_i) = \langle x_f, t_f | x_i, t_i\rangle = \mathcal{N} \sum_{n=1}^{N} e^{iS[\text{path}(n)]/\hbar}. \tag{4.24}$$

One can successively remove the barriers between the initial and final positions of the quantum particle and, as shown in Figure 4.5, there will be a great proliferation of possible paths. When there are no longer any slits, one has the limit of $N \to \infty$ or what is the same thing, there are infinitely many indistinguishable paths.

The transition amplitude is given by the sum over *all possible determinate paths*, going from the initial position x_i at time t_i to the final state x_f at time t_f, as shown in Figure 4.6, and yields

$$K(x_f, t_f; x_i, t_i) = \mathcal{N} \sum_{\text{all paths}} e^{iS[\text{path}]/\hbar} \quad : \text{Feynman path integral.} \tag{4.25}$$

The sum in Eq. 4.25 looks more figurative than a precise mathematical expression. After all, how are we supposed to actually perform a sum over infinitely many paths? Eq. 4.25 is re-cast into a precise and mathematical expression in Section 4.7.

In summary, from time t_i to time t_f, no measurement is performed on the particle's evolution and hence all the paths going x_i to final state x_f are *indistinguishable*. The total probability amplitude to make a transition from initial state x_i to final state x_f is equal to the sum of probability amplitudes $e^{iS/\hbar}$ over all the individual (indistinguishable) determinate paths that go from initial state x_i to final state x_f.

4.7 Path integral for evolution kernel

To render the sum over all paths in continuous space, namely $\sum_{\text{all paths}}$ given in Eq. 4.25, into a well-defined mathematical quantity, a derivation is given below of the path integral starting from the Schrödinger equation. A corollary result is to show that the definition of $K(x_i, x_f; t)$ given in Eq. 4.5 is equivalent to the one derived in Eq. 4.25.

Consider the Hamiltonian

$$H = \frac{\hbar}{2m}p^2 + V(x), \quad p = -i\hbar\frac{\partial}{\partial x}$$

and its evolution kernel, which from Eq. 4.5 is defined by

$$K(x_i, x_f; t) = \langle x_f | e^{-itH/\hbar} | x_i \rangle.$$

To simplify the notation, set $\hbar = 1$.

Note that in general $[p^2, V] \neq 0$; it is this noncommutativity that poses the main problem in evaluating the evolution kernel and makes quantum mechanics nontrivial. Ignoring the noncommutativity yields

$$\begin{aligned} e^{-itH} &\simeq e^{-it\frac{p^2}{2m}} e^{-itV} \\ \Rightarrow K(x, x'; t) &\simeq \langle x | e^{-it\frac{p^2}{2m}} e^{-itV} | x' \rangle \\ &\simeq e^{-itV(x')} \langle x | e^{-it\frac{p^2}{2m}} | x' \rangle. \end{aligned} \tag{4.26}$$

To evaluate the evolution kernel K in this approximation requires only the evolution kernel for the free particle Hamiltonian $p^2/2m$, which is given in Eq. 5.72.

Note the remarkable fact that for noncommuting operators A and B

$$e^A e^B = e^{A+B+\frac{1}{2}[A,B]+\cdots} \tag{4.27}$$

and hence, for infinitesimal time, namely $t = \epsilon$, we obtain

$$\begin{aligned} e^{-i\epsilon\hat{H}} &= e^{-i\epsilon(\frac{\hat{p}^2}{2m}+\hat{V})} \\ &= e^{\frac{-i\epsilon\hat{p}^2}{2m}} e^{-i\epsilon\hat{V}} + O(\epsilon^2). \end{aligned} \tag{4.28}$$

Hence for infinitesimal time ϵ, from Eq. 4.26 the transition amplitude $K(x, x^{'}; \epsilon)$ can be evaluated exactly to $O(\epsilon^2)$.

The path integral approach is employed fundamentally to build up the *finite time* transition amplitude by composing the *infinitesimal time* transition amplitude by repeatedly using the resolution of the identity operator given in Eq. 3.6.

The evolution kernel (transition amplitude) for a particle to go from initial position x_i to final position x_f in time t is written as

$$K(x_f, x_i; \tau) = \langle x_f|e^{-itH}|x_i\rangle = \langle x|(e^{-i\frac{t}{N}H})^N|x'\rangle,$$

where, for $\epsilon = \frac{t}{N}$, we have

$$K = \langle x_f|\underbrace{e^{-i\epsilon H}e^{-i\epsilon H}\cdots e^{-i\epsilon H}}_{N-times}|x_i\rangle. \tag{4.29}$$

Inserting, $N-1$ times, the completeness equation given in Eq. 3.6,

$$\mathbb{I} = \int_{-\infty}^{\infty} dx|x\rangle\langle x|,$$

yields

$$\begin{aligned} K(x_f, x_i; \tau) = \int dx_1 dx_2 \ldots dx_{N-1} \langle x_f|e^{-i\epsilon H}|x_{N-1}\rangle\langle x_{N-1}|e^{-i\epsilon H}|x_{N-2}\rangle \\ \cdots \langle x_{n+1}|e^{-i\epsilon H}|x_n\rangle \cdots \langle x_1|e^{-i\epsilon H}|x_i\rangle. \end{aligned} \tag{4.30}$$

Consider the matrix element

$$\langle x|e^{-i\epsilon H}|x'\rangle = \int_{-\infty}^{\infty} \frac{dp}{2\pi}\langle x|e^{-i\epsilon H}|p\rangle\langle p|x'\rangle. \tag{4.31}$$

Since $\langle x|p\rangle = e^{ipx}$, one has from Eq. 4.28

$$\begin{aligned} \langle x|e^{-i\epsilon H}|x'\rangle &= e^{-i\epsilon V(x')} \int \frac{dp}{2\pi} e^{-\frac{i\epsilon p^2}{2m}} e^{ip(x-x')} \\ &= \sqrt{\frac{m}{2\pi i\epsilon}} e^{i\frac{m}{2\epsilon}(x-x')^2 - i\epsilon V(x')}. \end{aligned} \tag{4.32}$$

Recall from Eq. 4.14, the Dirac–Feynman formula, for each infinitesimal time step ϵ, is given by

$$\langle x_{n+1}; t_{n+1}|x_n; t_n\rangle = \mathcal{N}(i\epsilon) \exp\left\{\epsilon\frac{i}{\hbar}\mathcal{L}(x_n, x_{n+1}; \epsilon)\right\}.$$

Recall from Eq. 4.4, the evolution kernel is defined by

$$K(x_f, t_f; x_i, t_i) = \langle x_f, t_f | x_i, t_i \rangle = \langle x_f | e^{-i(t_f - t_i)H/\hbar} | x_i \rangle.$$

Hence, from Eqs. 4.4 and 4.14, and simplifying the notation, we obtain the Lagrangian, defined for infinitesimal Euclidean time ϵ, given by

$$\langle x | e^{-i\epsilon H} | x' \rangle = \mathcal{N}(\epsilon) e^{j\epsilon \mathcal{L}(x,x';\epsilon)} \quad : \text{Dirac–Feynman formula.} \tag{4.33}$$

Hence from Eq. 4.32

$$\mathcal{N}(\epsilon) e^{i\epsilon \mathcal{L}(x,x';\epsilon)} = \langle x | e^{-i\epsilon H} | x' \rangle = \sqrt{\frac{m}{2\pi i \epsilon}} e^{i\frac{m}{2\epsilon}(x-x')^2 - i\epsilon V(x')}. \tag{4.34}$$

In summary, the particle degree of freedom has the Hamiltonian given by

$$H = -\frac{\hbar^2}{2m}\frac{\partial^2}{\partial x^2} + V(x),$$

and its Lagrangian is given by Eq. 4.34, for discrete time $t = n\epsilon$, by

$$\mathcal{L} = \frac{m}{2}\left(\frac{x - x'}{\epsilon}\right)^2 - V(x). \tag{4.35}$$

The Lagrangian is sometimes written more symmetrically as

$$\mathcal{L} = \frac{m}{2}(\frac{x - x'}{\epsilon})^2 - \frac{1}{2}[V(x) + V(x')] \tag{4.36}$$

and to $O(\epsilon)$ is the same as the one given in Eq. 4.35.

Hence the transition amplitude, restoring $\hbar$ in the last equation below, is given by

$$\begin{aligned} K(x_f, x_i; \tau) &= \int \prod_{n=1}^{N-1} dx_n \prod_{n=0}^{N-1} \langle x_{n+1} | e^{-i\epsilon H} | x_n \rangle \\ &= \int Dx e^{i\epsilon \sum_{n=0}^{N-1} \mathcal{L}(x_{n+1}, x_n)} \\ &= \int Dx \exp\{\frac{i}{\hbar} S[x]\}, \end{aligned} \tag{4.37}$$

$$\text{Boundary conditions} : x_0 = x_i, \quad x_N = x_f. \tag{4.38}$$

The lattice action and path integral integration measure is given by

$$S[x] = \epsilon \sum_{n=0}^{N-1} \frac{m}{2}\left(\frac{x_{n+1} - x_n}{\epsilon}\right)^2 - \epsilon \sum_{n=0}^{N-1} V(x_n),$$

$$\int Dx = \left(\frac{m}{2\pi i \epsilon}\right)^{\frac{N}{2}} \prod_{n=1}^{N-1} \int_{-\infty}^{+\infty} dx_n.$$

In the continuum limit $\epsilon \to 0$ one obtains

$$S[x] = \int_0^\tau \mathcal{L}dt, \quad \mathcal{L} = \frac{m}{2}\left(\frac{dx}{dt}\right)^2 - V(x), \tag{4.39}$$

$$\int Dx = \mathcal{N}\prod_{t=0}^{\tau}\int dx(t).$$

The continuum path integral for the evolution kernel is given by

$$K(x_f, x_i; \tau) = \langle x_f | e^{-itH/\hbar} | x_i \rangle = \int_{B.C.} Dx \exp\{\frac{i}{\hbar}S[x]\}$$

$$: \text{Feynman path integral}, \tag{4.40}$$

$$\text{Boundary condition}: \ x(0) = x_i \ , \ x(\tau) = x_f.$$

All paths between the initial and final position, figuratively shown in Figure 4.6, are summed over in the $\int Dx e^{iS/\hbar}$ path integration given in Eq. 4.40. The figurative summation over all paths $\sum_{\text{all paths}} e^{iS/\hbar}$ given in Eq. 4.25 is given a mathematical realization in Eq. 4.40, which is a functional integration over all indistinguishable paths from the initial to the final state.

At each instant, the position degree of freedom takes all its values; at instant $t \in [t_i, t_f]$, the degree of freedom is equal to the real line $\Re_t$; the total space of all paths is given by a tensor product over all instants and yields the total space of all paths equal to $\otimes_t \Re_t$. In general, for degree of freedom space given by $\mathcal{F}$, the path space is given by $\otimes_t \mathcal{F}_t$.

In summary, the *Feynman path integral* is an efficient mathematical instrument for evaluating the finite time matrix elements of the Euclidean continuation of the unitary operator $U(t)$, namely of $\langle x_f | e^{-itH/\hbar} | x_i \rangle$.

4.8 Composition rule for probability amplitudes

Consider the case of a particle going through N-slits, as shown in Figure 4.3, with all the paths being *indistinguishable*. Equation 4.12 yields the probability amplitude

$$\langle x_f, t_f | x_i, t_i \rangle = \sum_{n=1}^{N} \langle x_f, t_f | x_n, t \rangle \langle x_n, t | x_i, t_i \rangle.$$

Suppose the slits have spacing a, so that $x_n = na$, with $n = 0, \pm 1, \pm 2, \ldots \pm\infty$, that is, the slits extend over the entire x-axis. The probability amplitude, extending

Eq. 4.12 to the entire x-axis, is given by

$$\langle x_f, t_f | x_i, t_i \rangle = \sum_{n=-\infty}^{+\infty} \langle x_f, t_f | x_n, t \rangle \langle x_n, t | x_i, t_i \rangle. \tag{4.41}$$

To take the continuum limit of Eq. 4.41, the bra and ket vectors $|x_n, t\rangle$, $\langle x_n, t|$ defined on a discrete set $x_n = na$ need to be written in continuum notation; for $a \to 0$, let $x_n \to z$. The connection of the continuous and discrete state vector is given by Eq. 3.2,

$$\lim_{a\to 0}: \quad |x_n, t\rangle \to \sqrt{a}|z, t\rangle, \quad x_n = na, \quad -\infty \le z \le \infty$$
$$\langle x_n, t| \to \sqrt{a}|\langle z, t|. \tag{4.42}$$

Also, let $t_i = 0$, $t = \tau$ and $t_f = \tau + \tau'$. The initial and final state vectors are defined for continuous initial and final positions and hence have the limit

$$|x_i, t_i\rangle \to |x, 0\rangle, \quad \langle x_f, t| \to \langle x', \tau + \tau'|. \tag{4.43}$$

As shown in Figure 4.7, taking the $a \to 0$ limit, from Eqs. 4.41, 4.42, and 4.43, yields

$$\langle x', \tau + \tau' | x, 0\rangle = a \sum_{n=-\infty}^{+\infty} \langle x', \tau + \tau' | x_n, \tau \rangle \langle x_n, \tau | x, 0 \rangle, \tag{4.44}$$

$$\Rightarrow \langle x', \tau + \tau' | x, 0\rangle \to \int_{-\infty}^{+\infty} dz \langle x', \tau + \tau' | z, \tau \rangle \langle z, \tau | x, 0 \rangle. \tag{4.45}$$

Writing the transition amplitude in Eq. 4.43 in terms of the evolution kernel given in Eq. 4.4, for Euclidean time, yields

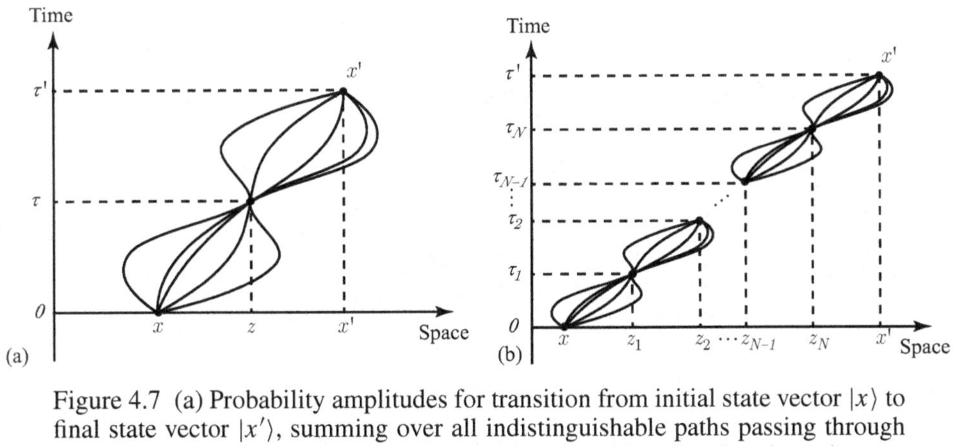

Figure 4.7 (a) Probability amplitudes for transition from initial state vector $|x\rangle$ to final state vector $|x'\rangle$, summing over all indistinguishable paths passing through position z at time τ. (b) The probability amplitude with path going through many intermediate positions $z_1, z_2, \ldots z_n$ at times $\tau_1, \tau_2, \ldots, \tau_N$.

$$K(x', x; \tau + \tau') = \int_{-\infty}^{+\infty} dz K(x', z; \tau') K(z, x; \tau). \tag{4.46}$$

Equation 4.46, illustrated in Figure 4.7, shows that the definition of the evolution kernel is consistent with the rules for the composition of probability amplitudes by summing over all intermediate indistinguishable paths.

In writing Eq. 4.46, only the property of the action was used. Consider a finite time slice $[0, \tau + \tau']$, as shown in Figure 4.7; due to the term $(dx/dt)^2$ in the action given in Eq. 4.39, one needs to specify the initial value x at $t = 0$ and a final value z at $t = \tau$, since two boundary conditions are required to specify paths going from x to z. The state space appears in the path integral via the boundary conditions imposed on the paths over which the path integration is defined.

The condition given in Eq. 4.46 is a fundamental property of probability amplitudes that allows one to define the state space – since the path integral can be interpreted as the matrix element connecting the initial and final state vector. The fundamental reason that the action satisfies the composition law is because, writing the action in terms of its initial and final boundary variables as $S[x_f, x_i]$, the action given in Eq. 4.39, in the notation of Figure 4.7, has the form

$$S[x', x] = S[x', z] + S[z, x]. \tag{4.47}$$

Interestingly enough, the above equation holds only for state space expressed in terms of coordinate state vectors $|x\rangle$. Unlike the Schrödinger equation that holds equally in momentum space, the composition law given in Eq. 4.46 does not hold when expressed in terms of Fourier transformed variables, essentially because Eq. 4.47 does not hold for Fourier transformed variables.

In many complicated cases such as quantum field theory on curved spacetime, the quantum theory is defined directly in terms of the action, and it may not be possible to derive a Hamiltonian; in such cases, one can directly base the existence of the state space on the properties of the Lagrangian and action.

For the case where there is a well defined Hamiltonian, Eq. 4.46 follows directly from the definition of the evolution kernel in terms of the Hamiltonian given in Eq. 4.40 and the completeness equation; more precisely, since $e^{-i(t+t')H} = e^{-itH}e^{-it'H}$, we have

$$\begin{aligned} K(x', x; t + t') &= \langle x'|e^{-i(t+t')\hat{H}}|x\rangle \\ &= \int_{-\infty}^{+\infty} dz \langle x|e^{-it\hat{H}}|z\rangle\langle z|e^{-it'\hat{H}}|x'\rangle \\ &= \int_{-\infty}^{+\infty} dz K(x', z; t) K(z, x; t'). \end{aligned}$$

4.9 Summary

The path integral is an independent formulation of quantum mechanics. To show the path integral's connection to the underlying foundations of indeterminacy of the quantum degree of freedom, the path integral has been derived from Schrödinger's state vector formulation. The probability amplitude for a finite determinate path was evaluated by breaking up the path into a series of infinitesimal steps. The Dirac–Feynman formula yields the probability amplitude for an infinitesimal time step; composing the infinitesimal paths yields that the probability amplitude for a finite determinate path is proportional to $\exp\{iS/\hbar\}$, where S is the action for the quantum degree of freedom.

The transition of a quantum entity from its initial to final state, without any observations during the interregnum, is made by the degree of freedom taking an indeterminate path, which is a collection of many indistinguishable determinate paths. The principle of quantum superposition yields the transition amplitude as the sum of the probability amplitudes of all the indistinguishable determinate paths and leads to the summing of $\exp\{iS/\hbar\}$ over all the determinate paths, and yields the Feynman path integral.

The evolution kernel was defined starting from the Hamiltonian of the degree of freedom. A path integral expression was obtained for the evolution kernel using methods based on the state space, and the Lagrangian was derived from the Hamiltonian.

In fields outside physics, the path integral is generalized and represents a random system with a probability distribution function for the different possible outcomes given by $\exp\{S/\hbar\}/Z$, where S is the analog of the action. For these random systems, the path integral is defined, similarly to quantum mechanics, by the sum over all possible outcomes.

5

Hamiltonian mechanics

Path integrals are by and large defined directly in terms of the configuration space representation of the quantum entity's degrees of freedom and employ the Lagrangian description of the quantum entity; most of the path integrals in this book follow this approach.

The Hamiltonian provides another independent approach for defining path integrals and is discussed in this chapter. Two important path integrals, which are directly based on the Hamiltonian, are the following: (a) one that is defined on the degree of freedom's *phase space*, defined as the tensor product of the degree of freedom space and its canonical conjugate momentum space; and (b) path integrals using the *coherent state basis* instead of the coordinate basis. Path integrals defined on phase space, or for coherent states, are both based on the Hamiltonian.

To put in the foreground the role of the Hamiltonian in quantum mechanics, the canonical equations connecting the Lagrangian to the Hamiltonian are discussed. A brief review of Hamiltonian mechanics, also called the canonical equations, is given in Section 5.1 and the connection of symmetries with conservation laws is discussed in Section 5.2. The Hamiltonian is derived from the Lagrangian in Section 5.3, for both Minkowski and Euclidean time. Phase space path integrals are defined in Section 5.4. Canonical quantization based on the Poisson brackets is discussed in Section 5.5, and Dirac brackets required for quantizing constrained systems are derived in Section 5.7. Coherent states and their path integrals are discussed in Sections 5.10 to 5.14.

5.1 Canonical equations

Consider for now Minkowski time denoted by t; a generic degree of freedom in configuration space is denoted by q and let $\dot{q} \equiv dq/dt$; the generalization to N degrees of freedom q_i, $i = 1, 2, \ldots N$ is straightforward and will be discussed when necessary. The Minkowski Lagrangian L and action are defined, from initial time t_i to final time t_f, by

$$L = \frac{1}{2}m\dot{q}^2 - V(q), \quad \dot{q} = \frac{dq}{dt}, \tag{5.1}$$

$$S[q] = \int_{t_i}^{t_f} dt L, \tag{5.2}$$

where $S[q]$ denotes the dependence of the action on the entire path taken by $q(t)$, from $q_i = q(t_i)$ to $q_f = q(t_f)$.

The classical equations of motion are defined by requiring that the classical trajectory, with the given boundary conditions, must minimize the action S, and this yields Lagrange's equation of motion

$$0 = \delta S[q] = \int_{t_i}^{t_f} dt \left[\frac{\partial L}{\partial q}\delta q + \frac{\partial L}{\partial \dot{q}}\delta \dot{q}\right] = \int_{t_i}^{t_f} dt \left[\frac{\partial L}{\partial q} - \frac{d}{dt}\left(\frac{\partial L}{\partial \dot{q}}\right)\right]\delta q$$

$$\Rightarrow \quad \frac{\partial L}{\partial q} = \frac{d}{dt}\left(\frac{\partial L}{\partial \dot{q}}\right). \tag{5.3}$$

The Lagrangian is based on the spacetime description of the entity and, in particular, depends on the degree of freedom q and its time derivative $\dot{q}$ and yields the following variation of the Lagrangian

$$dL = \frac{\partial L}{\partial q}dq + \frac{\partial L}{\partial \dot{q}}d\dot{q}. \tag{5.4}$$

For many applications, such as studying a system that conserves energy, it is more suitable to have a formulation that does not refer to time directly. Hence one would like to change the independent variables from $q, \dot{q}$ to another set of independent variables p, q and transform the Lagrangian to a new function, namely the Hamiltonian H; the variation of the Hamiltonian is given by

$$dH = \frac{\partial H}{\partial q}dq + \frac{\partial H}{\partial p}dp. \tag{5.5}$$

The canonical momentum p is defined below and yields, from Eq. 5.3, the following expression for the equation of motion:

$$p = \frac{\partial L}{\partial \dot{q}} \quad \Rightarrow \quad \frac{\partial L}{\partial q} = \dot{p}. \tag{5.6}$$

The Hamiltonian is defined by the Legendre transformation of L and, from Eqs. 5.4 and 5.6, is given by

$$dL = \frac{\partial L}{\partial q}dq + \frac{\partial L}{\partial \dot{q}}d\dot{q} = \dot{p}dq + pd\dot{q} = \dot{p}dq - \dot{q}dp + d(p\dot{q})$$

$$\Rightarrow \quad dH \equiv d(p\dot{q} - L) = -\dot{p}dq + \dot{q}dp. \tag{5.7}$$

The above equation yields the Hamiltonian H as well as the dynamical equations for q, p. From Eqs. 5.5 and 5.7, one obtains the canonical equations

$$H = p\dot{q} - L \Rightarrow \quad -\dot{p} = \frac{\partial H}{\partial q}, \ \dot{q} = \frac{\partial H}{\partial p}. \tag{5.8}$$

The canonical equations given in Eq. 5.8 are two first order equations and are equivalent to the single second order equation of motion given by the Lagrangian in Eq. 5.3.

5.2 Symmetries and conservation laws

One of the most fundamental symmetries of a Lagrangian is that it does not explicitly depend on time parameter t; what this means is that the system is homogeneous in time, having the same dynamics for each instant of time; in particular, there is no special value of time that affects the evolution of the system; all changes in time are effected via the dynamical variables $p(t), q(t)$. For such a Lagrangian homogeneous in time, energy conservation follows directly from the equations of motion.

Let the dynamical variables of the system be q_i, p_i with $i = 1, 2, \ldots N$. Using the classical equations of motion given in Eq. 5.3 yields[1]

$$\begin{aligned} dL &= \sum_i \left[\frac{\partial L}{\partial q_i} dq_i + \frac{\partial L}{\partial \dot{q}_i} d\dot{q}_i \right] = \sum_i \left[\frac{d}{dt} \left(\frac{\partial L}{\partial \dot{q}_i} \right) dq_i + \frac{\partial L}{\partial \dot{q}_i} d\dot{q}_i \right] \\ &= \frac{d}{dt} \left[\sum_i \left(\frac{\partial L}{\partial \dot{q}_i} \right) dq_i \right] \Rightarrow \frac{dL}{dt} = \frac{d}{dt} \left[\sum_i \left(\frac{\partial L}{\partial \dot{q}_i} \right) \dot{q}_i \right]. \end{aligned} \tag{5.9}$$

Hence, it follows from Eq. 5.9 that energy E, defined below, is a conserved quantity,

$$E = \sum_i \left(\frac{\partial L}{\partial \dot{q}_i} \right) \dot{q}_i - L \Rightarrow \frac{dE}{dt} = 0. \tag{5.10}$$

Note that although the definition of E looks very similar to the Hamiltonian H, the two are quite different: E is defined only for the classical trajectory that fixes, for each instant, both the position and momentum; in contrast, the Hamiltonian $H = H(p_i, q_i)$ is defined for all values of the canonical momenta and coordinates p_i, q_i. The equation $H(p_i, q_i) = E$ fixes a surface in phase space p_i, q_i; for a given fixed value of $E = E_0$, the classical trajectory, as it evolves in time, moves in phase space on the surface of constant E_0.

[1] An explicit dependence on time of L would yield an extra term $dt(\partial L/\partial dt)$ in dL that would spoil the conservation law for energy.

Suppose again that the Lagrangian L does not have any explicit dependence on time t. From Eqs. 5.4 and 5.6

$$dL = \sum_i \left[\frac{\partial L}{\partial q_i} dq_i + \frac{\partial L}{\partial \dot{q}_i} d\dot{q}_i \right] = \sum_i [\dot{p}_i dq_i + p_i d\dot{q}_i]$$
$$= \frac{d}{dt} \left[\sum_i p_i dq_i \right]. \tag{5.11}$$

Consider a general time independent transformation of the coordinates q_i that leaves the Lagrangian invariant, namely

$$\delta q_i = f_i(p, q)\delta \ \Rightarrow \ \delta L = 0. \tag{5.12}$$

Hence, from Eqs. 5.11 and 5.12

$$\delta L = 0 = \frac{d}{dt} \left[\sum_i p_i \delta q_i \right] = \delta \frac{d}{dt} \left[\sum_i p_i f_i(p, q) \right],$$
$$\sum_i p_i f_i(p, q) = \text{constant}. \tag{5.13}$$

The canonical equations lead to a remarkable conclusion: for every symmetry of the Lagrangian there is a conserved quantity that remains constant over time. This result continues to hold in quantum mechanics, with every symmetry of the Lagrangian resulting in a conserved quantity.

To illustrate the relation between symmetries and conservation laws, consider the Lagrangian for two uncoupled harmonic oscillators, namely

$$L = \frac{1}{2} m[\dot{x}^2 + \dot{y}^2] + \frac{1}{2} m\omega^2 [x^2 + y^2]. \tag{5.14}$$

It can be verified that under the following symmetry transformation, which is an infinitesimal rotation of the coordinates about the z-axis,

$$\delta x = -y\delta, \ \ \delta y = x\delta \ \Rightarrow \ f_x = -y, \ \ f_y = x, \tag{5.15}$$

the Lagrangian is invariant. Hence, the general result given in Eq. 5.13 yields, for the transformation given in Eq. 5.15, the conserved quantity

$$\sum_i p_i f_i(p, q) = -yp_x + xp_y : \ \text{constant}.$$

The conserved quantity is seen to be the z-component of angular momentum.

5.3 Euclidean Lagrangian and Hamiltonian

Euclidean time τ is defined as $\tau = it$. The coordinate degree of freedom remains unchanged and is q; Euclidean velocity $\dot{q}_E$ is defined as follows:

$$\tau = it, \tag{5.16}$$

$$\dot{q}_E = \frac{dq}{d\tau} = -i\frac{dq}{dt} = -i\dot{q}. \tag{5.17}$$

Hence, the Euclidean Lagrangian, from Eq. 5.1, is given by

$$L_E(q, \dot{q}_E) = -\frac{1}{2}m(\dot{q}_E)^2 - V(q) \tag{5.18}$$

and Euclidean action, similar to Eq. 5.2, is given by

$$S_E[q] = \int_{\tau_i}^{\tau_f} d\tau L_E. \tag{5.19}$$

The equations of motion for Euclidean time are obtained by minimizing the action S and, similarly to Eq. 5.3, yield Lagrange's equation of motion,

$$\delta S_E[q] = 0 \;\Rightarrow\; \frac{\partial L_E}{\partial q} = \frac{d}{d\tau}\left(\frac{\partial L_E}{\partial \dot{q}_E}\right). \tag{5.20}$$

Denoting Euclidean momentum by p_E, the Lagrangian yields

$$p_E = \frac{\partial L_E}{\partial \dot{q}_E} = -m\dot{q}_E. \tag{5.21}$$

Note the minus sign in the definition of p_E in terms of Euclidean velocity $\dot{q}_E$, which is due to Euclidean time switching the sign of the kinetic in the Euclidean Lagrangian as in Eq. 5.18. Since $\dot{q}_E = -i\dot{q}$, in terms of Minkowski momentum, the Euclidean momentum is given by

$$p_E = im\dot{q} = ip,$$
$$p \in [-\infty, +\infty] \;\Rightarrow\; p_E \in [-i\infty, +i\infty]. \tag{5.22}$$

Note the important fact that, since Minkowski momentum p is real, Euclidean momentum p_E is purely imaginary.

The Hamiltonian is defined by the Legendre transformation of L_E; from Eqs. 5.20 and 5.21

$$dL_E = \frac{\partial L_E}{\partial q}dq + \frac{\partial L_E}{\partial \dot{q}_E}d\dot{q}_E = \dot{p}_E dq + p_E d\dot{q}_E$$
$$\Rightarrow \quad dH_E \equiv d(p_E\dot{q}_E - L_E) = -\dot{p}_E dq + \dot{q}_E dp_E. \tag{5.23}$$

The above equation yields the Hamiltonian H_E as well as the dynamical equations for q, p_E. In particular, similarly to Eq. 5.23, the Hamilton equations of motion are given by

$$H_E = p_E \dot{q}_E - L_E \quad \Rightarrow \quad -\dot{p}_E = \frac{\partial H_E}{\partial q}, \quad \dot{q}_E = \frac{\partial H_E}{\partial p_E}. \tag{5.24}$$

The Euclidean Hamiltonian, from Eq. 5.29, is given by

$$H_E(p_E, q) = p_E \dot{q}_E - L_E = -\frac{1}{2m}(p_E)^2 + V(q). \tag{5.25}$$

The kinetic energy term of the Euclidean H_E has the opposite sign to the Minkowski kinetic term, reflecting the result obtained in Eq. 5.22.

Once the Euclidean Lagrangian and Hamiltonian are defined, all the results obtained earlier for Hamiltonian mechanics are valid for the Euclidean case.

5.4 Phase space path integrals

Path integral quantization discussed in Chapter 4 is based on the Lagrangian and the configuration space representation of the degree of freedom, as expressed in Eq. 4.40. In contrast, phase space path integration is based on the Hamiltonian H, which is a function of p_i, q_i. The canonical coordinate q_i and momentum p_i, taken together, define the phase space of the system, which for N particles (degrees of freedom) in three-dimensional space is the $3N$-dimensional Euclidean space $\Re^{3N}$.

The phase space path integrals for both Minkowski and Euclidean time are discussed here. The phase space path is defined as the sum over all possible values of $\mathbf{p}, \mathbf{q}$ from initial time t_i to final time t_f, with the boundary condition $\mathbf{q}_i, \mathbf{q}_f$ being imposed on only coordinate $\mathbf{q}$. For notational simplicity, unless required, the vector index is suppressed and phase space is represented by p, q.

Recall from Eq. 4.40 that the path integral quantization of the classical entity yields the transition amplitude given by

$$Z \equiv K(q_i, q_f; t_f, t_i) = \int Dq \exp\left\{\frac{i}{\hbar} S[q]\right\}.$$

Let us consider how this path integral is related to the phase space path integral. To distinguish between the Minkowski and Euclidean cases, the subscripts M and E are employed for the various quantities.

Minkowski path integral

The phase space action integral S_M for Minkowski time is given by

$$S_M[p, q] = \int_{t_i}^{t_f} dt \Big[p\dot{q} - H_M(p, q) \Big]. \tag{5.26}$$

Similarly to Eq. 4.40, the path integral is given by integrating over all possible configurations of the degrees of freedom in phase space and yields

$$Z_M = \mathcal{N} \int DqDp \exp\left\{\frac{i}{\hbar}\int_{t_i}^{t_f} dt\Big[p\dot{q} - H_M(p,q)\Big]\right\}. \tag{5.27}$$

For the Lagrangian $L_M = \frac{1}{2}m\dot{q}^2 - V(q)$ given in Eq. 5.1, the canonical momentum is given by

$$p = \frac{\partial L_M}{\partial \dot{q}} = m\dot{q} \tag{5.28}$$

and hence the Hamiltonian is

$$H_M = p\dot{q} - L = \frac{1}{2m}p^2 + V(q). \tag{5.29}$$

Hamiltonians that do not have velocity-dependent potentials and that are free from any constraint have the generic form given in Eq. 5.29.

In general, the integration over canonical momentum cannot be done explicitly. For the Hamiltonian given in Eq.5.29 that is quadratic in the canonical momentum, one can exactly perform the Gaussian path integral over the momentum variables in Eq. 5.27, which yields the Lagrangian

$$Z_M = \mathcal{N}' \int Dq \exp\left\{\frac{i}{\hbar}\int_{t_i}^{t_f} dt L_M(q,\dot{q})\right\}, \tag{5.30}$$

and this is the result given earlier in Eq. 4.40.

Euclidean path integral

The Euclidean path integral can be obtained by an analytic continuation of the Minkowski path integral. From Eq. 5.26, *define* the action for Euclidean time $S_E[p_E, q]$ by

$$S_E[p_E, q] = \int_{\tau_i}^{\tau_f} d\tau\Big[p_E\dot{q}_E - H_E(p_E, q)\Big]. \tag{5.31}$$

Continuing $t = -i\tau$ in Eq. 5.27 to imaginary Euclidean time yields

$$\frac{i}{\hbar}S_M = \frac{-i^2}{\hbar}S_E = \frac{1}{\hbar}S_E[p_E, q]. \tag{5.32}$$

Path integration is given by Eq. 4.40 and, from Eq. 5.22, because Euclidean momentum is pure imaginary, yields the Euclidean path integral

$$Z_E = \mathcal{N}_E \int Dq \int Dp_E \exp\left\{\frac{1}{\hbar} S_E[p_E, q]\right\}$$
$$= \mathcal{N}_E \int_{-\infty}^{+\infty} Dq \int_{-i\infty}^{+i\infty} Dp_E \exp\left\{\frac{1}{\hbar}\int_{\tau_i}^{\tau_f} d\tau\left[p_E \dot{q}_E - H_E(p_E, q)\right]\right\}.$$

The Euclidean Hamiltonian $\tilde{H}_E$, in terms of *real momentum* $p = -ip_E$ as in Eq. 5.22, is given by

$$\tilde{H}_E(p, q) = H_E(ip, q) = \frac{1}{2m}p^2 + V(q), \quad S_E[p, q] \equiv S_E[-ip_E, q]. \tag{5.33}$$

Hence

$$Z_E = \mathcal{N}_E \int Dq \int Dp \ \exp\{\frac{1}{\hbar} S_E[p, q]\} \tag{5.34}$$
$$= \mathcal{N}'_E \int_{-\infty}^{+\infty} Dq \int_{-\infty}^{+\infty} Dp \ \exp\left\{\frac{1}{\hbar}\int_{\tau_i}^{\tau_f} d\tau\left[ip\dot{q}_E - \tilde{H}_E(p, q)\right]\right\}. \tag{5.35}$$

Since the kinetic energy is quadratic in momentum, performing the Gaussian path integral over $\int Dp$ yields

$$Z_E = \mathcal{N}''_E \int Dq \exp\left\{\frac{1}{\hbar}\int_{\tau_i}^{\tau_f} d\tau L_E(q, \dot{q}')\right\}, \tag{5.36}$$

where the Euclidean Lagrangian L_E is given by

$$L_E(q, \dot{q}_E) = -\frac{1}{2}m(\dot{q}_E)^2 - V(q)$$

and agrees with the result obtained earlier in Eq. 5.18.

5.5 Poisson bracket

Consider arbitrary functions $f(p, q; t)$ and $g(p, q; t)$ of the canonical variables p, q. The Poisson bracket $\{f, g\}_P$ is defined as

$$\{f, g\}_P = \frac{\partial f}{\partial q}\frac{\partial g}{\partial p} - \frac{\partial f}{\partial p}\frac{\partial g}{\partial q}. \tag{5.37}$$

The Poisson bracket for the canonical conjugate variables is given by

$$\{q, p\}_P = \frac{\partial q}{\partial q}\frac{\partial p}{\partial p} - \frac{\partial q}{\partial p}\frac{\partial p}{\partial q} = 1. \tag{5.38}$$

The time evolution of the function f, using the dynamical equations given in Eq. 5.8, is given by

$$\frac{df}{dt} = \frac{\partial f}{\partial q}\dot{q} + \frac{\partial f}{\partial p}\dot{p} + \frac{\partial f}{\partial t} = \frac{\partial f}{\partial q}\frac{\partial H}{\partial p} - \frac{\partial f}{\partial p}\frac{\partial H}{\partial q} + \frac{\partial f}{\partial t}$$
$$\Rightarrow \frac{df}{dt} \equiv \{f, H\}_P + \frac{\partial f}{\partial t}. \tag{5.39}$$

For the special case of the Hamiltonian, one has in general that

$$\frac{dH}{dt} = \frac{\partial H}{\partial t} + \{H, H\}_P = \frac{\partial H}{\partial t}. \tag{5.40}$$

The Poisson bracket $\{f, g\}_P$ defined above can be generalized – for the case of N dynamical variables q_ℓ, p_ℓ; $\ell = 1, 2, \ldots N$ – to

$$\{f, g\}_P = \sum_{\ell=1}^{N} \left(\frac{\partial f}{\partial q_\ell}\frac{\partial g}{\partial p_\ell} - \frac{\partial f}{\partial p_\ell}\frac{\partial g}{\partial q_\ell} \right). \tag{5.41}$$

Poisson brackets have the following general properties:

$$\{f, g\}_P = -\{g, f\}_P, \quad \{f, gh\}_P = \{f, g\}_P h + g\{f, h\}_P \tag{5.42}$$

and obey the Jacobi identity

$$\{f, \{g, h\}_P\}_P + \{h, \{g, h\}_P\}_P + \{g, \{f, g\}_P\}_P = 0. \tag{5.43}$$

The Euclidean Hamiltonian and canonical variables obey the Poisson bracket relations. Similarly to Eq. 5.38,

$$\{q, p_E\}_P = \frac{\partial q}{\partial q}\frac{\partial p_E}{\partial p_E} - \frac{\partial q}{\partial p_E}\frac{\partial p_E}{\partial q} = 1. \tag{5.44}$$

Furthermore, from Eq. 5.24 and similarly to Eq. 5.39,

$$\frac{df}{d\tau} = \{f, H_E\}_P + \frac{\partial f}{\partial \tau}. \tag{5.45}$$

5.6 Commutation equations

Path integral quantization, as discussed in Section 5.4, is based on functional integration over the phase space of the degree of freedom. Another method for quantizing the degree of freedom is to introduce the Heisenberg commutation equations for the canonical dynamical variables q, p.

Minkowski case

Consider the single degree of freedom q; for Minkowski time t, the Minkowski momentum is $p = mdq/dt$. The procedure for canonical quantization is to elevate the dynamical variables q, p to Hermitian operators $\hat{q}, \hat{p}$ and replace the Poisson bracket of the classical dynamical variables q, p by the Heisenberg commutation equations for Hermitian operators $\hat{q}, \hat{p}$.

The degree of freedom q is quantized by *postulating* the following quantization condition [Weinberg (2013)]:

$$[\hat{q}, \hat{p}] = i\hbar\mathbb{I}\{q, p\}_P. \tag{5.46}$$

Note that the factor i on the right hand side of Eq. 5.46 is due to operators $\hat{q}, \hat{p}$ and the operator rendition of $\{q, p\}_P$ all being Hermitian.

The right hand side of Eq. 5.46 above is to be interpreted as follows: a) *first* evaluate the Poisson bracket considering p, q as classical dynamical variables; and b) *then* generalize p, q to Hermitian operators, with a *symmetrical ordering* of the terms in $\{q, p\}_P$ chosen to ensure that the right hand side is a Hermitian operator.

The Poisson bracket for the canonical conjugate variables given in Eq. 5.38 yields the well-known Heisenberg commutation equation[2]

$$\begin{aligned} [\hat{q}, \hat{p}] &= i\hbar\mathbb{I} \ \text{ since } \ \{q, p\}_P = 1 \\ \Rightarrow p &= -i\frac{\partial}{\partial q}. \end{aligned} \tag{5.47}$$

Euclidean case

Euclidean momentum p_E for Euclidean time $\tau = it$ is given, from Eq. 5.21

$$p_E = ip. \tag{5.48}$$

Note the Hermitian operators are related in the following manner:

$$\hat{p}_E = i\hat{p} \ \Rightarrow \hat{p}_E^\dagger = -\hat{p}_E : \ \text{anti-Hermitian.} \tag{5.49}$$

Hence, for Euclidean time, the rule for canonical quantization is given by

$$[\hat{q}, \hat{p}_E] = -\hbar\mathbb{I}\{q, p_E\}_P.$$

The Euclidean commutation equation does not have a factor of i since the Euclidean momentum operator $\hat{p}_E$ is *anti-Hermitian.* Since $\{q, p_E\}_P = 1$,

$$[\hat{q}, \hat{p}_E] = -\hbar\mathbb{I} \tag{5.50}$$

$$\Rightarrow p_E = \frac{\partial}{\partial q}. \tag{5.51}$$

The result obtained in Eq. 5.50 above is verified by an independent path integral derivation given in Eq. 6.58.

[2] The hat symbol is often dropped from the operators for simplicity of notation.

5.7 Dirac bracket and constrained quantization

The transition from classical dynamical variables to quantum operators, as in the commutation equations given in Eqs. 5.46 and 5.50, needs to be modified for the case of when the dynamical variables are constrained.

Consider a dynamical system in Euclidean time with L constraints that restrict the degree of freedom $\mathbf{q} = (q_1, \ldots q_N)$ to lie in an $N-L$ dimensional subspace of $\Re^N$ specified by

$$f_1(\mathbf{q}) = f_2(\mathbf{q}) \ldots = f_L(\mathbf{q}) = 0. \tag{5.52}$$

The question one needs to address is: what are the commutation equations of $\mathbf{p}_E$ and $\mathbf{q}$ (in boldface vector notation)? The commutation equations given in Eq. 5.50 are invalid for the constrained system as they do not respect the constraint equations given in Eq. 5.52.

Dirac (1964) has developed a procedure for quantizing constrained systems and in particular finding the commutation equations; the procedure is generalized to the case of Euclidean time.

The primary constraints $\phi_i = f_i(\mathbf{q}),\ i = 1, 2 \ldots L$ have to be supplemented by secondary constraints $\phi_{(L+1)}, \phi_{(L+2)}, \ldots \phi_K$ required for all the primary constraints to be conserved in time. All the constraints are *weakly* zero, namely

$$\phi_i(\mathbf{p}_E, \mathbf{q}) \approx 0, \quad i = 1, 2, \ldots K,$$

where the symbol $\approx$ means that the constraints ϕ_i are set to zero after performing the differentiations required for evaluating all the Poisson brackets.

The modified Hamiltonian is defined, similar to the procedure of defining constraints using Lagrange multipliers in the Lagrangian, by

$$H(\mathbf{p}_E, \mathbf{q}) = H_E(\mathbf{p}_E, \mathbf{x}) + \sum_{j=1}^{K} u_j(\mathbf{p}_E, \mathbf{q})\phi_j(\mathbf{p}_E, \mathbf{q}), \tag{5.53}$$

where H_E is the Euclidean Hamiltonian.

The coefficient functions u_i are determined by the requirement that the constraints are weakly conserved over time; this leads, from Eq. 5.45, to solving the following equations:

$$\frac{\partial \phi_I}{\partial \tau} = \{\phi_I, H\}_P = \{\phi_I, H_E\}_P + \sum_{j=1}^{K} \left(\{\phi_I, u_j\}_P \phi_j + u_j\{\phi_I, \phi_j\}_P\right) \approx 0. \tag{5.54}$$

The constraints $\phi_\ell, \ell = 1, 2 \ldots, K$ define the anti-symmetric constraint matrix

$$C_{\ell m} = \{\phi_\ell, \phi_m\}_P. \tag{5.55}$$

The Dirac brackets are defined, for arbitrary function $h(q, p), \; g(q, p)$, by

$$\{h, g\}_D = \{h, g\}_P - \sum_{\ell,m=1}^{K} \{h, \phi_\ell\}_P C^{-1}_{\ell m} \{\phi_m, g\}_P .$$

The Dirac brackets satisfy all the relations fulfilled by the Poisson bracket given in Eq. 5.42, and in particular can be shown to satisfy the crucial Jacobi identity given in Eq. 5.43.

The rule for defining the Heisenberg commutation equations for the constrained degree of freedom $\mathbf{q}$ is a generalization of Eq. 5.50, and for Euclidean time is given by

$$[\hat{q}_i, \hat{p}_{Ej}] = -\hbar \mathbb{I} \{q_i, p_{Ej}\}_D . \tag{5.56}$$

The advantage of using the Dirac bracket for quantization is that the commutation equation holds for all values of the degree of freedom $\mathbf{q}$ and is not restricted by the condition of weak equality.

5.7.1 Dirac bracket for two constraints

Consider the case of a single primary constraint [Weinberg (2013)] on the dynamical variable $\mathbf{x}$ given by

$$\phi_1 = f(\mathbf{x}) = 0.$$

The Euclidean path integral provides a straightforward method for quantization by constraining the degrees of freedom in the path integral and yields

$$Z = \int Dx \exp\{\int dt \mathcal{L}_E\} \prod_t \delta\Big(f(\mathbf{x}(t))\Big) \tag{5.57}$$

$$= \int Dx D\lambda \exp\{\int dt [\mathcal{L}_E + i\lambda(t) f(\mathbf{x}(t))]\}. \tag{5.58}$$

The Euclidean Hamiltonian, from Eq. 5.25, is given by

$$H_E(\mathbf{p}_E, \mathbf{x}) = \mathbf{p}_E \cdot \frac{d\mathbf{x}}{d\tau} - \mathcal{L}_E - i\lambda f(\mathbf{x})$$

$$= -\frac{1}{2m}(\mathbf{p}_E)^2 + V(\mathbf{x}) - i\lambda f(\mathbf{x}). \tag{5.59}$$

The primary constraint needs to be conserved over time and leads to the requirement

$$\frac{\partial \phi_1}{\partial \tau} = \{\phi_1, H_E\}_P = -\frac{1}{m}\mathbf{p}_E \cdot \nabla f \approx 0. \tag{5.60}$$

Hence, the second constraint is chosen to be

$$\phi_2 = \mathbf{p}_E \cdot \nabla f \approx 0, \tag{5.61}$$

which yields, from Eqs. 5.60 and 5.61, that constraint ϕ_1 is conserved over time. As required by Eq. 5.53, choose the modified Hamiltonian

$$\begin{aligned} H(\mathbf{p}_E, \mathbf{q}) &= H_E(\mathbf{p}_E, \mathbf{x}) + i\lambda\phi_1 \\ &= -\frac{1}{2m}(\mathbf{p}_E)^2 + V(\mathbf{x}). \end{aligned} \tag{5.62}$$

The time dependence of the constraint ϕ_2 is given by

$$\frac{\partial \phi_2}{\partial \tau} = \{\phi_2, H\}_P = -\frac{1}{m} p_{Ei} \frac{\partial}{\partial x_i} \left(\mathbf{p}_E \cdot \nabla f\right) \approx 0. \tag{5.63}$$

The last equality follows from the weak equality $\phi_2 = \mathbf{p}_E \cdot \nabla f \approx 0$ given in Eq. 5.61; since it holds for all $\mathbf{x}$, this implies that $\partial(\mathbf{p}_E \cdot \nabla f)/\partial x_i \approx 0$.

Since both the constraints ϕ_1, ϕ_2 are (weakly) conserved in time, no new constraints are required. The constraint matrix, from Eq. 5.55, is given by

$$C_{12} = \{\phi_1, \phi_2\}_P = \sum_i \frac{\partial \phi_1}{\partial x_i} \frac{\partial \phi_2}{\partial p_{Ei}} = (\nabla f)^2 = -C_{21}. \tag{5.64}$$

The inverse is given by

$$C_{12}^{-1} = -\frac{1}{(\nabla f)^2} = -C_{21}^{-1}. \tag{5.65}$$

The Dirac bracket, from Eq. 5.56, is given by

$$\begin{aligned} \{x_i, p_{Ej}\}_D &= \{x_i, p_{Ej}\}_P - \{x_i, \phi_2\}_P C_{21}^{-1} \{\phi_1, p_{Ej}\}_P \\ &= \delta_{i-j} - \frac{\partial f}{\partial x_i} \frac{1}{(\nabla f)^2} \frac{\partial f}{\partial x_j}. \end{aligned} \tag{5.66}$$

The other Dirac brackets have been shown by Weinberg (2013) to be zero, namely

$$\{x_i, x_j\}_D = 0 = \{p_{Ei}, p_{Ej}\}_D.$$

Hence, for Euclidean time, the Heisenberg commutation equations for a constrained degree of freedom, from Eqs. 5.56, 5.66, and 5.67, are given by

$$[\hat{x}_i, \hat{p}_{Ej}] = -\hbar\mathbb{I}\{x_i, p_{Ej}\}_D = -\hbar\mathbb{I}\left(\delta_{i-j} - \frac{\partial f}{\partial x_i} \frac{1}{(\nabla f)^2} \frac{\partial f}{\partial x_j}\right), \tag{5.67}$$

$$[x_i, x_j] = 0 = [p_{Ei}, p_{Ej}], \tag{5.68}$$

with the Hamiltonian given from Eq. 5.62 by

$$H(\mathbf{p}_E, \mathbf{q}) = -\frac{1}{2m}(\mathbf{p}_E)^2 + V(\mathbf{x}).$$

Note an important general feature of quantum mechanics is that the state space has a structure that is specified *independently* of the Hamiltonian. This is what comes to the fore in the case of a constrained degree of freedom.

For the constrained system that is being considered, the Hamiltonian is the same as the unconstrained case but the commutation equation given in Eq. 5.67 is changed from the unconstrained case to account for the constraint. The commutation equation, in turn, is a reflection of the structure of the state space.

For a more complex constrained system such as the one for the acceleration Hamiltonian considered in Section 13.4, both the Hamiltonian and the commutation equation are modified to account for the constraints imposed on the degrees of freedom.

5.8 Free particle evolution kernel

Consider the Minkowski and Euclidean Hamiltonian of a free quantum particle given by

$$H = \frac{1}{2m}\mathbf{p}^2 = -\frac{1}{2m}(\mathbf{p}_E)^2 = -\frac{\hbar^2}{2m}\frac{\partial^2}{\partial \mathbf{x}^2}. \tag{5.69}$$

The Hamiltonian for both the Minkowski and Euclidean time yields the same differential operator in terms of the coordinate degree of freedom $\mathbf{x} \in \Re^d$. The eigenstates of H are given by the plane wave eigenstates $|\vec{p}\rangle$,

$$H|\mathbf{p}\rangle = \frac{\mathbf{p}^2}{2m}|\mathbf{p}\rangle, \tag{5.70}$$

that yields, from Eq. 3.34, the completeness equation

$$\int_{-\infty}^{\infty} \frac{d\mathbf{p}}{(2\pi\hbar)^d}|\mathbf{p}\rangle\langle\mathbf{p}| = \mathbb{I}, \quad \langle\mathbf{p}|\mathbf{p}'\rangle = (2\pi\hbar)^d\delta(p - p'). \tag{5.71}$$

The free particle Minkowski evolution kernel, from Eq. 4.4, is given by

$$\begin{aligned} K(\mathbf{x}, \mathbf{x}'; t) &= \langle\mathbf{x}|e^{-it\frac{\mathbf{p}^2}{2m\hbar}}|\mathbf{x}'\rangle = \int \frac{d\mathbf{p}}{(2\pi\hbar)^d} e^{-it\frac{\mathbf{p}^2}{2m\hbar}} e^{i\mathbf{p}\cdot(\mathbf{x}-\mathbf{x}')/\hbar} \\ &= \left(\sqrt{\frac{m}{2\pi i\hbar t}}\right)^{d/2} \exp\left\{\frac{i}{\hbar}\frac{m(\mathbf{x}-\mathbf{x}')^2}{2t}\right\}. \end{aligned} \tag{5.72}$$

The evolution kernel for Euclidean time τ is given by

$$\begin{aligned} K(\mathbf{x}, \mathbf{x}'; \tau) &= \langle\mathbf{x}|e^{-\tau\frac{(\mathbf{p}_E)^2}{2m\hbar}}|\mathbf{x}'\rangle \\ &= \left(\sqrt{\frac{m}{2\pi\hbar\tau}}\right)^{d/2} \exp\left\{-\frac{1}{\hbar}\frac{m(\mathbf{x}-\mathbf{x}')^2}{2\tau}\right\}. \end{aligned} \tag{5.73}$$

As can be seen by comparing Eqs. 5.72 and 5.73, the Euclidean evolution kernel is a well defined Gaussian distribution whereas the Minkowski case is an oscillating exponential that needs to be defined using the theory of distributions.

5.9 Hamiltonian and path integral

The phase space path integral is defined by summing over all possible indeterminate evolutions of the degree of freedom and its canonical conjugate variable, and is one of the ways of defining a quantized Hamiltonian. Another approach, which leads to the Schrödiner equation, is to first quantize the classical Hamiltonian based on the commutation equations, and then use state space methods to determine the indeterminate evolution of the quantum degree of freedom encoded in its state vector.

In this section, the evolution kernel for a general Hermitian Hamiltonian $H(p, x)$ is studied in Euclidean time. The fundamental reason that analytic continuation is well defined is because the Hamiltonians for all physical systems are bounded from below. For all physical systems the lowest energy state, namely the ground state, has finite energy (and not infinite negative energy). Since the eigenspectrum of H is bounded from below, the operator $\exp\{-\tau H\}$ is well defined with none of the matrix elements having an ill-defined value.

The Euclidean time evolution kernel is given by[3]

$$K(x_N, x_0; \tau) = \langle x_N | \underbrace{e^{-\epsilon H} \cdots e^{-\epsilon H}}_{N-\text{factors}} | x_0 \rangle,$$

where $x_i = x_0,\ x_f = x_N$ for notational convenience. The general matrix elements of H are given by

$$\langle p | e^{-\epsilon H} | x \rangle = \langle x | e^{-\epsilon \hat{H}^\dagger} | p \rangle^* = \langle x | e^{-\epsilon \hat{H}} | p \rangle^* = e^{-\epsilon \hat{H}(p,x)} e^{-ipx}.$$

Using the completeness equation $\int dx |x\rangle\langle x| = \mathbb{I} = \int \frac{dp}{2\pi} |p\rangle\langle p|$ yields

$$\begin{aligned} K(x_N, x_0; \tau) &= \prod_{i=1}^{N-1} \int dx_i \langle x_N | e^{-\epsilon H} | x_{N-1} \rangle \cdots \langle x_{n+1} | e^{-\epsilon H} | x_n \rangle \cdots \langle x_1 | e^{-\epsilon H} | x_0 \rangle \\ &= \int Dx \prod_{i=1}^{N} \int \frac{dp_i}{2\pi} \langle x_N | p_N \rangle \langle p_N | e^{-\epsilon H} | x_{N-1} \rangle \cdots \\ &\quad \cdots \langle x_{n+1} | p_{n+1} \rangle \langle p_{n+1} | e^{-\epsilon H} | x_n \rangle \langle x_n | p_n \rangle \quad \cdots \langle x_1 | p_1 \rangle \langle p_1 | e^{-\epsilon H} | x_0 \rangle \end{aligned}$$

[3] Set $\hbar = 1$.

$$= \int DxDpe^{i\sum_{n=1}^{N} p_n x_n} e^{-\epsilon \sum_{n=1}^{N} H(p_n, x_{n-1})} e^{-i\sum_{n=1}^{N} p_n x_{n-1}},$$

$$= \int DxDpe^{S}, \tag{5.74}$$

where $\int DxDp = \prod_{n=1}^{N-1} \int_{-\infty}^{+\infty} dx_n \prod_{i=1}^{N} \int_{-\infty}^{+\infty} \frac{dp_i}{2\pi}.$

Hence, from Eq. 5.74, the limit of $\epsilon \to 0$ yields

$$S = i\sum_{n=1}^{N} p_n(x_n - x_{n-1}) - \epsilon \sum_{n=1}^{N} H(p_n, x_{n-1}) \to i\int_0^{\tau} dtp\dot{x} - \int_0^{\tau} H(p_t, x_t)dt$$

$$\Rightarrow S = \int_0^{\tau} dt\{ip\dot{x} - H(p, x)\}.$$

For the special case of a Euclidean Hamiltonian given by Eq. 5.25, namely

$$H = \frac{p^2}{2m} + V(x),$$

we obtain

$$K(x_N, x_0; \tau) = \int DxDpe^{-\int dtV} e^{-\int dt[\frac{p^2}{2m} - ip\dot{x}]} = \mathcal{N} \int Dxe^{-\int dt\left(\frac{m}{2}\dot{x}^2 + V\right)}$$

$$\Rightarrow K(x_N, x_0; \tau) = \int Dxe^{S} \text{ with } Dx = (\frac{m}{2\pi\epsilon})^{\frac{N}{2}} \prod_{n=1}^{N-1} dx_n.$$

Hence

$$S = \int_0^{\tau} dt\mathcal{L}, \quad \mathcal{L} = -\frac{m}{2}\dot{x}^2 - V$$

and the earlier result for the Euclidean Lagrangian given in Eq. 5.18 is recovered.

5.10 Coherent states

Coherent states are the natural basis states for studying systems that are defined in terms of the creation and destruction operators and many computations can be done in a transparent manner in this basis. Coherent states also provide a mathematical framework for the formulation of path integrals that, unlike many cases that require a Lagrangian, requires only a Hamiltonian.

Consider a Hamiltonian H that is expressed in terms of the annihilation and creation operators a, $a^\dagger$ and has the form $H = H(a, a^\dagger)$. The oscillator basis is defined by

$$[a, a^\dagger] = 1, \tag{5.75}$$

$$a|0\rangle = 0, \quad |n\rangle = \frac{(a^\dagger)^n}{\sqrt{n!}}|0\rangle, \; n = 1, 2, 3 \ldots$$
$$\langle n|m\rangle = \delta_{m-n}.$$

A coherent state is defined for complex number z by

$$|z\rangle = e^{za^\dagger}|0\rangle \;\Rightarrow\; |z\rangle = \sum_{n=0} \frac{z^n}{\sqrt{n!}}|n\rangle. \tag{5.76}$$

Note that the $|z\rangle$ states are over-complete since z is complex and hence there are many more states $|z\rangle$ than the integer valued oscillator basis $\{|n\rangle; n = 0, 1, 2 \cdots \infty\}$, which is complete.

The commutation equation given in Eq. 5.75 yields

$$a|n\rangle = \sqrt{n}|n-1\rangle, \quad a^\dagger|n\rangle = \sqrt{n+1}|n+1\rangle.$$

Hence

$$a|z\rangle = \sum_n \frac{z^n}{\sqrt{n!}} a|n\rangle = \sum \frac{z^n}{\sqrt{n!}}\sqrt{n}|n-1\rangle = z|z\rangle.$$

Similarly,

$$\langle z| = \langle 0|e^{z^* a} \;\Rightarrow\; \langle z|a^\dagger = \langle z|z^*.$$

Consider the scalar product

$$\langle z_2|z_1\rangle = \langle 0|e^{z_2^* a}e^{z_1 a^\dagger}|0\rangle.$$

Note that

$$\begin{aligned} e^{z_2^* a}e^{z_1 a^\dagger} &= e^{z_1 a^\dagger}e^{-z_1 a^\dagger}e^{z_2^* a}e^{z_1 a^\dagger} = e^{z_1 a^\dagger}\exp\left(z_2^* e^{-z_1 a^\dagger} a e^{z_1 a^\dagger}\right) \\ &= e^{z_1 a^\dagger}e^{z_2^*(a - z_1[a^\dagger, a])} = e^{z_1 a^\dagger}e^{z_2^* a}e^{z_2^* z_1}. \end{aligned}$$

Hence

$$\langle z_2|z_1\rangle = e^{z_2^* z_1}\langle 0|e^{z_2^* a^\dagger}e^{z_1 a}|0\rangle = e^{z_2^* z_1}. \tag{5.77}$$

5.11 Coherent state vector

The normalized coherent state function is defined by

$$|\psi_z\rangle = \mathcal{C}|z\rangle,$$

with the normalization given by

$$\langle\psi_z|\psi_z\rangle = 1 = \mathcal{C}^2 e^{+|z|^2} \;\Rightarrow \mathcal{C} = e^{-\frac{1}{2}|z|^2}.$$

One can use state functions $|\psi_z\rangle$ to define the completeness equation instead of $|z\rangle$; this just leads to a change of normalization and does not yield any changes in the results obtained.

The expectation value of the number operator $\hat{N} = a^\dagger a$ is given by

$$N = \langle\psi_z|\hat{N}|\psi_z\rangle = \langle\psi_z|a^\dagger a|\psi_z\rangle = |z|^2.$$

Since $N = |z|^2$, the likelihood of finding the coherent state in the ℓth excited state of the oscillator is given by

$$P_\ell(N) = |\langle\ell|\psi_z\rangle|^2 = \left|e^{-\frac{N}{2}}\frac{|z|^\ell}{\sqrt{\ell!}}\right|^2 = e^{-N}\frac{N^\ell}{\ell!}.$$

The result shows that the probability is given by the Poisson distribution. This is a well-known property of the photon field, in that the probability of a classical source emitting N photons of a given energy is given by the Poisson distribution.

Consider the harmonic oscillator Hamiltonian given by

$$H = \omega a a^\dagger.$$

The evolution of the coherent state is given by

$$\begin{aligned} e^{-\tau\omega a^\dagger a}|z\rangle &= e^{-\tau\omega a^\dagger a}e^{za^\dagger}|0\rangle = e^{-\tau\omega a^\dagger a}e^{za^\dagger}e^{\tau\omega a^\dagger a}e^{-\tau\omega a^\dagger a}|0\rangle \\ &= \exp\left(ze^{-\tau\omega a^\dagger a}a^\dagger e^{\tau\omega a^\dagger a}\right)|0\rangle. \end{aligned}$$

Note that

$$e^{-\lambda a^\dagger a}a^\dagger e^{\lambda a^\dagger a} = e^{-\lambda}a^\dagger.$$

To prove this consider

$$\begin{aligned} w(\lambda) &= e^{-\lambda a^\dagger a}a^\dagger e^{\lambda a^\dagger a} \\ \Rightarrow \frac{dw}{d\lambda} &= e^{-\lambda a^\dagger a}[a^\dagger, a^\dagger a]e^{\lambda a^\dagger a} = -w(\lambda), \end{aligned}$$

and integrating the above equation yields

$$w(\lambda) = e^{-\lambda}w(0) = e^{-\lambda}a^\dagger. \tag{5.78}$$

Hence

$$e^{-\tau\omega a^\dagger a}|z\rangle = \exp\left(ze^{-\omega\tau}a^\dagger\right)|0\rangle = |e^{-\omega\tau}z\rangle. \tag{5.79}$$

For the simple harmonic oscillator, the coherent state's time evolution results in a rescaling of its label z; this is the reason that coherent states are so useful in the study of quantum optics, where the photons are equivalent to a collection of harmonic oscillators.

5.12 Completeness equation: over-complete

For $z = x + iy$, $z^* = x - iy$, the completeness equation for the coherent states can be written as

$$\mathbf{1} = \int_{-\infty}^{+\infty} \frac{dxdy}{\pi} |z\rangle e^{-z^* z} \langle z| \equiv \int \frac{dzdz^*}{2\pi i} |z\rangle e^{-z^* z} \langle z|. \tag{5.80}$$

Note the new feature that there is a *metric* on the state space given by $e^{-z^* z}$, which is due to the fact that the basis $|z\rangle$ are over-complete; in other words, there are far more basis states $|z\rangle$ than are required for spanning the state space. The over-completeness is to be expected since the oscillator state space is spanned by the oscillator basis $|n\rangle$, with $n \geq 0$, whereas the $|z\rangle$ basis states are indexed by two continuous variables x, y.

To prove the completeness equation, note that consistency requires that the scalar product given in Eq. 5.77 be reproduced if the completeness equation is used for forming the scalar product. Hence consider

$$\begin{aligned} \langle z_2 | z_1 \rangle &= \int \frac{dxdy}{\pi} \langle z_2 | z \rangle e^{-z^* z} \langle z | z_1 \rangle = \int \frac{dxdy}{\pi} e^{z_2^* z} e^{-z^* z} e^{z^* z_1} \\ &= \int \frac{dxdy}{\pi} e^S. \end{aligned} \tag{5.81}$$

Since $z = x + iy$ and $z^* = x - iy$,

$$\begin{aligned} S &= z_2^*(x + iy) - (x^2 + y^2) + (x - iy) z_1 \\ &= -(x^2 + y^2) + x(z_1 + z_2^*) + iy(z_2^* - z_1). \end{aligned} \tag{5.82}$$

Therefore

$$\begin{aligned} \langle z_2 | z_1 \rangle &= \frac{1}{\pi} \int dxdy e^{-x^2 + x(z_1 + z_2^*)} e^{-y^2 + iy(z_2^* - z_1)} \\ &= \frac{1}{\pi} \cdot \sqrt{\pi} e^{\frac{1}{4}(z_1 + z_2^*)^2} \cdot \sqrt{\pi} e^{-\frac{1}{4}(z_2^* - z_1)^2} \\ &= e^{z_2^* z_1} \end{aligned} \tag{5.83}$$

as expected, hence verifying the completeness equation.

5.13 Operators; normal ordering

The matrix element of an operator $\mathcal{Q}(a^\dagger, a)$ is in general an arbitrary function of $a, a^\dagger$. Since a and $a^\dagger$ do not commute, the operator $\mathcal{Q}(a^\dagger, a)$ has to be brought into a standard form, which is chosen to be all the destruction operators a to the right of all the creation operators $a^\dagger$. The procedure for bringing an operator $\mathcal{Q}(a^\dagger, a)$

into the standard format is called *normal ordering*, and to emphasize the standard normal ordered form of the operator, it is represented by $: \mathcal{Q}(a^\dagger, a) :$.

One can then obtain the matrix elements of the operator to be given by

$$\langle z| : \mathcal{Q}(a^\dagger, a) : |z\rangle =: \mathcal{Q}(z^*, z) : \langle z|z\rangle =: \mathcal{Q}(z^*, z) : e^{z^* z}, \tag{5.84}$$

where, in the operator $\mathcal{Q}(a^\dagger, a)$, one has the replacement

$$a \to z, \quad a^\dagger \to z^*. \tag{5.85}$$

An example of normal ordering is

$$H = a^2 a^\dagger \quad \Rightarrow \quad : H :=: a^2 a^\dagger :\equiv a^\dagger a^2.$$

The relation between H and $: H :$ is

$$\begin{aligned} H &= a(aa^\dagger) = a(a^\dagger a + 1) = (a^\dagger a + 1)a + a \\ &=: H : +2a \quad \Rightarrow \quad : H := a^\dagger a^2. \end{aligned} \tag{5.86}$$

One then obtains, from Eq. 5.86,

$$\langle z_2| : H(a^\dagger, a) : |z_1\rangle = \langle z_2| : H(z_2^*, z_1) : |z_1\rangle = z_2^* z_1^2 e^{z_2^* z_1}. \tag{5.87}$$

The general matrix element for the normal ordered operator can be similarly determined.

5.14 Path integral for coherent states

In general, for an arbitrary Hamiltonian, using the completeness equation $N - 1$ times yields the evolution kernel

$$\begin{aligned} K(z_f, z_i; \tau) &= \langle z_f | e^{-\tau H(a^\dagger, a)} | z_i \rangle \\ &= \int DzDz^* \langle z_N | e^{-\epsilon H} | z_{N-1} \rangle e^{-z_{N-1}^* z_{N-1}} \langle z_{N-1} | \\ &\quad \cdots e^{-z_{n+1}^* z_{n+1}} \langle z_{n+1} | e^{-\epsilon H} | z_n \rangle \cdots \langle z_1 | e^{-\epsilon H} | z_0 \rangle, \end{aligned}$$

$$\text{where} \quad DzDz^* = \prod_{i=1}^{N-1} \frac{dz_i dz_i^*}{2\pi i} = \prod_{i=1}^{N-1} \frac{dx_i dy_i}{\pi}, \quad z_f = z_N, \; z_i = z_0.$$

To find the matrix elements of $\langle z_{n+1} | e^{-H(a, a^\dagger)} | z_n \rangle$ we first assume that the Hamiltonian $H(a, a^\dagger)$ has been normal ordered by moving all the destruction operators to the right. Hence, from the general property of the normal ordered operator given in Eq. 5.84,

$$\langle z_{n+1} | e^{-\epsilon \hat{H}(a^\dagger, a)} | z_n \rangle = e^{-\epsilon H(z_{n+1}^*, z_n)} \langle z_{n+1} | z_n \rangle = e^{-\epsilon H(z_{n+1}^*, z_n)} e^{z_{n+1}^* z_n}.$$

The evolution kernel is given by

$$K(z_f, z_i; \tau) = \int DzDz^* e^S,$$

$$\text{where} \quad S = -\epsilon \sum_{n=0}^{N-1} H(z^*_{n+1}, z_n) + \sum_{n=0}^{N-1} z^*_{n+1} z_n - \sum_{n=1}^{N-1} z^*_n z_n$$

$$= -\epsilon \sum_{n=0}^{N-1} H(z^*_{n+1}, z_n) + \sum_{n=0}^{N-1} (z^*_{n+1} - z^*_n) z_n + z^*_0 z_0.$$

The action has a new feature: the appearance of a boundary term, namely $z^*_0 z_0$; the term z_0 is fixed by the boundary condition whereas z^*_0 is an integration variable. The continuum limit is given by

$$S = -\int_0^\tau dt H(z^*(t), z(t)) + \int_0^\tau dt \frac{dz^*}{dt} z + z^*(0)z(0).$$

Since

$$\int_0^\tau dt \frac{dz^*}{dt} z = \int_0^\tau dt \left[\frac{d}{dt}(zz^*) - z^* \frac{dz}{dt}\right]$$

$$= z^*(\tau)z(\tau) - z^*(0)z(0) - \int_0^\tau z^* \frac{dz}{dt}$$

one obtains

$$S = -\int_0^\tau dt \left[H(z^*, z) + z * \frac{dz}{dt}\right] + z^*(\tau)z(\tau). \tag{5.88}$$

A symmetric expression for the action yields

$$S = -\int_0^\tau dt H(z^*(t), z(t)) + \frac{1}{2} \int_0^\tau dt \left(\frac{dz^*}{dt} z - z^* \frac{dz}{dt}\right)$$

$$+ \frac{1}{2}\left(z^*_f z(\tau) + z^*(0) z_i\right),$$

$$\text{Boundary conditions} : z_f = z(\tau), \; z_i = z(0) : \text{fixed}.$$

Note both $z(\tau)$ and $z^*(0)$ are integration variables and couple the boundary terms z_i, z_f via the time derivative terms in the action S to the other integration variables of the path integral.

The discrete coherent state path integral is well defined for the linear as well as for the nonlinear case. However, for the nonlinear case the continuum limit of the discretized coherent state path integral apparently has anomalies and ambiguities, as is the case when studying a spin system [Zinn-Justin (2005)]. The ambiguities can be resolved by using the fact that the coherent basis states are over-complete.

5.14.1 Simple harmonic oscillator

The Hamiltonian is given by

$$H = \omega a^\dagger a,$$

and from Eq. 5.79

$$e^{-\tau\omega a^\dagger a}|z\rangle = |e^{-\omega\tau}z\rangle.$$

Hence

$$\begin{aligned} K(z_f, z_i; \tau) &= \langle z_f|e^{-\tau\omega a^\dagger a}|z_i\rangle = \langle z_f|e^{-\omega\tau}z_i\rangle \\ &= \exp\left(e^{-\omega\tau}z_f^* z_i\right). \end{aligned} \tag{5.89}$$

The evolution kernel for the simple harmonic oscillator can also be obtained by performing the path integral. In the coherent state representation, from the definition of the matrix elements of operators given in Eq. 5.84, the matrix elements of the Hamiltonian are given by

$$\langle z|H(a^\dagger, a)|z\rangle = \omega\langle z|a^\dagger a|z\rangle = \omega z^* z e^{z^* z} \quad \Rightarrow \quad H(z^*, z) = \omega z^* z.$$

From Eq. 5.88, the action, for $\dot{z} \equiv dz/dt$, is given by

$$S = -\int_0^\tau (z^*\dot{z} + \omega\, z^* z) + z_f^* z(\tau).$$

The classical equation of motion is given by

$$\frac{\delta S}{\delta z^*} = \dot{z}_c + \omega z_c = 0.$$

For $z = x + iy$, one has $z^* = x^* - iy^*$, and x, y yields the complex classical trajectory of the particle,

$$z_c(t) = e^{-\omega\, t}z(0) \quad \Rightarrow \quad z_c^*(t) = e^{-(\tau - t)\omega}z_f^*.$$

Consider the change of variables

$$z = z_c + \varsigma, \quad z^* = z_c^* + \varsigma^*,$$

where

$$\varsigma(0) = 0 \quad , \quad \varsigma^*(\tau) = 0.$$

Hence

$$S = -\int_0^\tau (\varsigma^*\dot{\varsigma} + \omega\dot{\varsigma}\varsigma) + z_f^* e^{-\omega\tau}z_i \tag{5.90}$$

and yields the evolution kernel

$$K(z_f^*, z_i; \tau) = A \ \exp\{z_f^* e^{-\omega\tau} z_i\},$$

with A being a constant.

Let us perform a trace of K using the coherent basis completeness equation, and obtain

$$\begin{aligned}
K &= \langle z|e^{-TH}|z\rangle \\
\Rightarrow \ trK &= \int \frac{dzdz^*}{2\pi i} e^{-z^*z} K(z^*, z) = A \int \frac{dzdz^*}{2\pi i} e^{-(1-e^{-w\tau})|z|^2} \\
&= \frac{A}{1 - e^{-w\tau}}.
\end{aligned} \tag{5.91}$$

Performing the trace directly on the Hamiltonian using the oscillator basis yields

$$trK = tr(e^{-\tau H}) = \frac{1}{1 - e^{-w\tau}} \quad \Rightarrow \quad A = 1,$$

and hence, as given in Eq. 5.89, the evolution kernel is given by

$$K(z_f^*, z_i; \tau) = \exp\{z_f^* e^{-\omega\tau} z_i\}.$$

5.15 Forced harmonic oscillator

Consider the Hamiltonian

$$\begin{aligned}
H &= \omega a^\dagger a + \lambda(a^\dagger + a) \\
&= \omega\left(a^\dagger + \frac{\lambda}{\omega}\right)\left(a + \frac{\lambda}{\omega}\right) - \frac{\lambda^2}{\omega^2}.
\end{aligned}$$

Define the shifted operators by

$$b = a + \frac{\lambda}{\omega}, \quad b^\dagger = a^\dagger + \frac{\lambda}{\omega}.$$

Then, for $a|0\rangle = 0$ define the vacuum state for the shifted operators $|\upsilon\rangle$ by

$$b|\upsilon\rangle = 0 = (a + \frac{\lambda}{\omega})|\upsilon\rangle \ \Rightarrow \ a|\upsilon\rangle = -\frac{\lambda}{\omega}|\upsilon\rangle. \tag{5.92}$$

The state $|\upsilon\rangle$ is given by

$$|\upsilon\rangle = e^{\frac{\lambda}{\omega}(a - a^\dagger)}|0\rangle.$$

To prove this consider the expression

$$w(t) = e^{-t(a-a^\dagger)} a e^{t(a-a^\dagger)}.$$

It can be readily shown, similarly to the derivation of Eq. 5.78, that

$$w(t) = -t + a.$$

Hence one obtains the result given in Eq. 5.92, namely

$$a|\upsilon\rangle = ae^{\frac{\lambda}{\omega}(a-a^\dagger)}|0\rangle = e^{\frac{\lambda}{\omega}(a-a^\dagger)}[-\frac{\lambda}{\omega}+a]|0\rangle = -\frac{\lambda}{\omega}|\upsilon\rangle.$$

Using the CBH formula $e^A e^B = e^{A+B+\frac{1}{2}[A,B]}$ for $[A, B] \propto \mathbb{I}$ yields

$$e^{-\frac{\lambda}{\omega}a^\dagger} e^{\frac{\lambda}{\omega}a} = e^{\frac{\lambda}{\omega}(a-a^\dagger)-\frac{\lambda^2}{\omega^2}[a^\dagger,a]} = e^{+\frac{\lambda^2}{\omega^2}} e^{\frac{\lambda}{\omega}(a-a^\dagger)}$$

$$\Rightarrow \qquad |\upsilon\rangle = e^{-\frac{\lambda^2}{\omega^2}} e^{-\frac{\lambda}{\omega}a^\dagger}|0\rangle.$$

The evolution kernel for the anharmonic oscillator, in the coherent basis states, is given by

$$\begin{aligned} K &= \langle z_2|e^{-\tau H}|z_1\rangle = \langle 0|e^{z_2 a}e^{-\tau H}e^{z_1 a^\dagger}|0\rangle \\ &= e^{2\frac{\lambda^2}{\omega^2}}\langle\upsilon|e^{+\frac{\lambda}{\omega}a}e^{z_2^* a}e^{-\tau H}e^{z_1 a^\dagger}e^{+\frac{\lambda}{\omega}a^\dagger}|\upsilon\rangle \\ &= e^{\frac{\lambda^2}{\omega^2}}{}_b\langle\upsilon|e^{a(\frac{\lambda}{\omega}+z_2^*)}e^{-\tau\omega b^\dagger b}e^{a^\dagger(z_1+\frac{\lambda}{\omega})}|\upsilon\rangle \\ &= e^{\frac{\lambda^2}{\omega^2}}e^{-\frac{\lambda}{\omega}(\frac{\lambda}{\omega}+z_2^*)}\, e^{-\frac{\lambda}{\omega}(z_1+\frac{\lambda}{\omega})}\langle\upsilon|e^{b(z_2^*+\frac{\lambda}{\omega})}\, e^{-\tau\omega b^\dagger b}e^{(z_1+\frac{\lambda}{\omega})b^\dagger}|\upsilon\rangle. \end{aligned} \tag{5.93}$$

The definition of a coherent state given in Eq. 5.76, namely $|z\rangle = e^{za^\dagger}|0\rangle$, yields for the $b, b^\dagger$-oscillator

$$|z\rangle_b = e^{zb^\dagger}|\upsilon\rangle, \quad b|\upsilon\rangle = 0.$$

Hence, from Eq. 5.93

$$\begin{aligned} K &= \langle z_2|e^{-\tau H}|z_1\rangle \\ &= e^{-\frac{\lambda^2}{\omega^2}-\frac{\lambda}{\omega}(z_1+z_2^*)}\ {}_b\langle z_2^*+\frac{\lambda}{\omega}|e^{-\tau\omega b^\dagger b}|z_1+\frac{\lambda}{\omega}\rangle_b \\ &= \exp\left\{-\frac{\lambda^2}{\omega^2}-\frac{\lambda}{\omega}(z_1+z_2^*)\right\}\ \exp\left\{\left(z_2^*+\frac{\lambda}{\omega}\right)e^{-\omega\tau}\left(z_1+\frac{\lambda}{\omega}\right)\right\}. \end{aligned}$$

The evolution kernel of the simple harmonic oscillator, given in Eq. 5.89, has been used to obtain the final result.

5.16 Summary

Hamiltonian mechanics shows the close connection of the Lagrangian and Hamiltonian formulations of the dynamics of a system. The Hamiltonian is obtained by a Legendre transformation and yields the dynamics of the system in phase space, defined by the canonical coordinate and the conjugate momentum of the system. Conservation laws and symmetries are seen to emerge in a transparent manner in the canonical formulation.

The cardinal results of canonical equations, including the relation between the Lagrangian and the Hamiltonian, are seen to hold for the quantum case, but with an appropriate generalization of all the underlying symbols. In particular, in making the transition to a quantum case, the Hamiltonian is elevated to a Hermitian operator and the Lagrangian determines the probability amplitude for infinitesimal quantum paths. The Hamiltonian operator obtained from Hamiltonian mechanics leads to the formulation of the path integral in phase space as well as to the formulation of coherent state quantization.

The Poisson and Dirac brackets provide a well-defined algorithm for making the transition from the classical to the quantum case, with Dirac brackets playing a central role is providing a Hamiltonian formulation for constrained quantum systems. The case of constrained systems shows that the quantum system's state space is an independent ingredient that needs to be specified in addition to the Hamiltonian. The constraint can be formulated directly in terms of the path integral; it is seen that the path integral has complete information about both the state space and the Hamiltonian.

The oscillator basis provides another perspective to the study of the Hamiltonian and path integral. Coherent states extend the concepts of basis states and completeness to over-complete states. A few calculations using the coherent basis were performed to illustrate some of the interesting and novel features of this approach.

6
Path integral quantization

The Dirac–Feynman approach leads to the Feynman path integral, which is based on the mathematical concept of functional integration. Besides its application in quantum mechanics, the mathematical formalism of functional integration is one of the pillars of quantum mathematics and can be applied to classical random systems that are outside the domain of quantum mechanics.

The starting point of path integral quantization is the degree of freedom. One *postulates* that the degree of freedom takes *all possible* determinate paths in making a quantum transition from its initial to its final state. The probability amplitude for each determinate path is given by the properties of the degree of freedom encoded in the action S, and this in turn entails knowledge of the degree of freedom's Lagrangian $\mathcal{L}$.

In particular, the result obtained in Eq. 4.40 can be taken to be the starting point of path integral quantization. Hence, instead of starting from a Hamiltonian, as was done in Chapter 4, the quantum phenomena are specified by postulating a Lagrangian.

In Section 6.1 the Hamiltonian is derived from the Lagrangian and path integral. In Section 6.2 it is shown how the classical limit emerges from the path integral and in Section 6.7 the Heisenberg commutation equation is derived from the path integral, both for Minkowski and Euclidean time.

For a continuous degree of freedom the Lagrangian consists of a kinetic term that is usually the same for a wide class of systems; one needs to choose an appropriate potential $V(x)$ to fully describe the system. For the sake of rigor, consider the Euclidean Lagrangian and action given by ($\hbar = 1$),

$$\mathcal{L} = -\frac{1}{2}\left(\frac{dx}{dt}\right)^2 - V(x), \quad S = \int_{t_i}^{t_f} dt \mathcal{L}.$$

The evolution kernel is given by the superposition of all the determinate (indistinguishable) paths and is equal to the sum of e^S over all possible paths; hence

$$\langle x_f|e^{-\tau H}|x_i\rangle = \int_{B.C.} DXe^S \ : \text{Euclidean Feynman path integral}, \tag{6.1}$$
$$\text{Boundary condition}: \ x(0) = x_i \ , \ x(\tau) = x_f.$$

Path integral quantization is more general than starting from the Schrödinger equation for the following reasons:

- As discussed in Section 2.7, the Schrödinger approach is based on the properties of state space *in addition* to the Hamiltonian driving the Schrödinger equation.
- The spacetime symmetries of the quantum system are explicit in the Lagrangian-based path integral approach, whereas in the Schrödinger approach these are implicit and need to be extracted using the properties of the Hamiltonian and state space. In particular, one has to derive the symmetry operators that commute with the Hamiltonian.
- Path integral quantization yields a transparent formulation of constrained systems, as for example discussed in Section 5.7.1 and later in Chapter 13. In the Schrödinger formulation, one needs *both* the Hamiltonian and commutation relations, which for a constrained system are far from obvious and require a fair amount of derivation.

These considerations come to the forefront for complicated systems like non-Abelian gauge fields, where the starting point is the Lagrangian, and path integral quantization turns out to be more efficient than the Schrödinger approach.

6.1 Hamiltonian from Lagrangian

Recall that in Section 4.6 the Lagrangian was derived from the Hamiltonian using the Dirac–Feynman formula. The question naturally arises that if the Lagrangian is known, how would one derive its Hamiltonian H? As discussed in Section 5.1, Hamiltonian mechanics provides one procedure for obtaining H from $\mathcal{L}$; the purpose of this section is to provide an alternate derivation using quantum mechanical techniques.

A Lagrangian that is more general than the one discussed in Section 4.6, and arises in the study of option theory in finance [Baaquie (2004)], is chosen to illustrate some new features. Option theory is based on classical random processes that are similar to the diffusion equation and hence the time parameter t in the path integral appears as "Euclidean time" t, which for option theory is in fact calendar time.

Let the degree of freedom be the real variable ϕ. Consider the following Lagrangian and action

$$\mathcal{L}(t) = -\frac{1}{2}\left[me^{-2\nu\phi}\left\{\frac{d\phi}{dt} + \alpha(\phi,t)\right\}^2 + V(\phi)\right],$$

$$S = \int_0^\tau dt\mathcal{L}(t) = -\frac{1}{2}\int_0^\tau dt\left[me^{-2\nu\phi}\left\{\frac{d\phi}{dt} + \alpha(\phi,t)\right\}^2 + V(\phi)\right]. \quad (6.2)$$

For greater generality, a ϕ dependent mass equal to $me^{-2\nu\phi}$ and a drift term $\alpha(\phi,t)$ have been included in $\mathcal{L}$.

The path integral is given by generalization of Eq. 4.39,

$$K(\phi_i, \phi_f; \tau) = \int D\phi e^{-\nu\phi} e^S, \quad (6.3)$$

$$\int D\phi e^{-\nu\phi} \equiv \prod_{t=0}^{\tau}\int_{-\infty}^{+\infty} d\phi(t) e^{-\nu\phi(t)},$$

Boundary conditions $\phi(\tau) = \phi_f,\ \phi(t=0) = \phi_i$.

Note that the path integral integration measure $\int D\phi$ has a factor of $e^{-\nu\phi}$ needed to obtain a well-defined Hamiltonian.

Recall from the discussion of the evolution kernel in Section 4.6, that the path integral is related to the Hamiltonian H by Eq. 4.40, namely

$$K(\phi_i, \phi_f; T) = \int D\phi e^{-\nu\phi} e^S = \langle\phi_f|e^{-\tau H}|\phi_i\rangle. \quad (6.4)$$

One needs to extract the Hamiltonian H from the path integral on the left hand side of Eq. 6.4.

The Hamiltonian propagates the system for infinitesimal time; the time index t is discretized into a lattice with spacing ϵ, where $t = n\epsilon$ with $N = T/\epsilon$ and $\phi(x) \rightarrow \phi_n$. The path integral reduces to a finite $(N-1)$-fold multiple integral, analogous to what was obtained in Eq. 4.30. Discretizing the time derivative $d\phi/dt \rightarrow (\phi_{n+1} - \phi_n)/\epsilon$ yields the following lattice action and Lagrangian:

$$\langle\phi_N|e^{-\epsilon NH}|\phi_0\rangle = \prod_{n=1}^{N-1}\int d\phi_n e^{-\nu\phi_n} e^{S(\epsilon)}, \quad (6.5)$$

$$S(\epsilon) = \epsilon\sum_{n=0}^{N-1} L(n),$$

$$L(n) = -\frac{me^{-2\nu\phi_n}}{2\epsilon^2}\left[\phi_{n+1} - \phi_n + \epsilon\alpha_n\right]^2 - \frac{1}{2}\left[V(\phi_{n+1}) + V(\phi_n)\right].$$

As in Section 4.6, the completeness equation $\int d\phi_n|\phi_n\rangle\langle\phi_n| = \mathbb{I}$ is used $N-1$ times to write out the expression for $e^{-\epsilon NH}$, and the Hamiltonian is identified as

$$\langle \phi_{n+1}|e^{-\epsilon H}|\phi_n\rangle = \mathcal{N}(\epsilon)e^{-\nu\phi_n}e^{\epsilon L_n}$$
$$= \mathcal{N}(\epsilon)e^{-\nu\phi_n}\exp\left\{-\frac{me^{-\nu\phi}}{2\epsilon}\left[\phi_{n+1}-\phi_n+\epsilon\alpha_n\right]^2 - \frac{\epsilon}{2}\left[V(\phi_{n+1})+V(\phi_n)\right]\right\}.$$

Since the Hamiltonian depends on the value of ϕ at two different instants, to simplify notation let

$$\phi_{n+1}=\phi,\quad \phi_n=\phi',\quad \alpha_n=\alpha.$$

Ignoring terms that are of $O(\epsilon)$ in Eq. 6.5, the matrix elements of the Hamiltonian are given by

$$\langle \phi|e^{-\epsilon H}|\phi'\rangle = \mathcal{N}(\epsilon)e^{-\nu\phi}\exp\left\{-\frac{me^{-2\nu\phi}}{2\epsilon}\left[\phi-\phi'+\epsilon\alpha\right]^2 - \epsilon V(\phi)\right\}. \tag{6.6}$$

Note that unlike Eq. 4.33, for which the Hamiltonian is known and from which the Lagrangian was derived, in Eq. 6.6 one needs to *derive* the Hamiltonian *from* the known Lagrangian. This derivation is the quantum mechanical analog of the derivation of H given by Hamiltonian mechanics in Section 5.1.

The key feature of the Lagrangian that in general allows one to derive its Hamiltonian is that the Lagrangian contains *only first order* time derivatives; hence on discretization the Lagrangian involves only ϕ_n that are nearest neighbours in time, thus allowing it to be represented as the matrix element of $e^{-\epsilon H}$, as in Eq. 6.6.

In contrast, for Lagrangians that contain *second order* or higher order time derivatives, as in Chapter 13, the derivation of the Hamiltonian from the Lagrangian and path integral is nontrivial since the entire framework of coordinate and canonical momentum, as discussed in Section 5.1, is no longer applicable. Instead, one has to employ the method required for quantizing constrained systems, and in particular, evaluate the Dirac brackets for the system in order to obtain the Hamiltonian and commutation relations.

In Eq. 6.6, the time derivatives appear in a quadratic form; hence one can use Gaussian integration to re-write Eq. 6.6 as[1]

$$\langle \phi|e^{-\epsilon H}|\phi'\rangle = e^{-\nu\phi}e^{-\epsilon V(\phi)}\int_{-\infty}^{+\infty}\frac{dp}{2\pi}\exp\left\{-\frac{\epsilon}{2m}p^2 + ip\left[\phi-\phi'+\epsilon\alpha\right]e^{-\nu\phi}\right\}$$
$$= e^{-\epsilon V(\phi)}\int_{-\infty}^{+\infty}\frac{dp}{2\pi}\exp\left\{-\frac{\epsilon e^{2\nu\phi}}{2m}p^2 + ip(\phi-\phi'+\epsilon\alpha)\right\}, \tag{6.7}$$

where the pre-factor of $e^{-\nu\phi}$ has been canceled by re-scaling the integration variable $p \to pe^{\nu\phi}$.

[1] Henceforth $\mathcal{N}(\epsilon)$ is ignored since it is an irrelevant constant contributing only to the definition of the zero of energy.

The Hamiltonian $H = H(\phi, \partial/\partial\phi)$ is a differential operator and acts on the dual coordinate ϕ, as is required for all differential operators, as mentioned earlier after Eq. 3.31. Hence, for the state function $|\psi\rangle$, which is an element of the state space, the Hamiltonian acts on the *dual basis state* $\langle\phi|$, and yields $\langle\phi|H|\psi\rangle = H(\phi, \partial/\partial\phi)\psi(\phi)$, similarly to the result given in Eq. 2.10.

The Hamiltonian is hence given by[2]

$$\langle\phi|e^{-\epsilon H}|\phi'\rangle = e^{-\epsilon H(\phi,\partial/\partial\phi)}\langle\phi|\phi'\rangle = e^{-\epsilon H(\phi,\partial/\partial\phi)}\int_{-\infty}^{+\infty}\frac{dp}{2\pi}e^{ip(\phi-\phi')}, \tag{6.8}$$

since $\langle\phi|\phi'\rangle = \delta(\phi-\phi')$. Ignoring overall constants and using the property of the exponential function under differentiation, one can re-write Eq. 6.7 as

$$\langle\phi|e^{-\epsilon H}|\phi'\rangle = \exp\left\{\frac{1}{2m}\epsilon e^{2\nu\phi}\frac{\partial^2}{\partial\phi^2} + \epsilon\alpha\frac{\partial}{\partial\phi} - \epsilon V(\phi)\right\}\int_{-\infty}^{+\infty}\frac{dp}{2\pi}e^{ip(\phi-\phi)}. \tag{6.9}$$

Comparing Eq. 6.9 with Eq. 6.8 yields the Hamiltonian

$$H = -\frac{1}{2m}e^{2\nu\phi}\frac{\partial^2}{\partial\phi^2} - \alpha(\phi)\frac{\partial}{\partial\phi} + V(\phi). \tag{6.10}$$

The Hamiltonian is quite general since both $V(\phi)$ and $\alpha(\phi)$ can be functions of the degree of freedom ϕ. Note that the Hamiltonian H is non-Hermitian – and is Hermitian only for $\nu = 0$ and a pure imaginary α. The path integral has a non-trivial integration measure $\exp\{-\nu\phi\}$ that needs to be specified in addition to the Hamiltonian.

6.2 Path integral's classical limit $\hbar \to 0$

It is well known that classical physics is the limit of the quantum theory in the limit that $\hbar \to 0$. In this section it is shown how the classical limit emerges from a completely quantum world; in particular it is shown how the classical limit emerges from the path integral. Hence, to consider the classical limit, the presence of $\hbar$ is restored in the path integral, and the behavior of the path integral is studied in physical Minkowski time and in the limit of $\hbar \to 0$.

Recall that the evolution kernel is given in Euclidean time by

$$K(x, x'; \tau) = \int DXe^{S/\hbar},$$

$$S = \int_0^\tau \mathcal{L}dt, \quad \mathcal{L} = -\frac{m}{2}\dot{x}^2 - V(x).$$

[2] From Eq. 3.36, the convention for scalar product is $\langle p|\phi_n\rangle = \exp(-ip\phi_n)$, and the sign of the exponential in Eq. 6.8 reflects this choice. The definition of H requires it to act on the dual state vector $\langle\phi|$; if one chose to write the Hamiltonian as acting on the state vector $|\phi\rangle$, $H^\dagger$ would then have been obtained instead. Since H is not Hermitian, this would lead to an incorrect result.

To discuss the $\hbar \to 0$ limit, analytically continue back to Minkowski time $t = i\tau$,

$$\mathcal{L}_M = \frac{m}{2}\left(\frac{dx}{dt}\right)^2 - V(x), \quad S \to iS_M,$$

$$K(x, x'; t) = \int DXe^{iS_M/\hbar}.$$

In the $\hbar \to 0$ limit, one expects only the classical path to contribute; to achieve the classical limit one expects

$$\lim_{\hbar\to 0} e^{iS_M/\hbar} \to \mathcal{N} e^{iS_c/\hbar} \prod_{t=0}^{T} \delta(x(t) - x_c(t)),$$

where the normalization $\mathcal{N}$ is required for dimensional consistency. The equation above should be thought of only in a heuristic sense, since there are many paths "near" the classical path that also contribute, and in fact are responsible for the precise determination of the normalization constant. One hence obtains

$$\begin{aligned} K &\to \mathcal{N} \int Dx\delta[x - x_c]e^{iS_c/\hbar} \\ &= \mathcal{N} e^{iS_c/\hbar}. \end{aligned} \tag{6.11}$$

A geometric proof of Eq. 6.11 is given here, following Feynman and Hibbs (1965). Recall that x_c is the solution to the equation of motion given by

$$\left.\frac{\delta S}{\delta x(t)}\right|_{x(t)=x_c(t)} = 0 \ \Rightarrow \delta S_c = 0. \tag{6.12}$$

In other words, as one varies the path $x(t)$, to first order S_c is stationary with respect to this variation. Consider the discrete time path integral

$$K = \int Dxe^{iS_M/\hbar} \cong \sum_{\text{paths}} e^{iS_{\text{path}}//\hbar} \tag{6.13}$$

$$= e^{iS_1/\hbar} + e^{iS_2/\hbar} + \cdots \tag{6.14}$$

Note that, since $|e^{iS_{\text{path}}//\hbar}| = 1$, each term in the summation for K is a complex number of modulus unity. Hence K can be represented by a sum of vectors with unit length in the complex plane, one vector for each term $e^{iS_{\text{path}}/\hbar}$. Figure 6.1 shows the addition of the $e^{iS_{\text{path}}/\hbar}$ for different paths.

Consider two neighboring paths; these contribute $e^{iS_a/\hbar} + e^{iS_b/\hbar}$ to K. Note that if $S_b \simeq S_a + \pi\hbar$ we have

$$e^{iS_a/\hbar} + e^{iS_b/\hbar} = e^{iS_a/\hbar} - e^{iS_a/\hbar} = 0.$$

In other words, if the action changes by about $\pi\hbar$ or more for neighboring paths, then these paths interfere *destructively* and give no contribution to K.

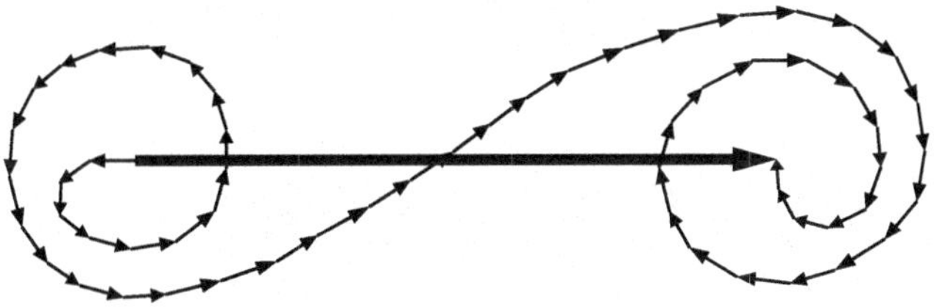

Figure 6.1 Classical path, indicated by the heavy arrow, results from the interference of quantum paths

Since S_c is stationary, we expect the paths close to $x_c(t)$ to have a value for the action S that is very close to S_c, and hence should interfere constructively, as shown in Figure 6.1. We expect that the paths in the neighborhood of the classical path all interfere constructively. Hence

$$K \simeq e^{iS_c/\hbar} + e^{iS'/\hbar} + \cdots + e^{iS''/\hbar} \tag{6.15}$$

$$\simeq \mathcal{N} e^{iS_c/\hbar}. \tag{6.16}$$

The classical path is indicated by the heavy arrowed vector in Figure 6.1.

Note that S' S'' ... are contributions to K from paths close to x_c, and for coherence (constructive interference) we expect

$$S', S'', \cdots \leq S_c + \pi\hbar. \tag{6.17}$$

6.2.1 Nonclassical paths and free particle

Consider a free particle of mass m traveling for a time interval T. The classical path is $x = vt$ and consider a nonclassical path $x = vt^2/T$. Let $v = 10^{-2}$ km s^{-1}, $T = 1$ s; this yields the following classical S_c and nonclassical S_{Nc} actions:

$$S_c = \frac{1}{2}mv^2 \int_0^T dt = \frac{1}{2}mv^2 T = \frac{1}{2}m 10^{-4} \text{ km}^2 \text{ s}^{-1},$$

$$S_{Nc} = \frac{2mv^2}{T^2} \int_0^T t^2 dt = \frac{2}{3}mv^2 T = \frac{2}{3}m\ 10^{-4} \text{ km}^2 \text{ s}^{-1},$$

$$\Delta S = S_{Nc} - S_c = \frac{1}{6}m \times 10^{-4} \text{ km}^2 \text{ s}^{-1}.$$

For a classical particle $m = 10^{-3}$ kg; using $\hbar = 1.05 \times 10^{-34}$ kg km^2 s^{-1} yields

$$\Delta S = \frac{1}{6} \times \frac{10^{-3} \times 10^{-4}}{1.05 \times 10^{-34}}\hbar$$

$$\simeq 1.6 \times 10^{26}\hbar \gg \pi\hbar. \tag{6.18}$$

Hence, for a classical particle the nonclassical path is not coherent with the classical path.

Consider the quantum case of the electron, with its mass given by

$$m_e = 9.1 \times 10^{-31} kg$$

and

$$\begin{aligned} \Delta S &= \frac{1}{6} \times \frac{9.1 \times 10^{-31} \times 10^{-4}}{1.05 \times 10^{-34}} \hbar \simeq \hbar/6 \\ \Rightarrow \quad & \Delta S/\hbar < \pi. \end{aligned} \tag{6.19}$$

We have the important conclusion that for a nonrelativistic electron (moving in the example at 1 cm/sec), in addition to the classical path, there are also *many* nonclassical paths that contribute to the transition amplitude.

Note that $\hbar$ sets the scale just right so that a nonrelativistic electron is required to be treated in a completely quantum mechanical manner. In fact, in most of chemistry and all of biology, the nuclei are treated as internally structure-less classical point charges, and it is the quantum mechanical nature of the nonrelativistic electrons that determines most of chemistry and biology.

6.3 Fermat's principle of least time

Consider light reflection off a mirror. In the limit of geometrical optics, where light can be considered as a ray,[3] the well-known law of reflection states that

$$\theta_i = \theta_r. \tag{6.20}$$

The light ray consists of photons. An approximate quantum action for the photons that arrive at B from source A after reflecting off the mirror, as shown in Figure 6.2, is given by

$$S_{\text{ray}}/\hbar \simeq \omega t_p = \theta_p, \tag{6.21}$$

where ω is the frequency of light and t_p is the time taken for light to travel the path from A to B. A path in this section always means the path of a ray in going from A to the *mirror* and then to B.

[3] Geometrical optics is a description of photons (light particles) when the wavelength and energy of individual photons are much smaller than the dimensions of the equipment and the experimental resolution of the photon's energy respectively.

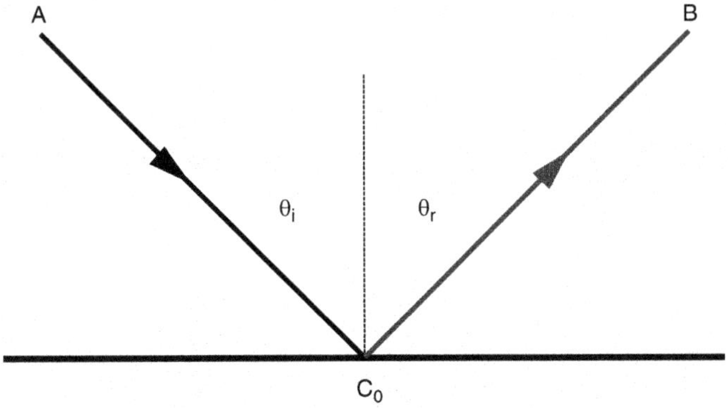

Figure 6.2 Reflection of light ray off a mirror. The classical path is the trajectory AC_0B.

The quantum mechanics principle of superposition states that light simultaneously takes all possible paths in going from A to B. The probability amplitude (A → B) is given by

$$\begin{aligned} K_{AB} &= \sum_{\{\text{all paths}\}} e^{iS/\hbar} \\ &= e^{i\theta_1} + e^{i\theta_2} + e^{i\theta_3} + e^{i\theta_4} + \cdots \end{aligned} \tag{6.22}$$

Note path AC_0B in Figure 6.3 is a classical path and consequently a minimum for the action S, since for this path $\delta S = 0$; hence, paths that are infinitesimally close to the classical path AC_0B all contribute coherently.

Each phase $e^{i\theta}$ in Eq. 6.22 can be considered as a two-dimensional unit vector in the two-dimensional plane and hence K_{AB}, which is a sum of complex numbers, can be obtained by adding the phases like vectors.

If the path is far from AC_0B the angle θ changes by large amounts, the reason being that the wavelength of light is much smaller than the distance from the source A to the mirror; hence, if one is not near the classical path, two paths differ by many wavelengths, giving rise to a large change in θ in going from one path to the other.

Adding all the phase contributions to K_{AB} yields the graphical representation given in Figure 6.3: one starts at the left end and starts to add paths starting from path AC_1B with phase $e^{i\theta_{AC_1B}}$, the path AC_2B contributes a phase $e^{i\theta_{AC_2B}}$ that adds destructively to the phase from path AC_1B and so on, and these are shown by arrows at the left end of Figure 6.1; similarly, phases from paths AC_3B and AC_4B cancel out.

What remains are phases that are close to the classical path AC_0B, and they all add coherently, as shown in Figure 6.1 by the heavy vector pointing straight from

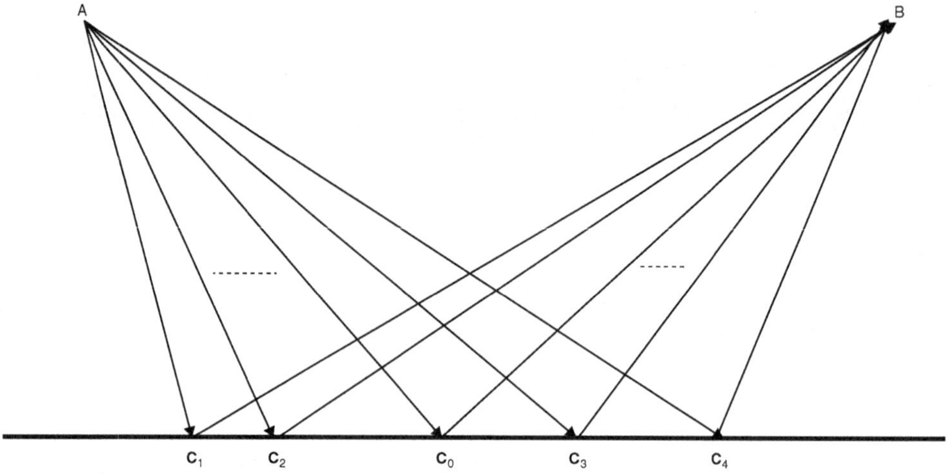

Figure 6.3 Many paths from source at A to detector at B reflected off the mirror at points $C_0, C_1, C_2, \ldots C_3, C_4$. The classical path is AC_0B.

left to right. The destructive and constructive interference yields the following the final result

$$K_{AB} = \mathcal{N} e^{iS_c/\hbar}. \tag{6.23}$$

In summary, all paths away from the classical path interfere destructively, whereas paths in the neighborhood of the classical paths interfere constructively yielding the final result that is dominated by the classical path. The fact that light taking the classical path contributes the most to light received at B can also be seen from the fact that if one removes most of the mirror except for a portion near the point C_0, the light received at B does not significantly change, showing that the other paths for light do not contribute significantly.

The path of least distance, which minimizes the action, yields the law that angle of incidence is equal to the angle of reflection,

$$\theta_i = \theta_r. \tag{6.24}$$

The path of least distance is also the path of least time and yields

$$K_{AB} = \mathcal{N} e^{iS_c/\hbar},$$

$\Rightarrow$ light takes path of least time in going from A to B: Fermat's principle.

Fermat's principle also yields Snell's law of refraction for light going from one medium to another having a different index of refraction.

6.4 Functional differentiation

Consider variables f_n, $n = 0, \pm 1, \pm 2 \ldots \pm N$ that satisfy

$$\frac{\partial f_n}{\partial f_m} = \delta_{n-m}.$$

Let $t = n\epsilon$, with $N \to \infty$. The limit $\epsilon \to 0$ yields

$$\frac{\partial f_n}{\partial f_m} \to \frac{\delta f(t)}{\delta f(t)'} \equiv \lim_{\epsilon \to 0} \frac{1}{\epsilon} \frac{\partial f_n}{\partial f_m}$$
$$\Rightarrow \frac{\delta f(t)}{\delta f(t')} = \lim_{\epsilon \to 0} \frac{1}{\epsilon} \delta_{n-m} \to \delta(t - t'). \tag{6.25}$$

In general, the *functional derivative* of $\Omega[f]$ – an arbitrary functional of the path $f(t)$ – is denoted by $\delta/\delta f(t)$ and is defined by

$$\frac{\delta \Omega[f]}{\delta f(t)} = \lim_{\epsilon \to 0} \frac{\Omega\left[f(t') + \epsilon\delta(t - t')\right] - \Omega[f]}{\epsilon}. \tag{6.26}$$

In the notation of state space one has

$$\left\langle f \left| \frac{\delta}{\delta f(t)} \right| \Omega \right\rangle = \frac{\delta}{\delta f(t)} \langle f | \Omega \rangle = \frac{\delta \Omega[f]}{\delta f(t)}.$$

Note that ϵ has the dimensions of $[f] \times [t]$.

Examples

- Consider the simplest function $\Omega[f] = f(t_0)$; then, from Eq. 6.26

$$\frac{\delta \Omega[f]}{\delta f(t)} = \frac{\delta f(t_0)}{\delta f(t)} = \lim_{\epsilon \to 0} \frac{f(t_0) + \epsilon\delta(t - t_0) - f(t_0)}{\epsilon} = \delta(t - t_0).$$

- Let $\Omega[f] = \int d\tau f^n(\tau)$; from above

$$\frac{\delta \Omega[f]}{\delta f(t)} = \int d\tau n f^{n-1}(\tau) \frac{\delta f(\tau)}{\delta f(t)}, \tag{6.27}$$
$$= \int d\tau n f^{n-1}(\tau) \delta(t - \tau) = n f^{n-1}(t). \tag{6.28}$$

6.4.1 Chain rule

The chain rule for the calculus of many variables has a generalization to functional calculus. Consider a change of variables from f_n to g_n; the chain rule of calculus yields

$$\frac{\partial}{\partial f_n} = \sum_{m=1}^{N} \frac{\partial g_m}{\partial f_n} \frac{\partial}{\partial g_m}.$$

As before, let $t = n\epsilon,\ t' = m\epsilon$; we can re-write the above expression as

$$\frac{1}{\epsilon}\frac{\partial}{\partial f_n} = \epsilon \sum_{m=1}^{N} \left[\frac{1}{\epsilon}\frac{\partial g_m}{\partial f_n}\right]\left[\frac{1}{\epsilon}\frac{\partial}{\partial g_m}\right].$$

Taking the limit of $N \to \infty$ and $\epsilon \to 0$ yields

$$\lim_{\epsilon\to 0}\frac{1}{\epsilon}\frac{\partial}{\partial f_n} \to \frac{\delta}{\delta f(t)} = \int dt' \frac{\delta g(t')}{\delta f(t)}\frac{\delta}{\delta g(t')} \quad : \ \text{chain rule.} \tag{6.29}$$

6.5 Equations of motion

For $T \to \infty$, we have

$$\begin{aligned} e^{-TH} &= e^{-TE_0}|\Omega\rangle\langle\Omega| + \cdots, \\ |\Omega\rangle &= \text{vacuum (ground) state.} \end{aligned} \tag{6.30}$$

Hence

$$\begin{aligned} Z &= \lim_{T\to\infty} tr(e^{-TH}) \\ &\simeq \langle\Omega|\Omega\rangle e^{-TE_0}. \end{aligned} \tag{6.31}$$

The path integral representation gives

$$Z = \int Dx e^{S}, \quad S = \int_{-\infty}^{+\infty} dt \mathcal{L}(t).$$

The equations of motion follow from the fact that the path integration measure is invariant under the displacement, namely

$$\begin{aligned} & x(t) \to x(t) + \epsilon(t) \\ \Rightarrow & \int Dx \to \int D(x+\epsilon) = \int Dx. \end{aligned} \tag{6.32}$$

On displacing the paths by $\epsilon(t)$ one obtains a functional Taylors expansion for the action given by

$$S[x(t) + \epsilon(t)] = S[x(t)] + \int_{-\infty}^{+\infty} dt \epsilon(t) \frac{\delta S[x(t)]}{\delta x(t)} + O(\epsilon^2). \tag{6.33}$$

Hence, in abbreviated notation

$$\begin{aligned} Z &= \int Dx e^{S[x]} = \int Dx e^{S[x+\epsilon]} \\ &= \int Dx e^{S[x]}\left[1 + \int \epsilon(t) \frac{\delta S[x]}{\delta x(t)} dt \cdots\right]. \end{aligned} \tag{6.34}$$

Note that the left hand side of Eq. 6.34 is *independent* of the ϵ parameter that appears in the right hand side of the same equation. In particular, the first order term in the Taylors expansion in ϵ of the right hand side of Eq. 6.34 must be zero, and yields

$$\Rightarrow 0 = \int dt\epsilon(t) \int Dx \frac{\delta S[x]}{\delta x(t)} e^S. \tag{6.35}$$

Since $\epsilon(t)$ is arbitrary, Eq. 6.35 yields that for each t the coefficient multiplying $\epsilon(t)$ must be zero, and hence

$$\int Dx \frac{\delta S[x]}{\delta x(t)} e^{S[x]} = 0 \quad : \quad \text{equations of motion.} \tag{6.36}$$

We have thus obtained the quantum mechanical generalization of the classical equations of motion; namely it is only the *average value* of $\delta S[x]/\delta x(t)$ that is equal to zero; denoting the functional average by

$$E[\mathcal{O}] \equiv \frac{1}{Z} \int Dx \mathcal{O} e^S, \tag{6.37}$$

the equations of motion can be succinctly written as

$$E[\delta S] = 0. \tag{6.38}$$

6.6 Correlation functions

Let $\tau_1 < \tau_2 < \tau_3 \ldots < \tau_k$; the k-point correlation is defined by

$$G(\tau_1, \tau_2 \ldots \tau_k) = \langle \Omega | \hat{x}_H(\tau_k) \ldots \hat{x}_H(\tau_1) | \Omega \rangle, \tag{6.39}$$

where $\hat{x}_H(\tau)$ are Heisenberg operators.

In general, time ordering operator T is defined so as to place the operators at the earliest time to the *right* of the operators at later time. More precisely, the time ordering operator is defined by

$$T(\hat{O}_t \hat{O}_{t'}) = \begin{cases} \hat{O}_{t'} \hat{O}_t & t' > t \\ \hat{O}_t \hat{O}_{t'} & t' < t \end{cases}. \tag{6.40}$$

To simplify the notation, denote $\hat{x}_H(\tau)$ by x_τ; time ordering yields

$$T(x_{\tau_k} x_{\tau_1} x_{\tau_2} \ldots) = x_{\tau_k} \ldots x_{\tau_2} x_{\tau_1}$$

and the correlation function is

$$G(\tau_1, \ldots \tau_k) = \langle \Omega | T(x_{\tau_k} x_{\tau_1} x_{\tau_2} \ldots) | \Omega \rangle. \tag{6.41}$$

Since all correlation functions are the time ordered vacuum expectation values of the Heisenberg operators,

$$|\Omega\rangle\langle\Omega| \simeq \lim_{T\to\infty} e^{-T(H-E_0)},$$

$$\begin{aligned}G(\tau_1\ldots\tau_k) &= e^{E_0 T} tr\left\{T(x_{\tau_k}x_{\tau_1}x_{\tau_2}\ldots)e^{-TH}\right\}\\ &= \frac{1}{Z}\int Dx\, x_{\tau_k}\ldots x_{\tau_1}e^S\\ &= \text{symmetric function of } \tau_1, \tau_2\ldots\tau_k.\end{aligned}$$

The correlation functions are interrelated through differential equations, called Ward identities. Consider for example

$$\begin{aligned}G(\tau_0) &= \frac{1}{Z}\int Dx x_{\tau_0}e^{S[x]}\\ &= \frac{1}{Z}\int Dx\left[x_{\tau_0}+\epsilon(\tau_0)\right]e^{S[x+\epsilon]}\\ &= \frac{1}{Z}\int Dx\left[x_{\tau_0}+\epsilon(\tau_0)\right]\left[1+\int dt\epsilon(t)\frac{\delta S}{\delta x(t)}\right]e^S,\end{aligned}$$

which yields

$$\begin{aligned}0 &= \int dt\epsilon(t)E\left[x_{\tau_0}\frac{\delta S}{\delta x(t)}+\delta(t-\tau_0)\right]\\ \Rightarrow E&\left[x_{\tau_0}\frac{\delta S}{\delta x(t)}\right] = -\delta(t-\tau_0).\end{aligned}$$

In general, for $\mathcal{L} = -\frac{1}{2}m\dot{x}^2 - V(x)$ all the correlations are required for defining the theory. The *only* special case is the Gaussian action with $V = \omega^2 x^2/2$ for which the two-point correlation function $E[x_\tau x_{\tau'}]$ is sufficient for obtaining all the correlation functions.

6.7 Heisenberg commutation equation

For a single degree of freedom, from Eq. 3.43, the Heisenberg commutation equation is given by

$$[\hat{x}, \hat{p}] = i\hbar\mathbb{I}. \tag{6.42}$$

Since the momentum operator $\hat{p}$ involves a time derivative, working in Euclidean time would yield an extra factor of i in the commutation equation. The calculation to obtain the Heisenberg commutation equation is carried out in physical Minkowski time so that the familiar expression is obtained.

Recall in the Schrödinger canonical quantization, the coordinate was taken to be a degree of freedom and it was then *postulated* in Eq. 3.31 that momentum is an operator conjugate to the coordinate degree of freedom and given by

$$\hat{p} = -i\hbar\frac{\partial}{\partial x} : \quad \text{canonical quantization.}$$

One postulates that the momentum of a degree of freedom is given by the "classical" looking expression for momentum, namely that

$$\hat{p} = m\frac{dx}{dt} : \quad \text{path integral quantization.} \tag{6.43}$$

However, Eq. 6.43 is *not* a classical expression since dx/dt is a degree of freedom and is indeterminate – having a determinate value only with a certain likelihood.

In the path integral quantization, only the matrix elements of the operators appear. Hence, one would like to take the expectation value of both sides of Eq. 6.42 to establish the Heisenberg commutation equation.

For action S_M in Minkowski time, one could try and evaluate the expectation value of the commutation equation, that simplistically is given by

$$E[([\hat{x},\hat{p}])] = \frac{1}{Z}\int Dx\,[\hat{x},\hat{p}]\,e^{iS_M/\hbar} = \frac{1}{Z}\int Dx\,(\hat{x}_t\hat{p}_t - \hat{p}_t\hat{x}_t)\,e^{iS_M/\hbar}. \tag{6.44}$$

If one takes $p = m\dot{x}$, then apparently

$$\int Dx\,(x_t\dot{x}_t - \dot{x}_t x_t)\,e^{iS_M/\hbar} = 0 \quad \text{(incorrect).} \tag{6.45}$$

The solution is to consider the product of operators as being *time ordered*, as is required when one goes from operators to correlation functions. Define $p_t = (x_{t+\epsilon} - x_t)/\epsilon$; time ordering then yields, for $\epsilon \to 0+$,

$$T(x_t p_t) = x_t p_{t-\epsilon} = \frac{m}{\epsilon}x_t\,(x_t - x_{t-\epsilon}),$$
$$T(p_t x_t) = p_t x_t = \frac{m}{\epsilon}\,(x_{t+\epsilon} - x_t)\,x_t,$$

from which it follows that the commutator has the time ordered form

$$E\,[([x,p])] \to E\,[T([x,p])] = E\,[T(x_t p_t - p_t x_t)], \tag{6.46}$$

and hence

$$T\,([x,p]) = x_t p_{t-\epsilon} - p_t x_t. \tag{6.47}$$

For discrete time $t = n\epsilon$, the (discretized) action is given by

$$S_M = \int dt \left(\frac{m}{2}\dot{x}^2 - V\right) \simeq \frac{m}{2\epsilon}\sum_n (x_{n+1} - x_n)^2 - \epsilon \sum_n V(x_n), \tag{6.48}$$

where it has been assumed that the potential $V(x)$ does not depend on velocity. The commutator yields the discrete expression

$$T([x, p]) = \frac{m}{\epsilon} x_t (2x_t - x_{t+\epsilon} - x_{t-\epsilon}). \tag{6.49}$$

Note that for any lattice variable x_ℓ, the action $S_M[x_\ell = \pm\infty] = \infty$ and yields

$$x_\ell e^{iS_M/\hbar}|_{x_\ell = \pm\infty} = 0.$$

An integration by parts for the specific coordinate x_ℓ yields

$$\int Dx \frac{\partial}{\partial x_\ell}\left(x_\ell e^{iS_M/\hbar}\right) = \int D'X \left[x_\ell e^{iS_M/\hbar}|_{x_\ell = +\infty} - x_\ell e^{iS_M/\hbar}|_{x_\ell = -\infty}\right] = 0, \tag{6.50}$$

where $\int D'X$ is the path integral excluding the integration variable x_ℓ. Using the chain rule for differentiation for the left hand side of Eq. 6.50 yields[4]

$$0 = E\left[1 + \frac{i}{\hbar} x_\ell \frac{\partial S_M}{\partial x_\ell}\right]. \tag{6.51}$$

Hence

$$\begin{aligned}
1 &= -\frac{i}{\hbar} E\left[x_\ell \frac{\partial S_M}{\partial x_\ell}\right] \\
&= -\frac{i}{\hbar} E\left[x_\ell \frac{m}{\epsilon}(2x_\ell - x_{\ell+1} - x_{\ell-1}) - \epsilon x_l \frac{\partial V(x_\ell)}{\partial x_\ell}\right] \\
&= -\frac{i}{\hbar} E\left[T([x_l, p_l])\right] + O(\epsilon) \\
\Rightarrow 1 &= -\frac{i}{\hbar}[x_l, p_l] + O(\epsilon).
\end{aligned}$$

Taking the limit of $\epsilon \to 0$ yields

$$[x, p] = i\hbar\mathbb{I} \quad : \quad \text{Heisenberg commutation equation.} \tag{6.52}$$

The commutation equation depends only on kinetic energy of action S, which contains the time derivative term. In particular, the commutation equation is independent of the potential energy $V(x)$ and of the boundary conditions.

[4] Recall $E[\mathcal{O}(x)] \equiv \int Dx \mathcal{O}(x) e^S / Z$.

6.7.1 Euclidean commutation equation

The commutation for Euclidean time has been derived in Section 5.6 using the Poisson bracket of the coordinate with its canonical momentum. Another derivation of the same result is given here using the path integral approach to verify the consistency of the two approaches.

Euclidean time τ and momentum p_E are given by

$$\tau = it, \quad p_E = -m\frac{dx(\tau)}{d\tau}. \tag{6.53}$$

The Euclidean action, for $\tau = n\epsilon$, is given by

$$\begin{aligned} S_E &= -\int d\tau \left\{\frac{m}{2}(\frac{dx(\tau)}{d\tau})^2 + V\right\} \\ \Rightarrow S_E &\simeq -\frac{m}{2\epsilon}\sum_n (x_{n+1} - x_n)^2 - \epsilon \sum_n V(x_n) \\ &= -\frac{m}{2\epsilon}\sum_n x_n\,(2x_n - x_{n+1} - x_{n-1}) - \epsilon\sum_n V(x_n). \end{aligned} \tag{6.54}$$

Similarly to the Minkowski case, the Euclidean commutator is given by

$$T([x, p_E]) = -\frac{m}{\epsilon}x_\tau\big(2x_\tau - x_{\tau+\epsilon} - x_{\tau-\epsilon}\big) = -\frac{m}{\epsilon}x_n\big(2x_n - x_{n+1} - x_{n-1}\big). \tag{6.55}$$

Note the difference of a minus sign between the Minkowski commutation equation given in Eq. 6.49 and the Euclidean case given in Eq. 6.55.

The discretized path integral, similarly to Eq. 6.44, yields the expectation value

$$E[T([x, p_E])] = -\frac{m}{\epsilon}\frac{1}{Z}\int Dx\;\; x_n\big(2x_n - x_{n+1} - x_{n-1}\big)e^{S/\hbar}. \tag{6.56}$$

Similarly to Eq. 6.51, for the Euclidean case

$$\begin{aligned} 0 &= E\left[1 + \frac{1}{\hbar}x_\ell\frac{\partial S}{\partial x_\ell}\right] \\ \Rightarrow 1 &= -\frac{1}{\hbar}E\left[x_\ell\frac{\partial S}{\partial x_\ell}\right] = -\frac{1}{\hbar}E\left[T([x, p_E])\right], \end{aligned} \tag{6.57}$$

and the Euclidean commutator is given by

$$[x, p_E] = -\hbar\mathbb{I} \quad : \quad \text{Euclidean commutation equation.} \tag{6.58}$$

The result obtained above confirms the earlier results obtained in Eq. 5.50 based on canonical quantization. Hence, the counterintuitive definition of Euclidean momentum $p_E = -m dx(\tau)/d\tau$ with an overall minus sign is confirmed by two independent derivations.

6.8 Summary

Path integral quantization is an independent formulation of quantum mathematics based on the concept of the Lagrangian and the action. The path integral measure has to be defined in *addition* to the Lagrangian in order to formulate the path integral. The Hamiltonian and state space are both incorporated in the path integral and the path integral measure arises from the properties of the state space. The Hamiltonian and the state space are independent structures of a quantum system that can both be derived from the path integral.

In the limit of $\hbar \to 0$, the classical limit emerges from the path integral due to the destructive interference of paths far from the classical path. A concrete illustration of this cancellation is displayed by light's reflection off a mirror and yields that the angle of incidence and reflection are equal.

The correlation functions of a quantum system obey a set of identities that have a transparent derivation using the invariance properties of the path integral's measure. The Heisenberg commutation equations are shown to follow from the properties of the correlation function of a large class of Lagrangians that depend only on the velocity squared of the degree of freedom. In particular, the commutation equation for Euclidean time was derived.

Part two

Stochastic processes

7

Stochastic systems

A quantum entity is intrinsically indeterminate and is not in a determinate (classical) state when it is not being observed. The quantum entity is found to be in a determinate state only if an experiment is carried out to observe it, as discussed in Baaquie (2013e). The indeterminate behavior of the quantum entity is described by the formalism of quantum mathematics discussed in Chapter 2. The measure of quantum indeterminacy and uncertainty is Planck's constant $\hbar$.

Randomness also exists in classical systems; however, a classical system's randomness is *in principle* different from quantum indeterminacy, since a priori, a classical random system is *intrinsically* in a determinate state. Classical randomness arises entirely due to our lack of complete knowledge of the determinate state. The theory of classical probability describes a classical system about which one has incomplete knowledge.

There are many ways of introducing randomness in a classical system, depending on its nature. A classical random system in thermal equilibrium is described by statistical mechanics. A classical system undergoing random time evolution can be described by taking the classical equations of motion and adding a random force to it. The evolution of the classical particle is then described by a stochastic process. The random force is usually represented by Gaussian white noise.

Classical random processes driven by white noise, such as those that arise in finance and other areas, form a wide class of problems that can be described by classical probability theory. It is shown in this chapter that the mathematical framework of quantum mechanics, and in particular, the Hamiltonian operator and path integrals, provide an appropriate mathematical formalism for describing processes that are driven by Gaussian white noise.

Many classical random systems are described by probability distribution functions [also called probability density functions] that obey linear partial differential equations, such as the Fokker–Planck equation [Risken (1988)]. These systems can be recast in terms of a Hamiltonian, which, in turn, leads to a path integral representation for conditional probabilities.

A more fundamental reason for the applicability of quantum mathematics to classical random systems is that random systems and processes have *many possible* determinate states, and all of them need to be taken into account: quantum mathematics is ideally suited for representing such systems.

Consider the classical equation of motion given by

$$m\frac{dv}{dt} + \gamma v + \Phi(v) = 0, \tag{7.1}$$

where γ is the coefficient of friction and $\Phi(v)$ is an arbitrary potential. For large objects, the solutions of Eq. 7.1 give an adequate description of the smooth trajectory that the body follows under the influence of the potential $\Phi(v)$.

Consider a tiny grain of pollen moving in a fluid medium with which it has many collisions per second. For example, a particle of air at room temperature and pressure has over 10^9 collisions per second. The trajectory of the grain or of the particle of air does indeed exist, but it becomes near to impossible to describe its motion in complete detail. The classical equation of motion given in Eq. 7.1 is not enough for describing it. To encode this lack of knowledge of the trajectory of the grain, classical randomness is introduced in the description of the motion of the particle.

Langevin proposed that the medium that is responsible for the random motion of the particle be modeled by a *random force*, and replaced Eq. 7.1 by a *stochastic differential equation* given by

$$m\frac{dv}{dt} + \gamma v + \Phi(v) = \sqrt{2A}R(t). \tag{7.2}$$

The random force $\sqrt{2A}R(t)$ is given by Gaussian white noise $R(t)$.

The introduction of a random force $\sqrt{2A}R(t)$ makes the trajectories of the particle random. For every choice of a random sample of values drawn for white noise $R(t)$, the trajectory changes. Hence, for a stochastic process one needs to specify the likelihood of different trajectories rather than the unique trajectory that is given in Eq. 7.1.

In Section 7.1 classical probability theory is briefly reviewed and in Section 7.3 Gaussian white noise is discussed. Ito calculus is discussed in Section 7.4 and its generalization given by the Wilson expansion is discussed in Section 7.5. In Section 7.6 the Langevin equation and its random paths are studied; Sections 7.7 and 7.8 examine the Langevin equation in an external potential and the nonlinear Langevin equation, respectively. In Section 7.9 stochastic quantization based on white noise is discussed. In Sections 7.10 and 7.11 the focus is on the Fokker–Planck Hamiltonian.

7.1 Classical probability: objective reality

Classical probability theory underpins the description and understanding of classical random phenomena, and is briefly reviewed here.

Consider a large collection of classical particles in a container. Each particle *objectively exists* in some determinate state and the *uncertainty in the knowledge* of the state of the particle is attributed to our *ignorance* of the microscopic state of a very large collection of molecules. In statistical mechanics, the large collection of particles is described by assuming that the position and velocity of each particle are classical random variables.

Classical probability is based on the concept of a random variable, which takes a range of values, and that exists objectively regardless of whether it is measured (sampled) or not. A unique probability, called the joint probability distribution, is assigned to a collection of random variables and predicts how frequently a collection of specific values will appear when the random variables are sampled.

Consider a random variable taking only two discrete values, say B (black) and W (white); each possible outcome has a probability, in this case, p_B and p_w. One can think of the random variable as an infinite collection of black and white balls with the relative frequency of B and W being given by their probability distribution. The probability of picking either B or W is given by p_B and p_w respectively, and is shown in Figure 7.1.

Following Kolmogorov, classical probability theory is defined by the following postulates:

- A collection of all possible allowed random sample values labeled by ω, forms a *sample space* Ω.
- A *joint probability distribution* function $P(\omega)$ determines the probability for the simultaneous occurrence for these random events and provides an exhaustive and complete description of the random system.

The events can be enumerated by *random variables*, say $\vec{X} = (X, Y, Z, \ldots)$, that map the random events ω of the sample space Ω to real numbers, namely

$$\begin{aligned} \vec{X} : \Omega \ &\rightarrow \ \Re^N, \\ X, Y, Z : \omega &\rightarrow \Re \otimes \Re \otimes \Re, \quad \omega \in \Omega, \\ P(X, Y, Z) \ &: \ \text{joint probability distribution.} \end{aligned}$$

Every element of the sample space Ω is assigned a likelihood of occurrence that is given by the joint probability distribution function $P(\omega)$; for the mapping of ω by random variables X, Y, Z to the real numbers, the joint probability distribution function is given by $P(X, Y, Z)$.

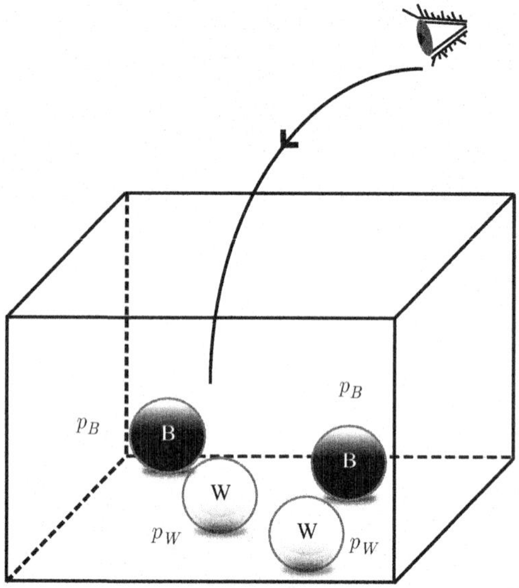

Figure 7.1 Classical random events: the number of black and white balls inside a closed box *exist objectively*, independent of being observed or not. Furthermore, there exists a probability p_B, p_W that is *intrinsic* to each possible outcome and is assigned to *each* black and white ball.

The assignment of a likelihood of occurrence $P(\omega)$ to *each element* of the sample space, namely to each $\omega \in \Omega$, is the *defining property* of classical probability theory; this assignment implicitly assumes that *each element* ω of Ω *exists objectively* – regardless of being observed or not – and an experiment finds it in its *pre-existing state* with probability specified by the probability distribution. It is precisely on this point that quantum probability is fundamentally different from classical probability. If a measurement is not made, the quantum degree of freedom is inherently indeterminate – having no objective existence, having no determinate value [Baaquie (2013e)]. This is the reason that the concept of the quantum degree of freedom replaces the classical concept of the random variable in the description of quantum phenomenon.

7.1.1 Joint, marginal and conditional probabilities

The joint probability distribution function obeys all the laws of classical probability. Consider random variables X, Y, Z. Their joint probability distribution is given by

$$P(X, Y, Z) \geq 0, \quad \int_{-\infty}^{\infty} dxdydzP(x, y, z) = 1.$$

In other words, $P(X = x, Y = y, Z = z)$ yields the probability for the *simultaneous occurrence* of the sample values x, y, and z of the random variables X, Y, Z. Consider a function H that depends on the random variables X, Y, Z; its (average) classical expectation value is given by

$$E_c[H] = \int dxdydzH(x, y, z)P(x, y, z).$$

If the random variables are independent, the joint probability distribution function factorizes and yields

$$P(x, y, z) = P_1(x)P_2(y)P_3(z).$$

For many random variables one can form various *marginal* and *conditional probability distributions*. The probability that random variables are observed having random values X, Y, regardless of the value of Z, is given by the marginal distribution for two random variables, namely

$$P(X, Y) = \int_{-\infty}^{\infty} dzP(X, Y, z), \quad \int_{-\infty}^{\infty} dxdyP(x, y) = 1. \tag{7.3}$$

The *conditional probability* for events A, B is defined as follows. Let $P(A, B)$ be the joint probability distribution that events A and B both occur. The conditional probability $P(A|B)$ that A occurs, given that B has definitely occurred, is given by conditional probability

$$P(A|B) = \frac{P(A, B)}{P(B)} \quad \Rightarrow \quad P(A|B)P(B) = P(B|A)P(A).$$

For the case of a classical random particle, the probability of finding the classical particle at point x, y, given that it has been definitely observed at z, is given by the *conditional probability*

$$P(X, Y|Z) = \frac{P(X, Y, Z)}{P(Z)} = \frac{P(X, Y, Z)}{\int_{-\infty}^{\infty} dxdyP(x, y, Z)}, \quad \int_{-\infty}^{\infty} dxdyP(x, y|Z) = 1.$$

7.2 Review of Gaussian integration

The mathematical framework of white noise and stochastic processes is based on the normal random variable, which in turn is described using Gaussian integration. Gaussian integration also plays a key role in studying path integrals in quantum mechanics.

The main results of Gaussian integration that will be employed later are derived below. For $\lambda > 0$ and complex parameter j, the basic Gaussian integral is given by

$$[j] = \int_{-\infty}^{+\infty} dx e^{-\frac{1}{2}\lambda x^2 + jx} = \sqrt{\frac{2\pi}{\lambda}} e^{\frac{1}{2\lambda} j^2}. \tag{7.4}$$

Normal (Gaussian) random variable

The normal, or Gaussian, random variable – denoted by $N(\mu, \sigma)$ – is a variable x that has a probability distribution given by

$$P(x) = \frac{1}{\sqrt{2\pi\sigma^2}} \exp\left\{-\frac{1}{2\sigma^2}(x-\mu)^2\right\}. \tag{7.5}$$

From Eq. 7.4

$$E[x] \equiv \int_{-\infty}^{+\infty} xP(x) = \mu \ \text{: mean},$$
$$E[(x-\mu)^2] \equiv \int_{-\infty}^{+\infty} (x-\mu)^2 P(x) = \sigma^2 \ \text{: variance}.$$

Any normal random variable is equivalent to the $N(0, 1)$ random variable via the following linear transformation

$$X = N(\mu, \sigma), \ Y = N(0, 1) \Rightarrow X = \mu + \sigma Y.$$

All the moments of the random variable $Y = N(0, 1)$ can be determined by the generating function given in Eq. 7.4; namely

$$E[y^n] = \frac{d^n}{dj^n} Z[j]|_{J=0}.$$

The cumulative distribution for the normal random variable $N(x)$ is defined by

$$\text{Prob}(-\infty \le z \le x) = N(x) = \frac{1}{\sqrt{2\pi}} \int_{-\infty}^{x} e^{-\frac{1}{2}z^2} dz. \tag{7.6}$$

A sum of normal random variables is also another normal random variable,

$$Z_1 = N(\mu_1, \sigma_1), Z_2 = N(\mu_2, \sigma_2), \ldots Z_n = N(\mu_n, \sigma_n)$$
$$\Rightarrow X = \sum_{i=1}^{n} Z_i = N(\mu, \sigma) \ \Rightarrow \ \mu = \sum_{i=1}^{n} \mu_i, \ \ \sigma^2 = \sum_{i=1}^{n} \sigma_i^2.$$

The result above can be proved using the generating function given in Eq. 7.4.

N-dimensional Gaussian integration

The moment generating function for the N-dimensional Gaussian random variable is given by

$$Z[J] = \int_{-\infty}^{+\infty} dx_1 \cdots dx_n e^S$$
$$\text{with} \quad S = -\frac{1}{2}\sum_{i,j=1}^{N} x_i A_{ij} x_j + \sum_i J_i x_i.$$

Let A_{ij} be a symmetric and positive definite matrix (that has only positive eigenvalues). A_{ij} can be diagonalized by an orthogonal matrix M,

$$A = M^T \begin{pmatrix} \lambda_1 & & \\ & \ddots & \\ & & \lambda_N \end{pmatrix} M$$
$$M^T M = \mathbb{I}.$$

Define new variables

$$z_i = M_{ij} x_j, \quad x_i = M^T_{ij} z_j$$
$$\prod_{i=1}^{N} dz_i = \det M \prod_{i=1}^{N} dx_i = \prod_{i=1}^{N} dx_i \equiv Dx.$$

Hence

$$Z[J] = \prod_i^N \int dz_i e^{-\frac{1}{2}\lambda_i z_i^2 + (JM^T)_i z_i} = \prod_{i=1}^{N}\left[\sqrt{\frac{2\pi}{\lambda_i}} e^{\frac{1}{2\lambda_i}(JM^T)_i (JM^T)_i}\right].$$

In matrix notation

$$\sum_i \frac{1}{\lambda_i}\left(JM^T\right)_i \left(JM^T\right)_i = J\frac{1}{A}J = JA^{-1}J, \quad \prod_{i=1}^{N}\sqrt{\frac{2\pi}{\lambda_i}} = (2\pi)^{N/2}\frac{1}{\sqrt{\det A}}.$$

Hence

$$Z[J] = \frac{(2\pi)^{N/2}}{\sqrt{\det A}} e^{\frac{1}{2}JA^{-1}J}, \tag{7.7}$$

where $Z[J]$ is called the generating function since all the correlators can be evaluated from it. In particular, the correlator of two of the variables is given by

$$E[x_i x_j] \equiv \frac{1}{Z}\int Dx\; x_i x_j e^S, \quad Z = Z[0]$$

$$E[x_i x_j] = \widehat{x_i x_j} = \frac{\partial^2}{\partial J_i \partial J_j} \ln Z[J]\Big|_{J=0}$$

$$= A_{ij}^{-1} \quad \text{: contraction.}$$

The correlation of N variables is given by

$$\langle x_{i_1} x_{i_2} x_{i_3} \ldots x_{i_{N-1}} x_{i_N} \rangle = \widehat{x_{i_1} x_{i_2}} \ldots \widehat{x_{i_{N-1}} x_{i_N}} + \text{all possible permutations.}$$

Let $t = n\epsilon$, $n = 0, \pm 1, \pm 2 \ldots \pm N$. Taking the limit of $N \to \infty$, $\epsilon \to 0$ yields

$$S_0 = -\frac{1}{2}\int_{-\infty}^{+\infty} dt dt' x_t A_{tt'} x_{t'},$$

$$Z = \int Dx e^S.$$

The generating functional for the simple harmonic oscillator is given by[1]

$$Z[j] = \frac{1}{Z}\int Dx e^{S_0 + \int dt j(t) x_t} = \exp\left\{\frac{1}{2}\int_{-\infty}^{+\infty} dt dt' j_t A_{tt'}^{-1} j_{t'}\right\}, \tag{7.8}$$

where

$$\int A_{tt'}^{-1} A_{t't''} dt' = \delta(t - t''). \tag{7.9}$$

7.3 Gaussian white noise

The properties of white noise are analyzed in this section. The fundamental properties of Gaussian white noise are that

$$E[R(t)] = 0, \quad E[R(t)R(t')] = \delta(t - t'). \tag{7.10}$$

Figure 7.2 shows how there is an independent (Gaussian) random variable $R(t)$ for each instant of time t.

Let us discretize time, namely $t = n\epsilon$, with $n = 1, 2 \ldots N$, and with $R(t) \to R_n$. The probability distribution function of white noise is given by

$$P(R_n) = \sqrt{\frac{\epsilon}{2\pi}} e^{-\frac{\epsilon}{2} R_n^2}. \tag{7.11}$$

1 The term generating functional is used instead of generating function as in Eq. 7.7 to indicate that one is considering a system with infinitely many variables.

Figure 7.2 One random variable $R(t)$ for each instant of time.

Hence, R_n is a Gaussian random variable with zero mean and $1/\sqrt{\epsilon}$ variance, and is denoted by $N(0, 1/\sqrt{\epsilon})$. The following result is essential in deriving the rules of Ito calculus:

$$R_n^2 = \frac{1}{\epsilon} + \text{random terms of } 0(1). \tag{7.12}$$

To prove the result stated in Eq.(7.12), it will be shown that, to leading order in $1/\epsilon$, the generating function of R_n^2 can be derived by considering R_n^2 to be deterministic. All the moments of R_n^2 can be determined from its generating function, namely

$$E\left[(R_n^2)^k\right] = \frac{d^k}{dt^k} E\left[e^{tR_n^2}\right]\Big|_{t=0}.$$

Note one needs to evaluate the generating function $E\left[e^{tR_n^2}\right]$ only in the limit of $t \to 0$. Hence, for ϵ small but fixed

$$\lim_{t\to 0} E\left[e^{tR_n^2}\right] = \int_{-\infty}^{+\infty} dR_n e^{tR_n^2} \sqrt{\frac{\epsilon}{2\pi}} \mathrm{e}^{-\frac{\epsilon}{2}R_n^2} = \frac{1}{\sqrt{1-\frac{2t}{\epsilon}}} \sim \exp\left(\frac{t}{\epsilon}\right) + O(1).$$

The probability distribution function $P(R_n^2)$ for R_n^2, which gives the above generating function, is given by

$$P\left(R_n^2\right) = \delta\left(R_n^2 - \frac{1}{\epsilon}\right) \Rightarrow E\left[e^{tR_n^2}\right] = \int_{-\infty}^{+\infty} dR_n e^{tR_n^2} \delta\left(R_n^2 - \frac{1}{\epsilon}\right) = \exp\left(\frac{t}{\epsilon}\right).$$

In other words, although R_n is a random variable, the quantity R_n^2 is **not** a random variable, but is instead fixed at the value of $1/\epsilon$.

To write the probability measure for $R(t)$, with $t_1 \le t \le t_2$ discretize continuous time, namely $t \to n\epsilon$. White noise $R(t)$ has the probability distribution given in Eq. 7.11. The probability measure for the white noise random variables in the interval $t_1 \leq t \leq t_2$ is given by

$$\mathcal{P}[R] = \prod_{n=1}^{N} P(R_n) = \prod_{n=1}^{N} e^{-\frac{\epsilon}{2}R_n^2}, \tag{7.13}$$

$$\int dR = \prod_{n=1}^{N} \sqrt{\frac{\epsilon}{2\pi}} \int_{-\infty}^{+\infty} dR_n.$$

Taking the continuum limit of $\epsilon \to 0$ yields, for $t_1 < t < t_2$,

$$\mathcal{P}[R, t_1, t_2] \to e^{S_0}, \quad S_0 = -\frac{1}{2}\int_{t_1}^{t_2} dt R^2(t), \tag{7.14}$$

$$Z = \int DRe^{S_0}, \quad \int dR \to \int DR.$$

The action functional S_0 is ultra-local with all the variables being decoupled. Gaussian integration, given in Eq. 7.7, yields

$$Z[j, t_1, t_2] = \frac{1}{Z}\int DRe^{\int_{t_1}^{t_2} dtj(t)R(t)} e^{S_0} = e^{\frac{1}{2}\int_{t_1}^{t_2} dtj^2(t)}. \tag{7.15}$$

The correlation functions are given by

$$E[R(t)] = 0, \quad E[R(t)R(t')] = \frac{1}{Z}\int DR\ R(t)R(t')e^{S_0} = \delta(t - t'),$$

and yield the result given in Eq. 7.10.

The results given in Eqs. 7.14 and 7.15 show that white noise is represented by a path integral with an ultra-local action S_0. Path integrals can be used to represent a great variety of white noise, as discussed by Kleinert (1990).

7.3.1 Integrals of white noise

Consider the integral of white noise

$$I = \int_t^T dt' R(t') \sim \epsilon \sum_{n=0}^{M} R_n, \quad M = \left[\frac{T - t}{\epsilon}\right],$$

where ϵ is an infinitesimal. For Gaussian white noise

$$R_n = N\left(0, \frac{1}{\sqrt{\epsilon}}\right) \quad \Rightarrow \epsilon R_n = N\left(0, \sqrt{\epsilon}\right).$$

The integral of white noise is a sum of normal random variables and hence, from Eq.(7.7) and above, is also a Gaussian random variable given by

$$I \sim N\left(0, \sqrt{\epsilon M}\right) \to N\left(0, \sqrt{T - t}\right). \tag{7.16}$$

In general, for

$$Y = \int_t^T dt' a(t')R(t') \Rightarrow Y = N\left(0, \sigma^2\right),\ \sigma^2 = \int_t^T dt' a^2(t').$$

Due to the Dirac delta function singularity in white noise $R(t)$, there is an ambiguity in discretizing white noise. Consider the integral of white noise

$$I = \int_0^\tau dt f(R(t); t)R(t) = \int_0^\tau f(W(t); t)dW(t),\quad dW(t) = R(t)dt,$$
$$E\left[W(t)W\left(t'\right)\right] = t\theta\left(t' - t\right) + t'\theta\left(t - t'\right) = \min\left(t, t'\right), \tag{7.17}$$

where the theta function is defined in Eq. 3.9.

Let $t_n = n\epsilon$ and with $\tau = N\epsilon$; $W(t) \to W_n$; there are two different and *inequivalent* ways of discretizing the integral, namely the Ito and Strantovich discretizations, which are given by

$$I_I = \sum_{n=0}^{N-1} f\left(W_n; t_n\right)\left(W_{n+1} - W_n\right),\quad E[W_n W_{n'}] = \left[\min(n, n')\right]\epsilon, \tag{7.18}$$
$$I_S = \sum_{n=0}^{N-1} f\left(\frac{W_{n+1} + W_n}{2}; \frac{t_{n+1} + t_n}{2}\right)\left(W_{n+1} - W_n\right).$$

The expectation value of the stochastic integrals for the case of $f = W(t)$ is used below to illustrate the ambiguity in the discretization of the stochastic integral. From Eq. 7.18

$$E[I_I] = \sum_{n=0}^{N-1} E\left[W_n(W_{n+1} - W_n)\right] = \sum_{n=0}^{N-1}(n\epsilon - n\epsilon) = 0. \tag{7.19}$$

In contrast

$$E[I_S] = \frac{1}{2}\sum_{n=0}^{N-1} E\left[(W_{n+1} + W_n)(W_{n+1} - W_n)\right]$$
$$= \frac{1}{2}\sum_{n=0}^{N-1} E\left[W_{n+1}^2 - W_n^2\right] = \frac{1}{2}\sum_{n=0}^{N-1}\{(n+1)\epsilon - n\epsilon\} = \frac{\tau}{2}. \tag{7.20}$$

Hence, Eqs. 7.19 and 7.20 show that the two methods for discretizing the stochastic integral are inequivalent.

Both the Ito and Strantovich discretizations are used extensively. It can be shown that the Ito discretization is the correct one for mathematical finance since it is consistent with the requirement that the market is free from arbitrage opportunities [Baaquie (2004)]; the Strantovich discretization requires a *future* value of $f(W)$, which in turn is in conflict with the requirement of no arbitrage. The Strantovich discretization is used in the study of the Fokker–Planck equation [Risken (1988)].

7.4 Ito calculus

A brief discussion is given below of Ito calculus and its relation to stochastic differential equations. Due to the singular nature of white noise $R(t)$, functions of white noise have new features. In particular, the infinitesimal behavior of such functions, as seen in their Taylor's expansions, acquire new terms.

Consider the stochastic differential equation that occurs in finance; the time derivative of a security $S(t)$ is generically expressed as

$$\frac{dS(t)}{dt} = \phi(t)S(t) + \sigma(t)S(t)R(t). \tag{7.21}$$

Let f be some arbitrary function of $S(t)$. Ito's definition of a derivative yields

$$\frac{df}{dt} = \lim_{\epsilon \to 0} \frac{f(t+\epsilon, S(t+\epsilon)) - f(t, S(t))}{\epsilon}$$

or, using Taylors expansion

$$\frac{df}{dt} = \frac{\partial f}{\partial t} + \frac{\partial f}{\partial S}\frac{dS}{dt} + \frac{\epsilon}{2}\frac{\partial^2 f}{\partial S^2}\left[\frac{dS}{dt}\right]^2 + 0\left(\epsilon^{1/2}\right). \tag{7.22}$$

The last term in Taylor's expansion is order ϵ for smooth functions, and goes to zero. However, due to the singular nature of white noise, Eq. 7.21 yields

$$\left[\frac{dS}{dt}\right]^2 = \sigma^2 S^2 R^2 + 0(1) = \frac{1}{\epsilon}\sigma^2 S^2 + 0(1). \tag{7.23}$$

Hence, from Eqs. 7.21, 7.22, and 7.23, for $\epsilon \to 0$

$$\frac{df}{dt} = \frac{\partial f}{\partial t} + \frac{\partial f}{\partial S}\frac{dS}{dt} + \frac{\sigma^2}{2}\frac{\partial^2 f}{\partial S^2} = \frac{\partial f}{\partial t} + \frac{1}{2}\sigma^2 S^2 \frac{\partial^2 f}{\partial S^2} + \phi S \frac{\partial f}{\partial S} + \sigma S \frac{\partial f}{\partial S} R. \tag{7.24}$$

Since Eq. 7.24 is of central importance for the theory of security derivatives, a derivation is given based on Ito-calculus. Let us rewrite Eq. 7.21 in terms of differentials as

$$dS = \phi S dt + \sigma S dz, \quad dz = R dt, \tag{7.25}$$

where dz is a Wiener process. Equation 7.12 yields $R^2(t) = 1/\epsilon = 1/dt$ and hence

$$(dz)^2 = R(t)^2 (dt)^2 = dt + 0(dt^{3/2})$$

and

$$(dS)^2 = \sigma^2 S^2 dt + 0\left(dt^{3/2}\right).$$

From the equations for dS and $(dS)^2$ given above,

$$\begin{aligned} df &= \frac{\partial f}{\partial t} dt + \frac{\partial f}{\partial S} dS + \frac{1}{2}\frac{\partial^2 f}{\partial S^2}(dS)^2 + 0\left(dt^{3/2}\right) \\ &= \left(\frac{\partial f}{\partial t} + \frac{1}{2}\sigma^2 S^2 \frac{\partial^2 f}{\partial S^2}\right) dt + \sigma S \frac{\partial f}{\partial S} dz, \end{aligned}$$

and Eq. 7.24 is recovered using $dz/dt = R$.

Suppose $g(t, S(t)) \equiv g_t$ is another function of the white noise $S(t)$. The abbreviated notation $\delta g_t \equiv g_{t+\epsilon} - g_t$ yields

$$\begin{aligned} \frac{d(fg)}{dt} &= \lim_{\epsilon \to 0} \frac{1}{\epsilon}\left[f_{t+\epsilon} g_{t+\epsilon} - f_t g_t\right] \\ &= \lim_{\epsilon \to 0} \frac{1}{\epsilon}\left[\delta f_t g_t + f_t \delta g_t + \delta f_t \delta g_t\right]. \end{aligned}$$

Usually the last term $\delta f_t \delta g_t$ is of order ϵ^2 and goes to zero. However, due to the singular nature of white noise

$$\frac{d(fg)}{dt} = \frac{df}{dt} g + f \frac{dg}{dt} + \frac{df}{\sqrt{dt}} \frac{dg}{\sqrt{dt}} \quad : \text{Ito's chain rule.} \tag{7.26}$$

Similarly to Eq. 7.26, in terms of infinitesimals, the Ito chain rule is given by

$$d(fg) = df g + f dg + df dg.$$

7.4.1 Stock price

To illustrate stochastic calculus, the stochastic differential equation, Eq. 7.21, is integrated. Consider the change of variable and the subsequent integration

$$x(t) = \ln[S(t)], \quad \Rightarrow \frac{dx}{dt} = \phi - \frac{\sigma^2}{2} + \sigma R(t), \tag{7.27}$$

$$\Rightarrow x(T) = x(t) + (\phi - \frac{\sigma^2}{2})(T - t) + \sigma \int_t^T dt' R(t'). \tag{7.28}$$

The random variable $\int_t^T dt' R(t')$ is a sum of normal random variables and is shown in Eq.7.16 to be equal to a normal $N(0, \sqrt{T-t})$ random variable. Hence

$$S(T) = S(t) e^{\left(\phi - \frac{\sigma^2}{2}\right)(T-t) + \left(\sigma\sqrt{T-t}\right)Y} \quad \text{with } Y = N(0, 1). \tag{7.29}$$

The stock price evolves randomly from its given value of $S(t)$ at time t to a whole range of possible values $S(T)$ at time T. Since the random variable $x(T)$ is a normal (Gaussian) random variable, the security $S(T)$ is a lognormal random variable.

Geometric mean of stock price

The probability distribution of the geometric mean of the stock price can be exactly evaluated. For $\tau = T - t$ and $m = x(t) + (\phi - \frac{\sigma^2}{2})\tau$, Eq. 7.28 yields

$$\begin{aligned}
S_{GM} &= e^G, \quad \tau = T - t, \\
G &\equiv \frac{1}{\tau} \int_t^T dt' x\left(t'\right) = m + \frac{\sigma}{\tau} \int_t^T dt' \int_t^{t'} dt'' R\left(t''\right) \\
&= m + \frac{\sigma}{\tau} \int_t^T dt' \left(T - t'\right) R\left(t'\right).
\end{aligned}$$

From Eq.7.16 the integral of white noise is a Gaussian random variable, which is completely specified by its means and variance. Hence, using $E[G] = m$ and Eq. 7.10 for $E[R(t)R(t')] = \delta(t - t')$ yields

$$\begin{aligned}
E\left[(G-m)^2\right] &= \left(\frac{\sigma}{\tau}\right)^2 \int_t^T dt' \left(T - t'\right) \int_t^T dt'' \left(T - t''\right) E\left[R\left(t'\right) R\left(t''\right)\right] \\
&= \left(\frac{\sigma}{\tau}\right)^2 \int_t^T dt' \left(T - t'\right)^2 = \frac{\sigma^2 \tau}{3}.
\end{aligned}$$

Hence

$$G = N\left(m, \frac{\sigma^2\tau}{3}\right). \tag{7.30}$$

The geometric mean of the stock price is lognormal with the same mean as the stock price, but with its volatility being one-third of the stock price's volatility.

7.5 Wilson expansion

The product of nonlinear (non-Gaussian) quantum fields is the subject matter of what is called the "short distance" Wilson expansion, discussed by Wilson (1969), Wilson and Zimmermann (1972), and Zinn-Justin (1993). The Wilson expansion of quantum fields is a very general technique that allows one to isolate the singularities in the product of quantum fields. In the context of stochastic systems, the Wilson expansion provides a generalization of Ito calculus to the case where the stochastic phenomenon is driven by the two-dimensional Gaussian quantum field $\mathcal{A}(t, x)$, as discussed by Baaquie (2009).

The time derivative of various quantities like the underlying security $S(t)$ or stochastic volatility are generically expressed as

$$\frac{dS(t)}{dt} = \mu(t) + \sigma(t)R(t).$$

Ito's stochastic calculus, for discrete time $t = n\epsilon$, is a result of the identity [Baaquie (2004)]

$$E\left[R(t)R(t)\right] = \delta(t - t') \Rightarrow R^2(t) = \frac{1}{\epsilon} + O(1). \tag{7.31}$$

The singular term for $R^2(t)$ in Eq. 7.31 is *deterministic*, namely, equal to $1/\epsilon$; all the random terms that occur for $R^2(t)$ are finite as $\epsilon \to 0$.

The two-dimensional quantum field $\mathcal{A}(t, x)$ is an independent degree of freedom for each t and each x. For Gaussian quantum fields such as $\mathcal{A}(t, x)$ that have a quadratic action, the full content of a Gaussian (free) quantum field, as discussed by Zinn-Justin (1993), is encoded in its *propagator*. Consider a propagator given by

$$E\left[\mathcal{A}(t, x)\mathcal{A}(t', x'\right] = \delta\left(t - t'\right) D\left(t, x; t', x'\right). \tag{7.32}$$

Similarly to white noise correlator $E[R(t)R(t)] = \delta(t - t')$, the correlation function $E[\mathcal{A}(t, x)\mathcal{A}(t', x')]$ given in Eq. 7.32 is infinite for $t = t'$. The singular product of two Gaussian quantum fields is the simplest case of the Wilson expansion. The singularity of the correlation function in the product of the quantum field $\mathcal{A}(t, x)$, similarly to Eq. 7.31, can be expressed as

$$\mathcal{A}(t, x)\,\mathcal{A}\left(t, x'\right) = \frac{1}{\epsilon}\mathcal{D}\left(x, x'; t\right) + O(1). \tag{7.33}$$

All the fluctuating components, which are contained in $\mathcal{A}(t, x)\mathcal{A}(t, x')$, are regular and finite as $\epsilon \to 0$. The correlation of $\mathcal{A}(t, x)$ is singular for $t = t'$, very much like the singularity of white noise $R(t)$.

Since $\mathcal{A}(t, x)$ is an indeterminate quantity, one may ask how one can assign it a deterministic numerical value as in Eq. 7.33? What Eq. 7.33 means is that in any correlation function, wherever a product of fields is at equal time, namely $\mathcal{A}(t, x)\mathcal{A}(t, x')$, then – to leading order in ϵ – the product can be replaced by the deterministic quantity $\mathcal{D}(x, x'; t)/\epsilon$. In terms of symbols, Eq. 7.33 states

$$E\left[\mathcal{A}(t_1,x_1)\mathcal{A}(t_2,x_2)\ldots\mathcal{A}(t,x_n)\mathcal{A}(t,x_{n+1})\ldots\mathcal{A}(t_N,x_N)\right]$$
$$=\frac{1}{\epsilon}E\left[\mathcal{A}(t_1,x_1)\mathcal{A}(t_2,x_2)\ldots\mathcal{A}(t_{n-1},x_{n-1})\mathcal{D}(x_n,x_{n+1};t)\right.$$
$$\left.\times\mathcal{A}(t_{n+2},x_{n+2})\ldots\mathcal{A}(t_N,x_N)\right]+O(1).$$

Equations 7.31 and 7.33 have a similar singularity structure, and Gaussian quantum fields yield a natural generalization of Ito calculus. Ito calculus is a special case of the Wilson expansion, and is given by taking the limit

$$\mathcal{A}(t,x)\to R(t),\ \ \mathcal{D}\left(x,x';t\right)\to 1, \tag{7.34}$$
$$\Rightarrow\ E\left[\mathcal{A}(t,x)\mathcal{A}(t',x')\right]\to E\left[R(t)R(t)\right]=\delta(t-t').$$

7.6 Linear Langevin equation

Consider the case with no external potential, namely $\Phi(v)=0$. One can rewrite the Langevin equation given in Eq. 7.2, for $\beta\equiv\gamma/m$, as

$$\frac{dv}{dt}=-\frac{\gamma v}{m}+\frac{\sqrt{2A}}{m}R\equiv-\beta v+\frac{\sqrt{2A}}{m}R. \tag{7.35}$$

The term R is white noise, with probability density given in Eq. 7.14 by

$$P[R]=\frac{1}{Z}\exp\left(-\frac{1}{2}\int R^2(t)dt\right).$$

Integrating the first order stochastic differential

$$\frac{dv}{dt}+\frac{\gamma}{m}v=\frac{\sqrt{2A}}{m}R=e^{-\frac{\gamma}{m}t}\frac{d}{dt}\left(e^{\frac{\gamma}{m}tv}\right)=\frac{\sqrt{2A}}{m}R.$$

yields for $v(t)$

$$v=v_0e^{-\gamma t/m}+\sqrt{2A}\frac{e^{-\frac{\gamma t}{m}}}{m}\int_0^t dt'e^{\frac{\gamma}{m}t'}R(t').$$

Define

$$u=v-v_0e^{-\gamma t/m}, \tag{7.36}$$

such that $E[u(t)]=0$; hence

$$E\left[u^2(t)\right]=\frac{2A}{m^2}\int_0^t d\eta d\tau e^{-\frac{\gamma}{m}(t-\eta)}e^{-\frac{r}{m}(t-\tau)}E\left[R(\eta)R(\tau)\right]$$
$$=\frac{2A}{m^2}\int_0^t d\eta e^{-2\beta(t-\eta)}=\frac{A}{m\gamma}\left(1-e^{-2\gamma t/m}\right)$$
$$\Rightarrow\ \lim_{t\to\infty}E\left[u^2(t)\right]=E[v^2(t)]=\frac{A}{m\gamma}.$$

The particle reaches equilibrium with its medium after a long time, namely for $t = \infty$. For the particle in a medium at temperature T, the Maxwell velocity distribution is given by

$$P[v] = \sqrt{\frac{m}{2\pi kT}} e^{-\frac{1}{kT}\cdot\frac{1}{2}mv^2} \Rightarrow E\left[v^2\right] = \frac{kT}{m}. \tag{7.37}$$

Hence, in equilibrium

$$\frac{A}{\gamma m} = \frac{kT}{m} \Rightarrow A = kT\gamma. \tag{7.38}$$

The velocity $v(t)$ is a random variable, and hence is determined by a probability distribution function $P[v; t]$. Solving a stochastic differential equation, unlike an ordinary differential, consists of determining the probability distribution $P[v; t; v_0]$ for the occurrence of different values for $v(t)$, given the initial velocity v_0.

Recall from Eq. 7.2 that the Langevin equation yields

$$v = v_0 e^{-\frac{\gamma t}{m}} + \int_0^t dt' \rho(t') R(t'). \tag{7.39}$$

The linear sum of Gaussian random variables is also a Gaussian random variable. Hence all we need to do is to determine the mean and variance of $v(t)$ to determine $P[v; t]$. Recall from Eq. 7.36 that $u = v - v_0 e^{-\gamma t/m}$ and hence

$$E[v] = v_0 e^{-\frac{\gamma}{m}t} \Rightarrow \sigma_v^2 = \sigma_u^2 = E[u^2(t)] = \frac{kT}{m}(1 - e^{-2\frac{\gamma}{m}t}). \tag{7.40}$$

Hence

$$\begin{aligned} P[v; t; v_0] &= \frac{1}{\sqrt{2\pi\sigma_v^2}} e^{-\frac{1}{2\sigma_v^2}\left(v - v_0 e^{-\frac{\gamma}{m}t}\right)^2} \\ &= \sqrt{\frac{m}{2\pi kT\left(1 - e^{-2\frac{\gamma}{m}t}\right)}} \exp\left\{-\frac{m}{2kT}\frac{\left(v - v_0 e^{-\frac{\gamma}{m}t}\right)^2}{\left(1 - e^{-\frac{\gamma}{m}t}\right)}\right\}. \end{aligned} \tag{7.41}$$

Note, as expected

$$\begin{aligned} &\lim_{t\to 0} P[v; t; v_0] \to \delta(v - v_0) \\ &\lim_{t\to\infty} P[v; t; v_0] \to \sqrt{\frac{m}{2\pi kT}} e^{-\frac{m}{2kT}v^2}. \end{aligned} \tag{7.42}$$

7.6.1 Random paths

From Eq. 7.35, the Langevin equation is given by

$$\frac{dv}{dt} = -\beta v + \frac{\sqrt{2A}}{m} R.$$

The random paths $x(t)$, as determined by the Langevin equation, are given by

$$\frac{dx(t)}{dt} = v(t), \tag{7.43}$$

$$x(t) = x_0 + \int_0^t d\xi v(\xi) \Rightarrow E[x(t)] = x_0 + \int_0^t d\xi E[v(\xi)],$$

$$\text{or,} \quad E[x(t)] = x_0 + v_0 \int_0^t d\xi e^{-\beta\xi} = x_0 + \frac{v_0}{\beta}(1 - e^{-\beta t}). \tag{7.44}$$

Let $x_0 = -v_0/\beta$, then

$$E[x(t)] = -\frac{v_0}{\beta} e^{-\beta t} \to 0 \qquad \text{as} \qquad t \to \infty.$$

Let $y(t) = x(t) + \frac{v_0}{\beta} e^{-\beta t}$, $y(0) = 0$. The change from $x(t)$ to $y(t)$ is made to remove unnecessary terms arising from the boundary conditions on $x(t)$ and $v(t)$. Defining $\dot{y} = dy/dt$ yields

$$\frac{dy}{dt} = v - e^{-\beta t} v_0 \quad \Rightarrow \quad \dot{y} = \sqrt{2A} \frac{e^{-\beta t}}{m} \int_0^t d\xi e^{\beta\xi} R(\xi).$$

Hence

$$\begin{aligned} E\left[\dot{y}(t)\dot{y}(t')\right] &= \frac{2A}{m^2} \int_0^t d\xi \int_0^{t'} d\eta e^{-\beta(t-\xi)} e^{-\beta(t'-\eta)} E\left[R(\xi)R(\eta)\right] \\ &= e^{-\beta(t+t')} \frac{2A}{m^2} \int_0^t d\xi \int_0^{t'} d\eta e^{\beta\xi} e^{\beta\eta} \delta(\xi - \eta). \end{aligned}$$

But

$$\int_0^{t'} d\eta \delta(\xi - \eta) = \begin{cases} 0 & \xi > t' \\ 1 & \xi \leqslant t' \end{cases} = \theta(t' - \xi).$$

Let $t \geq t'$, then

$$E\left[\dot{y}(t)\dot{y}\left(t'\right)\right] = \frac{A}{m^2} e^{-\beta(t+t')} \int_0^{t'} d\xi e^{2\beta\xi} = \frac{A}{2\beta m^2} e^{-\beta(t+t')} \left[-1 + e^{2\beta t'}\right].$$

Due to time ordering expressed in $t \geq t'$, the correlator is given by

$$E\left[y(t)y\left(t'\right)\right] = \frac{A}{2\beta m^2} \int_0^t d\tau \int_0^{t'} d\tau' \left[-e^{-\beta(\tau+\tau')} + e^{-\beta|\tau-\tau'|}\right]. \tag{7.45}$$

For $t = t'$,

$$E\left[y^2(t)\right] = \frac{A}{2\beta m^2}\int_0^t d\tau d\tau'\left[-e^{-\beta(\tau+\tau')} + e^{-\beta|\tau-\tau'|}\right]. \tag{7.46}$$

Integrating the equation above yields

$$E\left[y^2(t)\right] = \frac{2A}{\beta^2}t + \frac{Am}{\beta^3}\left[-3 + 4e^{-\frac{\gamma}{m}t} - e^{-\frac{2\gamma}{m}t}\right]. \tag{7.47}$$

Note that $E[y^2(0)] = 0$ as expected. In the limit of $t \to \infty$ we have that $y(t) \to x(t)$, and this yields

$$\sqrt{E\left[x^2(t)\right]} = \sqrt{\frac{2A}{\gamma^2}}\cdot\sqrt{t}, \tag{7.48}$$

where $\sqrt{t}$ is the characteristic signal of a random walk. Equivalently

$$E\left[x^2(t)\right] = \frac{2A}{\gamma^2}t \equiv 2Dt, \tag{7.49}$$

where D is the diffusion constant; hence

$$D = \frac{A}{\gamma^2} = \frac{kT}{\gamma} \quad : \quad \text{Einstein relation.} \tag{7.50}$$

Note $E[x(t)] = 0$ and hence $\sqrt{E[x^2(t)]}$ measures the degree of dispersal of the randomly moving particle.

For a uniformly moving particle $x \sim vt$. In contrast, for a randomly walking particle $x \sim \sqrt{t}$, which is much slower than a freely moving particle, due to the frequent random collisions that the particle has with the medium it is in.

7.7 Langevin equation with potential

In the presence of an external force due to a potential field B, the Langevin equation is

$$\frac{d\vec{v}}{dt} = -\beta\vec{v} + \frac{1}{m}\vec{K} + \frac{1}{m}\vec{F}, \tag{7.51}$$

where $\vec{K} = -\vec{\nabla}B$.

For the one-dimensional simple harmonic oscillator, $B = \frac{1}{2}m\omega^2x^2$ and

$$\frac{dv}{dt} = -\beta v - \omega^2 x + \frac{1}{m}F, \tag{7.52}$$

or

$$\frac{d^2x}{dt^2} + \beta\frac{dx}{dt} + \omega^2 x = \frac{1}{m}F. \tag{7.53}$$

Let $D = d/dt$, then

$$\left(D^2 + \beta D + \omega^2\right) x = \frac{1}{m} F, \tag{7.54}$$

or

$$(D + r_1)(D + r_2)x = \sigma R \quad \left(\sigma = \frac{\sqrt{2A}}{m}\right), \tag{7.55}$$

$$r_{1,2} = \frac{\beta}{2} \pm \sqrt{\frac{\beta^2}{4} - \omega^2}.$$

Let x_H be the homogeneous solution of Eq.7.53 with

$$(D + r_1)(D + r_2)x_H = 0. \tag{7.56}$$

Note the boundary values that specify a unique solution for Eq. 7.53 are carried by x_H. The complete solution of Eq. 7.53 is then given by

$$\begin{aligned} x &= x_H + \frac{1}{(D + r_1)(D + r_2)} \sigma R \\ &= x_H + \frac{\sigma}{r_2 - r_1} \left[\frac{1}{D + r_1} - \frac{1}{D + r_2} \right] R \\ &= x_H + \sigma \int_0^t B(t, t') R(t') dt', \end{aligned} \tag{7.57}$$

where $x_H = Ae^{-r_1 t} + Be^{-r_2 t}$.
Using

$$\frac{1}{D + r_1} R = \int_0^t e^{-r_1(t-\xi)} R(\xi) d\xi$$

yields

$$B(t, t') = \frac{1}{\sqrt{\beta^2 - 4\omega^2}} \left\{ e^{-r_2(t-t')} - e^{-r_1(t-t')} \right\}. \tag{7.58}$$

The coefficients A, B are determined from the initial conditions using $x(0) = x_0$, $\dot{x}(0) = v_0$, or any other specifications of the boundary conditions.

7.7.1 Correlation functions

Consider the unequal time correlator given by

$$C_{t,t'} = E[x(t)x(t')] = x_H(t)x_H(t') + \frac{2A}{m^2} \int_0^t d\xi \int_0^{t'} d\xi' B(t, \xi) B(t', \xi') \delta(\xi - \xi').$$

Choose time ordering $t > t'$, and for $C_H = x_H(t)x_H(t')$, one obtains

$$C_{t,t'} = C_H + \frac{2A}{m^2}\int_0^{t'} d\xi' B\left(t,\xi'\right)B\left(t',\xi'\right). \tag{7.59}$$

For the *under damped case*

$$\lambda^2 = \omega^2 - \frac{\beta^2}{4} > 0 \tag{7.60}$$

and hence

$$r_1 = -\frac{\beta}{2} + i\lambda, \qquad r_2 = -\frac{\beta}{2} - i\lambda \tag{7.61}$$

$$\Rightarrow C_{tt'} = C_H + \frac{2A}{m^2}\frac{1}{2\beta\omega^2}e^{-\frac{\beta}{2}(t-t')}\left\{\cos\lambda\left(t-t'\right) + \frac{\beta}{2\lambda}\sin\lambda\left(t-t'\right)\right\}. \tag{7.62}$$

Similar expressions can be obtained for the *over damped case* for which $\lambda^2 > 0$.

7.8 Nonlinear Langevin equation

Consider the nonlinear Langevin equation given by

$$\frac{dv}{dt} = \beta v - gv^3 + \sigma R, \tag{7.63}$$

with $\beta, g > 0$ and $v(0) = v_0$. Note that the sign of β is positive, and is the *opposite* of the sign of the viscosity term γ in the Langevin equation as given in Eq. 7.2.

There are two competing effects in the nonlinear Langevin equation, namely:

- the β term acts to exponentially increase v;
- the dissipative term g damps and reduces the value of v.

It is not known how to exactly solve Eq. 7.63. Instead, a self-consistent *mean field approximation* is used to find a solution. The accuracy of this approximate solution is discussed later.

Let $f(t) = E[v^2(t)]$ and $f(0) = v_0^2$. The mean field approximation consists of (self-consistently) linearizing the equations of motion, namely

$$\frac{dv}{dt} \to \beta v - gvE[v^2] + \sigma R$$

$$\Rightarrow \frac{dv}{dt} = \beta v - gvf + \sigma R, \quad \text{mean field approximation.} \tag{7.64}$$

Note that

$$\frac{df}{dt} = 2E\left[v(t)\frac{dv}{dt}(t)\right] = 2E\left[v(t)\left\{\beta v(t) - gv^3(t) + \sigma R(t)\right\}\right]. \tag{7.65}$$

Using the mean field approximation for the equations of motion given in Eq. 7.64 yields

$$E\left[v(t)^4\right] \simeq E\left[v(t)^2\right]^2 \equiv f^2(t).$$

Hence, in the mean field approximation

$$\frac{df}{dt} = 2\beta f - 2gf^2 + 2\sigma E\left[v(t)R(t)\right].$$

Defining $h(t) \equiv g\int_0^t d\tau f(\tau) - \beta t$, Eq. 7.64 yields

$$v(t) = e^{-h(t)}v_0 + e^{h(t)}\sigma \int_0^t e^{-h(\tau)}R(\tau)d\tau$$

and hence

$$E\left[v(t)R(t)\right] = e^{-h(t)}\sigma \int_0^t e^{h(\tau)}\delta(t-\tau)d\tau = \frac{1}{2}\sigma,$$

where for consistency $\int_0^t \delta(t-\tau)d\tau = \frac{1}{2}$. Hence, from Eq. 7.65

$$\frac{df}{dt} = 2\beta f - 2gf^2 + \sigma^2 \quad : \quad \text{Riccati equation.} \tag{7.66}$$

Let

$$f = \frac{1}{2g}\frac{\dot{u}}{u} \Rightarrow \dot{f} = \frac{1}{2g}\left(\frac{\ddot{u}}{u} - \left(\frac{\dot{u}}{u}\right)^2\right).$$

Substituting the expression for f, $\dot{f}$ on both sides of the Riccati equation given in Eq. 7.66 yields

$$\frac{1}{2g}\left(\frac{\ddot{u}}{u} - \frac{\dot{u}^2}{u^2}\right) = 2\beta \cdot \frac{1}{2g}\frac{\dot{u}}{u} - \frac{1}{2g}\left(\frac{\dot{u}}{u}\right)^2 + \sigma^2.$$

Note that the nonlinear term $(\dot{u}/u)^2$ cancels in the above equation, and yields a linear equation for $u(t)$, namely

$$\ddot{u} - 2\beta\dot{u} - 2g\sigma^2 u = 0.$$

Using the ansatz $u(t) \simeq e^{\alpha t}$ yields

$$\alpha^2 - 2\beta\alpha - 2g\sigma^2 = 0 \Rightarrow \alpha = \beta \pm \sqrt{\beta^2 + 2\sigma^2 g} = \beta \pm \lambda.$$

Hence the solution is given by

$$u(t) = ce^{\beta t}\left(e^{\lambda t} + ke^{-\lambda t}\right),$$

where c, k are two integration constants. The superfluous constant c cancels out, as is required, for the "physical" quantity $f(t)$, in the following manner:

$$E\left[v^2(t)\right] = f = \frac{1}{2g}\frac{\dot{u}}{u}$$

$$\Rightarrow E\left[v^2(t)\right] = \frac{1}{2g}\frac{(\beta+\lambda)e^{\lambda t} + k(\beta-\lambda)e^{-\lambda t}}{e^{\lambda t} + ke^{-\lambda t}}$$

$$v_0^2 = \frac{1}{2g}\frac{\beta+\lambda+k(\beta-\lambda)}{1+k}, \tag{7.67}$$

or, equivalently

$$k = \frac{2gv_0^2 - \beta - \lambda}{\beta - \lambda - 2gv_0^2}.$$

Note that

$$\lambda \equiv \beta\left(1 + \frac{2\sigma^2 g}{\beta^2}\right)^{1/2} \simeq \beta\left(1 + \frac{\sigma^2 g}{\beta^2}\right) \simeq \beta.$$

Consider $1 \ll \beta t \ll \infty$, that is $\frac{1}{\beta} \ll t \ll \infty$. Let $v_0 \ll 1$, then, for infinite time, the average fluctuation for $v^2(t)$ is given by

$$\Rightarrow \lim_{t\to\infty} E\left[v^2(t)\right] \cong \frac{1}{2g}\left[\frac{2\beta e^{\lambda t}}{e^{\lambda t} + ke^{-\lambda t}}\right] = \frac{\beta}{g}\frac{\frac{1}{k}e^{2\lambda t}}{1 + \frac{1}{k}e^{2\lambda t}} \to \frac{\beta}{g}.$$

The cross-over of the velocity $v(t)$ from its initial value v_0 to its equilibrium value β/g takes place at t_0 such that $E[v^2(t)] \simeq \beta/g$, as shown in Figure 7.3.

Hence, $1 + e^{2\lambda t_0}/k \simeq e^{2\lambda t_0}/k \simeq 1$, and this yields

$$\frac{1}{k}e^{2\lambda t_0} \simeq \mathbf{1}, \quad t_0 \sim \ln(k)/2\lambda.$$

At time t_0 dissipative structures far from equilibrium are formed due to the synergy between nonlinear dissipation driven by g and random fluctuations due to σ^2/β. As time flows the dissipative structures evolve into the final equilibrium state. Figure 7.3 shows the evolution of the particle from its initial state at $t = 0$, via the formation of dissipative structures t_0, to equilibrium at t_E.

To check whether the mean field result is consistent, note that on reaching equilibrium

$$\frac{dv}{dt} = 0 = \beta v - gv^3 + R,$$

$$\Rightarrow \beta E[v] = gE[v^3] \simeq gE[v]E[v^2] \quad : \quad \text{mean field,}$$

$$\Rightarrow E[v^2] \simeq \frac{\beta}{g} \quad : \quad \text{consistent.}$$

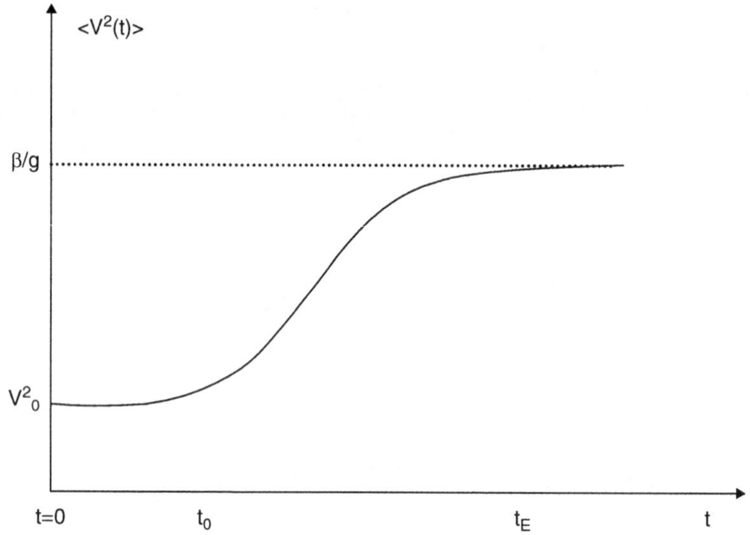

Figure 7.3 The value of $E[v^2(t)]$ as a function of time, which starts at the initial value v_0^2 and follows a nonlinear trajectory to its final value, with cross-over at $t_0 \simeq \ln(k)/(2\lambda)$.

How good is the mean field approximation? Can one systematically improve it? There is no small parameter in terms of which the mean field approximation is defined, and hence one cannot go to higher and higher order in the mean field approximation.

The mean field approximation can correctly ascertain whether, for example, a system undergoes a phase transition, but cannot give an accurate estimate of the physical parameters that describe the transition. Rather, mean field approximation is a guide to the qualitative behaviour of a system and can capture many features like the system having instabilities or other global properties.

7.9 Stochastic quantization

The stochastic Langevin differential equation expresses the evolution of the fundamental random degree of freedom, say the random velocity $v(t)$ itself, and is analogous to the Heisenberg operator equations. The solution of a stochastic differential equation entails determining the probability distribution function for every time $P(v; t; v_0)$, given the initial condition at $t = 0$ by $v(0) = v_0$. $P(v; v_0; t)$ is the evolution kernel for a nonequilibrium and dissipative system, driven by the Fokker–Planck Hamiltonian.

As a warm-up to a more detailed analysis of the Fokker–Planck equation, a quick and heuristic derivation is given for $P(v; t; v_0)$ for the (linear) Langevin equation. Recall

$$\frac{dv}{dt} = -\beta v + \Phi(v) + \sigma R, \qquad v(0) = v_0.$$

Since both v_t and R_t are random variables, stochastic quantization defines a path integral by integrating over all possible values for v_t and R_t.

The conditional probability $P(v; v_0; T)$ is the likelihood of the final value of the velocity $v(T)$ having a value equal to v, given that its initial value was v_0. The Langevin equation is a first order differential and needs only one boundary condition, which has been taken to be the initial condition; the second condition $v(T) = v$ is a constraint on the allowed random paths and is put in as a delta-function constraint in the path integral. The path integral is given by

$$\begin{aligned} &P(v; v_0; T) \\ &= \int DvDR\left[\prod_t \delta\left(\frac{\dot{v}_t + \beta_t v_t - \Phi(v)}{\sigma} - R(t)\right)\right]\delta(v(T) - v)e^{-\frac{1}{2}\int R^2(t)dt} \\ &= \int Dv\delta\left(v(T) - v\right)e^{-\frac{1}{2\sigma^2}\int dt(\dot{v}+\beta v-\Phi(v))^2}, \quad \text{B.C.}: \ v(0) = v_0. \end{aligned}$$

Once the white noise path integral has been performed, one can write the path integral for finite time by directly incorporating the two boundary conditions on the velocity, namely $v(0) = v_0, \ v(T) = v$; the path integral can be written more compactly as

$$P(v; v_0; T) = \int Dve^{-\frac{1}{2\sigma^2}\int dt(\dot{v}+\beta v-\Phi(v))^2}. \tag{7.68}$$

For the nonlinear Langevin equation given in Eq. 7.63 one obtains the action

$$S = -\frac{1}{2\sigma^2}\int_0^T dt\left(\dot{v} - \beta v + gv^3\right)^2. \tag{7.69}$$

The path integral for the nonlinear Langevin equation in principle is exact and can be used for systematically generating a power series in any small parameter by expanding the action and doing the path integral term by term. In particular, one could for example compute the expectation values $E[v^3], \ E[v^2], \ E[v]$ as a perturbation expansion in $g << 1$, and then compare $E[v^3]$ with $E[v^2]E[v]$ to determine the accuracy of the mean field approximation.

7.9.1 Linear Langevin path integral

Consider the linear Langevin equation for which $\Phi(v) = 0$. The stochastic differential equation is given by

$$\frac{dv}{dt} = -\beta v + \sigma R$$

with volatility σ=constant. The action is given by

$$S = -\frac{1}{2\sigma^2}\int_0^T dt\,(\dot{v} + \beta v)^2 .$$

Consider change of variables

$$w = \dot{v} + \beta v = e^{-\beta t}\frac{d}{dt}\left(e^{\beta t}v\right).$$

For incorporating the initial condition $v(0) = v_0$, consider

$$v(T) = e^{-\beta T}v_0 + e^{-\beta T}\int_0^T dt' e^{\beta t'} w_{t'}. \qquad (7.70)$$

As discussed in deriving Eq. 7.68, the second condition $v(T) = v$ is put in as a delta-function constraint in the path integral and yields the following path integral for the conditional probability

$$P\,(v;\, v_0;\, T) = N\int DW\delta\,(v - v(T))\, e^{-\frac{1}{2\sigma^2}\int w^2}.$$

Using the Fourier representation for the δ-function and using Eq. 7.70 for $v(T)$ yields

$$\begin{aligned}
P(v;\, v_0;\, T) &= N\int \frac{d\xi}{2\pi} e^{i\xi\left(v - v_0 e^{-\beta T}\right)} \int DW e^{-i\xi e^{\beta T}\int dt' e^{\beta t'} w_{t'}} e^{-\frac{1}{2\sigma^2}\int w^2} \\
&= \int \frac{d\xi}{2\pi} e^{i\xi\left(v - v_0 e^{-\beta T}\right)} \exp\left\{-\frac{\sigma^2}{2} e^{-2\beta T}\xi^2 \int_0^T dt' dt'' e^{\beta t'}\delta(t' - t'') e^{\beta t''}\right\} \\
&= \int \frac{d\xi}{2\pi} e^{i\xi\left(v - v_0 e^{-\beta T}\right)} \exp\left\{-\frac{\sigma_v^2}{2}\xi^2\right\},
\end{aligned}$$

where

$$\sigma_v^2 = \sigma^2 e^{-2\beta T}\int_0^T dt e^{2\beta t} = \frac{\sigma^2}{2\beta}\left(1 - e^{-2\beta T}\right)$$

and, as expected from Eq. 7.40,

$$\frac{\sigma^2}{2\beta} = \frac{2A}{2m^2}\cdot\frac{m}{\gamma} = \frac{A}{m\gamma} = \frac{kT}{m}.$$

Performing the ξ integration yields the final result

$$P\,(v;\, v_0;\, T) = \frac{1}{\sqrt{2\pi\sigma_v^2}} e^{-\frac{1}{2\sigma_v^2}\left(v - e^{-\beta T}v_0\right)^2}, \qquad (7.71)$$

as was obtained earlier in Eq. 7.41.

7.10 Fokker–Planck Hamiltonian

The Fokker–Plank equation is a partial differential equation for evolving the conditional probability $P(v; v_0; t)$ [Risken (1988)].

Consider the general Langevin equation

$$\frac{dv(t)}{dt} = a(v,t) + \sigma(v,t)R(t). \tag{7.72}$$

There are two possible boundary conditions, namely

- Initial condition $v(t_0) = v_0$;

  or
- Final condition $v(T) = v_T$.

The Langevin equation has the forward Euler discretization given by $t = n\epsilon$ and

$$v(t+\epsilon) = v(t) + \epsilon\left[a(v,t) + \sigma(v,t)R(t)\right], \tag{7.73}$$

$$v = v' + \epsilon\left\{a\left(v'\right) + \sigma(v')R_n\right\}. \tag{7.74}$$

For discrete time, white noise has the correlators

$$E[R_n] = 0, \qquad E\left[R_n^2\right] = \frac{1}{\epsilon}.$$

The forward conditional probability is given by

$$\begin{aligned}
P_F(v, v_0, t+\epsilon) &= \int dv' E\left[\delta\left(v - v' - \epsilon[a\left(v'\right) + \sigma\left(v'\right)R_t]\right)\right] P_F\left(v', v_0, t\right) \\
&= \int dv' E\left[\delta\left(v - v'\right) + \epsilon\left(a\left(v'\right) + \sigma\left(v'\right)R\right)\frac{\partial}{\partial v'}\delta\left(v - v'\right)\right. \\
&\quad \left. + \frac{\epsilon^2}{2}\left(a + \sigma R\right)^2 \frac{\partial^2}{\partial v'^2}\delta\left(v - v'\right)\right] P_F\left(v'; v_0, t\right) \\
&= P_F(v, v_0, t) - \epsilon\frac{\partial}{\partial v}(aP_F) + \frac{\epsilon}{2}\frac{\partial^2}{\partial v^2}\left(\sigma^2 P_F\right) + O\left(\epsilon^2\right).
\end{aligned}$$

Hence

$$\begin{aligned}
\frac{\partial P_F}{\partial t} &\equiv \frac{1}{\epsilon}\left[P_F(v, v_0, t+\epsilon) - P_F(v, v_0, t)\right] \\
&= \left[\frac{1}{2}\frac{\partial^2}{\partial v^2}\sigma^2 - \frac{\partial}{\partial v}a\right] P_F \\
&\equiv -H_F P_F,
\end{aligned}$$

where the forward Fokker–Planck Hamiltonian is given by

$$H_F = -\frac{1}{2}\frac{\partial^2}{\partial v^2}\sigma^2(v) + \frac{\partial}{\partial v}a = -\frac{1}{2}\frac{\partial^2}{\partial v^2}\sigma^2(v) + a\frac{\partial}{\partial v} + \frac{\partial a}{\partial v} \tag{7.75}$$

and

$$P_F(v, v_0; t) = \langle v \left| e^{-(t-t_0)H} \right| v_0 \rangle .$$

Note that in the discretization of the Langevin equation

$$v = v' + \epsilon \left(a\left(v'\right) + \sigma\left(v'\right) R\right) \neq v' + \epsilon(a(v) + \sigma(v)R),$$

due to singular nature of the white noise.

Consider for example

$$a\left(v'\right) = a\left(v + \epsilon\sigma\left(v'\right) R + 0(\epsilon)\right) = a(v) + \epsilon\sigma\left(v'\right) Ra'(v) + 0(\epsilon)$$
$$\simeq a(v) + \epsilon^{1/2}\sigma(v) + \cdots$$

The $0(\epsilon^{1/2})$ term changes the Langevin equation since

$$\left(\frac{v - v'}{\epsilon}\right) = \left(a\left(v'\right) + \sigma\left(v'\right) R + \epsilon\sigma Ra'(v)\right)^2$$
$$\Rightarrow \left(\frac{dv}{dt}\right)^2 = (a + \sigma R)^2 + 2\epsilon\sigma^2 a' R^2 + 0\left(\epsilon^2\right)$$
$$\neq (a + \sigma R)^2.$$

The backward probability distribution is given by the backward Fokker–Planck equation results from the same discretization of the Langevin equation as the forward case. The only difference being that for the backward case one integrates over the later velocity $v(t+\epsilon) \equiv v$, which yields

$$P_B(v_T, v'; t) = \int dvE\left[\delta\left[v - v' - \epsilon(a(v') + \sigma(v')R_t)\right]\right] P(v_T, v, ; t + \epsilon)$$
$$= P\left(v_T, v', t + \epsilon\right) + \epsilon\left[a(v)\frac{\partial P}{\partial v} + \frac{1}{2}\sigma^2(v)\frac{\partial^2 P}{\partial v^2}\right].$$

The backward Fokker–Planck Hamiltonian is hence

$$\frac{\partial P_B}{\partial t} = H_B P_B, \quad H_B = -\frac{\sigma^2}{2}\frac{\partial^2}{\partial v^2} - a\frac{\partial}{\partial v}.$$

Note that for the backward FP time is given by $-t$, that is, time is running backward, and consequently

$$\frac{1}{\epsilon}\left[P_B(v_T, v; t) - P_B(v_T, v; t + \epsilon)\right] = -\frac{\partial P_B}{\partial(-t)} = -H_B P_B$$
$$\Rightarrow \quad \frac{\partial P_B}{\partial t} = H_B P_B.$$

To write P_B as an evolution kernel, note that in Dirac's notation |initial⟩ always represents the |starting state⟩ and the ending state is given by ⟨final| = ⟨ending state|. Hence

$$P_B(v_T, v; t) = \left\langle v \left| e^{-(T-t)\hat{H}} \right| v_T \right\rangle, \quad P_B(v_T, v; T) = \delta(v - v_T),$$

where

$$H = H_B = -\frac{\sigma^2}{2}\frac{\partial^2}{\partial v^2} - a\frac{\partial}{\partial v}.$$

7.11 Pseudo-Hermitian Fokker–Planck Hamiltonian

For non-Hermitian operators, the rule for the action of operators is that all operators act on the *dual vectors*; that is

$$\left\langle v|\hat{H}|\psi\right\rangle = \hat{H}\left(v, \frac{\partial}{\partial v}\right)\langle v|\psi\rangle = H(v, \partial_v)\psi(v).$$

The definition of Hermitian conjugation is given by

$$\left\langle \psi|\hat{H}|v\right\rangle = \left(\left\langle v|\hat{H}^\dagger|\psi\right\rangle\right)^* = (H^\dagger(v, \partial_v)\psi(v))^*.$$

The forward Fokker–Planck Hamiltonian is given by the Hermitian conjugation of the backward Fokker–Planck Hamiltonian as follows:

$$H_F = H_B^\dagger = -\frac{1}{2}\frac{\partial^2}{\partial v^2}\sigma^2 + \frac{\partial}{\partial v}a \neq H_F \tag{7.76}$$
$$: \text{non-Hermitian.}$$

H_F has two sources of non-Hermiticity, namely [Risken (1988)]:

1. The volatility depends on velocity, that is $\sigma = \sigma(v)$.
2. The first partial derivative $a\,\partial/\partial v$ term is not Hermitian.

An effective Hermitian Fokker–Planck Hamiltonian can be obtained for one dimension in which volatility $\sigma(v)$ is made into a constant by a change of variables and the first derivative term is removed by a similarity transformation.

Consider the backward Fokker–Planck Hamiltonian given by

$$H = -\frac{1}{2}\sigma^2(v)\partial_v^2 - a(v)\partial_v, \quad \partial_v \equiv \frac{\partial}{\partial v}, \tag{7.77}$$

where both σ and a depend on the velocity v. Change variables from v to w such that

$$H = -\frac{1}{2}\sigma_0^2\partial_w^2 + \alpha(w)\partial_w,$$

where σ_0^2 is chosen to be a constant, independent of v.

Using the chain rule for differentiation yields

$$\frac{\partial^2}{\partial v^2} = \frac{\partial w}{\partial v}\frac{\partial}{\partial w}\left(\frac{\partial w}{\partial v}\frac{\partial}{\partial w}\right) = \left(\frac{\partial w}{\partial v}\right)^2 \frac{\partial^2}{\partial w^2} + \frac{\partial w}{\partial v}\left[\frac{\partial}{\partial w}\left(\frac{\partial w}{\partial v}\right)\right]\frac{\partial}{\partial w}. \tag{7.78}$$

Choose constant σ_0^2,

$$\sigma_0^2 = \left(\frac{\partial w}{\partial v}\right)^2 \sigma^2,$$

which yields the definition for the change of variables from v to w as

$$\frac{dw}{dv} = \frac{\sigma_0}{\sigma(v)} \Rightarrow w = \sigma_0 \int_0^v \frac{dv'}{\sigma(v')}. \tag{7.79}$$

Hence, the Hamiltonian in Eq. 7.77 yields from Eqs. 7.78 and 7.79

$$\begin{aligned} H &= -\frac{1}{2}\sigma^2(v)\partial_v^2 - a(v)\partial_v \\ &= -\frac{1}{2}\sigma_0^2\frac{\partial^2}{\partial w^2} - \frac{1}{2}\sigma^2 \cdot \frac{\sigma_0}{\sigma(v)} \cdot \left[\frac{\partial}{\partial w}\left(\frac{\sigma_0}{\sigma(v)}\right)\right]\frac{\partial}{\partial w} - a(v)\frac{\partial w}{\partial v}\frac{\partial}{\partial w} \\ &= -\frac{\sigma_0^2}{2}\partial_w^2 + \alpha(w)\partial_w, \end{aligned} \tag{7.80}$$

where

$$\alpha(w) = \frac{\sigma_0}{\sigma}\left[\frac{1}{2}\sigma_0\frac{\partial\sigma}{\partial w} - a\right].$$

The volatility σ_0 is constant and hence the first term in the Hamiltonian given in Eq. 7.80 is Hermitian. A similarity transformation is now defined that removes the first order derivative term in Eq. 7.80.

For an abitrary function $\Phi = \Phi(w)$, note that

$$\begin{aligned} e^{-\Phi}\left(-\frac{1}{2}\sigma_0^2\partial_w^2\right)e^{\Phi} &= -\frac{\sigma_0^2}{2}\left(\Phi'' + \Phi'^2 + 2\Phi'\partial_w + \partial_w^2\right), \\ -\frac{\sigma_0^2}{2}\partial_w^2 - \frac{\sigma_0^2}{2}(2\Phi'\partial_w) &= e^{-\Phi}\left(-\frac{1}{2}\sigma_0^2\partial_w^2 + \frac{\sigma_0^2}{2}\Phi'' + \frac{\sigma_0^2}{2}\Phi'^2\right)e^{\Phi}, \end{aligned} \tag{7.81}$$

where $\Phi' = d\Phi/dw$. Choose Φ such that

$$\begin{aligned} -\frac{\sigma_0^2}{2}\left(2\Phi'\partial_w\right) &= \alpha(w)\partial_w \\ \Rightarrow \Phi' = -\frac{1}{\sigma_0^2}\alpha \;\Rightarrow\; \Phi &= -\frac{1}{\sigma_0^2}\int_0^w \alpha(w)dw. \end{aligned} \tag{7.82}$$

Hence from Eqs. 7.81 and 7.82

$$-\frac{\sigma_0^2}{2}\partial_w^2 + \alpha(w)\partial_w = e^{-\Phi}\left(-\frac{1}{2}\sigma_0^2\partial_w^2 + \frac{\sigma_0^2}{2}\Phi'' + \frac{\sigma_0^2}{2}\Phi'^2\right)e^{\Phi}$$
$$= e^{-\Phi}H_{\text{eff}}e^{\Phi}, \tag{7.83}$$

where the effective Hermitian Hamiltonian is given by

$$H_{\text{eff}} = H_{\text{eff}}^\dagger \;:\; \text{Hermitian}$$
$$= -\frac{1}{2}\sigma_0^2\partial_w^2 + \frac{\sigma_0^2}{2}\Phi'' + \frac{\sigma_0^2}{2}\Phi'^2. \tag{7.84}$$

In summary, the backward and forward Fokker–Planck Hamiltonians are given by

$$H = H_B = -\frac{\sigma_0^2}{2}\partial_w^2 + \alpha(w)\partial_w = e^{-\Phi}H_{\text{eff}}e^{\Phi},$$
$$H_F = H^\dagger = e^{\Phi}H_{\text{eff}}e^{-\Phi} \neq H.$$

Hamiltonian H is *pseudo-Hermitian* since

$$H^\dagger = e^{2\Phi}He^{-2\Phi}.$$

Pseudo-Hermitian Hamiltonians are a special class of operators with many interesting properties, and are studied in great detail in Chapter 14.

The effective Hamiltonian given in Eq. 7.84 can be further simplified using Eq. 7.82 as follows:

$$H_{\text{eff}} = \frac{1}{2}\sigma_0^2\left(-\partial_w^2 - \frac{1}{\sigma_0^2}\alpha' + \frac{1}{\sigma_0^4}\alpha^2\right)$$
$$= \frac{1}{2}\sigma_0^2\left(-\partial_w + \frac{1}{\sigma_0^2}\alpha\right)\left(\partial_w + \frac{1}{\sigma_0^2}\alpha\right)$$
$$= \frac{1}{2}\sigma_0^2 Q^\dagger Q, \tag{7.85}$$

where

$$Q = \partial_w + \frac{1}{\sigma_0^2}\alpha, \quad Q^\dagger = -\partial_w + \frac{1}{\sigma_0^2}\alpha.$$

The eigenfunction equation for H_{eff} is given by

$$H_{\text{eff}}|\psi_n\rangle = E_n|\psi_n\rangle.$$

The eigenvalues E_n of H_{eff} are nonnegative since, from Eq. 7.85

$$E_n = \langle\psi_n|H_{\text{eff}}|\psi_n\rangle = \frac{1}{2}\sigma_0^2\langle\psi_n|Q^\dagger Q|\psi_n\rangle$$
$$= \frac{1}{2}\sigma_0^2||Q|\psi_n\rangle||^2 \geq 0. \tag{7.86}$$

The ground state can be obtained exactly. Note, for $|\psi_0\rangle \equiv |\Omega\rangle$

$$H_{\text{eff}}|\Omega\rangle = 0, \quad E_0 = 0$$
$$\Rightarrow Q\Omega = \left(\partial_w + \frac{1}{\sigma_0^2}\alpha\right)\Omega = 0, \tag{7.87}$$
$$\Omega(w) = \exp\left(-\frac{1}{\sigma_0^2}\int_0^w \alpha\left(w'\right)dw'\right), \tag{7.88}$$

: stationary (time independent) solution.

Since $H_{\text{eff}}|\psi_n\rangle = E_n|\psi_n\rangle$, the eigenfunctions of H are given by

$$H\left[e^{-\Phi}|\psi_n\rangle\right] = e^{-\Phi}H_{\text{eff}}|\psi_n\rangle = E_n\left[e^{-\Phi}|\psi_n\rangle\right].$$

Hence the eigenfunctions and eigenvalues of H are given by

$$e^{-\Phi}|\psi_n\rangle, \quad E_n.$$

In particular, the conditional transition probability is given by

$$P_B(v_T, v; t) = \left\langle v|e^{-(T-t)H}|v_T\right\rangle = e^{\Phi(v)-\Phi(v_T)}\sum_n c_n e^{-E_n(T-t)}, \tag{7.89}$$

where $c_n = \langle v|\psi_n\rangle\langle\psi_n|v_T\rangle$.

Note that, since $E_n \geq 0$, the conditional probability $P_B(v_T, v; t)$ given in Eq. 7.89 is convergent. This is a reflection, in the Hamiltonian framework, of the fact that the path integral expression for $P_B(v_T, v; t)$ is convergent – in spite of the Fokker–Planck Hamiltonian not being Hermitian – since the Lagrangian is a squared quantity due to Gaussian white noise driving the random system.

For $T \to \infty$, since $E_0 = 0$,

$$P_B(v_T; v) \simeq \Omega_v^*(v)\Omega(v_T)e^{\Phi(v)-\Phi(v_T)}. \tag{7.90}$$

7.12 Fokker–Planck path integral

Consider the case $\sigma = \sigma_0 =$ constant; the path integral, from Eq. 7.68, is given by

$$Z_F = \int Dv e^{S_F}. \tag{7.91}$$

The forward Fokker–Planck Hamiltonian, from Eq. 7.75, is given by

$$H_F = -\frac{1}{2}\frac{\partial^2}{\partial v^2}\sigma^2(v) + a\frac{\partial}{\partial v} + \frac{\partial a}{\partial v}.$$

Hence

$$\begin{aligned}\langle v|e^{-\epsilon H_F}|\tilde{v}\rangle &= \int \frac{dp}{2\pi}\langle v\,|e^{-\epsilon H_F}|\,p\rangle\langle p|\tilde{v}\rangle \\ &= N\int \frac{dp}{2\pi}e^{-\epsilon\left(\frac{\sigma^2}{2}p^2+iap+a'\right)}e^{ip(v-v')} = e^{-\frac{\epsilon}{2\sigma^2}(v-\tilde{v}-\epsilon a)}e^{-\epsilon a'} \\ &= Ne^{\epsilon L} \;\Rightarrow\; L = -\frac{1}{2}(\dot{v}-a)^2 - \frac{\partial a}{\partial v}.\end{aligned}$$

Hence, the forward action is given by

$$S_F = -\frac{1}{2}\int dt\frac{1}{\sigma^2}\left(\frac{dv}{dt}-a\right)^2 - \int dt\frac{\partial a}{\partial v}. \tag{7.92}$$

The backward path integral Z_B has the intuitive stochastic quantization discussed in Section 7.9,

$$Z_B = \int DrDR\prod_t \delta\left[\frac{dv}{dt} - a - \sigma R\right]e^{-\frac{1}{2}\int R^2(t)}.$$

Rescaling the white noise variable $R \to R/\sigma$ and doing the path integral over white noise yields

$$Z_B = \int DvDR\delta\left[\dot{v}-a-R\right]e^{-\frac{1}{2}\int\frac{1}{\sigma^2}R^2} = \int Dve^{-\frac{1}{2}\int\frac{1}{\sigma^2}(\dot{v}-a)^2}. \tag{7.93}$$

Recall that

$$H_B = -\frac{1}{2}\sigma\frac{\partial^2}{\partial v^2} - a\frac{\partial}{\partial v}, \quad H_F = H_B^\dagger$$

and yields

$$Z_B = \int Dve^{-\frac{1}{2\sigma_0^2}\int(\dot{v}-a)^2dt} = \langle v|e^{-(T-t)H_B}|v_F\rangle, \tag{7.94}$$

which is the expected result.

Similarly to the arbitrariness in choosing complete basis states, as discussed in Section 3.2, the choice of the coordinates for the integration variables in the path integral is arbitrary. Just as one can make a change in the basis states of Hilbert space by a unitary transformation, one can make a change of the integration variables; the analog of the requirement of unitarity in changing the basis states is that the change of the path integration variables needs to be invertible, and this in turn yields a positive Jacobian of the transformation. More precisely, let the new integration variables be defined by

$$y = y(x) \Rightarrow dy(t) = \int dt' C(t, t') dx(t') \Rightarrow Dy = J[x] DX,$$

where $J = \det[C]$ is the Jacobian of the transformation.

The forward and backward Fokker–Planck Lagrangians differ by the term $\partial a(r)/\partial r$. The difference lies in the manner in which the path integral $\int DR$ is carried out. For the forward Lagrangian, if one repeats the calculation expressed in Eq. 7.93, the integration yields an extra Jacobian term that precisely gives the extra term in Eq. 7.92, namely $\partial a(r)/\partial r$, that is present in the forward Lagrangian [Zinn-Justin (1993)].

7.13 Summary

A classical random process driven by white noise was addressed and shown to be described by the formalism of path integrals and Hamiltonians.

Gaussian white noise, that forms the backbone of the study of stochastic processes, is based on Gaussian integration, and this was briefly reviewed. Ito calculus and the Wilson expansion were seen to follow from the properties of Gaussian stochastic processes, with the Wilson expansion discussed in this chapter being the property of Gaussian quantum fields.

The linear and nonlinear Langevin equation was analyzed as a stochastic differential equation and a number of exact and approximate results were obtained. In particular, the mean field approximation led to many interesting results for the nonlinear Langevin equation, including the formation of structures far from equilibrium.

Stochastic quantization was defined based on Gaussian white noise and the path integral for the Langevin equation was derived. The rest of the chapter was focused on the Fokker–Planck equation, and the Fokker–Planck Hamiltonian was shown to be a pseudo-Hermitian Hamiltonian.

Part three

Discrete degrees of freedom

8
Ising model

The simplest possible quantum degree of freedom has only *two discrete* possible values and is called an Ising spin. This could be a spin system having only two possible values, namely "up" or "down," or this could be a quantum particle that can occupy only two possible positions.

To simplify the problem to its bare essentials, a two-state system is studied in zero-dimension space and in Euclidean time; the two-state system occupies a *single point* and, as it evolves in time, it can point either up or down. Furthermore, time evolution is simplified by discretizing time into a finite time lattice.

The two-state degree of freedom in zero-dimension space propagating on a time lattice is called the Ising model. The one-dimensional Ising model is a toy model that is studied to develop the essential ideas of quantum mechanics. The concepts of state space, Hamiltonian, evolution kernel, and path integration have a discrete and simple representation in the Ising model.

The Ising model can be viewed as a statistical mechanical system in thermal equilibrium, with the Ising spins occupying a one-dimensional space lattice, and hence is called the one-dimensional Ising model. The Ising model can be defined on higher dimensional lattices and forms one of the bedrocks of theoretical statistical mechanics.

In Sections 8.1 to 8.5 the degree of freedom, state space, Hamiltonian, path integral and correlator for the Ising model are discussed. In Section 8.6, the technique of spin decimation is illustrated using an Ising model with coupling depending on the lattice site. The last Section 8.7 discusses how the one-dimensional can be generalized by considering the Ising model on a $2 \times N$ lattice.

8.1 Ising degree of freedom and state space

The Ising degree of freedom is studied in some detail to motivate the results for more complicated degrees of freedom.

The electron has an *intrinsic* spin 1/2 angular momentum; the electron's spin in three dimensions can point in any direction, which is specified by two angular variables; the (z-component of the) spin of the electron can either point up or down, and hence forms a double-valued system. If one observes only the z-component of the spin, then the spin 1/2 degree of freedom can be represented by the Ising spin degree of freedom. In quantum information theory, the Ising degree of freedom is called a qubit since it is the quantum generalization of the classical "bit" of information science.

The quantum state of the Ising degree of freedom is mathematically described by specifying the state vector $|\psi\rangle$, and belongs to the state space $\mathcal{V}$. The general state vector of the Ising spin can be described by basis states $|\mu\rangle$ with $\mu = \pm 1$ and are sometimes called "up" $= +1$ or "down" $= -1$ states. The two distinct states corresponding to the two possible values of the Ising spin (degree of freedom) should be "orthogonal" to each other, as in Eq. 2.3, so that being in one basis state is completely different from being in the other basis state,

$$|\psi_{\text{up}}\rangle = |u\rangle = |+1\rangle, \quad |\psi_{\text{down}}\rangle = |d\rangle = |-1\rangle, \quad \langle u|d\rangle = 0.$$

A general state vector of the Ising spin can be represented by a collection of complex valued two-dimensional vectors. One should note that the state space $\mathcal{V}$ has no relation with physical space, but rather is a mathematical construction for describing the state vector of the Ising spin.

One possible representation of the Ising basis states is to consider $|u\rangle$ and $|d\rangle$ as the basis vectors of a two-dimensional vector space, as shown in Figure 8.1. The distinct basis state vectors and their dual basis vectors have the representation

$$|u\rangle = \begin{bmatrix} 1 \\ 0 \end{bmatrix}, \quad |d\rangle = \begin{bmatrix} 0 \\ 1 \end{bmatrix}$$
$$\langle u| = \begin{bmatrix} 1 & 0 \end{bmatrix}, \quad \langle d| = \begin{bmatrix} 0 & 1 \end{bmatrix}. \tag{8.1}$$

The *inner product* of two vectors is defined as

$$\langle d|u\rangle = \begin{bmatrix} 0 & 1 \end{bmatrix} \cdot \begin{bmatrix} 1 \\ 0 \end{bmatrix} = 0. \tag{8.2}$$

Figure 8.1 The Ising spin is represented by up and down vectors.

The completeness equation expresses the fact that a collection of basis vectors forms a complete basis of a linear vector space and is given in Eq. 3.2 for a degree of freedom having infinitely many discrete values. For the Ising spin, the completeness equation for the two-dimensional state space is realized, as follows, by the *outer product* of two vectors

$$\begin{aligned} |u\rangle\langle u| + |d\rangle\langle d| &= \begin{bmatrix} 1 \\ 0 \end{bmatrix} \otimes \begin{bmatrix} 1 & 0 \end{bmatrix} + \begin{bmatrix} 0 \\ 1 \end{bmatrix} \otimes \begin{bmatrix} 0 & 1 \end{bmatrix} \\ &= \begin{bmatrix} 1 & 0 \\ 0 & 0 \end{bmatrix} + \begin{bmatrix} 0 & 0 \\ 0 & 1 \end{bmatrix} = \begin{bmatrix} 1 & 0 \\ 0 & 1 \end{bmatrix} = \mathbb{I}. \end{aligned} \tag{8.3}$$

8.1.1 Ising spin's state space $\mathcal{V}$

A spin pointing purely up $|u\rangle$ or down $|d\rangle$ is the closest that, in quantum mechanics, one can get to a classical object since, on being observed, it is certain to be in either the up or down. The general state vector of the Ising spin is more subtle: the two basis states can be *superposed*, as discussed after Eq. 2.13, yielding an indeterminate state that simultaneously points up and down, but with only a certain *likelihood*.[1] Superposing a quantum spin pointing up with one pointing down yields a state vector given by

$$\begin{aligned} |\psi\rangle &= \alpha|u\rangle + \beta|d\rangle, \quad \langle\psi| = \alpha^*\langle u| + \beta^*\langle d| \\ \langle\psi|\psi\rangle &= |\psi|^2 = |\alpha|^2 + |\beta|^2, \end{aligned} \tag{8.4}$$

where α and β are *complex numbers*.

For a probabilistic interpretation of $|\psi\rangle$, the total probability has to be unity, and one obtains the normalization of the state vector given by

$$\langle\psi|\psi\rangle = |\psi|^2 = |\alpha|^2 + |\beta|^2 = 1. \tag{8.5}$$

The coefficients now have the physical interpretation

$$\begin{aligned} |\alpha|^2 &= \text{probability that the spin is pointing up,} \\ |\beta|^2 &= \text{probability that the spin is pointing down.} \end{aligned}$$

The coefficients α, β parameterize the state space $\mathcal{V}_2$ of the two-state system. Note that state vector $|\psi\rangle$ given in Eq. 8.4 cannot be in general represented by a two-dimensional unit vector in two-dimensional Euclidean space $\Re^2$, since only vectors with real coefficients can be drawn in $\Re^2$. In contrast, α, β are complex numbers and, from Eq. 8.5, it can be seen that their possible values constitute a three-dimensional sphere S^3.

[1] The concept of quantum superposition is discussed in-depth in Section 4.3.

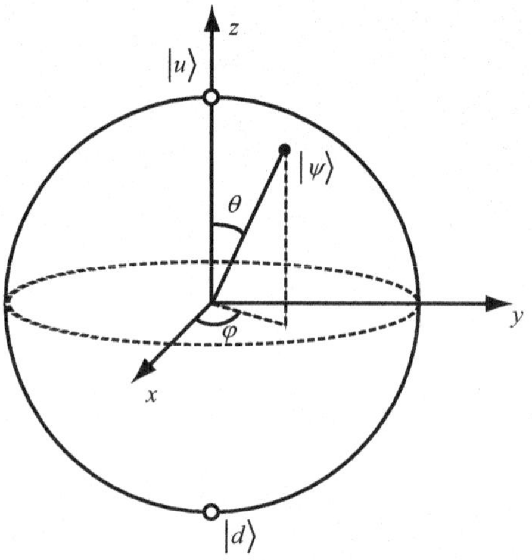

Figure 8.2 Bloch sphere for the two-dimensional state space.

There is, however, a redundancy in this description since the coefficients α, β can both be re-scaled by the same constant phase and yield

$$|\tilde{\psi}\rangle = e^{i\phi}|\psi\rangle \ \Rightarrow \ \langle\tilde{\psi}|\tilde{\psi}\rangle = \langle\psi|\psi\rangle. \tag{8.6}$$

Hence, both state vectors $|\psi\rangle$ and $|\tilde{\psi}\rangle$ provide the same description; in other words states linked by a global pure phase, namely $|\psi\rangle \to e^{i\phi}|\psi\rangle$, are *equivalent*.[2]

The phase ϕ forms a space that is isomorphic to a circle $\mathcal{S}^1$. Hence, one needs to "divide out" $\mathcal{S}^3$ by $\mathcal{S}^1$ to form equivalence classes of state vectors, yielding $\mathcal{S}^3/\mathcal{S}^1 = \mathcal{S}^2$, which is a two-dimensional surface.[3]

In summary, all unique state vectors of the Ising spin are parameterized by the angles of the Bloch sphere $\mathcal{S}^2$; each point on the Bloch sphere corresponds to a state vector of the two-state system state space $\mathcal{V}$.

8.1.2 Bloch sphere

The state space $\mathcal{V}$ of the Ising spin is isomorphic to the Bloch sphere $\mathcal{S}^2$, shown in Figure 8.2. The Bloch sphere allows for a specific representation of the most general state vector for the two-state system.

Consider the spherical polar coordinates θ, ϕ shown in Figure 8.2. An arbitrary normalized two-state ket vector $|\psi\rangle$, shown in Figure 8.2, is given by the following mapping of the Bloch sphere into two-dimensional state space:

[2] ϕ being a pure phase means that it is real.

[3] To prove this result, one needs to construct $\mathcal{S}^3$ by a Hopf fibration, using the mathematics of fiber bundles, by fibrating the base manifold $\mathcal{S}^2$ with fibers given by $\mathcal{S}^1$.

$$|\psi(\theta,\phi)\rangle = \cos\left(\frac{\theta}{2}\right)\begin{bmatrix}1\\0\end{bmatrix} + e^{i\phi}\sin\left(\frac{\theta}{2}\right)\begin{bmatrix}0\\1\end{bmatrix}, \tag{8.7}$$

$$0 \le \theta \le \pi, \quad 0 \le \phi \le 2\pi.$$

The basis states have been chosen to reflect the fact that $\theta = 0$ points in the "up" direction and $\theta = \pi$ points in the "down" direction, as shown in Figure 8.2; more precisely, Eq. 8.7 yields

$$|\psi(0,\phi)\rangle = \begin{bmatrix}1\\0\end{bmatrix} = |u\rangle,$$

$$|\psi(\pi,\phi)\rangle = e^{i\phi}\begin{bmatrix}0\\1\end{bmatrix} \equiv \begin{bmatrix}0\\1\end{bmatrix} = |d\rangle. \tag{8.8}$$

Note that the ambiguity in the definition of the state vector $|\psi(\theta,\phi)\rangle$ for $\theta = \pi$ is automatically removed for quantum states since states differing by a pure phase are equivalent, as shown in Eq. 8.6.

The presence of the phase $e^{i\phi}$ in Eq. 8.7 shows that vector $|\psi(\theta,\phi)\rangle$ does not correspond to a two-dimensional unit vector in Euclidean space, which has only real-valued components. In fact, every state vector $|\psi(\theta,\phi)\rangle$ corresponds to a (unique) unit *three-dimensional* Euclidean vector, as shown in Figure 8.2, and is uniquely parameterized by the coordinates $\theta \in [0,\pi]$ and $\phi \in [0, 2\pi]$. In summary, every point on the surface of the Bloch sphere $\mathcal{S}^2$ corresponds to a unique state vector in $\mathcal{V}_2$.[4]

8.2 Transfer matrix

The state function $|\psi_t\rangle$ is evolved in time by the Hamiltonian operator H and is given by

$$|\psi_t\rangle = e^{-itH/\hbar}|\psi_0\rangle.$$

It is generally more efficient to analytically continue Minkowski time t to imaginary Euclidean time τ by the mapping $t = -i\hbar\tau$, which yields

$$|\psi_\tau\rangle = e^{-\tau H}|\psi_0\rangle.$$

To reduce the system to its bare essential ingredients consider discrete Euclidean time $\tau = n\epsilon$, with $n = 0, 1, 2, \ldots M$, shown in Figure 8.3. There is an Ising spin μ_n for lattice site n; the spin variable takes two values ± 1, indicated by an arrow pointing up or down. There are in total 2^{M+1} possible spin configurations for an $M + 1$ lattice, and Figure 8.4 shows some of the possible Ising configurations for five lattice sites.

[4] The ambiguity for the special value of $\theta = \pi$ is removed by Eq. 8.8.

Figure 8.3 Ising lattice.

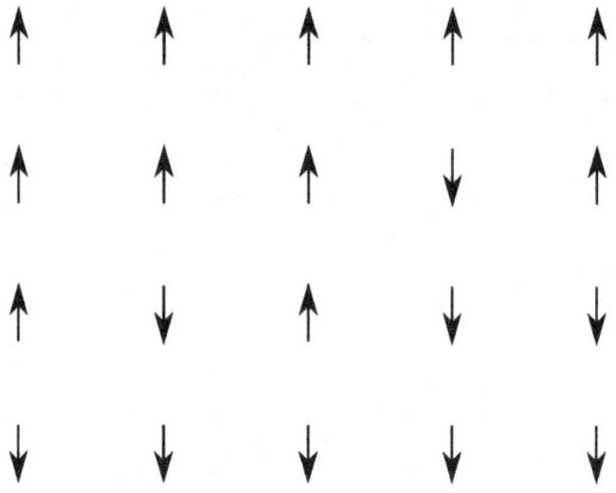

Figure 8.4 A few Ising spin configurations for five sites.

The system evolves in time through jumps over a discrete time interval ϵ. The time evolution of the state function $|\psi_\tau\rangle \to |\psi_{n\epsilon}\rangle \equiv |\psi_n\rangle$ is hence given by

$$|\psi_{n+1}\rangle = e^{-\epsilon H}|\psi_n\rangle = L|\psi_n\rangle$$
$$\Rightarrow L \equiv e^{-\epsilon H} \quad : \quad \text{transfer matrix.}$$

In statistical mechanics L is called the *transfer matrix*.

The $|\mu\rangle$ basis yields

$$\begin{aligned}\langle\mu|\psi_{n+1}\rangle &= \langle\mu|L|\psi_n\rangle \\ &= \sum_{\mu'=\pm}\langle\mu|L|\mu'\rangle\langle\mu'|\psi_n\rangle.\end{aligned}$$

To define the dynamics of the system $\langle\mu|L|\mu'\rangle$ needs to be specified. Since $\mu^2 = 1$ one has the most general expansion for a Hermitian L given by

$$\begin{aligned}L\left(\mu,\mu'\right) &= \langle\mu|L|\mu'\rangle \\ &= a + b(\mu+\mu') + c\mu\mu'.\end{aligned}$$

A more convenient representation is given by (discarding an overall constant)

$$L\left(\mu,\mu'\right) = e^{K\mu\mu' + \frac{h}{2}(\mu+\mu')}.$$

The constant K quantifies the strength of interaction of nearest neighboring spins, and the constant h is the strength of the external magnetic field that the spins are subjected to. Note all the quantities in the Ising model are *dimensionless*, and the size of the lattice spacing (or time step ϵ) is encoded in the coupling constant K; this aspect of the lattice model is discussed later in Section 9.6.

The Heisenberg operators $X_H(t)$ satisfy the evolution equation

$$\frac{\partial \hat{X}_H(t)}{\partial t} = \frac{i}{\hbar}\left[H, \hat{X}_H(t)\right]. \tag{8.9}$$

From Eq. 8.9, the Heisenberg operators have the evolution

$$\hat{X}_H(t) = e^{itH/\hbar}\hat{X}e^{-itH/\hbar}, \tag{8.10}$$

where $\hat{X}$ is the Schrödinger operator. Euclidean time $\tau = it/\hbar$ yields

$$\hat{X}_H(\tau) = e^{\tau H}\hat{X}e^{-\tau H}. \tag{8.11}$$

For discrete time, $\tau \to k\epsilon$ and $L = e^{-\epsilon H}$. The Heisenberg spin operator $\hat{\mu}_k$ obeys the lattice version of Eq. 8.11 given by

$$\hat{\mu}_k = L^{-k}\hat{\mu}L^k. \tag{8.12}$$

The operator $\hat{\mu}$ is the co-ordinate Schrödinger operator for the Ising discrete degree of freedom; its explicit representation, using the completeness equation given in Eq. 8.3, is given by

$$\hat{\mu} = \hat{\mu}\mathbb{I} = \sum_{\mu} \hat{\mu}|\mu\rangle\langle\mu|. \tag{8.13}$$

Since $\hat{\mu}|\mu\rangle = \mu|\mu\rangle$, one has

$$\begin{aligned}\hat{\mu} &= \sum_{\mu} \mu|\mu\rangle\langle\mu| = |+\rangle\langle+| - |-\rangle\langle-| \\ &= \begin{pmatrix}1 & 0\\ 0 & 0\end{pmatrix} - \begin{pmatrix}0 & 0\\ 0 & 1\end{pmatrix} = \begin{pmatrix}1 & 0\\ 0 & -1\end{pmatrix}.\end{aligned} \tag{8.14}$$

Note that as expected $\hat{\mu}^2 = \mathbb{I}$.

8.3 Correlators

A typical quantity of interest is the correlation function, called correlator for brevity. If a disturbance is created at time k, one would like to know for how long does the disturbance propagate. In general, for a d-dimensional system, one would like to know what is the system's correlation length, which quantifies the extent to which any disturbance propagates in the system.

For the open chain lattice the normalized correlator is defined by

$$C_r = \frac{1}{Z}\langle\mu|L^{N/2}\hat{\mu}_{k+r}\hat{\mu}_k L^{N/2}|\mu'\rangle, \tag{8.15}$$

where Z is a normalization constant and $\hat{\mu}_k$ is the Heisenberg operator at time k. The correlator is given in Figure 8.5.

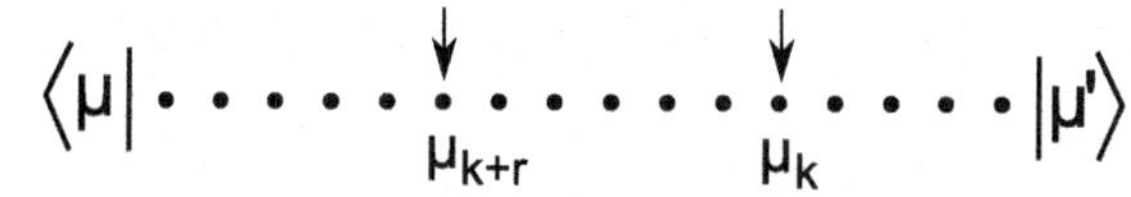

Figure 8.5 Ising correlation function with open boundary conditions.

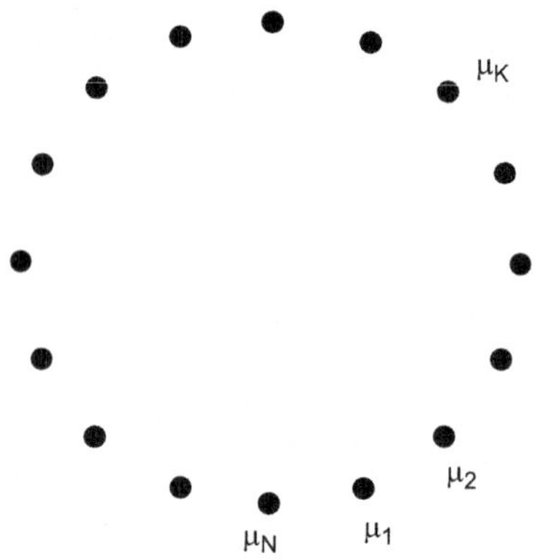

Figure 8.6 Periodic lattice with $\mu_{N+1} = \mu_1$.

For $r = 0$, since $\hat{\mu}_k^2 = 1$ one expects $C_0 = 1$ to be the maximum value of the self-correlation. Hence

$$C_0 = 1 \;\Rightarrow\; Z = \langle\mu|L^N|\mu'\rangle.$$

From Eq. 8.12, the Heisenberg spin operators yield

$$C_r = \frac{1}{Z}\langle\mu|L^{N/2-(k+r)}\hat{\mu}L^r\hat{\mu}L^{N/2+k}|\mu'\rangle, \tag{8.16}$$

: open chain with μ,μ' end-points.

8.3.1 Periodic lattice

To simplify the computation consider the case of periodic boundary conditions such that $\mu = \mu'$ and a sum is carried out over the states μ with the periodic boundary condition that $\mu_1 = \mu_{N+1}$. The periodic lattice is shown in Figure 8.6. The correlator simplifies since the trace is cyclic, that is $tr(ABC) = tr(CAB)$.

The correlator is defined by

$$\begin{aligned} C_r &= \frac{1}{Z}\sum_{\mu=\pm 1}\langle\mu|L^{N/2}\hat{\mu}_{k+r}\hat{\mu}_k L^{N/2}|\mu\rangle \\ &= \frac{1}{Z}tr\left(\hat{\mu}_{k+r}\hat{\mu}_k L^N\right). \end{aligned} \tag{8.17}$$

Using the cyclicity of the trace, the correlator is given by

$$C_r = \frac{1}{Z} tr\left(L^{-(k+r)} \hat{\mu} L^r \hat{\mu} L^{N+k}\right)$$
$$= \frac{1}{Z} tr\left(L^{N-r} \hat{\mu} L^r \hat{\mu}\right). \tag{8.18}$$

The normalization is given by $C_0 = 1$. Hence

$$Z = tr\left(L^N\right) : \text{ partition function,} \tag{8.19}$$

and the normalized correlator for the periodic lattice is

$$C_r = \frac{1}{Z} tr\left(L^N \hat{\mu}_{k+r} \hat{\mu}_k\right). \tag{8.20}$$

8.4 Correlator for periodic boundary conditions

The correlator is explicitly evaluated for the case of $h = 0$. One has that

$$L\left(\mu, \mu'\right) = e^{K\mu\mu'}, \tag{8.21}$$

with

$$L = \begin{pmatrix} L_{11} & L_{1-1} \\ L_{-11} & L_{-1-1} \end{pmatrix}$$
$$= \begin{pmatrix} e^K & e^{-K} \\ e^{-K} & e^K \end{pmatrix}. \tag{8.22}$$

The eigenfunction equation is given by

$$L|\lambda\rangle = \lambda|\lambda\rangle, \tag{8.23}$$
$$\det|L - \lambda| = 0, \tag{8.24}$$

and yields

$$\lambda_1 = e^K + e^{-K}, \quad \lambda_2 = e^K - e^{-K}. \tag{8.25}$$

Hence the partition function, from Eq. 8.19, is given by

$$Z = tr\left(L^N\right) = \lambda_1^N + \lambda_2^N. \tag{8.26}$$

The eigenvectors are

$$|\lambda_1\rangle = \frac{1}{\sqrt{2}} \begin{pmatrix} 1 \\ 1 \end{pmatrix} \equiv |1\rangle, \tag{8.27}$$

$$|\lambda_2\rangle = \frac{1}{\sqrt{2}} \begin{pmatrix} 1 \\ -1 \end{pmatrix} \equiv |2\rangle, \tag{8.28}$$

$$\mathbb{I} = \sum_{i=1}^{2} |i\rangle\langle i| \quad : \quad \text{completeness equation.} \tag{8.29}$$

Hence the transfer matrix is given by

$$L = \sum_{i=1}^{2} \lambda_i |i\rangle\langle i|. \tag{8.30}$$

The correlator is given by

$$\begin{aligned} C_r &= \frac{1}{Z} tr\left(L^{N-r}\hat{\mu}L^r\hat{\mu}\right) \\ &= \frac{1}{Z}\sum_{i,j=1}^{2} \lambda_i^{N-r}\lambda_j^r tr\left(|i\rangle\langle i|\hat{\mu}|j\rangle\langle j|\hat{\mu}\right) \\ &= \frac{1}{Z}\sum_{i,j=1}^{2} \lambda_i^{N-r}\lambda_j^r \langle i|\hat{\mu}|j\rangle\langle j|\hat{\mu}|i\rangle, \end{aligned} \tag{8.31}$$

since $\mathrm{tr}(M|a\rangle\langle b|) = \langle b|M|a\rangle$.

The coordinate operator is Hermitian; hence

$$\begin{aligned} \langle j|\hat{\mu}|i\rangle &= \langle j\hat{\mu}|i\rangle^* = \langle i|\hat{\mu}^\dagger|j\rangle \\ &= \langle i|\hat{\mu}|j\rangle. \end{aligned} \tag{8.32}$$

From Eqs. 8.14, 8.27, and 8.28

$$\langle 1|\mu|1\rangle = 0 = \langle 2|\mu|2\rangle, \tag{8.33}$$

$$\langle 1|\mu|2\rangle = 1 = \langle 2|\mu|1\rangle. \tag{8.34}$$

The correlator is given by

$$C_r = \frac{1}{Z}\sum_{ij}\left(\langle i|\hat{\mu}|j\rangle\right)^2 \lambda_i^{N-r}\lambda_j^r. \tag{8.35}$$

Hence

$$\begin{aligned} C_r &= \frac{1}{Z}\left\{\lambda_1^{N-r}\lambda_2^r + \lambda_2^{N-r}\lambda_1^r\right\} \\ &= \frac{\lambda_1^N}{\lambda_1^N + \lambda_2^N}\left[\tanh^r K + \tanh^{N-r} K\right] \\ &= \frac{\tanh^r K + \tanh^{N-r} K}{1 + \tanh^N K}. \end{aligned} \tag{8.36}$$

Note that $C_0 = 1$ as expected. For $N \to \infty$, $\tanh^N K \to 0$; hence

$$C_r = \tanh^r K = e^{-r/\xi}, \tag{8.37}$$

where the correlation length(time) ξ is given by

$$\xi = -\frac{1}{\ln(\tanh K)}, \tag{8.38}$$

$$\sim \begin{cases} \frac{1}{2}e^{2K} & K \gg 1 \\ \frac{1}{-\ln(K)} & K \sim 0 \end{cases}. \tag{8.39}$$

From the above it can be seen that for $K \gg 1$ the system becomes strongly correlated, whereas for $K \to 0$ the system has short-time correlation.

8.4.1 Correlator as vacuum expectation values

Since $\lambda_1 > \lambda_2$, the state $|1\rangle$ has the lower energy. Hence, for $N \to \infty$, and labeling $|1\rangle = |\Omega\rangle =$ the vacuum state, yields

$$L^N = \lambda_1^N \left(|\Omega\rangle\langle\Omega| + O\left(\left(\frac{\lambda_2}{\lambda_1}\right)^N\right)\right). \tag{8.40}$$

Hence, for $N \to \infty$, the periodic correlator given in Eq. 8.17 reduces to

$$\begin{aligned} C_r &= \frac{1}{Z}\lambda_1^N tr(|\Omega\rangle\langle\Omega\langle|\hat{\mu}_r\hat{\mu}_0) \\ &= \langle\Omega|\hat{\mu}_r\hat{\mu}_0|\Omega\rangle. \end{aligned} \tag{8.41}$$

Recall from Eq. 6.40 that time ordering of operators is given by

$$T(\hat{O}_t\hat{O}_{t'}) = \begin{cases} \hat{O}_{t'}\hat{O}_t & t' > t \\ \hat{O}_t\hat{O}_{t'} & t' < t \end{cases}.$$

Hence, it follows, in general, that

$$C_r = \langle\Omega|T(\hat{\mu}_{k+r}\hat{\mu}_k)|\Omega\rangle. \tag{8.42}$$

8.5 Ising model's path integral

The derivation given in Section 8.4 is for the partition function and correlator of the periodic chain L. Another derivation is now given (for a periodic lattice of size N) of the partition function and the correlator by summing over all the possible 2^N configurations. This derivation is an example of evaluating a discrete path integral.

8.5.1 Ising partition function

The partition function is given by $Z = tr(L^N)$; one can evaluate the trace for Z by inserting, N times, the complete set of states given in Eq. 8.1, yielding

$$\sum_{\mu=\pm 1} |\mu\rangle\langle\mu| = \mathbb{I}, \quad |\mu\rangle\,, \quad \mu = -1, +1.$$

Introducing an index μ_i on μ to distinguish the different completeness equations that are inserted into the trace to evaluate Z, one can write L^N as an N-fold matrix product and the trace is achieved by having periodic boundary conditions on the spin variables, namely $\mu_i = \mu_{i+N}$; in particular $\mu_1 = \mu_{N+1}$. Hence the partition function is given by

$$\begin{aligned} Z &= tr(L^N) \\ &= \sum_{\mu_N} \langle\mu_N| \underbrace{LLL\cdots LLL\cdots LLL}_{N-times} |\mu_N\rangle \\ &= \Big[\prod_{i=1}^{N} \sum_{\mu_i=\pm 1}\Big]\Big[\prod_{i=1}^{N} \langle\mu_{i+1}|L|\mu_i\rangle\Big] = \Big[\prod_{i=1}^{N} \sum_{\mu_i=\pm 1}\Big] \prod_{i=1}^{N} e^{K\mu_{i+1}\mu_i} \\ &= \sum_{\{\mu\}} e^S \qquad : \text{periodic boundary conditions} \quad \mu_1 = \mu_{N+1}. \end{aligned}$$

The action S is given by

$$S = K\sum_{i=1}^{N} \mu_{i+1}\mu_i. \tag{8.43}$$

Note $\mu_1, \mu_2, \ldots \mu_N$ are discrete random variables taking values of ± 1. There are 2^N possible spin configurations, and $e^{S[\mu_1,\ldots\mu_N]}/Z$ is the (normalized) probability distribution for the occurrence of a particular spin configuration (also called a fluctuation).

The action can be re-written using the special property of the spin variables that $(\mu_n\mu_{n+1})^2 = 1$; this fact yields

$$\begin{aligned} e^S &= e^{K\sum_{n=1}^{N}\mu_{n+1}\mu_n} = \prod_{n=1}^{N}\left[e^{K\mu_n\mu_{n+1}}\right] \\ &= \prod_{n=1}^{N}[\cosh K + \sinh K\mu_n\mu_{n+1}] \\ &= \cosh^N K \prod_{n=1}^{N}[1 + \tanh K\mu_n\mu_{n+1}]\,. \end{aligned} \tag{8.44}$$

Representing $\mu_n \mu_{n+1}$ by a bond connecting the lattice site n with $n+1$, one can see that e^S has an expansion that consists of the sum of all possible combinations of bonds, with the term 1 arising from no bond to the term $\prod_{n=1}^{N} \mu_n$ that has one contribution from every bond.

Consider the simple product

$$
\begin{aligned}
(1+a)(1+b) &= 1+a+b+ab \\
(1+a)(1+b)(1+c) &= 1+a+b+ab+c+ac+bc+abc \\
\ldots\ldots &= \ldots\ldots..
\end{aligned}
\tag{8.45}
$$

Since $\sum_\mu \mu = 0$, we conclude that, at each lattice site n, only even or no spin variables can contribute to the sum for Z. There are only two terms in the expansion for S that satisfy this condition, namely the term with no bonds and the term with all the bonds. Every sum contributes a factor of 2, since $\sum_{u_i=\pm 1} = 2$, and hence $\sum_{\{\mu\}} = 2^N$. The partition function is

$$
\begin{aligned}
Z &= tr\left(L^N\right) = 2^N \cosh^N K[1+\tanh^N K] \\
&= \lambda_1^N + \lambda_2^N
\end{aligned}
$$

and we obtain the expected result given in Eq. 8.26.

8.5.2 Path integral calculation of C_r

The correlator, given by Eq. 8.17, can be written as

$$
\begin{aligned}
C_r &= \frac{1}{Z} tr\left(\hat{\mu}_{k+r}\hat{\mu}_k L^N\right) \\
&= \frac{1}{Z} tr\left(L^{N-r}\hat{\mu} L^r \hat{\mu}\right) \\
&= \frac{1}{Z} \sum_{\mu_k=\pm 1} \langle \mu_k | L^{N-r}\hat{\mu} L^r \hat{\mu} | \mu_k \rangle .
\end{aligned}
$$

Inserting a complete set of states yields

$$
\begin{aligned}
C_r &= \frac{1}{Z} \sum_{\mu_k} \sum_{\mu_{k+r}} \langle \mu_k | L^{N-r}\hat{\mu} | \mu_{k+r} \rangle \langle \mu_{k+r} | L^r | \mu_k \rangle \mu_k \\
&= \frac{1}{Z} \sum_{\mu_k} \sum_{\mu_{k+r}} \mu_{k+r}\mu_k \langle \mu_k | L^{N-r} | \mu_{k+r} \rangle \langle \mu_{k+r} | L^r | \mu_k \rangle \\
&= \frac{1}{Z} \sum_{\{\mu\}} e^{K\sum_{n=1}^{N} \mu_n \mu_{n+1}} \mu_{k+r}\mu_k ,
\end{aligned}
$$

where $\sum_{\{\mu\}}$ is a sum over all possible configurations. Hence

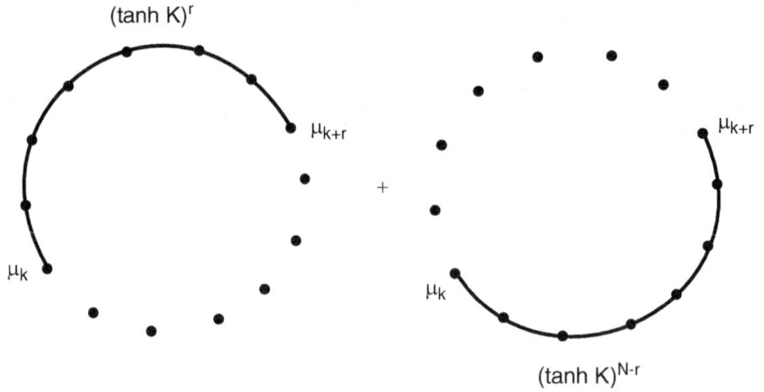

Figure 8.7 The two bond configurations that contribute to the correlation function $\sum_{\{\mu\}} e^S \mu_{k+r}\mu_k/Z$. Note that the correlation function is independent of k.

$$C_r = \frac{1}{Z}\sum_{\{\mu\}} e^S \mu_{k+r}\mu_k \quad \text{: path integral,}$$

where the action is given in Eq. 8.43. Since the lattice is periodic, the correlator C_r does not depend on the index k and hence

$$C_r = \frac{1}{Z}\sum_{\{\mu\}} e^S \mu_r\mu_0.$$

Using the expression Eq. 8.44 for the Ising action yields

$$C_r = \frac{\cosh^N K}{Z}\sum_{\{\mu\}}\prod_n (1 + \tanh K \mu_n\mu_{n+1})\mu_r\mu_0. \tag{8.46}$$

The μ_r and μ_0 terms must be canceled by bonds. Only two terms survive from the product, namely a product of bonds clockwise going from μ_r and μ_0 and another term going counter-clockwise, as shown in Figure 8.7. Hence performing the sum yields, as expected

$$\begin{aligned} C_r &= \frac{2^N \cosh^N K}{Z}\left[\tanh^r K + \tanh^{N-r} K\right] \\ &= \frac{1}{1 + \tanh^N K}\left[\tanh^r K + \tanh^{N-r} K\right]. \end{aligned} \tag{8.47}$$

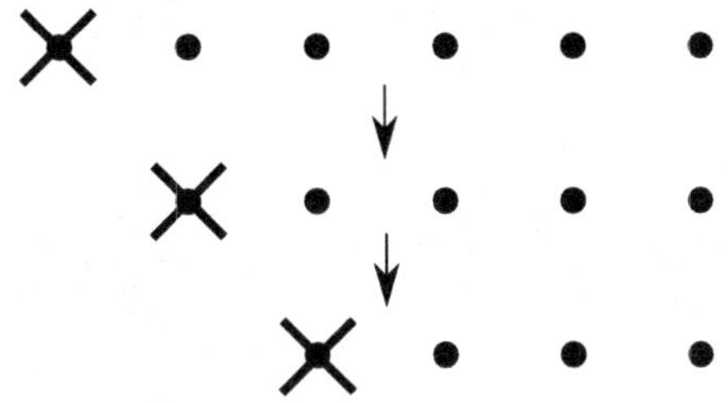

Figure 8.8 The spins are decimated (summed over) from the boundary.

8.6 Spin decimation

Consider an open chain one-dimensional Ising model with coupling constants K_n and the action given by

$$S = \sum_{n=1}^{N-1} K_n \mu_n \mu_{n+1}, \tag{8.48}$$

$$Z = \sum_{\{\mu\}} e^S. \tag{8.49}$$

The partition function Z can be evaluated by summing over the spin μ_N at the boundary, called spin decimation. Then decimate spin μ_{N-1}, μ_{N-2},... all the way to spin μ_1. In symbols one has

$$\begin{aligned} S &= \sum_{n=1}^{N-1} K_n \mu_n \mu_{n+1}, \\ Z &= \prod_{n=1}^{N} \sum_{\mu_n=\pm 1} e^{K_n \mu_n \mu_{n+1}} \\ &= \sum_{\mu_N} e^{K_{N-1}\mu_{N-1}\mu_N} \sum_{\mu_{N-1}} e^{K_{N-2}\mu_{N-2}\mu_{N-1}} \dots \\ &= \left(e^{K_{N-1}\mu_{N-1}} + e^{-K_{N-1}\mu_{N-1}}\right) \sum_{\mu_{N-1}} e^{K_{N-2}\mu_{N-2}\mu_{N-1}} \dots \\ &= 2\cosh(K_{N-1}) \sum_{\mu_{N-1}} e^{K_{N-2}\mu_{N-2}\mu_{N-1}} \dots \end{aligned} \tag{8.50}$$

Recursively decimating the spins from the boundary inwards, as shown in Figure 8.8, yields

$$Z = \left[\prod_{n=1}^{N-1} 2\cosh(K_n)\right] \sum_{\mu_1}$$
$$= 2\left[\prod_{n=1}^{N-1} 2\cosh(K_n)\right].$$

To evaluate $\langle \mu_n \mu_{n+r} \rangle$ note that the correlation function can be re-written by differentiating the action with respect to the coupling constants K_m; one starts with the coupling constants K_n corresponding to the term in the action $K_n \mu_n \mu_{n+1}$ and keeps differentiating on the next lattice site until one reaches site $n + r - 1$. In symbols one has

$$E[\mu_n \mu_{n+r}] = \frac{1}{Z} \sum \mu_n \mu_{n+r} e^S$$
$$= \frac{1}{Z} \frac{\partial}{\partial K_n} \frac{\partial}{\partial K_{n+1}} \cdots \frac{\partial}{\partial K_{n+r-1}} \sum e^S, \tag{8.51}$$
$$= \prod_{\ell=n}^{n+r-1} \frac{\sinh(K_\ell)}{\cosh(K_\ell)} = \prod_{\ell=n}^{n+r-1} \tanh(K_\ell). \tag{8.52}$$

As expected, the limit $K_\ell = K =$ constant yields

$$E[\mu_n \mu_{n+r}] \to \tanh^r K.$$

8.7 Ising model on 2 x N lattice

Consider the Ising model on a two site by N steps in the time direction. At each step there are two spin variables μ_n and μ'_n. Suppose the time lattice is a periodic lattice, that is $\mu_{N+1} = \mu_1$ and $\mu'_{N+1} = \mu'_1$. The action is given by

$$S = K \sum_{n=1}^{N} \mu_n \mu'_n + K \sum_{n=1}^{N} \mu_n \mu_{n+1} + K \sum_{n=1}^{N} \mu'_n \mu'_{n+1} = \sum_{n=1}^{N} \mathcal{L}(n).$$

The action is defined on the Ising "ladder", as shown in Figure 8.9, and yields the Lagrangian

$$\mathcal{L}(n) = \frac{K}{2}\left(\mu_n \mu'_n + \mu_{n+1}\mu'_{n+1}\right) + K\mu_n \mu_{n+1} + K\mu'_n \mu'_{n+1}. \tag{8.53}$$

To simplify the notation, let

$$\mu_{n+1} = \mu, \quad \mu'_{n+1} = \mu'$$
$$\mu_n = \lambda, \quad \mu'_n = \lambda'.$$

Figure 8.9 Ising "ladder" for $2 \times N$ lattice.

This yields

$$\mathcal{L}\left(\mu, \mu'; \lambda\lambda'\right) = \frac{K}{2}\left(\lambda'\lambda + \mu\mu'\right) + K\left(\lambda\mu + \lambda'\mu'\right). \tag{8.54}$$

The partition function is given by

$$Z = \sum_{\{\mu;\mu'\}} e^{S[\mu;\mu']} = \sum_{\{\mu,\lambda\}} e^{S[\mu,\lambda]}.$$

One has to choose a set of basis vectors to write out the matrix elements of the transfer matrix L. Since there are two spins at each time n, the natural basis states for each n are given by the tensor product of the basis states of the two lattice sites, namely $|\mu_n; \mu'_n\rangle \equiv |\mu_n\rangle \otimes |\mu'_n\rangle$. In the simplified notation introduced above, the matrix elements of L are given by $\langle\mu; \mu'|L|\lambda; \lambda'\rangle$.

The explicit construction of the basis vectors $|\lambda\rangle \otimes |\lambda'\rangle$, using the rules of tensor product of state vectors [Baaquie (2013e)], yields

$$|1\rangle = |+\rangle \otimes |+\rangle = \begin{pmatrix}1\\0\end{pmatrix} \otimes \begin{pmatrix}1\\0\end{pmatrix} = \begin{pmatrix}1\\0\\0\\0\end{pmatrix},$$

$$|2\rangle = |+\rangle \otimes |-\rangle = \begin{pmatrix}1\\0\end{pmatrix} \otimes \begin{pmatrix}0\\1\end{pmatrix} = \begin{pmatrix}0\\1\\0\\0\end{pmatrix},$$

$$|3\rangle = |-\rangle \otimes |+\rangle = \begin{pmatrix}0\\1\end{pmatrix} \otimes \begin{pmatrix}1\\0\end{pmatrix} = \begin{pmatrix}0\\0\\1\\0\end{pmatrix},$$

$$|4\rangle = |-\rangle \otimes |-\rangle = \begin{pmatrix}0\\1\end{pmatrix} \otimes \begin{pmatrix}0\\1\end{pmatrix} = \begin{pmatrix}0\\0\\0\\1\end{pmatrix}.$$

The matrix elements of L are defined with respect to the four basis vectors, and the matrix can be written as $L_{ij} \equiv \langle i|L|j\rangle$, where $|i\rangle, i = 1, 2, 3, 4$ as given above and $\langle j|$ is the transpose of the $|i\rangle$ vectors. Hence

$$L = \sum_{i,j=1}^{4} L_{ij}|i\rangle\langle j| = \begin{pmatrix} L_{11} & L_{12} & L_{13} & L_{14} \\ .. & L_{22} & L_{23} & L_{24} \\ .. & .. & L_{33} & L_{34} \\ .. & .. & .. & L_{44} \end{pmatrix}, \quad L_{ij} = L_{ji}, \tag{8.55}$$

with the other elements being fixed by the fact that L is a symmetric matrix, for example $L_{11} = \langle 1|L|1\rangle$ and so forth. Thus the transfer matrix is given by

$$L_{ij} = e^{\mathcal{L}(\mu,\mu';\lambda\lambda')},$$

where the matrix elements are given by

$$\begin{aligned}
&L_{11} = e^{\mathcal{L}(1,1;1,1)} = e^{3K}; \;\; L_{12} = \; L_{13} = e^{\mathcal{L}(1,1;-1,1)} = 1, \\
&L_{14} = e^{\mathcal{L}(1,1;-1,-1)} = e^{-K}, L_{22} = e^{\mathcal{L}(1,-1;1,-1)} = e^{K}, \\
&L_{23} = e^{\mathcal{L}(1,-1;-1,1)} = e^{-3K}, L_{24} = e^{\mathcal{L}(1,-1;-1,-1)} = 1, \\
&L_{33} = e^{\mathcal{L}(-1,1;-1,1)} = e^{K}, L_{34} = e^{\mathcal{L}(-1,1;-1,-1)} = 1, \\
&L_{44} = e^{\mathcal{L}(-1,-1;-1,-1)} = e^{3K}.
\end{aligned}$$

Collecting all the results yields the symmetric transfer matrix L given by

$$L = \begin{pmatrix} e^{3K} & 1 & 1 & e^{-K} \\ 1 & e^{K} & e^{-3K} & 1 \\ 1 & e^{-3K} & e^{K} & 1 \\ e^{-K} & 1 & 1 & e^{3K} \end{pmatrix},$$

eigenvalues $\lambda_1 > \lambda_2 > \lambda_3 > \lambda_4$ (decreasing in magnitude) are given by

$$\begin{aligned}
&\lambda_1 = e^{-3k}(-1 + e^{4k}), \;\; \lambda_2 = \;\; e^{-k}(-1 + e^{4k}), \\
&\lambda_3 = \frac{1}{2}e^{-4k}\left(e^{k} + e^{3k} + e^{5k} + e^{7k} - e^{k}(1 + 2^{2k})\sqrt{1 - 4e^{2k} + 10e^{4k} - 4e^{6k} + e^{8k}}\right), \\
&\lambda_4 = \frac{1}{2}e^{-4k}\left(e^{k} + e^{3k} + e^{5k} + e^{7k} + e^{k}(1 + 2^{2k})\sqrt{1 - 4e^{2k} + 10e^{4k} - 4e^{6k} + e^{8k}}\right).
\end{aligned}$$

The partition function is given by

$$Z = \sum_{\{\mu,\lambda\}} e^{S[\mu,\lambda]} = tr(L^N) = \sum_{i=1}^{4} \lambda_i^N.$$

The Ising model can be extended to a square lattice by adding lattice sites to extend the Ising "ladder" and create an $N \times N$ two-dimensional lattice. The transfer matrix is $2^N \times 2^N$ and has 2^N eigenvalues λ_I with the partition function given by

$$Z = \sum_{I=1}^{2^N} [\lambda_I]^N.$$

The limit of $N \to \infty$ of the two-dimensional Ising model was exactly solved by Onsager in a landmark derivation and exhibits a second order phase transition [Papon *et al.* (2002)].

Similarly, one can go to higher dimensions by defining Ising spins on an N^d lattice given by $N \times N \times N \ldots \times N$. The transfer matrix rapidly becomes more and more complicated.

8.8 Summary

The Ising model is based on the simplest possible degree of freedom, namely one having only two possible values. The one-dimensional Ising model is a toy model that has all the mathematical structures of quantum mechanics, from the state space to operators and onto path integrals and correlation functions.

The periodic lattice was studied for explicitly deriving the partition function and the correlation function, and these derivations illustrate the general features of the computations that are carried out for more complex systems.

One significant feature of the Ising model is the expansion of the action in a power series. This is possible since the Ising spin is a *compact variable*, taking values in a finite range. This property does not hold for Gaussian degrees of freedom and hence is a special feature of the Ising model.

The generalization of the one-dimensional Ising model to the $2 \times N$ Ising ladder can be extended to defining the Ising model on an $N \times N$ lattice and shows the procedure that is required for defining the model in higher dimensions.

9

Ising model: magnetic field

The Ising model in a magnetic field is a continuation of the discussions of Chapter 8. The equations have to be generalized to include the presence of a magnetic field. A number of new features are studied, in particular, the effect of the Ising model's boundary conditions on the partition function and correlation functions.

The Ising model in a magnetic field is introduced in Section 9.1 and the evolution for this system is obtained. The magnetization of the Ising models is an important manifestation of the magnetic field and is discussed for a periodic lattice in Section 9.3. The correlation function is discussed in Section 9.4 using the concept of linear regression. Magnetization for an open chain is discussed in Section 9.5, and block spin renormalization is discussed in Section 9.6.

9.1 Periodic Ising model in a magnetic field

The earlier calculation of the Ising model for a periodic lattice is generalized to the case with a nonzero magnetic field. The calculation is more complex than the earlier case showing new features of the transfer matrix.

The partition function for an N size periodic lattice is defined by

$$Z_N = \sum_{\{\mu\}} e^S, \quad \sum_{\{\mu\}} \equiv \prod_{n=1}^{N} \sum_{\mu_n = \pm 1}, \quad \mu_{N+1} = \mu_1,$$

where

$$S \equiv \sum_{j=1}^{N} \left(K\mu_j\mu_{j+1} + h\mu_j \right). \tag{9.1}$$

Writing the partition function as

$$Z_N = tr\left(L^N\right) \tag{9.2}$$

yields the Hermitian (symmetric) transfer matrix

$$L(\mu, \mu') = \exp\left\{K\mu\mu' + \frac{h}{2}(\mu + \mu')\right\}. \tag{9.3}$$

The eigenvalues and eigenfunctions of the transfer matrix are

$$L = \begin{pmatrix} e^{K+h} & e^{-K} \\ e^{-K} & e^{K-h} \end{pmatrix}, \quad L|\phi_i\rangle = \lambda_i|\phi_i\rangle, \quad i = 1, 2, \tag{9.4}$$

$$\lambda_1 = e^K \cosh h + \left(e^{2K}\sinh^2(h) + e^{-2K}\right)^{1/2},$$

$$\lambda_2 = e^K \cosh h - \left(e^{2K}\sinh^2(h) + e^{-2K}\right)^{1/2}, \tag{9.5}$$

$$|\phi_1\rangle = \frac{1}{\sqrt{1+a^2}}\begin{pmatrix} 1 \\ a \end{pmatrix}, \tag{9.6}$$

$$|\phi_2\rangle = \frac{1}{\sqrt{1+a^2}}\begin{pmatrix} a \\ -1 \end{pmatrix}, \tag{9.7}$$

$$a = e^K\left(\lambda_1 - e^{K+h}\right).$$

The completeness equation is given by

$$\sum_{i=1,2}|\phi_i\rangle\langle\phi_i| = 1.$$

From Eqs. 9.6, 8.14, and 9.7

$$\langle\phi_1|\hat{\mu}|\phi_1\rangle = -\langle\phi_2|\hat{\mu}|\phi_2\rangle = \frac{1-a^2}{1+a^2}, \quad \langle\phi_1|\hat{\mu}|\phi_2\rangle = \langle\phi_2|\hat{\mu}|\phi_1\rangle = \frac{2a}{1+a^2}. \tag{9.8}$$

The eigenvalues yield

$$L = \lambda_1|\phi_1\rangle\langle\phi_1| + \lambda_2|\phi_2\rangle\langle\phi_2| \ \Rightarrow\ L^N = \lambda_1^N|\phi_1\rangle\langle\phi_1| + \lambda_2^N|\phi_2\rangle\langle\phi_2|. \tag{9.9}$$

Hence, the partition function is given by

$$Z_N = tr\left(L^N\right) = \lambda_1^N + \lambda_2^N. \tag{9.10}$$

The periodic Ising model with negative coupling, namely $-K$, is given by

$$S' \equiv \sum_{j=1}^{N}\left(-K\mu_j\mu_{j+1} + h\mu_j\right).$$

By a change of variables for only the odd spins, that is, $\mu_{2n+1} \to -\mu_{2n+1}$, one can change the sign of K for all the bonds. This yields

$$S' \equiv \sum_{j=1}^{N}\left(K\mu_j\mu_{j+1} + h(-1)^n\mu_j\right).$$

Hence, one can see that $-K$ corresponds to anti-ferromagnetic coupling, with the nearest neighbor spins tending to be oppositely aligned. The alternating sign of the magnetic field drives this anti-alignment. All the formulas derived for $+K$ can be applied to the case for $-K$ by a simple change of sign.

9.2 Ising model's evolution kernel

From Eq. 9.4, the transfer matrix of the Ising model in a magnetic field is given by

$$L = \begin{pmatrix} e^{K+h} & e^{-K} \\ e^{-K} & e^{K-h} \end{pmatrix}.$$

The eigenvalues and eigenvectors of L, given in Eqs. 9.5, 9.6, and 9.7, yield

$$\begin{aligned} L^N &= \frac{1}{1+a^2}\begin{pmatrix} 1 & a \\ a & -1 \end{pmatrix}\begin{pmatrix} \lambda_1^N & 0 \\ 0 & \lambda_2^N \end{pmatrix}\begin{pmatrix} 1 & a \\ a & -1 \end{pmatrix} \\ &= \frac{1}{1+a^2}\begin{pmatrix} \lambda_1^N + a^2\lambda_2^N & a\left(\lambda_1^N - \lambda_2^N\right) \\ a\left(\lambda_1^N - \lambda_2^N\right) & a^2\lambda_1^N + \lambda_2^N \end{pmatrix}. \end{aligned} \tag{9.11}$$

The evolution kernel for the Ising model is given by

$$K\left(\mu', \mu; N\right) = \langle \mu' | L^N | \mu \rangle = \exp\left\{\beta + \alpha\mu\mu' + \frac{\gamma}{2}\left(\mu + \mu'\right)\right\}.$$

In matrix notation

$$K = e^{\beta}\begin{pmatrix} e^{\alpha+\gamma} & e^{-\alpha} \\ e^{-\alpha} & e^{\alpha-\gamma} \end{pmatrix}. \tag{9.12}$$

From Eqs. 9.11 and 9.12, after some algebra

$$\begin{aligned} e^{2\gamma} &= \frac{\lambda_1^N + a^2\lambda_2^N}{a^2\lambda_1^N + \lambda_2^N}, \quad e^{4\alpha} = \frac{\left(\lambda_1^N + a^2\lambda_2^N\right)\left(a^2\lambda_1^N + \lambda_2^N\right)}{a^2\left(\lambda_1^N - \lambda_2^N\right)^2}, \\ e^{\beta} &= \frac{\sqrt{a}}{1+a^2}\left(\lambda_1^N - \lambda_2^N\right)^{1/2}\left(\lambda_1^N + a^2\lambda_2^N\right)^{1/4}\left(a^2\lambda_1^N + \lambda_2^N\right)^{1/4}. \end{aligned}$$

In the limit of $h \to 0$

$$a \to 1, \quad \lambda_1 \to \kappa_1, \quad \lambda_2 \to \kappa_2,$$

with

$$\kappa_1^N = 2^N \cosh^N(K), \quad \kappa_2^N = 2^N \sinh^N(K).$$

The evolution kernel for zero magnetic field $h \to 0$ is hence given by the parameters

$$\alpha \to \frac{1}{2}\ln\left(\frac{\kappa_1^N + \kappa_2^N}{\kappa_1^N - \kappa_2^N}\right); \quad \beta \to \frac{1}{2}\ln\left(\kappa_1^{2N} - \kappa_2^{2N}\right) - \ln 2; \quad \gamma \to 0.$$

9.3 Magnetization

The magnetization is defined by

$$\mathcal{M}_N = E[\mu_n] = \frac{1}{Z}\sum_{\{\mu\}} \mu_n e^S.$$

Due to the periodicity of the lattice, $\langle \mu_n \rangle = \langle \mu_{n+N} \rangle$ and hence $\mathcal{M}_N$ is independent of n. Hence

$$\begin{aligned}\mathcal{M}_N &= \frac{1}{N}\sum_{n=1}^{N} E[\mu_n] = \frac{1}{N}\cdot\frac{1}{Z}\frac{\partial}{\partial h}\sum_{\{\mu\}} e^S \\ &= \frac{1}{N}\frac{\partial}{\partial h}\ln(Z_N) \\ &= \frac{\partial}{\partial h}\ln(\lambda_1) + \frac{1}{N}\frac{\partial}{\partial h}\ln\left(1 + \left(\frac{\lambda_2}{\lambda_1}\right)^N\right).\end{aligned}$$

Consider the limit of $N \to \infty$, called the thermodynamic limit. Since $\lambda_1 > \lambda_2$, the partition function is given by

$$Z_N = \lambda_1^N \left[1 + O\left(e^{-N}\right)\right].$$

Hence, the magnetization is given by

$$\begin{aligned}\mathcal{M} = \lim_{N\to\infty} \mathcal{M}_N &= \frac{\partial}{\partial h}\ln(\lambda_1) + O(e^{-N}) \\ &= \frac{1}{\lambda_1}\left[e^K \sinh h + \frac{e^{2K}\sinh h \cosh h}{\left(e^{2K}\sinh^2(h) + e^{-2K}\right)^{1/2}}\right].\end{aligned} \tag{9.13}$$

The magnetization can also be derived using the transfer matrix. For a periodic lattice

$$\begin{aligned}\mathcal{M}_N = \langle \mu_n \rangle &= \frac{1}{Z_N}\sum_{\{\mu\}} \mu_n e^S \\ &= \frac{1}{Z_N} tr\left(L^N \hat{\mu}\right).\end{aligned} \tag{9.14}$$

From Eqs. 9.8, 9.9, and 9.10

$$\begin{aligned}\mathcal{M}_N &= \frac{1}{Z_N}\sum_{i=1}^{2}\langle \phi_i | L^N \hat{\mu} | \phi_i \rangle \\ &= \frac{1}{Z_N}\left[\lambda_1^N \langle \phi_1 | \hat{\mu} | \phi_1 \rangle + \lambda_2^N \langle \phi_2 | \hat{\mu} | \phi_2 \rangle\right]\end{aligned}$$

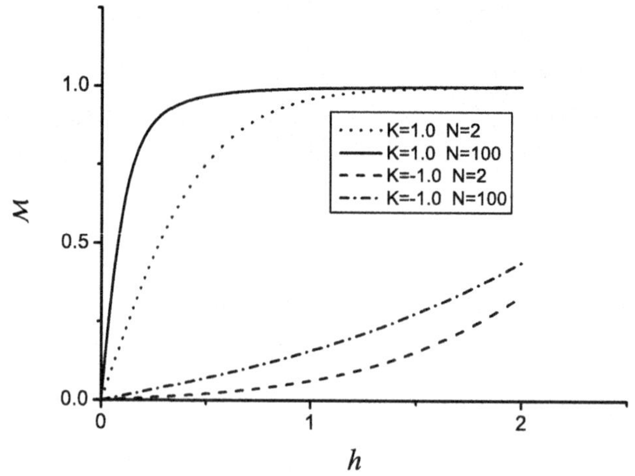

Figure 9.1 Ising magnetization for the case of $+K$ and $-K$.

$$
\begin{aligned}
&= \frac{1}{Z_N} \cdot \frac{(1-a^2)}{(1+a^2)} \cdot (\lambda_1^N - \lambda_2^N) \\
&= \frac{1-a^2}{1+a^2} \cdot \frac{\lambda_1^N - \lambda_2^N}{\lambda_1^N + \lambda_2^N}.
\end{aligned} \tag{9.15}
$$

Taking the limit of $N \to \infty$ yields the magnetization

$$
\mathcal{M} = \lim_{N \to \infty} \mathcal{M}_N = \frac{1-a^2}{1+a^2}, \tag{9.16}
$$

and it can be shown to be equal to the result given in Eq. 9.13.

Figure 9.1 plots the magnetization given in Eq. 9.15 as a function of h with $K = \pm 1$ and for two cases, namely $N = 2$ and $N = 100$. It can be seen that the magnetization converges very fast to its large lattice value. For $-K$, the magnetization has a smaller value since, due to the anti-ferromagnetic coupling, the spins tend to anti-align.

9.3.1 Correlator

The periodic chain correlation function is given by

$$
E[\mu_k \mu_{k+r}] = \frac{1}{Z_N} \sum_{\{\mu\}} \mu_k \mu_{k+r} e^S,
$$

where the partition function is given by Eq. 9.10. From Eq. 8.17, in terms of the transfer matrix, the correlator is given by

$$
C_r = \frac{1}{Z_N} tr \left(L^{N-r} \hat{\mu} L^r \hat{\mu} \right).
$$

The expression given above for the nonzero magnetic field looks the same as for the case of zero magnetic field, the only difference between the two cases being the values of the eigenvalues and eigenvectors of the transfer matrix L. The correlation function can then be written, similarly to Eq. 8.31, as

$$C_r = \frac{1}{Z_N} \sum_{i,j=1}^{2} \lambda_i^{N-r} \lambda_j^r \left(\langle \phi_i | \hat{\mu} | \phi_j \rangle \right)^2.$$

A straightforward but long derivation, using the result above and Eqs. 9.5 and 9.10, yields

$$C_r = \frac{1}{(1+a^2)^2} \left[\left(1-a^2\right)^2 + 4a^2 \frac{\left(\frac{\lambda_2}{\lambda_1}\right)^r + \left(\frac{\lambda_2}{\lambda_1}\right)^{N-r}}{1 + \left(\frac{\lambda_2}{\lambda_1}\right)^N} \right]. \tag{9.17}$$

As expected, $C_0 = 1$. Recall from Eq. 9.16

$$\mathcal{M} = \frac{1-a^2}{1+a^2}$$

and hence

$$C_r = \mathcal{M}^2 + \frac{4a^2}{\left(1+a^2\right)^2} \frac{\left(\frac{\lambda_2}{\lambda_1}\right)^r + \left(\frac{\lambda_2}{\lambda_1}\right)^{N-r}}{1 + \left(\frac{\lambda_2}{\lambda_1}\right)^N}. \tag{9.18}$$

Note that since $\lambda_1 > \lambda_2$, we can define correlation length ξ by

$$\left(\frac{\lambda_2}{\lambda_1}\right)^r \equiv \exp(-r/\xi), \quad \lim_{N\to\infty} \left(\frac{\lambda_2}{\lambda_1}\right)^{N-r} \simeq \left(\frac{\lambda_2}{\lambda_1}\right)^N \to 0.$$

Hence, for an infinitely large lattice, $N \to \infty$, from Eq. 9.18

$$C_r = \mathcal{M}^2 + \frac{4a^2}{\left(1+a^2\right)^2} e^{-r/\xi} + O\left(e^{-N}\right). \tag{9.19}$$

The limit of $h = 0$ leads to $a \to 0$ and results in $\mathcal{M} = 0$; the value of C_r converges to the value of the correlator for zero magnetic field given in Eq. 8.36.

9.4 Linear regression

For two random variables to be correlated means that the variables take their random values in tandem, that is, the value of one of them takes a predictable range of values if the other takes certain values and vice versa.

The concept of correlation is very different from causation, in which the cause determines the effect: heat and fire are correlated as well as causally linked, with heat causing fires.

Consider the height and weight of a person in a given age group; although both are random variables they are nevertheless related (correlated) since the taller the person is, the greater is the likelihood that the person is heavier and vice versa. However, although they are correlated, height is not taken to be the cause of weight and vice versa. Another example is intelligence and say height – which have no correlation – again without one being the cause of the other.

Consider random variables X and Y. Suppose they are linearly related and let

$$Y = p + qX. \tag{9.20}$$

The equation above is written so that the random variable X is considered as the independent random variable and Y the dependent random variable. For the case of height and weight, the height can be taken to be the *independent* random variable with the weight being the *dependent* random variable. One uses the average value and the correlation of X, Y and variance of X to fix the parameters p, q in the following manner. Let

$$E[XY]_c = E[XY] - E[X]E[Y],$$
$$\sigma^2(X) = E[X^2] - E[X]^2.$$

Then

$$q = \frac{E[XY]_c}{\sigma^2(X)}, \quad p = E[Y] - qE[X]. \tag{9.21}$$

The estimate of the variance of Y from above is given by

$$\sigma^2_{\text{Est}}(Y) = q^2\sigma^2(X) = \frac{\langle XY \rangle_c^2}{\sigma^2(X)}. \tag{9.22}$$

The exact variance of Y is $\sigma^2(Y)$. Hence, one measure of the accuracy of the linear regression between X and Y is given by the fractional error

$$FE = \sqrt{\frac{\sigma^2(Y) - \sigma^2_{\text{Est}}(Y)}{\sigma^2(Y)}}$$
$$= \sqrt{1 - \frac{E[XY]_c^2}{\sigma^2(X)\sigma^2(Y)}}. \tag{9.23}$$

From the point of view of probability theory, the spins μ_0 and μ_r are two random variables; let

$$X = \mu_r, \quad Y = \mu_0.$$

The average value of the random variable is given by the magnetization, namely

$$E[\mu_r] = \frac{1}{Z_N} \sum_{\{\mu\}} e^S \mu_r. \tag{9.24}$$

For the periodic lattice, the magnetization is given by

$$E[\mu_r] = E[\mu_0] = \mathcal{M}_N. \tag{9.25}$$

The expectation value of the product of two spin variables, namely μ_0 and μ_r, is given by

$$E[\mu_r \mu_0] = \frac{1}{Z_N} \sum_{\{\mu\}} e^S \mu_r \mu_0. \tag{9.26}$$

The *connected* two spin correlation function is given by

$$E[XY]_c = G_r = E[\mu_r \mu_0] - E[\mu_r]E[\mu_0]. \tag{9.27}$$

For the case of the Ising spin on a finite periodic lattice, from Eqs. 9.15 and 9.18, the correlator is

$$\begin{aligned} G_r &= C_r - \mathcal{M}_N^2 \\ &= \frac{(1-a^2)^2}{(1+a^2)^2}\left[\frac{2\lambda_2^N}{\lambda_1^N + \lambda_2^N}\right] + \frac{4a^2}{(1+a^2)^2}\left[\frac{\left(\frac{\lambda_2}{\lambda_1}\right)^r + \left(\frac{\lambda_2}{\lambda_1}\right)^{N-r}}{1+\left(\frac{\lambda_2}{\lambda_1}\right)^N}\right]. \end{aligned} \tag{9.28}$$

Furthermore, the variance of μ_r is given by

$$\sigma^2(\mu_r) = E\left[\mu_r^2\right] - (E\left[\mu_r\right])^2 = 1 - \mathcal{M}_N^2 = \sigma^2(\mu_0), \tag{9.29}$$

where the last equation follows due to the periodic lattice.

From the point of view of probability theory all the spins in the lattice, namely μ_n with $n = 1, 2, \ldots N$ are random variables. The behavior of all the spins can be fully described by specifying how the different random variables are correlated.

One can start with the simplest relation between the random variables' as given by the linear regression. From Eqs. 9.20 and 9.21

$$\begin{aligned} \mu_r &\simeq p + q\mu_0, \\ q &= \frac{G_r}{\sigma^2(\mu_r)} = \frac{G_r}{1-\mathcal{M}_N^2}, \\ p &= \langle\mu_r\rangle - q\langle\mu_0\rangle = [1-q]\mathcal{M}_N. \end{aligned}$$

From Eq. 9.23 the fractional error for the Ising spins is given by

$$FE = \sqrt{1 - \left[\frac{G_r}{1 - \mathcal{M}_N^2}\right]^2}$$
$$= \begin{cases} 0 & r = 0 \\ 1 & r >> 1 \end{cases} . \tag{9.30}$$

For simplicity, consider an infinitely large lattice $N \to \infty$ that yields, from Eq. 9.19,

$$\mu_r \simeq p + e^{-r/\xi}\mu_0, \quad p = \left[1 - e^{-r/\xi}\right]\mathcal{M},$$
$$FE = \sqrt{1 - e^{-2r/\xi}}.$$

For $r \gg \xi$ we have to leading order that the random variables μ_r and μ_0 are decorrelated and take values *independently* of each other; that is

$$r \gg \xi \;\Rightarrow\; \mu_r \simeq \mathcal{M} \quad : \text{ independent of } \mu_0 .$$

The fractional error FE is small when r, the distance of the two spins μ_0 and μ_r, is less than ξ and hence the linear regression is a good fit; when $r \gg \xi$ the error FE ~ 1, which is equal to the magnitude of the spin, and hence the linear regression is no longer a good fit.

In summary, the correlation length ξ is the crucial quantity in the validity of linear regression. The correlation length ξ specifies the distance within which the spins are correlated. In fact, in statistical mechanics and quantum field theory, the correlation length ξ plays a central role in describing physical phenomena.

To fully specify the theory one needs to have a generalized nonlinear regression that expresses all the product spins, namely all possible combination of bonds, in terms of the other remaining spins. Note that one cannot arbitrarily assign values to the regression of various products of spins since these must satisfy consistency conditions. At most, one can try and define a theory perturbatively by specifying the regression coefficients for the product of a few random variables.

The assignment of the probability distribution – in statistical mechanics and quantum theory – is given by the expression $e^{S[\mu_0,\ldots\mu_{N-1}]}/Z$ for the (normalized) probability distribution for a particular spin configuration. This completely and uniquely defines the theory, as well as determining the regression and correlation of all the random spin variables in a self-consistent manner.

Figure 9.2 Open lattice with boundary spins given by μ_1, μ_N.

9.5 Open chain Ising model in a magnetic field

Consider the one-dimensional Ising model for an open chain of length N with arbitrary boundary values for μ_1, μ_N as shown in Figure 9.2. The open chain partition function is defined by

$$Z_N = \sum_{\{\mu\}} e^S$$

and the action is given by

$$S \equiv K \sum_{j=1}^{N-1} \mu_j \mu_{j+1} + h \sum_{j=1}^{N} \mu_j,$$

$$\sum_{\{\mu\}} \equiv \prod_{n=1}^{N} \sum_{\mu_n=\pm 1}.$$

The partition function can be written in terms of the transfer matrix as

$$Z_N = \sum_{\{\mu\}} \exp\left[K \sum_{j=1}^{N-1} \mu_j \mu_{j+1} + \frac{h}{2} \sum_{j=1}^{N-1} (\mu_j + \mu_{j+1})\right] \exp\left[\frac{h}{2}(\mu_N + \mu_1)\right]. \tag{9.31}$$

The sum for the action runs from 1 to $N - 1$ since it extends to the boundaries. The term in the action

$$\exp\left\{\frac{h}{2}(\mu_N + \mu_1)\right\}$$

is left over due to the open chain condition. The transfer matrix is given by

$$L(\mu, \mu') = \exp\left\{K\mu\mu' + \frac{h}{2}\left(\mu + \mu'\right)\right\}, \tag{9.32}$$

which is the same as the one given in Eq. 9.3 for the periodic lattice.

From Eq. 9.31, the partition function Z_N can be written as

$$Z_N = \sum_{\mu_1} \cdots \sum_{\mu_N} \left[\prod_{j=1}^{N-1} \langle \mu_j | L | \mu_{j+1} \rangle\right] \exp\left[\frac{h}{2}(\mu_N + \mu_1)\right]$$

$$= \sum_{\mu_1} \cdots \sum_{\mu_N} \langle\mu_1| \left[\prod_{j=2}^{N-2} L|\mu_j\rangle\langle\mu_j| \right] L|\mu_N\rangle e^{\frac{h}{2}(\mu_N+\mu_1)}$$

$$= \sum_{\mu_1} \sum_{\mu_N} \langle\mu_1|L^{N-2} \cdot L|\mu_N\rangle e^{\frac{h}{2}(\mu_N+\mu_1)}$$

$$= tr \left(\sum_{\mu_1} \sum_{\mu_N} L^{N-1}|\mu_N\rangle\langle\mu_1|e^{\frac{h}{2}(\mu_N+\mu_1)} \right).$$

In matrix notation, the partition function is given by

$$Z_N = tr(L^{N-1}\mathcal{B}), \tag{9.33}$$

where the boundary condition $\mathcal{B}$ is given by

$$\mathcal{B} = \sum_{\mu_1} \sum_{\mu_N} |\mu_N\rangle\langle\mu_1|e^{\frac{h}{2}(\mu_N+\mu_1)}.$$

In components

$$\langle\mu_N|\mathcal{B}|\mu_1\rangle = e^{\frac{h}{2}\mu_N} e^{\frac{h}{2}\mu_1} \ \Rightarrow \ \mathcal{B} = \begin{pmatrix} e^h & 1 \\ 1 & e^{-h} \end{pmatrix}. \tag{9.34}$$

Hence, from Eqs. 9.6 and 9.7

$$\Rightarrow Z_N = \lambda_1^{N-1}\langle\phi_1|\mathcal{B}|\phi_1\rangle + \lambda_2^{N-1}\langle\phi_2|\mathcal{B}|\phi_2\rangle$$

$$= \lambda_1^{N-1} \frac{\left(e^{\frac{h}{2}} + ae^{-\frac{h}{2}}\right)^2}{1+a^2} + \lambda_2^{N-1} \frac{\left(e^{-\frac{h}{2}} - ae^{\frac{h}{2}}\right)^2}{1+a^2}. \tag{9.35}$$

The case of $h = 0$ yields the partition function

$$Z_N = 2(2\cosh K)^{N-1}. \tag{9.36}$$

9.5.1 Open chain magnetization

Consider the expectation value $\langle\mu_k\rangle$ of single spin variable μ_k; note that for an open chain $\langle\mu_k\rangle$ depends on how far the lattice site k is from the boundaries. In particular

$$\langle\mu_k\rangle = \frac{1}{Z_N} \sum_{\mu_1} \cdots \sum_{\mu_N} \left(\prod_{j=1}^{N-1} \langle\mu_j|L|\mu_{j+1}\rangle \right) \mu_k e^{\frac{h}{2}(\mu_1+\mu_N)}$$

$$= \frac{1}{Z_N} tr\left(L^{k-1}\hat{\mu}L^{N-k}B\right), \tag{9.37}$$

where, from Eq. 8.14

$$\hat{\mu} = \begin{pmatrix} 1 & 0 \\ 0 & -1 \end{pmatrix}.$$

A long calculation yields

$$\langle \mu_k \rangle = \frac{1}{Z_N} \frac{1}{(1+a^2)^2} \Big[\lambda_1^{N-1} \left(e^{\frac{h}{2}} + a e^{-\frac{h}{2}} \right)^2 \left(1 - a^2\right) - \lambda_2^{N-1} \left(a e^{\frac{h}{2}} - e^{-\frac{h}{2}} \right)^2 \left(1 - a^2\right) + 2a \left\{ \lambda_1^{N-k} \lambda_2^{k-1} + \lambda_1^{k-1} \lambda_2^{N-k} \right\} \left(a e^{h} - 1 + a^2 - a e^{-h} \right) \Big]. \tag{9.38}$$

Note that for $h \to 0$, one recovers the expected result

$$E[\mu_k] = 0.$$

9.6 Block spin renormalization

Wilson's (1983) concept of renormalization plays a central role in quantum field theory. The essential idea is that at each length scale, there is an effective theory that completely describes the physics at that length scale. The one-dimensional Ising model provides a toy model to examine the essential features of renormalization theory; the concept of length scale in the case of the Ising model is the distance between two spins.

The large distance action is related to the short distance action by the procedure of renormalization. In Wilson's (1983) approach, the short distance degrees of freedom are summed over (integrated out) to generate the action appropriate for describing the longer distance physics.

Let the lattice spacing for the Ising model be denoted by a, described by action S_0. Consider the odd and even lattice sites, and on them the spins of the one-dimensional Ising model. The distance between the even spins is twice the distance between adjacent spins. Hence if all the odd spins are summed over, the new lattice will have a spacing of $a_1 = 2a$ and describe a new effective action S_1; the transformation relating the action for the original lattice S_0 to the action of the new lattice S_1 is the renormalization transformation $\mathcal{R}$ and $S_1 = \mathcal{R}[S_0]$ [Kardar (2007)].

Successive transformations form a group, and hence the name renormalization group for this procedure for studying the distinct scales of a system. If one repeats the renormalization transformation l times, then the lattice spacing of the final lattice is $a_l = 2^l a$ and the action S_l is given by $S_l = \mathcal{R}^l[S_0]$.

The correlation length on the original lattice is say $\xi_0 = L/a$; hence, the correlation on the lattice with spacing a_l is given by $\xi_l = 2^{-l}\xi_0$, where ξ_l is the correlation length measured by the scale of the final lattice, namely $a_l = 2^l a$ and $\xi_l = L/a_l$.

Divide the lattice into odd and even sites and we sum over all the spins residing at the *even* sites, thus generating a lattice of double the original lattice. The partition function remains invariant. The renormalization transformation is given by

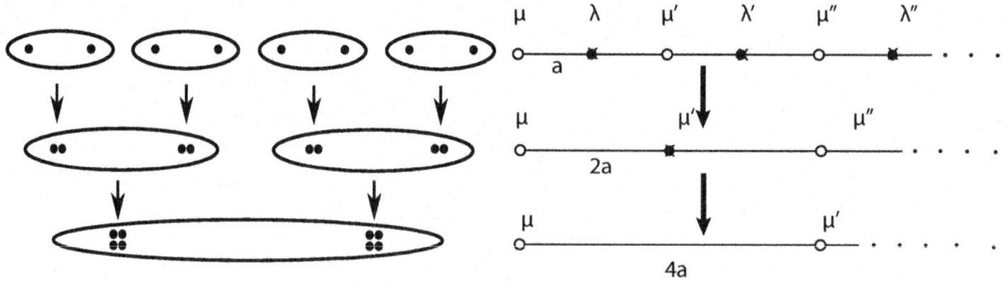

Figure 9.3 Block-spin renormalization.

$$Z = \sum_{\{\mu:\text{all}\}} e^{S_0} = \sum_{\{\mu:\text{odd}\}} e^{S_1}, \tag{9.39}$$

$$S_0 = g_0 + K_0 \sum_n \mu_n \mu_{n+1}. \tag{9.40}$$

Let the even spins be labeled by λ_n as shown in Figure 9.3; the renormalization transformation is given by

$$e^{S_1} = \sum_{\{\mu:\text{even}\}} e^{S_0} = \sum_{\{\lambda\}} e^{S_0}.$$

One can think of integrating over the even spins as combining two spins together, as shown in Figure 9.3, to generate a new block-spin that is defined on the new lattice and described by its own effective action. For a lattice of infinite size, this process can be repeated indefinitely.

Each λ_n spin is coupled only to its nearest neighbor; hence the sum over the λ_n spins can be done separately over each λ_n spin. As shown in Figure 9.3, the Ising spin at the even site labeled by $\lambda = \pm 1$ couples to its nearest neighbor spins μ, μ'; the two bonds $\lambda\mu$ and $\mu'\lambda$ contain the spin variable λ; summing over λ yields the fundamental renormalization group transformation

$$e^{2g_0} \sum_{\lambda=\pm 1} e^{K_0 \lambda(\mu+\mu')} = e^{g_1 + K_1 \mu\mu'}, \tag{9.41}$$

where it is postulated that the effective bonds coupling the lattice at double the lattice spacing are given by the right hand side of Eq. 9.41. The new coupling constants g_1, K_1 are now determined in terms of the original coupling constants g_0, K_0. Consider the following cases for Eq. 9.41:

- $\mu = 1 = \mu'$ and $\mu = -1 = \mu'$ both yield the same equation, namely

$$2e^{2g_0}\left[e^{2K_0} + e^{-2K_0}\right] = e^{g_1 + K_1}. \tag{9.42}$$

- $\mu = 1 = -\mu'$ and $\mu = -1 = -\mu'$ both yield the same equation

$$2e^{2g_0} = e^{g_1 - K_1}. \tag{9.43}$$

Equations. 9.42 and 9.43 yield

$$e^{g_1} = 2e^{2g_0} \cosh^{1/2} 2K_0, \quad e^{K_1} = \cosh^{1/2} 2K_0,$$

and we obtain the renormalization group transformation

$$K_1 = \frac{1}{2} \ln\cosh 2K_0, \quad g_1 = 2g_0 + \frac{1}{2} \ln\cosh 2K_0 + \ln 2. \tag{9.44}$$

The lattice spacing for S_1 is twice the lattice spacing of S_0; hence, the correlation lengths ξ as computed from the two actions should scale accordingly, namely

$$\xi_1 = \frac{1}{2}\xi_0 \Rightarrow \xi(K_1) = \frac{1}{2}\xi(K_0). \tag{9.45}$$

To prove Eq. 9.45, recall that Eq. 8.38 yields the correlation lengths

$$\xi(K_0) = -\frac{1}{\ln(\tanh K_0)} \quad \text{and} \quad \xi(K_1) = -\frac{1}{\ln(\tanh K_1)}.$$

The expression for K_1 in Eq. 9.44 yields

$$2\left(e^{2K_1} \pm 1\right) = \left[e^{K_0} \pm e^{-K_0}\right]^2 \quad \Rightarrow \quad \tanh(K_1) = [\tanh(K_0)]^2.$$

Hence, the result given in Eq. 9.45 is realized as

$$\xi(K_1) = -\frac{1}{\ln(\tanh K_1)} = -\frac{1}{2}\frac{1}{\ln(\tanh K_0)} = \frac{1}{2}\xi(K_0). \tag{9.46}$$

Note the remarkable result that the new coupling constant K_1 on the larger lattice "remembers" that it is the result of summing over the smaller lattice with coupling constant K_0. One of the fundamental insights that the renormalization group provides about the different length scales in a problem is the following: that each scale 2^ℓ of the system has its own corresponding coupling constant K_ℓ.

The recursion equation connecting coupling constants g_l, K_l defined for a lattice of size $2^l a$, with coupling constants g_{l+1}, K_{l+1} defined for a lattice of size $2^{l+1}a$, is given by

$$g_{l+1} = 2g_l + \frac{1}{2} \ln\cosh 2K_l + \ln 2 = \mathcal{R}(g_l), \quad K_{l+1} = \frac{1}{2} \ln\cosh(2K_l) = \mathcal{R}(K_l).$$

One can write the renormalization group recursion equations more generally as

$$(K_l, g_l) = \mathcal{R}(K_{l-1}, g_{l-1}) = \mathcal{R}^\ell(K_0, g_0), \tag{9.47}$$

where $\mathcal{R}$ is a 2×2 matrix.

The coupling constant K_l is the renormalized coupling constant, describing the physics at length scale $2^l a$ and is obtained from the initial (bare) coupling constant

K_0 by the repeated application of the renormalization group transformation $\mathcal{R}$. The correlation length is given by

$$\xi_\ell = \frac{1}{2}\xi_{\ell-1} = \frac{1}{2^\ell}\xi_0.$$

The *fixed point* of the renormalization group transformation is given by

$$K^* = \mathcal{R}(K^*) \Leftrightarrow \xi^* = \frac{1}{2}\xi^*$$

and hence leads to the result

$$\xi^* = \frac{1}{2}\xi^* \quad \Rightarrow \xi^* = 0 \quad or \quad \xi^* = \infty.$$

The case of $\xi^* = 0$ corresponds to a decoupled system. A system is said to be critical – undergoing a phase transition – if its correlation length is infinite, namely $\xi^* = \infty$; hence the fixed point of the renormalization transformation corresponds to the system being critical. Since ξ^* is unchanged under a change of lattice spacing, the critical system is seen to be scale invariant.

Note the remarkable fact that the fixed point of the renormalization group transformation $\mathcal{R}$ depends only on $\mathcal{R}$ and not on the details of the (Ising) system we start from; this feature of phase transitions leads to the property of *universality* in that many different types of phase transitions are described by the same critical system defined by $\mathcal{R}$.

The fixed point is analyzed as follows:

- To find the fixed point of K_l, consider the limit $\lim_{l\to\infty} K_l \to K^*$.
- Fixed point of $K_0 = \infty = K^*$ result from the fact that $K_l = K^* = \infty$. The initial Ising system (fixed by K_0) is strongly correlated with the correlation length being $\xi^* = \infty$. Hence, under renormalization, the correlation length remains infinite no matter how large the effective lattice spacing.
- For any $K_l \simeq 0$,

  $$K_{l+1} \simeq \frac{1}{2}\ln\left(1 + \frac{4K_l^2}{2}\right) \approx K_l^2 < K_l.$$

  Hence $\lim_{l\to\infty} K_l \to K^* = 0$. The system is decoupled at large enough distances for any initial value $K_0 < 0$. In fact, for any $K_0 < \infty$, $\lim_{l\to\infty} K_l \to 0 = K^*$. For any initial finite correlation length, the large distance correlation is always zero, namely $\xi^* = 0$.

The flow of the coupling constant K_l is shown in Figure 9.4; if one starts with any finite value for $K_0 < \infty$, then the effective coupling K_l for larger and larger lattice size flows towards zero, with the one-dimensional Ising model becoming decoupled at large distances.

Figure 9.4 Flow of coupling constant under renormalization.

Near $K^* \simeq \infty$,

$$K_{l+1} \simeq \frac{1}{2}\ln\left(\frac{e^{2K_l}}{2}\right) \simeq K_l - \frac{1}{2}\ln 2 < K_l.$$

Hence, for any $K_l < \infty$

$$K_{l+1} < K_l, \quad \lim_{l\to\infty} K_l \to 0.$$

9.6.1 Block spin renormalization: magnetic field

One can introduce a magnetic field h, as in Eq. 9.1, and repeat the block spin calculation starting with the initial Ising model given by

$$S_0 \equiv \sum_{j=-\infty}^{+\infty} (K\mu_j\mu_{j+1} + h\mu_j).$$

The renormalization group transformation is given by

$$\begin{aligned}&\exp\left[K'\mu\mu' + \frac{h'}{2}(\mu+\mu') + g'\right]\\ &= \sum_{\lambda=\pm 1}\exp\left[K\lambda\left(\mu+\mu'\right) + \frac{h}{2}\left(\mu+\mu'\right) + h\lambda + 2g\right].\end{aligned} \tag{9.48}$$

To solve for the renormalized interaction it is convenient to set

$$\begin{cases} x = e^K,\ y = e^h,\ z = e^g \\ x' = e^{K'},\ y' = e^{h'},\ z' = e^{g'}. \end{cases} \tag{9.49}$$

Similarly to the case for zero magnetic field, the four possible configurations of the bond this time yield only three equations in three unknowns. The solution is given by Kardar (2007)

$$\begin{cases} z'^4 = z^8\left\{x^2y + x^{-2}y^{-1}[x^{-2}y + x^2y^{-1}(y+y^{-1})^2]\right\} \\ y'^2 = y^2\frac{x^2y+x^{-2}y^{-1}}{x^{-2}y+x^2y^{-1}} \\ x'^4 = \frac{(x^2y+x^{-2}y^{-1})(x^{-2}y+x^2y^{-1})}{(y+y^{-1})^2}. \end{cases} \tag{9.50}$$

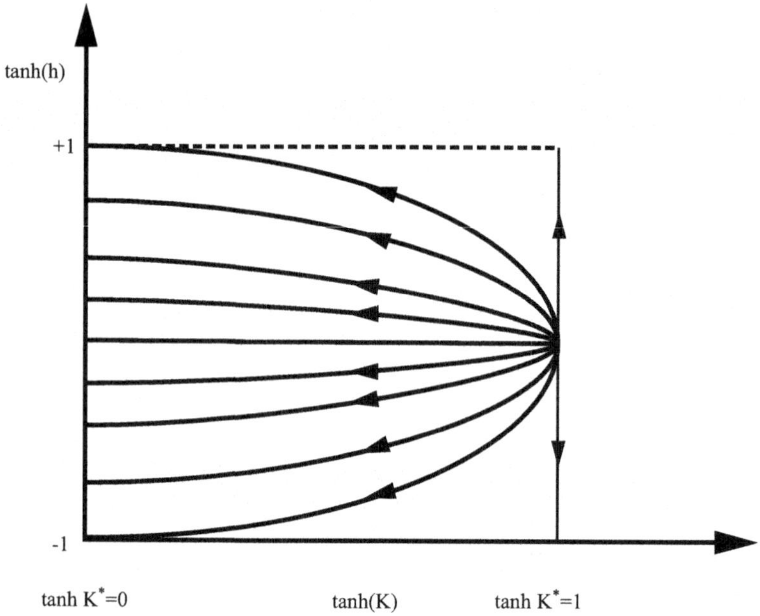

Figure 9.5 Flow of coupling constant K and magnetic field h under renormalization, from an initial value of K_0, h_0.

The renormalization of coupling constants with three equations is a generalization of the renormalization group equations given in Eq. 9.47; taking the logarithm of Eq. 9.50 yields

$$(K_{l+1}, g_{l+1}, h_{l+1}) = \mathcal{R}(K_l, g_l, h_l), \tag{9.51}$$

where $\mathcal{R}$ is now a 3×3 matrix.

Similarly to the case of $h_l = 0$, the flow of g_l is irrelevant for the critical properties of the Ising system in the presence of a magnetic field. The coupling constants K_l, h_l have flows as shown in Figure 9.5; the system is critical for $K^* = \infty$ and has two unstable directions [Kardar (2007)].

For $K_0 < \infty$, as one recurses and goes to long distance, the system decouples with the coupling constant going to zero, namely $K_l \to 0$; the magnetic field flows to $h_l \to h^*$, namely a decoupled spin system with magnetic field h^*, the value of h^* being fixed by the initial values K_0, h_0. Figure 9.5 shows the renormalization flow for $K_0 = K^* = \infty$ and $h_0 = 0$, as well as for the flows that originate from the fixed point in (K, h) space.

9.7 Summary

The Ising model with a magnetic field has many new features. An exact solution of the partition function, of the magnetization, and of the correlation function can be obtained using the transfer matrix. Unlike the zero magnetic field case, the expansion e^S in a power series does not lead to a tractable method for calculation due to the difficulty in doing the combinatorics in the presence of a magnetic field.

Linear regression in the presence of a magnetic field provides an intuitive and physical interpretation of the concept of the correlation function.

An important feature of all lattice systems is that the lattice spacing a does not appear explicitly. The consequence of this plays a central role in the concept of renormalization and of the renormalization group.

Let the physical correlation *length* of a system described by the Ising model be L, in units of say meters; let the lattice spacing be a meters; then on the lattice, since all quantities are dimensionless, the dimensionless correlation length is given by $\xi(K_0) = L/a$. The correlation for the lattice of spacing $2a$ with coupling constant K_1 is $\xi(K_1)$ and hence the dimensionless correlation is given by $\xi(K_1) = L/(2a)$.

The dimensionless numerical value of the correlation $\xi(K_1)$ is seen to be half the value of the correlation for lattice a with coupling constant K_0 given by $\xi(K_0)$, namely $\xi(K_1) = L/(2a) = \xi(K_0)/2$. The result of the renormalization transformation leads to the nontrivial result that the effective lattice spacing, instead of appearing explicitly, appears through the value of the coupling constant K_l which corresponds to a lattice of spacing $2^l a$, and yields $\xi_l = 2^{-l}\xi_0$.

10

Fermions

The degrees of freedom studied so far have been either real or complex variables. These variables commute under multiplication, in the sense that two numbers a, b satisfy $ab = ba$; commuting variables are generically called bosonic variables, or bosonic degrees of freedom. Typical of the bosonic case are the degrees of freedom for a collection of quantum mechanical particles.

Interactions of fundamental particles are generally mediated by bosonic fields such as the Maxwell electromagnetic field, whereas mass is usually carried by particles that are fermions, the most familiar being the electron.

Two key features distinguish fermions from bosons:

- Fermions obey the Pauli exclusion principle, which states that no two fermions can occupy the same quantum state. This is the reason the concept of intensity does not apply to a fermion. A high intensity electric field is a reflection of the presence of a large number of photons, which are bosons, in the same quantum state; for photons, any number of photons can be in the same quantum state. In contrast, an electron is either in a quantum state or it is not; in particular, ignoring spin, an electron exists at a point or there is no electron there.
- The state function of a multi-bosonic system is totally symmetric in that the exchange of any two bosonic degrees of freedom yields the same state function. In contrast, a multi-fermion system is totally anti-symmetric: the exchange of any two fermion degrees of freedom gives the same state – *but* with a *negative sign.*

The Pauli exclusion principle implies a *discrete* nature for fermions since a fermion degree of freedom has only two possibilities, either occupying a state or not occupying a state. In discussing the fermion Hilbert space in Section 10.3, it will be seen that the Hilbert space of a single fermion is identical (isomorphic) to the Hilbert space of a *discrete degree of freedom* that takes only two values.

Both the key features of fermions, namely obeying the Pauli exclusion principle and the state function being anti-symmetric, can be mathematically realized by introducing a new type of variable, namely fermionic variables; similarly to the bosonic case, fermionic degrees of freedom can be described by either real or complex fermionic variables.

Fermionic variables are defined in Section 10.1 and fermion integration is discussed in Section 10.2. The fermionic Hilbert space as well as its dual space are defined in Section 10.3. The concept of the antifermionic Hilbert space is discussed in Section 10.4.

Gaussian integration for real and complex fermions is discussed in Section 10.6. The fermionic path integral is defined for the particle and anti-particle system in Section 10.8, and the transition probability amplitude is obtained in Section 10.11. A simple one-dimensional toy model is constructed in Section 10.12 to show how quark confinement can arise by coupling the fermions to a gauge field.

10.1 Fermionic variables

The defining property of fermions is the Pauli exclusion principle, namely that at most only one fermion can occupy a quantum state. In other words, for fermions, a state either has no fermions, or at most one fermion.

Let$|0\rangle_F$ be the fermion vacuum state, and let $a_F^\dagger$ be the fermion creation operator. Then

$$\begin{aligned} |0\rangle_F &: \text{ground state; no fermions,} \\ a_F^\dagger|0\rangle_F &: \text{one fermion,} \\ (a^\dagger)_F^2|0\rangle = 0 &: \text{null state.} \end{aligned} \tag{10.1}$$

The second defining property of fermions is that two different fermions must give an anti-symmetric state function on being exchanged; hence two distinct fermionic creation operators, represented by say $a_1^\dagger, a_2^\dagger$, must satisfy the anti-symmetry

$$a_1^\dagger a_2^\dagger|0\rangle = -a_2^\dagger a_1^\dagger|0\rangle,$$

which is realized by imposing the anti-commutation relation

$$a_1^\dagger a_2^\dagger = -a_2^\dagger a_1^\dagger \quad \Rightarrow \quad \{a_1^\dagger, a_2^\dagger\} = 0,$$

where the anti-commutator is defined for any two quantities A, B by

$$\{A, B\} \equiv AB + BA. \tag{10.2}$$

Instead of working with fermionic creation and annihilation operators acting on the ground state $|0\rangle$, one can instead describe fermions using a calculus distinct from the calculus based on real numbers that is used for describing bosons.

An independent and self-contained formalism for realizing all the defining properties of fermions is provided by a set of *anti-commuting fermionic variables* $\psi_1, \psi_2, \ldots \psi_N$ and its conjugate $\bar{\psi}_1, \bar{\psi}_2, \ldots \bar{\psi}_N$, defined by the properties

$$\{\psi_i, \psi_j\} = -\{\psi_i, \psi_j\},$$
$$\{\bar{\psi}_i, \psi_j\} = -\{\psi_j, \bar{\psi}_i\},$$
$$\{\bar{\psi}_i, \bar{\psi}_j\} = -\{\bar{\psi}_j, \bar{\psi}_i\}.$$

Hence, it follows that

$$\psi_i^2 = 0 = \bar{\psi}_i^2.$$

Fermionic differentiation is defined by

$$\frac{\delta}{\delta\psi_i}\psi_j = \delta_{i-j}, \quad \frac{\delta}{\delta\psi_i}\bar{\psi}_j = 0$$

and

$$\frac{\delta^2}{\delta\psi_i\delta\psi_j} = -\frac{\delta^2}{\delta\psi_j\delta\psi_i} \Rightarrow \frac{\delta^2}{\delta\psi_i^2} = 0 = \frac{\delta^2}{\delta\bar{\psi}_i^2}.$$

Similarly, all the fermionic derivative operators $\delta/\delta\psi_i$,$\delta/\delta\bar{\psi}_i$ anti-commute.

10.2 Fermion integration

Similarly to the case of $\int_{-\infty}^{+\infty} dx f(x)$ which is invariant under $x \to x + a$, that is $\int_{-\infty}^{+\infty} dx f(x) = \int_{-\infty}^{+\infty} dx f(x + a)$, we *define* fermion integration by

$$\int d\bar{\psi}\, f(\bar{\psi}) = \int d\bar{\psi}\, f(\bar{\psi} + \bar{\eta}). \tag{10.3}$$

Since $\bar{\psi}^2 = 0$, Taylors expansion shows that the most general function of the variable $\bar{\psi}$ is given by

$$f = a + b\bar{\psi}.$$

It follows that rules of fermion integration that yield Eq. 10.3 are given by

$$\int d\bar{\psi} = 0 = \int d\psi,$$
$$\int d\bar{\psi}\bar{\psi} = 1 = \int d\psi\psi,$$
$$\int d\bar{\psi} d\psi \psi\bar{\psi} = 1 = -\int d\bar{\psi} d\psi \bar{\psi}\psi. \tag{10.4}$$

For N fermionic variables ψ_i, with $i = 1, 2, \ldots N$, one has the generalization

$$\left[\prod_{n=1}^{N} \int d\psi_n\right] \psi_{i_1}\psi_{i_2}\ldots\psi_{i_n} = \epsilon_{i_1,i_2,\ldots i_n}, \tag{10.5}$$

where $\epsilon_{i_1,i_2,\ldots i_n}$ is the completely anti-symmetric epsilon tensor.

Consider a change variable, namely

$$\psi = a\chi + \zeta,$$

where a is a constant and ζ is a constant fermion. From Eq. 10.4, the non-zero fermion integral yields

$$1 = \int d\psi\psi = \int d\psi(a\chi + \zeta) = \int d\psi a\chi \Rightarrow d\psi = \frac{1}{a}d\chi. \tag{10.6}$$

Note that this is the inverse for the case of real variables, since $x = ay$ yields $dx = ady$.

For the case of N fermions, the anti-symmetric matrix $M_{ij} = -M_{ji}$ yields the change of variables

$$\psi_i = \sum_{j=1}^{N} M_{ij}\chi_j \Rightarrow \psi = M\chi.$$

Similarly to Eq. 10.6, it follows that

$$\prod_{i=1}^{N} d\psi_i = \frac{1}{\det M}\prod_{j=1}^{N} d\chi_j \Rightarrow D\psi = \frac{1}{\det M}D\chi, \tag{10.7}$$

where $D\psi = \prod_{i=1}^{N} d\psi_i$ and so on.

10.3 Fermion Hilbert space

The discussion in Section 2.4 on the bosonic state space and its dual is valid for fermions as well. There are two distinct fermionic variables, namely the variable ψ and its dual $\bar{\psi}$. Comparing with the bosonic case, if one takes the fermionic variable $\bar{\psi}$ to be the analog of the coordinate x, then ψ is the analog of the conjugate momentum variable p. The two different state spaces $V_{\bar{\psi}}$ and V_ψ are based on the coordinate variable $\bar{\psi}$ and its conjugate ψ, respectively. There are consequently *two* Hilbert spaces, namely $V_{\bar{\psi}}$ and V_ψ, that are dual to each other, as shown in Figure 2.3.

In analogy with bosonic variables x for which $\langle\phi|\phi\rangle = \int dx\phi^*(x)\phi(x) \geq 0$, the norm for the fermionic Hilbert space needs to be defined so as to yield a positive norm fermionic Hilbert space.

Choose $V_{\bar{\psi}}$ to be the state space of the fermionic degree of freedom $\bar{\psi}$. A fermion state function is then given by the scalar product of the dual coordinate state vector $\langle\bar{\psi}| \in V_{\psi}$ with $|f\rangle \in V_{\bar{\psi}}$ and yields

$$f(\bar{\psi}) = \langle\bar{\psi}|f\rangle = a + b\bar{\psi}.$$

The dual state of $|f\rangle$, denoted by $\langle f|$, is defined such that

$$\langle f|f\rangle = |a|^2 + |b|^2 > 0. \tag{10.8}$$

Physical (normalizable) state functions have $\langle f|f\rangle = 1$ and yield the interpretation

$$\begin{aligned}\langle f|f\rangle = 1 \quad &\Rightarrow \quad |a|^2 + |b|^2 = 1,\\ |a|^2 &= \text{probability, there is no fermion,}\\ |b|^2 &= \text{probability, there is one fermion.}\end{aligned}$$

Since $a,\ b$ are complex numbers, the space of all physical state functions is equal to a three-dimensional sphere S^3. Note that a fermion state function, similarly to the boson case, is equivalent to all state functions related to it by a global phase $e^{i\phi}$. Factoring out the phase from the physically distinct state functions yields, as discussed in Section 8.1.1, the Hilbert space

$$V_{\bar{\psi}} \equiv S^3/S^1 = S^2 : \text{Bloch sphere.}$$

Hence, similar to a spin $1/2$ system, the distinct physical states of a single fermion Hilbert space are parameterized by the points of a two-dimensional sphere. Or more formally, each state vector of the single fermion space of states corresponds to one point on a two-dimensional sphere.

The single fermion state space is seen to be isomorphic to the state space of the Ising spin discussed in Section 8.1.1, and shows that in essence a fermion is a *discrete* degree of freedom.

Define the dual state by

$$\bar{\psi} \rightarrow \psi \quad \Rightarrow \quad f^D(\psi) = \langle f|\psi\rangle = a^* + b^*\psi. \tag{10.9}$$

To achieve the required scalar product, note that

$$\begin{aligned}\langle f|f\rangle &= \int d\bar{\psi}d\psi f^D(\psi)e^{-\bar{\psi}\psi} f(\psi)\\ &= \int d\bar{\psi}d\psi(a^* + b^*\psi)e^{-\bar{\psi}\psi}(a + b\bar{\psi})\\ &= \int d\bar{\psi}d\psi\Big(|a|^2 + |b|^2\psi\bar{\psi} + a^*b\bar{\psi} + ab^*\psi\Big)(1 - \bar{\psi}\psi).\end{aligned} \tag{10.10}$$

Using the rules for fermion integration given in Eq. 10.4, the positive definite scalar product is given by

$$\langle f|f\rangle = |a|^2 + |b|^2.$$

10.3.1 Fermionic completeness equation

The fermionic variables have a phase space representation with $(\bar{\psi}, \psi)$ being analogous to the coherent state representation of the creation and destruction operators $(a^\dagger, a)$. Let $|\psi\rangle$ be the fermionic eigenstate and $\langle\bar{\psi}|$ be the dual fermionic eigenstate. The completeness equation for the fermion degree of freedom is given by

$$\mathbb{I} = \int d\bar{\psi} d\psi |\psi\rangle e^{-\bar{\psi}\psi} \langle\bar{\psi}|. \tag{10.11}$$

The fermion completeness equation is similar to the completeness equation for coherent states discussed in Section 5.12, and to Eq. 5.80 in particular. The fermion basis states are over-complete – as is the case for the bosonic coherent basis states – and the metric on the fermion state space that accounts for the over-completeness is given by $\exp\{-\bar{\psi}\psi\}$.

The inner product of the basis with its dual is given by the self-consistency equation that follows from the completeness equation and yields

$$\langle\bar{\psi}|\psi\rangle = e^{\bar{\psi}\psi}. \tag{10.12}$$

To verify that Eq. 10.12 is indeed consistent with the completeness equation given in Eq. 10.11, let $\bar{\zeta}, \zeta$ be fermionic variables; then

$$\begin{aligned}
\langle\bar{\zeta}|\zeta\rangle &= \int d\bar{\psi} d\psi \langle\bar{\zeta}|\psi\rangle e^{-\bar{\psi}\psi} \langle\bar{\psi}|\zeta\rangle = \int d\bar{\psi} d\psi \left(1 + \bar{\zeta}\psi\right)\left(1 - \bar{\psi}\psi\right)\left(1 + \bar{\psi}\zeta\right) \\
&= \int d\bar{\psi} d\psi \left(\bar{\zeta}\psi\bar{\psi}\zeta - \bar{\psi}\psi\right) = \int d\bar{\psi} d\psi \left(1 + \bar{\zeta}\zeta\right)\psi\bar{\psi} = e^{\bar{\zeta}\zeta}
\end{aligned}$$

as expected.

The inner product of the basis states and completeness is self-consistently determined since one requires the other: to prove completeness one needs the inner product. In particular, a proof of the resolution of the identity as given in Eq. 10.11 is

$$\begin{aligned}
\mathbb{I}^2 &= \int d\bar{\psi} d\psi d\bar{\zeta} d\zeta |\psi\rangle e^{-\bar{\zeta}\psi} \langle\bar{\zeta}|\zeta\rangle e^{-\bar{\psi}\zeta} \langle\bar{\psi}| \\
&= \int d\bar{\psi} d\psi d\bar{\zeta} d\zeta |\psi\rangle e^{-\bar{\zeta}\psi} e^{\bar{\zeta}\zeta} e^{-\bar{\psi}\zeta} \langle\bar{\psi}| \\
&= \int d\bar{\psi} d\psi |\psi\rangle e^{-\bar{\psi}\psi} \langle\bar{\psi}| = \mathbb{I},
\end{aligned}$$

where the result follows from the inner product given in Eq. 10.12 and performing the $\bar{\zeta}, \zeta$ integrations.

10.3.2 Fermionic momentum operator

The state space depends on $\bar{\psi}$, so we need to determine the representation of the dual coordinate ψ on the state space. The fermion coordinate operator $\hat{\psi}$ has the fermionic coordinate eigenstate given by

$$\hat{\psi}|\psi\rangle = \psi|\psi\rangle,$$

where the coordinate eigenvalue ψ is fermionic. The scalar product yields

$$\langle\bar{\psi}|\hat{\psi}|\psi\rangle = \psi\langle\bar{\psi}|\psi\rangle = \psi e^{\bar{\psi}\psi} = \frac{\delta}{\delta\bar{\psi}}(e^{\bar{\psi}\psi}),$$

$$\Rightarrow \psi\langle\bar{\psi}|\psi\rangle = \frac{\delta}{\delta\bar{\psi}}e^{\bar{\psi}\psi} \quad \Rightarrow \quad \psi = \frac{\delta}{\delta\bar{\psi}}.$$

Hence, on state space $V_{\bar{\psi}}$ the dual coordinate ψ yields

$$\psi f(\bar{\psi}) = \frac{\delta}{\delta\bar{\psi}} f(\bar{\psi}). \tag{10.13}$$

Hence, as mentioned in Section 10.3, the variable ψ is the analog of the momentum operator p and is evidenced by its action on state space, as given in Eq. 10.13.

In terms of the fermionic variables the creation and annihilation operators have the realization

$$a^\dagger = \bar{\psi}, \quad a = \frac{\delta}{\delta\bar{\psi}} \quad \text{with} \quad \{a^\dagger, a\} = 1.$$

The identification of the fermion variables with the fermion creation and destruction operators is consistent with the identification made earlier, in the discssion on Eq. 10.11, of the fermion completeness equation with the completeness equation for the coherent state space.

10.4 Antifermion state space

Let χ, $\bar{\chi}$ be a set of fermionic variables; the dual coordinate eigenstate is defined by $\langle\chi| \in V_\chi$; hence

$$\langle\chi|f\rangle = a + b\chi.$$

This change of definition for the coordinate of the state space will lead to the conclusion that V_χ is the state space for antifermions. The completeness equation continues to be

$$\mathbb{I} = \int d\bar{\chi} d\chi |\bar{\chi}\rangle e^{-\bar{\chi}\chi} \langle \chi|. \tag{10.14}$$

Consistency with the completeness equation for the antifermionic variables requires that the inner product continues to be the same as the particle case, namely

$$\langle \chi|\bar{\chi}\rangle = \exp\{\bar{\chi}\chi\}. \tag{10.15}$$

The following derivation provides a consistency check for the above completeness equation

$$\begin{aligned}
\langle \eta|\bar{\eta}\rangle &= \int d\bar{\chi} d\chi e^{\bar{\chi}\eta} e^{-\bar{\chi}\chi} e^{\bar{\eta}\chi} \\
&= \int d\bar{\chi} d\chi \,(1+\bar{\chi}\eta)\,(1-\bar{\chi}\chi)\,(1+\bar{\eta}\chi) \\
&= \int d\bar{\chi} d\chi \,(-\bar{\chi}\chi + \bar{\chi}\eta\bar{\eta}\chi) = \int d\bar{\chi} d\chi \,(-\bar{\chi}\chi)\,(1+\bar{\eta}\eta) \\
&= \exp\{\bar{\eta}\eta\}.
\end{aligned}$$

In the completeness equation given in Eq. 10.14, the variables $\chi, \bar{\chi}$ have been interchanged for the state space vectors – as compared to the completeness equation for variables ψ, $\bar{\psi}$, as in Eq. 10.11 – but with the metric being unchanged, namely $\exp\{-\bar{\chi}\chi\}$ and $\exp\{-\bar{\psi}\psi\}$.

To compensate for the difference in these two cases, an extra minus sign needs to introduce the conjugation of the state space vector $f(\chi)$. Conjugation is defined by $\chi \rightarrow -\bar{\chi}$.[1] Hence

$$f^D(\bar{\chi}) = f^*(-\bar{\chi}) \;=\; a^* - b^*\bar{\chi}.$$

It is verified that the rule for conjugation yields a positive definite norm for the state space V_χ,

$$\begin{aligned}
\langle f|f\rangle &= \int d\bar{\chi} d\chi f^D(\bar{\chi}) e^{-\bar{\chi}\chi} f(\chi) \\
&= \int d\bar{\chi} d\chi (a^* - b^*\bar{\chi})(a + b\chi)(1 - \bar{\chi}\chi) \\
&= \int d\bar{\chi} d\chi (|a|^2 + |b|^2)\chi\bar{\chi} = |a|^2 + |b|^2.
\end{aligned}$$

The fermion momentum operator is defined by

$$\langle \chi|\bar{\chi}|\bar{\chi}\rangle = \bar{\chi} e^{\bar{\chi}\chi} \;=\; -\frac{\delta}{\delta\chi} e^{\bar{\chi}\chi}$$

and yields

[1] This is because the order of integration in the scalar product is *reversed* compared to the fermion case, for which, under conjugation $\psi \rightarrow \bar{\psi}$.

$$\bar{\chi} f(\chi) = -\frac{\delta}{\delta \chi} f(\chi).$$

The fact that anticommuting variables $\chi, \bar{\chi}$ represent antifermions becomes clear when the antifermions are combined with fermions, as is done in the following section.

10.5 Fermion and antifermion Hilbert space

One can choose either of the Hilbert spaces $V_{\bar{\psi}}$ or V_{χ} to be fermionic state space; once a choice is made for the fermion state space, it automatically allows for the introduction of the concept of antifermions. The representation of antifermions is fixed by the choice that is made for the fermion state space. The normal convention is to choose $V_{\bar{\psi}}$ to be the fermionic state space and $\bar{\psi}$ the fermionic degree of freedom and V_{χ} to be the antifermionic state space and χ to be the antifermion degree of freedom.

The state space and path integral for the fermion and antifermion system is discussed below. A system containing both particle and anti-particle has a state space given by the tensor product of the fermion and antifermion state spaces, namely $V_{\bar{\psi}} \otimes V_{\chi}$.

The most general state vector is given by

$$\begin{aligned} f(\bar{\psi}, \chi) &= \langle \bar{\psi}, \chi | f \rangle = a + b\bar{\psi} + c\chi + d\bar{\psi}\chi, \\ |a|^2 &+ |b|^2 + |c|^2 + |d|^2 = 1. \end{aligned} \tag{10.16}$$

The interpretation of the state vector is the following:

- $|a|^2$ = probability of the system having no fermion or antifermion,
- $|b|^2$ = probability of the system having one fermion,
- $|c|^2$ = probability of the system having one antifermion,
- $|d|^2$ = probability of the system having one fermion and one antifermion.

Hermitian conjugation for the fermion and antifermion state space is defined by the operations:

1. Complex conjugate all the coefficients;
2. Reverse the order of all the fermion variables;
3. Make the substitution

$$\begin{pmatrix} \psi \\ \chi \end{pmatrix} \rightarrow \begin{pmatrix} 1 & 0 \\ 0 & -1 \end{pmatrix} \begin{pmatrix} \bar{\psi} \\ \bar{\chi} \end{pmatrix}. \tag{10.17}$$

The completeness for the fermion–antifermion degrees of freedom is

$$\mathbb{I} = \int d\bar{\psi} d\psi d\bar{\chi} d\chi |\psi, \bar{\chi}\rangle e^{-\bar{\psi}\psi - \bar{\chi}\chi} \langle \bar{\psi}, \chi|.$$

Consider a state vector $|f\rangle$ given in Eq. 10.16; its dual state vector, using the rules for fermion and antifermion conjugation, is given by

$$f^D(\psi, \bar{\chi}) = \langle f|\psi, \bar{\chi}\rangle = a^* + b^*\bar{\psi} - c^*\bar{\chi} - d^*\bar{\chi}\psi.$$

The rule for conjugation yields the positive definite inner product

$$\begin{aligned}
\langle f|f\rangle &= \int d\bar{\psi} d\psi d\bar{\chi} d\chi \, \langle f|\psi, \bar{\chi}\rangle \, e^{-\bar{\psi}\psi - \bar{\chi}\chi} \left\langle \bar{\psi}, \chi|f\right\rangle \\
&= \int d\bar{\psi} d\psi d\bar{\chi} d\chi \, \left(a^* + b^* - c^*\bar{\chi} - d^*\bar{\chi}\psi\right)\left(1 - \bar{\psi}\psi - \bar{\chi}\chi + \bar{\psi}\psi\bar{\chi}\chi\right) \\
&\quad \times \left(a + b + c\chi + d\bar{\psi}\chi\right) \\
&= \int d\bar{\psi} d\psi d\bar{\chi} d\chi \, \left(|a|^2 + |b|^2 + |c|^2 + |d|^2\right) \bar{\psi}\psi\bar{\chi}\chi \\
&= |a|^2 + |b|^2 + |c|^2 + |d|^2 .
\end{aligned}$$

10.6 Real and complex fermions: Gaussian integration

The rules of fermion integration are used for evaluating real and complex fermion Gaussian integration. Consider N real fermion variables χ_n with $n = 1, 2, \ldots, N$. Define the partition function

$$Z[J] = \prod_{n=1}^{N} \int d\chi_n \exp\left\{-\frac{1}{2}\chi_n M_{nm} \chi_m + J_n \chi_n\right\}. \tag{10.18}$$

The external source J_n is fermionic.

For N odd, $Z[J]$ is always zero for the case of $J = 0$; on expanding the exponential, one always has powers of the even product of fermions; hence, term by term, the partition function is given by terms that are all zero, namely

$$\prod_{i=1}^{N} \int d\chi_i \left(\text{even products of } \chi_j\right) = 0.$$

The term M is a real anti-symmetric matrix. One way of evaluating the partition function is to try and diagonalize M. However, one can only use a real transformation since all the fermions are real. Matrix algebra yields the result that every real anti-symmetric matrix M can be brought into a block-diagonal form by an orthogonal transformation O in the following manner:

$$M = O^T \begin{pmatrix} 0 & \lambda_1 & & & & & \\ -\lambda_1 & 0 & & & & & \\ & & 0 & \lambda_2 & & & \\ & & -\lambda_2 & 0 & & & \\ & & & & \ddots & & \\ & & & & & 0 & \lambda_{\frac{N}{2}} \\ & & & & & -\lambda_{\frac{N}{2}} & 0 \end{pmatrix} O = O^T \Lambda O, \qquad (10.19)$$

where

$$O^T O = 1 : \text{orthogonal}.$$

Let us perform a change of variables, and from Eq. 10.7

$$O\chi = \xi : \text{real}$$

$$\Rightarrow \prod_{i=1}^{N} d\chi_i = \det(O) \prod_{i=1}^{N} d\xi_i = \prod_{i=1}^{N} d\xi_i.$$

The partition function factors into a product of $N/2$ terms and yields

$$\begin{aligned} Z[0] &= \prod_{i=1}^{N} \int d\xi_i \exp\left\{-\frac{1}{2}\xi_n \Lambda_{nm} \xi_m\right\} \\ &= \prod_{i=1}^{N/2} \left[\int d\xi_1 d\xi_2 \exp\left\{-\frac{1}{2}\begin{pmatrix}\xi_1 & \xi_2\end{pmatrix}\begin{pmatrix} 0 & \lambda_i \\ -\lambda_i & 0\end{pmatrix}\begin{pmatrix}\chi_1 \\ \chi_2\end{pmatrix}\right\}\right] \\ &= \prod_{i=1}^{N/2} \int d\xi_1 d\xi_2 e^{+\lambda_i \xi_1 \xi_2} \;=\; \prod_{i=1}^{N/2} \lambda_i. \end{aligned} \qquad (10.20)$$

From Eq. 10.19

$$\det M = \prod_{i=1}^{N/2} \det \begin{pmatrix} 0 & \lambda_i \\ -\lambda_i & 0 \end{pmatrix} = \prod_{i=1}^{N/2} \lambda_i^2.$$

Hence, from Eq. 10.20, the partition function is given by

$$Z[0] = \sqrt{\det M}. \qquad (10.21)$$

The square root of an antisymmetric matrix is known as a Pfaffian and has many remarkable mathematical properties.

The general case of the partition function in the presence of an external fermionic source J_i can be derived from the results obtained. Rewriting the action given in Eq. 10.18 yields

$$Z[J] = \int D\chi \exp\left\{-\frac{1}{2}\left(\chi + JM^{-1}\right) M \left(\chi + M^{-1}J\right) + \frac{1}{2}J^{-1}MJ\right\}. \quad (10.22)$$

Using the fundamental invariance of fermion integration under the shift of the fermion integration variable given in Eq. 10.3 allows us to shift the integration variable,

$$\chi \to \chi - JM^{-1},$$

and yields, from Eqs. 10.22 and 10.21

$$Z[J] = \int D\chi e^{-\frac{1}{2}\chi M\chi} e^{\frac{1}{2}J^T M^{-1} J} = Z[0] e^{\frac{1}{2}J^T M^{-1} J}$$
$$\Rightarrow Z[J] = \sqrt{\det M} e^{\frac{1}{2}J^T M^{-1} J}.$$

The propagator can be obtained by fermionic differentiations on J_i,

$$\langle \chi_i \chi_j \rangle = \frac{\delta^2}{\delta J_i \delta J_j} \cdot \frac{1}{Z} \int D\chi e^{S+J\chi}\bigg|_{J=0} = \frac{\delta^2}{\delta J_i \delta J_j} \exp\left\{\frac{1}{2}\sum_{mn=1}^{N} J_m^T M_{mn}^{-1} J_n\right\}\bigg|_{J=0}$$
$$= \frac{\delta}{\delta J_i}\left[\left(\frac{1}{2}M_{jk}^{-1}J_k - \frac{1}{2}J_k M_{kj}^{-1}\right)e^S\right]\bigg|_{J=0} = \frac{1}{2}M_{ji}^{-1} - \frac{1}{2}M_{ij}^{-1}$$
$$\Rightarrow \langle \chi_i \chi_j \rangle = M_{ji}^{-1}.$$

10.6.1 Complex Gaussian fermion

Consider the N-dimensional Gaussian integral for fermions ψ_n and $\bar{\psi}_n$,

$$Z[J] = \prod_{n=1}^{N} \int d\bar{\psi}_n d\psi_n \exp\left\{-\bar{\psi}_n M_{nm} \psi_m + \bar{J}_n \psi_n + \bar{\psi}_n J_n\right\},$$

where $M_{nm} = -M_{mn}$ is an antisymmetric matrix. For real fermions $\psi = \psi^*$. For complex fermions $\psi = \psi_1 + i\psi_2$.

An antisymmetric matrix $M = -M^T$ can be diagonalized by a *unitary* transformation

$$M = U^\dagger \begin{pmatrix} \lambda_1 & & \\ & \ddots & \\ & & \lambda_N \end{pmatrix} U, \; U^\dagger U = 1.$$

In matrix notation

$$M = U^{\dagger}\Lambda U, \qquad UU^{\dagger} = 1, \qquad \det(UU^{\dagger}) = 1, \tag{10.23}$$

where $\Lambda = diag(\lambda_1, \ldots \lambda_N)$. Since the fermions ψ, $\bar{\psi}$ are complex, let us define the change of variables using the unitary matrix U, and from Eq. 10.7[2]

$$\begin{aligned} \bar{\psi}U^{\dagger} &= \bar{\eta}, \qquad \eta = U\psi, \\ D\eta D\bar{\eta} &= \frac{1}{\det(U)}D\psi \times \frac{1}{\det(U^{\dagger})}D\bar{\psi} = D\psi D\bar{\psi}. \end{aligned} \tag{10.24}$$

Hence, the fermion integrations completely factorize and yield

$$\begin{aligned} Z[0] &= \prod_n \int d\bar{\eta}_n d\eta_n \exp\{-\sum_{n=1}^{N} \lambda_n \bar{\eta}_n \eta_n\} \\ &= \prod_n \left[\int d\bar{\eta}_n d\eta_n e^{-\lambda_n \bar{\eta}_n \eta_n}\right] = \prod_n \lambda_n = \det M. \end{aligned} \tag{10.25}$$

The partition function with an external source is given as before by a shift of fermion integration variables. Write the partition function as

$$Z[J] = \int D\bar{\psi}D\psi \exp\left\{-\left(\bar{\psi} - \bar{J}M^{-1}\right)M\left(\psi - M^{-1}J\right) + \bar{J}M^{-1}J\right\}.$$

Using the fundamental property of fermion integration that it is invariant, a constant shift of fermion variables as given in Eq. 10.3 yields

$$\bar{\psi} \to \bar{\psi} + \bar{J}M^{-1}, \quad \psi \to \psi + M^{-1}J$$

and hence

$$Z[J] = \int D\bar{\psi}D\psi \exp\left\{-\bar{\psi}M\psi + \bar{J}M^{-1}J\right\} = (\det M)\exp\left\{\bar{J}M^{-1}J\right\}.$$

The fermion Gaussian integration obtained in Eq. 10.25 can be directly done using the rules of fermion integration. On expanding the exponential term $\exp\{\bar{\psi}M\psi\}$, only *one term* – $\left(\bar{\psi}M\psi\right)^N/N!$ containing the product of all the fermion variables – is nonzero inside the integrand. Using the notation of summing over repeated indexes, Eq. 10.5 yields

$$\begin{aligned} \int D\bar{\psi}D\psi e^{\bar{\psi}M\psi} &= \int D\bar{\psi}D\psi\left[\frac{1}{N!}\left(\bar{\psi}M\psi\right)^N\right] \\ &= \frac{1}{N!}M_{i_1j_1}M_{i_2j_2}\cdots M_{i_Nj_N}\int D\bar{\psi}D\psi\bar{\psi}_{i_1}\psi_{j_1}\bar{\psi}_{i_2}\psi_{j_2}\cdots\bar{\psi}_{i_N}\psi_{j_N} \end{aligned}$$

[2] Note that for real fermions one could not use a unitary transformation for a change of variables as this would lead to the transformed fermions being complex.

$$= \frac{1}{N!} M_{i_1 j_1} M_{i_2 j_2} \cdots M_{i_N j_N} \epsilon_{i_1 i_2 \cdots i_N} \epsilon_{j_1 j_2 \cdots j_N}$$
$$= \det M.$$

10.7 Fermionic operators

Operators for fermions, such as the Hamiltonian and momentum have a representation in terms of fermionic variables. Given that the dual of the fermionic coordinate $\bar{\psi}$ is given by ψ, all operators for fermions are expressed in terms of both the fermion coordinate and its dual. In this sense, the operators of the fermion degree of freedom are defined analogously to bosonic operators that are defined on phase space, as in Section 5.4.

Consider a fermionic operator $\mathcal{O}$; the operator is a mapping from $V_{\bar{\psi}}$ to itself and hence, similar to the bosonic case given in Eq. 2.5, $\mathcal{O}$ is an element of the tensor product space $V_{\bar{\psi}} \otimes V_{\psi}$,

$$\mathcal{O} \in \mathcal{V} \otimes \mathcal{V}_D \equiv V_{\bar{\psi}} \otimes V_{\psi}.$$

The matrix elements of $\mathcal{O}$ are given by

$$\langle \bar{\psi} | \mathcal{O} | \psi \rangle = \mathcal{O}(\bar{\psi}, \psi) \langle \bar{\psi} | \psi \rangle = \mathcal{O}(\bar{\psi}, \psi) e^{\bar{\psi}\psi}.$$

In particular, the Hamiltonian operator H is defined by

$$\langle \bar{\psi} | H | \psi \rangle = H(\bar{\psi}, \psi) \langle \bar{\psi} | \psi \rangle = H(\bar{\psi}, \psi) e^{\bar{\psi}\psi} \tag{10.26}$$

and yields, from Eq. 10.26,

$$\langle \bar{\psi} | e^{-\epsilon H} | \psi \rangle = \langle \bar{\psi} | [1 - \epsilon H] \psi \rangle = [1 - \epsilon H(\bar{\psi}, \psi)] e^{\bar{\psi}\psi} + O(\epsilon^2)$$
$$\simeq e^{-\epsilon H(\bar{\psi}, \psi)} e^{\bar{\psi}\psi} + O(\epsilon^2). \tag{10.27}$$

10.8 Fermionic path integral

One way of understanding the difference between a fermion and an antifermion is to examine the evolution of fermions in (Euclidean) time. For clarity, consider a time lattice with spacing ϵ and let lattice time be denoted by $n\epsilon$. The fermion degrees of freedom are defined on the lattice and denoted by $\bar{\psi}_n; \psi_n$. Consider a typical action for the lattice fermions given by

$$S_P = -\sum_n \bar{\psi}_n \psi_n + 2K \sum_n \bar{\psi}_{n+1} \psi_n$$
$$= \sum_n \mathcal{L}(n),$$

where the fermion Lagrangian is given by

$$\begin{aligned}\mathcal{L}(n) &= -\bar{\psi}_n\psi_n + 2K\bar{\psi}_{n+1}\psi_n \\ &= 2K\left[\bar{\psi}_{n+1} - \bar{\psi}_n\right]\psi_n + (2K-1)\bar{\psi}_n\psi_n.\end{aligned} \tag{10.28}$$

The path integral is given by

$$Z_P = \prod_n \int d\bar{\psi}_n d\psi_n e^S p. \tag{10.29}$$

The path integral is written in the fermion coherent state basis, similarly to the coherent basis path integral for bosons (real and complex degrees of freedom) as discussed in Section 5.14.

Note that the fermion variable ψ_n at time n propagates to variable $\bar{\psi}_{n+1}$ at time $n+1$. The partition function is re-written by repeatedly using the completeness equation given in Eq. 10.11, and yields

$$Z_P = \int D\bar{\psi}D\psi \langle\bar{\psi}_{n+1}|e^{-\epsilon H}|\psi_n\rangle e^{-\bar{\psi}_n\psi_n}\langle\bar{\psi}_n|e^{-\epsilon H}|\psi_{n-1}\rangle e^{-\bar{\psi}_{n-1}\psi_{n-1}}\ldots$$

Hence, from the definition of fermionic operators given in Eq. 10.27

$$\begin{aligned}e^{2K\bar{\psi}_{n+1}\psi_n} &= \langle\bar{\psi}_{n+1}|e^{-\epsilon H}|\psi_n\rangle \\ &\simeq e^{-\epsilon H(\bar{\psi}_{n+1},\psi_n)}\langle\bar{\psi}_{n+1}|\psi_n\rangle = e^{-\epsilon H(\bar{\psi}_{n+1},\psi_n)}e^{\bar{\psi}_{n+1}\psi_n}.\end{aligned} \tag{10.30}$$

Dropping the index of time on the fermion variables yields the particle Hamiltonian

$$\begin{aligned}&-\epsilon H_P(\bar{\psi},\psi) \simeq 2K\bar{\psi}\psi - \bar{\psi}\psi \\ &\Rightarrow H_P(\bar{\psi},\psi) \simeq \left(\frac{1-2K}{\epsilon}\right)\bar{\psi}\psi = \langle\bar{\psi}|H|\psi\rangle.\end{aligned}$$

Note that, as is the case for coherent states (Section 5.13), the Hamiltonian is automatically normal ordered.

The limit of continuous time is taken by defining the following continuum fermions $\bar{\psi}(t)$, $\psi(t)$ and Hamiltonian

$$\begin{aligned}&\bar{\psi}_t = \sqrt{2K}\bar{\psi}_n, \quad \psi_t = \sqrt{2K}\psi_n, \quad t = n\epsilon, \\ &H_P\left(\bar{\psi},\psi\right) = \left(\frac{1-2K}{2K\epsilon}\right)\bar{\psi}_t\psi_t = \omega\bar{\psi}_t\frac{\delta}{\delta\bar{\psi}_t}, \quad \omega = \frac{1-2K}{2K\epsilon}.\end{aligned} \tag{10.31}$$

The continuum Lagrangian $\mathcal{L}(t)$ and action, from Eqs. 10.28 and 10.31, are given by

$$\mathcal{L}(n) = -\epsilon\left[\frac{\partial\bar{\psi}_t}{\partial t}\psi_t + \frac{2K-1}{2\epsilon K}\bar{\psi}_t\psi_t\right] \equiv \epsilon\mathcal{L}(t),$$

$$S_P = \int dt\mathcal{L}(t), \quad \mathcal{L}(t) = -\left[\frac{\partial\bar{\psi}_t}{\partial t}\psi_t + \omega\bar{\psi}_t\psi_t\right].$$

The term S_P is the action for a particle propagating forward in time. Consider another action that describes the time evolution of *anti-particles*, namely

$$S_A = -\sum_n \bar{\chi}_n\chi_n + 2K\sum_n \bar{\chi}_n\chi_{n+1}$$

with

$$Z_A = \int D\bar{\chi}D\chi e^{S_A}. \tag{10.32}$$

The antifermion variable $\bar{\chi}_n$ propagates from time n to the variable χ_{n+1} at time $n+1$. If one thinks of the variable χ_{n+1} as representing a particle, then one can think of the anti-particle as being equivalent to a particle propagating **backwards** in time; although this way of thinking is not required, it does help to develop some physical intuition about the peculiarities of anti-particles.

Given the nature of the state space of the anti-particles, Z_A can be decomposed as

$$Z_A = \int D\bar{\chi}D\chi \ldots \langle\chi_{n+1}|e^{-\epsilon H_A}|\bar{\chi}_n\rangle e^{-\bar{\chi}_n\chi_n}\langle\chi_n|e^{-\epsilon H_A}|\bar{\chi}_{n-1}\rangle \ldots \tag{10.33}$$

and by inspection

$$\begin{aligned}\langle\chi_{n+1}|e^{-\epsilon H_A}|\bar{\chi}_n\rangle &= e^{2K\bar{\chi}_n\chi_{n+1}}\\ &= e^{-\epsilon H_A(\bar{\chi}_n\chi_{n+1})}e^{\bar{\chi}_n\chi_{n+1}}\\ \Rightarrow H_A &= \left(\frac{1-2K}{\epsilon}\right)\bar{\chi}\chi = \langle\chi|H|\bar{\chi}\rangle.\end{aligned}$$

Note that the order of the matrix element of the antifermions, namely $\langle\chi|H|\bar{\chi}\rangle$, is the reverse of the fermion case. We define continuum fermion variables by

$$\bar{\chi}_t = \sqrt{2K}\,\bar{\chi}_n, \quad \chi_t = \sqrt{2K}\,\chi_n, \quad \omega = \left(\frac{1-2K}{2K\epsilon}\right), \quad t = n\epsilon.$$

Anti-commuting the fermionic variables and ignoring an irrelevant constant yields, similarly to Eq. 10.31, the continuum antifermion Hamiltonian

$$H_A = -\omega\chi_t\bar{\chi}_t. \tag{10.34}$$

After normal ordering, interpreting H_A as an operator yields, using $\bar{\chi} = -\delta/\delta\chi$,

$$H_A = \omega\chi_t\frac{\delta}{\delta\chi_t}.$$

10.9 Fermion–antifermion Hamiltonian

From Eq. 10.31, a Hamiltonian for fermions is given by

$$H_P = \omega\bar{\psi}\psi = \omega\bar{\psi}\frac{\delta}{\delta\bar{\psi}}. \tag{10.35}$$

The eigenstates and eigenenergies of the Hamiltonian are given by

$$\begin{aligned} H_P\Phi_n &= E_n\Phi_n \\ \Phi_0 &= 1 \qquad E_0 = 0 \\ \Phi_1 &= \bar{\psi} \qquad E_1 = \omega. \end{aligned}$$

From Eq. 10.34, a typical antifermion Hamiltonian is given by

$$H_A = \tilde{\omega}\chi\frac{\delta}{\delta\chi} = -\tilde{\omega}\chi\bar{\chi} \tag{10.36}$$

with eigenstates and eigenenergies

$$\begin{aligned} \Phi_0^A &= 1 \qquad\qquad E_0 = 0 \\ \Phi_1^A &= \chi \qquad\qquad E_1 = \omega. \end{aligned}$$

For a single fermion degree of freedom, the Hamiltonian H has the form $\omega\bar{\psi}\psi$ and this is all that one can construct. Hence, it is necessary to look at systems coupling fermion and antifermion as well as coupling fermions to bosons.

Consider a fermion and antifermion system. The Hilbert state space is four-dimensional since there are four possible states for the system, as enumerated in Eq. 10.16. A simple Hamiltonian for the fermion and antifermion system is a sum of the fermion and antifermion system with a coupling term, namely

$$\begin{aligned} H &= \omega\bar{\psi}\frac{\delta}{\delta\bar{\psi}} + \omega'\chi\frac{\delta}{\delta\chi} + \lambda\bar{\psi}\chi \\ &\equiv \omega\bar{\psi}\psi - \omega'\chi\bar{\chi} + \lambda\bar{\psi}\chi \end{aligned} \tag{10.37}$$

with real values for ω, ω', λ.

The normalized eigenfunctions and eigenvalues are given by

$$H\Phi_n = E_n\Phi_n.$$

One can directly verify that

$$\begin{aligned} \Phi_1 &= \chi \qquad E_1 = \omega' \\ \Phi_2 &= \bar{\psi} \qquad E_2 = \omega \\ \Phi_3 &= \bar{\psi}\chi \qquad E_3 = \omega + \omega'. \end{aligned}$$

Note that the eigenstates do not form a complete set since the Hilbert space is four-dimensional, the reason being that this Hamiltonian is not Hermitian. Using the rules of conjugation given in Eq. 10.17,

$$\bar{\psi} \to \psi, \chi \to -\bar{\chi}$$
$$\psi \to \bar{\psi}, \bar{\chi} \to -\chi$$

and reversing the order of the fermion variables shows that

$$H^{\dagger} = \omega\bar{\psi}\psi - \omega'\chi\bar{\chi} - \lambda\bar{\chi}\psi \neq H. \tag{10.38}$$

A Hermitian Hamiltonian for the fermion–antifermion system is given by

$$\begin{aligned} H &= \omega\bar{\psi}\frac{\delta}{\delta\bar{\psi}} + \omega'\chi\frac{\delta}{\delta\chi} + \lambda\bar{\psi}\chi\frac{\delta^2}{\delta\bar{\psi}\delta\chi} \\ &= \omega\bar{\psi}\psi + \omega'\bar{\chi}\chi - \lambda\bar{\psi}\chi\psi\bar{\chi}. \end{aligned}$$

Hermitian conjugation, explicitly shown below, shows that the Hamiltonian is Hermitian,

$$\begin{aligned} H^{\dagger} &= \omega\bar{\psi}\psi + \omega'\bar{\chi}\chi - \lambda\bar{\chi}^c\psi^c\chi^c\bar{\psi}^c \\ &= \omega\bar{\psi}\psi + \omega'\bar{\chi}\chi - \lambda(-\chi)\bar{\psi}(-\bar{\chi})\psi \\ &= \omega\bar{\psi}\psi + \omega'\bar{\chi}\chi - \lambda\chi\bar{\psi}\bar{\chi}\psi \\ &= \omega\bar{\psi}\psi + \omega'\bar{\chi}\chi - \lambda\bar{\psi}\chi\psi\bar{\chi} \\ &= H. \end{aligned}$$

The eigenfunctions can be read off by inspection and are

$$\begin{aligned} &\Phi_0 = 1 \quad E_0 = 0 \\ &\Phi_1 = \bar{\psi} \quad E_1 = \omega \\ &\Phi_2 = \chi \quad E_2 = \omega' \\ &\Phi_3 = \bar{\psi}\chi \quad E_3 = \omega + \omega' - \lambda. \end{aligned}$$

The Hamiltonian given in Eq. 10.37 can be made Hermitian by adding the term required, and yields

$$H = \omega\bar{\psi}\frac{\delta}{\delta\bar{\psi}} + \omega'\chi\frac{\delta}{\delta\chi} + i\lambda(\bar{\psi}\chi + \bar{\chi}\psi). \tag{10.39}$$

Two of the four orthogonal eigenstates can be obtained by inspection, and this yields

$$\begin{aligned} &\Phi_1 = \chi \quad E_1 = \omega' \\ &\Phi_2 = \bar{\psi} \quad E_2 = \omega. \end{aligned}$$

For the other two eigenfunctions consider the ansatz

$$\Phi = \bar{\psi}\chi + ic.$$

Applying the Hamiltonian given in Eq. 10.39 on Φ yields

$$H\Phi = \left[\omega + \omega' - c\lambda\right]\left(\bar{\psi}\chi - i\frac{\lambda}{\omega + \omega' - c\lambda}\right).$$

One obtains the eigenvalue condition

$$c = -\frac{\lambda}{\omega + \omega' - c\lambda},$$

which has two solutions, namely

$$c_{\pm} = \frac{1}{2\lambda}\left(\omega + \omega' \pm \sqrt{(\omega + \omega')^2 + 4\lambda^2}\right), \quad c_+c_- = -1. \tag{10.40}$$

Hence, the remaining two eigenfunctions and eigenvalues are given by

$$\Phi_3 = \frac{1}{\sqrt{1 + c_+^2}}\left[\bar{\psi}\chi + ic_+\right], \quad E_3 = \omega + \omega' - \lambda c_+$$

$$\Phi_4 = \frac{1}{\sqrt{1 + c_-^2}}\left[\bar{\psi}\chi + ic_-\right], \quad E_4 = \omega + \omega' - \lambda c_-. \tag{10.41}$$

The interpretation of the state Φ_3 is that $|c_+|^2/\sqrt{1 + |c_+|^2}$ is the likelihood that the system has no particles and $1/\sqrt{1 + |c_+|^2}$ is the likelihood of having a fermion–antifermion pair, with a similar interpretation for Φ_4.

10.9.1 Orthogonality and completeness

To illustrate the workings of fermion calculus, the orthogonality of the states Φ_3, Φ_4 is explicitly computed. Using the rules for forming the conjugate state function yields

$$\begin{aligned}
\langle\Phi_3|\Phi_4\rangle &= \int d\bar{\psi}d\psi d\bar{\chi}d\chi\, \langle\Phi_3|\psi;\bar{\chi}\rangle e^{-\bar{\psi}\psi-\bar{\chi}\chi}\langle\bar{\psi};\chi|\Phi_4\rangle \\
&= \int d\bar{\psi}d\psi d\bar{\chi}d\chi\, \Phi_3^{\dagger}[\psi;\bar{\chi}\,]\Phi_4[\bar{\psi};\chi]e^{-\bar{\psi}\psi-\bar{\chi}\chi} \\
&= \int d\bar{\psi}d\psi d\bar{\chi}d\chi\, (-\bar{\chi}\psi - ic_+)(\bar{\psi}\chi + ic_-)e^{-\bar{\psi}\psi-\bar{\chi}\chi} \\
&= (1 + c_+c_-)\int d\bar{\psi}d\psi d\bar{\chi}d\chi\;\; \bar{\psi}\psi\bar{\chi}\chi \\
&= (1 + c_+c_-) = 0,
\end{aligned}$$

since $c_+c_- = -1$.

The completeness equation can be expressed in terms of the eigenfunctions by

$$\mathbb{I} = \sum_{i=1}^{4} |\Phi_i\rangle\langle\Phi_i|. \tag{10.42}$$

To verify Eq. 10.42, one needs to prove that

$$\sum_{i=1}^{4} \langle\bar{\psi}, \chi|\Phi_i\rangle\langle\Phi_i|\psi, \bar{\chi}\rangle = \exp\{\bar{\psi}\psi + \bar{\chi}\chi\}.$$

Using the explicit form of the state functions derived above yields

$$\begin{aligned}\sum_{i=1}^{4} \langle\bar{\psi}, \chi|\Phi_i\rangle\langle\Phi_i|\psi, \bar{\chi}\rangle &= \bar{\chi}\chi + \bar{\psi}\psi + N_+ \left[\bar{\psi}\chi + ic_+\right][-\bar{\chi}\psi - ic_+] \\ &\quad + N_- \left[\bar{\psi}\chi + ic_-\right][-\bar{\chi}\psi - ic_-] \\ &= \bar{\chi}\chi + \bar{\psi}\psi + \left[c_+^2 N_+ + c_-^2 N_-\right] + [N_+ + N_-]\,\bar{\psi}\psi\bar{\chi}\chi \\ &\quad - i[c_+ N_+ + c_- N_-][\bar{\psi}\chi + \bar{\chi}\psi] \\ \text{where } N_+ &= \frac{1}{1+c_+^2}, \quad N_- = \frac{1}{1+c_-^2}.\end{aligned}$$

From Eq. 10.40 it follows that

$$\begin{aligned}c_+^2 N_+ + c_-^2 N_- &= 1 = N_+ + N_-, \\ c_+ N_+ + c_- N_- &= 0.\end{aligned}$$

Hence

$$\sum_{i=1}^{4} \langle\bar{\psi}, \chi|\Phi_i\rangle\langle\Phi_i|\psi, \bar{\chi}\rangle = 1 + \bar{\chi}\chi + \bar{\psi}\psi + \bar{\psi}\psi\bar{\chi}\chi = \exp\left\{\bar{\psi}\psi + \bar{\chi}\chi\right\},$$

thus verifying Eq. 10.42.

10.10 Fermion–antifermion Lagrangian

Consider the propagation of a fermion–antifermion system given by the Lagrangian

$$\epsilon\mathcal{L}_n = -\bar{\psi}_n\psi_n - \bar{\chi}_n\chi_n + 2K\left(\bar{\psi}_{n+1}\psi_n + \bar{\chi}_n\chi_{n+1}\right). \tag{10.43}$$

The term $\mathcal{L}_n$ consists of a fermion propagating forward in time and its antifermion (since K is the same for both) propagating "backward" in time.

Define a two-component spinor

$$\psi_n = \begin{pmatrix}\psi_n \\ \chi_n\end{pmatrix} \tag{10.44}$$

and

$$\bar{\psi}_n = \begin{pmatrix}\bar{\psi}_n & \bar{\chi}_n\end{pmatrix} \tag{10.45}$$

and let

$$\gamma_0 = \begin{pmatrix}1 & 0\\ 0 & -1\end{pmatrix}. \tag{10.46}$$

Then

$$\mathcal{L}_n = -\bar{\psi}_n\psi_n + K\left\{\bar{\psi}_{n+1}(1+\gamma_0)\psi_n + \bar{\psi}_n(1-\gamma_0)\psi_{n+1}\right\}. \tag{10.47}$$

The fermion action is

$$S = \epsilon\sum_n \mathcal{L}_n. \tag{10.48}$$

Consider the continuum limit by defining $t = n\epsilon$ and take the limit of $\epsilon \to 0$, yielding

$$\begin{aligned} K\bar{\psi}_{n+1}\psi &= K(\bar{\psi}_{n+1} - \bar{\psi}_n)\psi_n + K\bar{\psi}_n\psi_n \\ &\cong K\epsilon\partial_0\bar{\psi}_n\psi_n + K\bar{\psi}_n\psi_n. \end{aligned} \tag{10.49}$$

Similarly

$$K\bar{\chi}_n\chi_{n+1} = \epsilon K\bar{\chi}_n\partial_0\chi_n + K\bar{\chi}_n\chi_n. \tag{10.50}$$

Hence

$$\epsilon\mathcal{L}_n = (-1+2K)\left[\bar{\psi}_n\psi_n + \bar{\chi}_n\chi_n\right] + 2K\epsilon\left[\partial_0\bar{\psi}_n\psi_n + \bar{\chi}_n\partial_0\chi_n\right]. \tag{10.51}$$

Let us define continuum fermionic variables by

$$\begin{aligned} \bar{\psi}_t &= (2K)^{1/2}\bar{\psi}_n, \quad \psi_t = (2K)^{1/2}\psi_n, \\ \bar{\chi}_t &= (2K)^{1/2}\bar{\chi}_n, \quad \chi_t = (2K)^{1/2}\chi_n. \end{aligned}$$

The continuum Lagrangian is given by

$$\epsilon\mathcal{L}(t) = \left(\frac{2K-1}{2K}\right)\left[\bar{\psi}_t\psi_t + \bar{\chi}_t\chi_t\right] + \epsilon\left[\bar{\psi}_t(-\partial_0)\psi_t + \bar{\chi}_t\partial_0\chi_t\right].$$

We define

$$\begin{aligned} &m\epsilon = -\left[\frac{2K-1}{2K}\right] \quad\Rightarrow\quad 2K = \frac{1}{1+m\epsilon} \\ &\Psi = \begin{pmatrix}\psi_t\\ \chi_t\end{pmatrix} \qquad \bar{\Psi} = \begin{pmatrix}\bar{\psi}_t & \bar{\chi}_t\end{pmatrix} \qquad \gamma_0 = \begin{pmatrix}1 & 0\\ 0 & -1\end{pmatrix}. \end{aligned} \tag{10.52}$$

Hence, the continuum action and Lagrangian are given by

$$S = \int_{-\infty}^{+\infty} dt \mathcal{L},$$
$$\mathcal{L}(t) = -\bar{\Psi}(\gamma_0 \partial_0 + m)\Psi$$
$$: \text{one-dimensional Dirac Lagrangian.}$$

10.11 Fermionic transition probability amplitude

The derivation done earlier for finding the eigenstates and eigenenergies of the fermionic Hamiltonian can be obtained directly working with the fermion transition probability amplitude. Recall for a fermion particle

$$\langle \bar{\psi} | e^{-\epsilon H} | \psi \rangle = e^{2K\bar{\psi}\psi}. \tag{10.53}$$

The eigenvalue and eigenstates that are given by $\Phi_0 \sim 1$ and $\Phi_1 \sim \bar{\psi}$ can be directly obtained by performing fermion integration in the following manner:

$$\begin{aligned}
\langle \bar{\psi} | e^{-\epsilon H} | \Phi_0 \rangle &= \int d\varsigma d\bar{\varsigma} \langle \bar{\psi} | e^{-\epsilon H} | \varsigma \rangle e^{-\bar{\varsigma}\varsigma} \langle \bar{\varsigma} | \Phi_0 \rangle \\
&= \int d\varsigma d\bar{\varsigma} e^{2K\bar{\psi}\varsigma} e^{-\bar{\varsigma}\varsigma} \\
&= \int d\varsigma d\bar{\varsigma} (1 + 2K\bar{\psi}\varsigma)(1 - \bar{\varsigma}\varsigma) \\
&= 1,
\end{aligned} \tag{10.54}$$
$$\Rightarrow \Phi_0(\bar{\psi}) = \langle \bar{\psi} | \Phi_0 \rangle = 1, \quad E_0 = 0.$$

For the eigenstate $\Phi_1 \sim \bar{\psi}$ consider the calculation

$$\begin{aligned}
\langle \bar{\psi} | e^{-\epsilon H} | \Phi_1 \rangle &= \int d\varsigma d\bar{\varsigma} e^{2K\bar{\psi}\varsigma} e^{-\bar{\varsigma}\varsigma} \bar{\varsigma} \\
&= 2K\bar{\psi} \int d\varsigma d\bar{\varsigma} \varsigma \bar{\varsigma} = 2K\bar{\psi} = e^{-\epsilon E_1} \bar{\psi},
\end{aligned}$$
$$\Rightarrow \langle \psi | \Phi_1 \rangle = \bar{\psi},$$

which yields the expected answer. The eigenenergy is given by

$$\begin{aligned}
\Rightarrow E_1 &= -\frac{1}{\epsilon} \ln(2K) \\
&= -\frac{1}{\epsilon} \ln\left(\frac{1}{1 + m\epsilon}\right) \\
&\simeq m + O(\epsilon).
\end{aligned} \tag{10.55}$$

Similarly for the antifermion, recall

$$\langle\chi|e^{-\epsilon H}|\bar{\chi}\rangle = e^{2K\bar{\chi}\chi}. \tag{10.56}$$

Hence, for the antifermion ground state, $e^{-\epsilon H}|\Phi_0\rangle = |\Phi_0\rangle$ since $\Phi_0 \sim 1$.

For $\langle\chi|\Phi_1\rangle = \chi$ one has

$$\begin{aligned}\langle\chi|e^{-\epsilon H}|\Phi_1\rangle &= \int \langle\chi|e^{-\epsilon H}|\bar{\varsigma}\rangle e^{-\bar{\varsigma}\varsigma}\langle\varsigma|\Phi_1\rangle \\ &= \int_{\bar{\varsigma}\varsigma} e^{2K\bar{\varsigma}\chi}e^{-\bar{\varsigma}\varsigma}\varsigma \\ &= \int_{\bar{\varsigma}\varsigma}(1+2K\bar{\varsigma}\chi)(1-\bar{\varsigma}\varsigma)\varsigma \\ &= 2K\chi\int(-\bar{\varsigma}\varsigma) = 2K\chi. \end{aligned} \tag{10.57}$$

Note that the transition amplitude automatically yields a normal ordered Hamiltonian with the energy given by

$$\begin{aligned} E_1 &= -\frac{1}{\epsilon}\ln(2K) \\ &\simeq m. \end{aligned} \tag{10.58}$$

Until now, fermions and antifermions have equal mass but have not been coupled, and hence their contrasting properties have not come into play. One can couple them by nonlinear terms in the Lagrangian such as $\lambda\bar{\psi}\psi\bar{\chi}\chi$, as well as by their coupling to gauge fields, and this is briefly explored below.

10.12 Quark confinement

Consider a one-dimensional toy model of quarks (complex fermions) and antiquarks (complex antifermions) given by

$$\mathcal{L}_n = -\bar{\psi}_n\psi_n + 2K\left(\bar{\psi}_{n+1}\psi_n + \bar{\chi}_n\chi_{n+1}\right) - \bar{\chi}_n\chi_n.$$

A gauge transformation on the fermions is defined by

$$\begin{aligned} \psi_n &\rightarrow e^{i\phi_n}\psi_n & \bar{\psi}_n &\rightarrow \bar{\psi}_n e^{-i\phi_n} \\ \chi_n &\rightarrow e^{i\phi_n}\chi_n & \bar{\chi}_n &\rightarrow \bar{\chi}_n e^{-i\phi_n}. \end{aligned}$$

To leave $\mathcal{L}_n$ invariant we need fix the nearest neighbor term. Let us introduce gauge field e^{iB_n} and modify $\mathcal{L}_n$ to

$$\mathcal{L}_n = -\bar{\psi}_n\psi_n - \bar{\chi}_n\chi_n + 2K\left(\bar{\psi}_{n+1}e^{-iB_n}\psi_n + \bar{\chi}_n e^{iB_n}\chi_{n+1}\right).$$

Under a gauge transformation, let B_n have the transformation

$$e^{iB_n} \rightarrow e^{i\phi_n} e^{iB_n} e^{-i\phi_{n+1}},$$

and hence the combined transformations on the fermions and gauge field leave the Lagrangian $\mathcal{L}$ invariant.

Note that B_n is an angular variable taking values in $[-\pi, +\pi]$. The partition function is

$$Z = \prod_n \int_{-\pi}^{\pi} \frac{dB_n}{2\pi} d\bar{\psi}_n d\psi_n d\bar{\chi}_n d\chi_n e^S.$$

Since e^{iB_n} couples nearest neighboring instants of time, one can derive as before the transition amplitude for quark–antiquark system, namely

$$\langle \bar{\psi}_{n+1}, \chi_{n+1} | e^{-\epsilon H} | \psi_n, \bar{\chi}_n \rangle,$$

by generalizing Eq. 10.30. One can choose to include the integration over the gauge field in the definition of the transition probability amplitude for the fermions, and this yields

$$\langle \bar{\psi}_{n+1}, \chi_{n+1} | e^{-\epsilon H} | \psi_n, \bar{\chi}_n \rangle = \int_{-\pi}^{\pi} \frac{dB_n}{2\pi} e^{\mathcal{L}_n}.$$

Dropping the subscripts from $\mathcal{L}_n$ yields

$$\begin{aligned}\langle \bar{\psi}, \chi | e^{-\epsilon H} | \psi, \bar{\chi} \rangle &= \int_{-\pi}^{\pi} \frac{dB}{2\pi} \left[1 + 2K\bar{\psi} e^{-iB} \psi\right]\left[1 + 2K\bar{\chi} e^{iB} \chi\right] \\ &= 1 + (2K)^2 \bar{\psi}\psi\bar{\chi}\chi.\end{aligned}$$

The transition probability amplitude acting on a general state function $|\Phi\rangle$ yields

$$\begin{aligned}\langle \bar{\psi}, \chi | \Phi \rangle &= \int d\bar{\zeta} d\zeta d\bar{\xi} d\xi \langle \bar{\psi}, \chi | e^{-\epsilon H} | \zeta, \bar{\xi} \rangle e^{-\bar{\zeta}\zeta - \bar{\xi}\xi} \langle \bar{\zeta}, \xi | \Phi \rangle \\ &= \int d\bar{\zeta} d\zeta d\bar{\xi} d\xi \left[1 + (2K)^2 \bar{\psi}\zeta\bar{\xi}\chi\right] e^{-\bar{\zeta}\zeta - \bar{\xi}\xi} \Phi(\bar{\zeta}, \xi). \qquad (10.59)\end{aligned}$$

The only nonzero integral is

$$\int d\bar{\zeta} d\zeta d\bar{\xi} d\xi \;\; \zeta\bar{\zeta}\xi\bar{\xi} = 1.$$

Solving the eigenvalue equation of the quark–antiquark Hamiltonian

$$e^{-\epsilon H} |\Phi_n\rangle = e^{-\epsilon E_n} |\Phi_n\rangle$$

using Eq. 10.59 yields the eigenfunctions and eigenvalues given in the table below.

$\Phi_n(\bar{\zeta}, \xi)$	E_n	State
1	0	vacuum
$\bar{\psi}$	∞	one quark
χ	∞	one antiquark
$\bar{\psi}\chi$	$2\left[-\frac{\ln(2K)}{\epsilon}\right]$	quark+antiquark

Note that single quark $\bar{\psi}$ and antiquark χ states are confined since they have infinite energy and cannot propagate in time; they are fixed at whatever moment in time they are created. Only the vacuum and paired states of quark–antiquark, namely state 1 and $\bar{\psi}\chi$, are finite energy eigenstates and hence can propagate in time.

10.13 Summary

The fermion variable takes only two values and is fundamentally different from a real variable. The fermion degree of freedom describes a system that is essentially discrete and at the same time quite distinct from the Ising variable, that belongs to the category of real variables.

A differential and integral calculus was developed for the fermion variable and the concepts applicable to a real variable were generalized to the fermion case. Gaussian integration was defined for both real and complex fermions and the results are similar to the real variable case, but with a few significant differences.

Fermion and antifermion variables emerge naturally, based on the manner in which conjugation of the state vector is defined. The state space and Hamiltonian for fermions and antifermions were derived and the state space was shown to behave in the manner that one intuitively expects for a discrete system. A few simple models of the fermion and antifermion path integral, Hamiltonian, and state space were discussed. A one-dimensional toy model based on fermion and antifermion degrees of freedom coupled to a gauge field was shown to exhibit quark confinement.

Part four

Quadratic path integrals

11

Simple harmonic oscillator

A simple harmonic oscillator, or oscillator in short, is one of the most important systems in physics as well as in quantum mechanics. The path integral and Hamiltonian of a simple harmonic oscillator are leading exemplars for the description of a wide class of physical systems. An oscillator is a theoretical model of great utility, primarily due to its simplicity that allows for the exact derivation of many results.

The simple harmonic oscillator can be generalized to the case of infinitely many oscillators, and is also the starting point for analysis of many nonlinear quantum systems.

Path integrals for the simple harmonic oscillator are all based on Gaussian integration, which has been discussed in Section 7.2. The harmonic oscillator is described by Gaussian *path integrals*, and is a bedrock of the general theory of path integration. Moreover, all perturbation expansions of nonlinear path integrals about the oscillator path integral, which includes the semi-classical expansion, are based on results of Gaussian integration.

The properties of the harmonic oscillator are analyzed in the coordinate space representation, and its path integral as well as its correlation functions are analyzed. All formulas and derivations are for Euclidean time.

In Sections 11.1 and 11.2 the Hamiltonian and state space of the oscillator are studied and the correlator is derived using state space methods. In Section 11.2 the infinite time oscillator is introduced and in Sections 11.3–11.5 the path integral for the oscillator is evaluated. In Sections 11.6–11.11 the oscillator path integral is studied on a finite lattice and the path integral is reduced to a finite-dimensional ordinary multiple integral. In Section 11.12 the finite lattice path integral is derived using the technique of the transfer matrix.

11.1 Oscillator Hamiltonian

The oscillator Hamiltonian H has a kinetic term equal to the free particle case – as in Eq. 5.69 – and which has an operator form valid both in Minkowski and Euclidean time; the oscillator Hamiltonian has a quadratic potential and hence is given by

$$H = -\frac{1}{2m}\frac{\partial^2}{\partial x^2} + \frac{1}{2}m\omega^2 x^2.$$

Let us define the creation and annihilation operators as

$$x = \frac{1}{\sqrt{2m\omega}}\left(a + a^\dagger\right), \quad \frac{\partial}{\partial x} = \sqrt{\frac{m\omega}{2}}\left(a - a^\dagger\right),$$
$$a = \frac{1}{\sqrt{2m\omega}}\left(m\omega x + \frac{\partial}{\partial x}\right), \quad a^\dagger = \frac{1}{\sqrt{2m\omega}}\left(m\omega x - \frac{\partial}{\partial x}\right),$$

and these yield the commutation equations

$$\left[x, \frac{\partial}{\partial x}\right] = -\mathbb{I} \ \Rightarrow \ \left[a, a^\dagger\right] = \mathbb{I}.$$

We define the oscillator ground state $|0\rangle$ by $a|0\rangle = 0$. The Hamiltonian and its eigenfunctions and eigenvalues are given by

$$H = \omega a^\dagger a + \frac{1}{2}\omega,$$
$$|n\rangle = \frac{(a^\dagger)^n}{\sqrt{n!}}|0\rangle, \quad \langle m|n\rangle = \delta_{n-m}, \quad a|0\rangle = 0,$$
$$H|n\rangle = E_n|n\rangle, \quad E_n = n\omega + \frac{1}{2}\omega.$$

Furthermore, the oscillator algebra $[a, a^\dagger] = \mathbb{I}$ yields the basis states

$$a^\dagger\,|n\rangle = \sqrt{n+1}\,|n+1\rangle, \quad a\,|n\rangle = \sqrt{n}\,|n-1\rangle,$$

which are complete, namely

$$\sum_n |n\rangle\langle n| = \mathbb{I}. \tag{11.1}$$

11.2 The propagator

The Heisenberg operator is given by

$$x_H(t) = e^{tH}\hat{x}e^{-tH} \equiv \hat{x}_t.$$

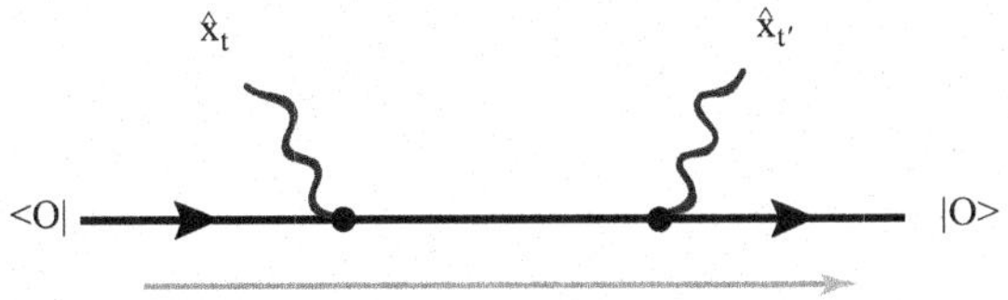

Figure 11.1 The correlator of two coordinate operators acting on the vacuum state.

Consider the correlation function, shown in Figure 11.1,

$$D(t,t') = \langle 0|T[x_H(t)x_H(t')]|0\rangle, \tag{11.2}$$

which is a measure of the time interval over which disturbances on the vacuum are correlated. Given the special role of $D(t,t')$ – the correlator of two coordinate operators – it is called the propagator.

Taking $t' = 0$ and $t > 0$, the correlator is given by

$$\begin{aligned} D(t) &= e^{\frac{\omega\tau}{2}}\langle 0|\hat{x}e^{-tH}\hat{x}|0\rangle, && (11.3)\\ &= \frac{1}{2m\omega}\langle 0|(a+a^\dagger)e^{-t\omega a^+ a}(a+a^\dagger)|0\rangle \\ &= \frac{1}{2m\omega}\langle 1|e^{-t\omega a^+ a}|1\rangle \\ &= \frac{1}{2m\omega}e^{-\omega t}. && (11.4)\end{aligned}$$

For the general case given in Eq. 11.2, it can readily be shown that

$$D(t,t') = \frac{1}{2m\omega}e^{-\omega|t-t'|}. \tag{11.5}$$

11.2.1 Finite time propagator

The propagator for finite time can be evaluated exactly for the case of the harmonic oscillator. The oscillator basis is used for the derivation and another derivation will be given later using the path integral. Figure 11.2 shows two coordinate operators acting at two different times, with a finite periodic time given by τ.

Recall that $x_H(t) \equiv \hat{x}_t$; the finite time correlator is

$$D\left(t,t',\tau\right) = \frac{1}{Z(\tau)}tr\left(T\left[\hat{x}_t\hat{x}_{t'}\right]e^{-\tau\hat{H}}\right),\quad Z(\tau) = tr\left(e^{-\tau\hat{H}}\right). \tag{11.6}$$

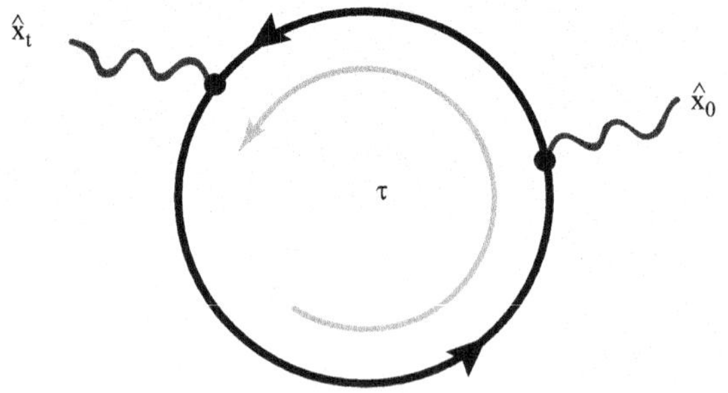

Figure 11.2 The correlator of two coordinate operators for periodic time.

Taking $t' = 0$ and $t > 0$ yields, for $D(t, 0, \tau) = D(t, \tau)$,

$$\begin{aligned} D(t, \tau) &\equiv \frac{1}{Z(\tau)} tr\left(\hat{x}_t \hat{x}_0 e^{-\tau \hat{H}}\right) = \frac{1}{Z(\tau)} tr\left(e^{-(\tau - t)\hat{H}} \hat{x} e^{-t\hat{H}} \hat{x}\right) \\ &= \frac{N}{Z(\tau)}, \end{aligned}$$

$$\text{where } N = tr\left(e^{-(\tau - t)\hat{H}} \hat{x} e^{-t\hat{H}} \hat{x}\right).$$

The numerator N is given by using the complete oscillator basis states $|n\rangle$ given by Eq. 11.1; hence

$$\begin{aligned} N &= \sum_{n=0}^{\infty} \langle n \left| e^{-(\tau - t)\hat{H}} \hat{x} e^{-t\hat{H}} \hat{x} \right| n\rangle \\ &= \frac{1}{2m\omega} \sum_{n=0}^{\infty} e^{-(\tau - t)\left(n + \frac{1}{2}\right)\omega} \langle n \left| \left(a + a^\dagger\right) e^{-t\hat{H}} \left(a + a^\dagger\right) \right| n\rangle. \end{aligned}$$

Using the oscillator algebra yields

$$\begin{aligned} N &= \frac{1}{2m\omega} \sum_{n=0}^{\infty} e^{-(\tau - t)\left(n + \frac{1}{2}\right)\omega} \Big\{ \left(\sqrt{n+1}\,\langle n+1| + \sqrt{n}\,\langle n-1|\right) e^{-t\hat{H}} \\ &\qquad \times \left(\sqrt{n+1}\,|n+1\rangle + \sqrt{n}\,|n-1\rangle\right) \Big\} \\ &= \frac{1}{2m\omega} \sum_{n=0}^{\infty} e^{-\tau\left(n + \frac{1}{2}\right)\omega} \left\{ (n+1) e^{-\omega t} + n e^{\omega t} \right\} \\ &= \frac{e^{-\frac{\omega\tau}{2}}}{2m\omega} \sum_{n=0}^{\infty} \left\{ n e^{-n\omega\tau} \left(e^{-\omega t} + e^{\omega t}\right) + e^{-n\omega\tau} e^{-\omega t} \right\}. \end{aligned}$$

Note the two summations

$$\sum_{n=0}^{\infty} e^{-n\omega\tau} = \frac{1}{1 - e^{-\omega\tau}}, \quad \sum_{n=0}^{\infty} n e^{-n\omega\tau} = -\frac{\partial}{\partial(\omega\tau)} \sum_{n=0}^{\infty} e^{-n\omega\tau} = \frac{e^{-\omega\tau}}{(1 - e^{-\omega\tau})^2}.$$

Hence, the partition function is given by

$$Z(\tau) = tr(e^{-\tau \hat{H}}) = e^{-\frac{\omega\tau}{2}} \sum_{n=0}^{\infty} e^{-n\omega\tau} = \frac{1}{2\sinh\left(\frac{\omega\tau}{2}\right)},$$

and the numerator simplifies to

$$\begin{aligned} N &= \frac{e^{-\frac{\omega\tau}{2}}}{2m\omega}\left[\frac{\left(e^{-\omega t} + e^{\omega t}\right)e^{-\omega\tau}}{(1-e^{-\omega\tau})^2} + \frac{e^{-\omega t}}{1-e^{-\omega\tau}}\right] \\ &= \frac{e^{-\frac{\omega\tau}{2}}}{2m\omega}\left[\frac{e^{-\omega t}e^{-\omega\tau} + e^{\omega t}e^{-\omega\tau} + e^{-\omega t} - e^{-\omega\tau}e^{-\omega t}}{(1-e^{-\omega\tau})^2}\right] \\ &= \frac{1}{2m\omega}\left[\frac{e^{\omega t}e^{-\frac{\omega\tau}{2}} + e^{-\omega t}e^{\frac{\omega\tau}{2}}}{\left(e^{\frac{\omega\tau}{2}} - e^{-\frac{\omega\tau}{2}}\right)^2}\right] \\ &= \frac{1}{2m\omega}\frac{\cosh\left(\frac{\omega\tau}{2} - \omega t\right)}{2\sinh^2\left(\frac{\omega\tau}{2}\right)}. \end{aligned}$$

Hence, the propagator is given by

$$D(t,\tau) = \frac{N}{Z(\tau)} = \frac{1}{2m\omega}\frac{\cosh\left(\omega t - \frac{\omega\tau}{2}\right)}{\sinh\left(\frac{\omega\tau}{2}\right)}. \tag{11.7}$$

In general, for two arbitrary times t and t', the correlator is given by time ordering the operators and yields

$$D\left(t,t';\tau\right) = \frac{1}{2m\omega}\frac{\cosh\left(\omega|t-t'| - \frac{\omega\tau}{2}\right)}{\sinh\left(\frac{\omega\tau}{2}\right)}. \tag{11.8}$$

Consider the limiting case of $\tau \to \infty$,

$$\lim_{\tau\to\infty} e^{-\tau H} \simeq e^{-\frac{\omega\tau}{2}}|0\rangle\langle 0|$$

and hence, from Eq. 11.6,

$$\begin{aligned} \lim_{\tau\to\infty} D(t,t';\tau) &\to \frac{tr\left(T\left[\hat{x}_t\hat{x}_{t'}\right]e^{-\frac{\omega\tau}{2}}|0\rangle\langle 0|\right)}{tre^{-\frac{\omega\tau}{2}}|0\rangle\langle 0|} \\ &= \langle 0|T\left[\hat{x}_t\hat{x}_{t'}\right]|0\rangle. \end{aligned}$$

Equation 11.2 shows that the finite time correlator reduces to the one for infinite time. Directly taking the limit of $\tau \to \infty$ in Eq. 11.8 yields the limiting value

$$\lim_{\tau\to\infty} D(t,t';\tau) \to \frac{1}{2m\omega}e^{-\omega|t-t'|},$$

as was obtained earlier in Eq. 11.5.

11.3 Infinite time oscillator

The action for the infinite time oscillator is given by

$$\begin{aligned}S &= -\frac{m}{2}\int_{-\infty}^{+\infty} dt\,(\dot{x}^2+\omega^2x^2)\\ &= -\frac{m}{2}\int dtx\left(-\frac{d^2}{dt^2}+\omega^2\right)x\\ &= -\frac{m}{2}\int dtdt'x_tx_{t'}\left(-\frac{d^2}{dt^2}+\omega^2\right)\delta(t-t'),\end{aligned}$$

and yields

$$A_{tt'} = \left(-\frac{d^2}{dt^2}+\omega^2\right)\delta\left(t-t'\right).$$

The inverse of $A_{tt'}$ is given by

$$\begin{aligned}A^{-1}_{tt'} &= \left(\frac{1}{-\frac{d^2}{dt^2}+\omega^2}\right)\delta(t-t') = \frac{1}{\left(-\frac{d^2}{dt^2}+\omega^2\right)}\int\frac{dp}{2\pi}e^{ip(t-t')}\\ &= \int\frac{dp}{2\pi}\frac{e^{ip(t-t')}}{p^2+\omega^2} = \frac{1}{2\omega}e^{-\omega|t-t'|}.\end{aligned} \tag{11.9}$$

Hence

$$\begin{aligned}E[x_\tau x_0] &= \frac{1}{Z}\int Dx\; x_\tau x_0 e^S\\ &= \frac{1}{m}A^{-1}_{\tau 0} = \frac{1}{2m\omega}e^{-\omega|\tau|},\end{aligned} \tag{11.10}$$

as expected from Eq. 11.4.

To interpret the meaning of the correlator one can repeat the earlier linear regression analysis discussed in Section 9.4. Since $E[x_\tau]=0$ and $E[x_0^2]=1/(2m\omega)$, the random variables are related by the linear regression

$$x_\tau \simeq \frac{A^{-1}_{\tau 0}}{E[x_0^2]}x_0 \simeq e^{-\omega|\tau|}x_0.$$

11.4 Harmonic oscillator's evolution kernel

The simple harmonic oscillator's action and Lagrangian is given by

$$\mathcal{L} = -\frac{m}{2}\dot{x}^2 - \frac{1}{2}m\omega^2x^2.$$

The Euclidean time path integral is given by

$$K(x_f, x_i; \tau) = \langle x'|e^{-\tau H}|x\rangle = \int Dx\, e^S .$$

The continuum action for finite time is given by

$$S = \int_0^\tau \mathcal{L} dt = -\frac{m}{2}\int_0^\tau dt\, (\dot{x}^2 + \omega^2 x^2) .$$

Consider the classical field equation

$$\frac{\delta S}{\delta x(t)} = m\ddot{x}_c(t) - m\omega^2 x_c(t) = 0, \tag{11.11}$$

$$\Rightarrow x_c(t) = Ae^{-\omega t} + Be^{\omega t}, \tag{11.12}$$

$$\text{boundary conditions:} \quad x_c(0) = x, \quad x_c(\tau) = x'. \tag{11.13}$$

Define new quantum variables $\xi(t)$ by

$$x(t) = \xi(t) + x_c(t)$$

that obey the boundary conditions

$$\xi(0) = 0 = \xi(\tau).$$

Hence, the path integral yields

$$\int Dx\, e^S = \int D\xi\, e^{S[\xi + x_c]},$$

where $Dx = D\xi$. The action, for $\dot{\xi} = d\xi/dt$, is

$$\begin{aligned} S[\xi + x_c] &= -\frac{m}{2}\int_0^\tau dt \left\{ \left(\dot{\xi} + \dot{x}_c\right)^2 + \omega^2 \left(\xi + x_c\right)^2 \right\} \\ &= -\frac{m}{2}\int_0^\tau dt \left\{ \dot{\xi}^2 + 2\dot{\xi}\dot{x}_c + \dot{x}_c^2 + \omega^2 \left(\xi^2 + 2\xi x_c + x_c^2\right) \right\} \\ &= -\frac{m}{2}\int dt\, (\dot{x}_c^2 + \omega^2 x_c^2) - m\int (\dot{\xi}\dot{x}_c + \omega^2 \xi x_c) - \frac{m}{2}\int dt\, (\dot{\xi}^2 + \omega^2\xi^2) . \end{aligned}$$

Note, since $\xi(0) = \xi(\tau) = 0$,

$$\int_0^\tau \dot{\xi}\dot{x}_c = \int_0^\tau \left\{ \frac{d}{dt}(\xi \dot{x}_c) - \xi \ddot{x}_c \right\} = \xi(0)\dot{x}_c(0) - \xi(\tau)\dot{x}_c(\tau) - \int \xi \ddot{x}_c dt$$
$$= - \int \xi \ddot{x}_c dt.$$

The cross-terms of x_c and ξ cancel using the classical equation, Eq. 11.11, and

$$S[x_c + \xi] = S[x_c] + S[\xi]. \tag{11.14}$$

Hence, the evolution kernel factorizes and yields

$$K(x', x; \tau) = \mathcal{N} e^{S_c},$$

with

$$\mathcal{N} = \int D\xi e^{S[\xi]} : \text{ independent of } x, x'. \tag{11.15}$$

It is shown in Eq. 11.24 that

$$\int D\xi e^{S[\xi]} = \mathcal{N} = \sqrt{\frac{m\omega}{2\pi \sinh \omega\tau}},$$

yielding the result

$$K(x', x; \tau) = \sqrt{\frac{m\omega}{2\pi \sinh \omega\tau}} e^{S_c}. \tag{11.16}$$

The classical action S_c is given by

$$S_c = -\frac{m}{2} \int_0^\tau dt \left(\dot{x}_c^2 + \omega^2 x_c^2 \right)$$
$$= -\frac{m}{2} \left(x_c \dot{x}_c \right) |_0^\tau - \frac{m}{2} \int_0^\tau dt x_c \left(-\ddot{x}_c + \omega^2 \ddot{x}_c \right)$$
$$= -\frac{m}{2} \left[x_c(\tau)\dot{x}_c(\tau) - x_c(0)\dot{x}_c(0) \right].$$

Equation 11.12 yields

$$\dot{x}_c(t) = \omega \left(-Ae^{-\omega t} + Be^{\omega t} \right).$$

The classical action is hence

$$S_c[x_c] = -\frac{m\omega}{2} \left[B^2 \left(e^{2\omega\tau} - 1 \right) + A^2 \left(1 - e^{-2\omega\tau} \right) \right]. \tag{11.17}$$

The boundary conditions, from Eq. 11.13, are

$$x_c(0) = A + B = x, \quad x_c(\tau) = Ae^{-\omega\tau} + Be^{\omega\tau} = x',$$

and yield

$$A = \frac{1}{2\sinh\omega\tau}\left(xe^{\omega\tau} - x'\right), \quad B = \frac{1}{2\sinh\omega\tau}\left(x' - xe^{-\omega\tau}\right), \tag{11.18}$$

and hence the classical solution for S_0, from Eqs. 11.12 and 11.18, is given by

$$x_c(t) = \frac{1}{2\sinh\omega\tau}\left[\left(xe^{\omega\tau} - x'\right)e^{-\omega t} + \left(x' - xe^{-\omega\tau}\right)e^{\omega t}\right]. \tag{11.19}$$

From Eq. 11.17, the classical action is

$$S_c = -\frac{m\omega}{2\sinh\omega\tau}\left[\left(x^2 + x'^2\right)\cosh\omega\tau - 2xx'\right], \tag{11.20}$$

yielding the final result for the evolution kernel

$$K(x', x; \tau) = \sqrt{\frac{m\omega}{2\pi\sinh\omega\tau}}\exp\left\{-\frac{m\omega}{2\sinh\omega\tau}\left[\left(x^2 + x'^2\right)\cosh\omega\tau - 2xx'\right]\right\}. \tag{11.21}$$

11.5 Normalization

Recall $K(x', x; \tau) = \langle x'|e^{-\tau\hat{H}}|x\rangle$. Consider the periodic trace of K, namely

$$trK = \int dx\langle x|e^{-\tau H}|x\rangle = \int dx K(x, x; \tau).$$

Using $K = \mathcal{N}\exp\{S_c\}$ yields

$$\begin{aligned} trK &= \mathcal{N}\int_{-\infty}^{+\infty} dx e^{S_c(x,x;\tau)} \\ &= \mathcal{N}\int dx \exp\left\{-\frac{m\omega}{\sinh\omega\tau}[\cosh\tau - 1]x^2\right\} \\ &= \mathcal{N}\sqrt{\frac{2\pi}{m\omega}\frac{\sinh\omega\tau}{2(\cosh\omega\tau - 1)}}. \end{aligned} \tag{11.22}$$

In the oscillator basis

$$H = \omega a^\dagger a + \frac{\omega}{2}$$

the trace yields, using this basis,

$$\begin{aligned} tr(K) &= tr\left(e^{-\tau H}\right) \\ &= e^{-\frac{\tau\omega}{2}} tr\left(e^{-\tau\omega a^\dagger a}\right) = e^{-\frac{\tau\omega}{2}}\sum_n\langle n|e^{-\tau\omega a^\dagger a}|n\rangle \\ &= e^{-\frac{\tau\omega}{2}}\sum_{n=0}^{\infty}\left(e^{-n\tau\omega}\right) = e^{-\frac{\tau\omega}{2}}\frac{1}{1 - e^{-\tau\omega}}. \end{aligned} \tag{11.23}$$

Hence, from Eqs. 11.22 and 11.23

$$\mathcal{N} = \sqrt{\frac{m\omega}{2\pi \sinh(\omega\tau)}}. \tag{11.24}$$

11.6 Generating functional for the oscillator

Consider the simple harmonic oscillator in the presence of an external source $j(t)$ given by

$$\mathcal{L} = -\frac{1}{2}m\dot{x}^2(t) - \frac{1}{2}m\omega^2 x^2(t) + j(t)x(t)$$

and with the finite time action

$$S = -\frac{1}{2}m\int_0^\tau \left(\dot{x}^2 + w^2x^2\right) + \int_0^\tau j(t)x(t) = S_0 + \int_0^\tau j(t)x(t). \tag{11.25}$$

The path integral for the normalized generating functional is given by

$$Z[j] = \frac{1}{Z}\int Dxe^S = \frac{1}{Z}\int Dxe^{S_0+\int j(t)x(t)},$$

$$Z = \int Dxe^S.$$

The generating functional contains the full content of the quantum system; all the correlation functions for the action S_0, in the presence of the boundary conditions, can be evaluated by functional differentiation. In particular, the first few correlators are given by

$$\frac{1}{Z}\frac{\delta Z[j]}{\delta j(t)}\bigg|_{j=0} = \frac{1}{Z}\int Dx\; x(t)e^{S_0},$$

$$\frac{1}{Z}\frac{\delta^2 Z[j]}{\delta j(t)\delta j(t')}\bigg|_{j=0} = \frac{1}{Z}\int Dx\; x(t)x(t')e^{S_0},$$

$$\frac{1}{Z}\frac{\delta^3 Z[j]}{\delta j(t)\delta j(t')\delta j(t'')}\bigg|_{j=0} = \frac{1}{Z}\int Dx\; x(t), x(t')x(t'')e^{S_0},$$

and so on.

11.6.1 Classical solution with source

Similar to the derivation in Section 11.4, one can use the classical equations of motion to evaluate the Gaussian path integral. The classical solution obeys the equation

$$\left(-\frac{d^2}{dt^2}+w^2\right)x_c = \frac{J(t)}{m},$$

$$x_c = x_H + x_I, \tag{11.26}$$

where x_H, x_I are the homogeneous and inhomogeneous solutions respectively. Note that the homogeneous solution depends on the boundary conditions. In particular,

$$\left(-\frac{d^2}{dt^2}+w^2\right)x_H = 0 \;\Rightarrow\; x_H = A'e^{-wt} + B'e^{wt}.$$

The inhomogeneous solution is given by

$$\begin{aligned} x_I(t) &= \frac{1}{m}\langle t|\frac{1}{\left(-\frac{d^2}{dt^2}+w^2\right)}|J\rangle = \frac{1}{2mw}\int_0^\tau dt' e^{-w|t-t'|}J(t') \\ &= \frac{1}{2mw}\left[\int_0^t dt' e^{-w(t-t')}J(t') + \int_t^\tau dt' e^{-w(t'-t)}J(t')\right], \end{aligned} \tag{11.27}$$

where

$$\left(-\frac{d^2}{dt^2}+w^2\right)^{-1}\delta(t-t') = \frac{e^{-w|t-t'|}}{2w}. \tag{11.28}$$

We fix boundary conditions by

$$x_c(0) = x = x_H(0) + x_I(0), \quad x_c(\tau) = x' = x_H(\tau) + x_I(\tau).$$

Hence

$$\begin{aligned} x(0) &= x = A' + B' - \frac{1}{2m\omega}\int_0^\tau dt e^{-\omega t}J(t), \\ x(\tau) &= x' = A'e^{-\omega\tau} + B'e^{\omega\tau} - \frac{1}{2m\omega}\int_0^\tau dt e^{-\omega(\tau-t)}J(t), \end{aligned}$$

which yields

$$\begin{aligned} A' &= \frac{1}{2\sinh(\omega\tau)}\left[x' - e^{-\omega\tau}x - \frac{1}{m}\int_0^\tau dt \sinh\omega(\tau-t)J(t)\right], \\ B' &= \frac{1}{2\sinh(\omega\tau)}\left[e^{-\omega\tau}x - x' - \frac{e^{-\omega\tau}}{m\omega}\int_0^\tau dt \sinh(\omega t)J(t)\right]. \end{aligned}$$

After considerable simplifications,

$$S[x_c; j] = -\frac{m\omega}{2\sinh\omega\tau}\left[(x^2 + x'^2)\cosh\omega\tau - 2xx'\right]$$
$$+ \frac{x}{\sinh w\tau}\int_0^\tau dt J(t)\sinh w(\tau - t) + \frac{x'}{\sinh w\tau}\int_0^\tau dt J(t)\sinh(wt)$$
$$+ \frac{1}{mw\sinh w\tau}\int_0^\tau\int_0^t dt' J(t)\sinh w(\tau - t)\sinh\left(wt'\right) J\left(t'\right). \tag{11.29}$$

The generating functional is hence given by the decomposition

$$x(t) = x_c(t) + \xi(t), \quad \xi(0) = 0 = \xi(\tau),$$

that yields

$$\int Dxe^S = \tilde{A}e^{S[x_c;j]}\int D\xi e^{S_0[\xi]},$$
$$\Rightarrow \tilde{A}\int D\xi e^{S_0[\xi]} = \mathcal{N} \ : \ \text{independent of } x, x', \text{and } j(t),$$

with normalization constant $\mathcal{N}$ given by Eq. 11.24. The generating functional, from Eq. 11.29, is given by

$$Z[j] = \frac{1}{Z}\int Dxe^S = e^{S[\xi;j]},$$

with the normalization constant $\mathcal{N}$ canceling out.

11.6.2 Source free classical solution

The classical solution yields the inhomogeneous term x_I, as in Eq. 11.26, arising from the source $j(t)$. To obtain the final answer given in Eq. 11.29 requires a fair amount of algebra.

Given the importance of the simple harmonic oscillator, another derivation of the same result is given here. This derivation turns out to be useful in situations that are more complicated than the simple one that we are considering.

For this derivation, consider the classical solution for the source free action, that is *only* $S_0[x]$ – the external source $j(t)$ is *not* included in $S_0[x]$ – and hence the classical solution does not have an inhomogeneous term x_I. The change of variables $x = x_c + \xi$ now yields a path integral for which the quantum variable is coupled to the source $j(t)$ – and one needs to perform a path integral over the quantum degree of freedom $\xi(t)$ to obtain the result.

Recall, from Eq. 11.25, that the simple harmonic action for the generating functional is given by

$$S = S_0 + \int_0^\tau j(t)x(t), \quad S_0 = -\frac{1}{2}m\int_0^\tau \left(\dot{x}^2 + w^2x^2\right).$$

The classical solution for S_0, from Eq. 11.19, is given by

$$x_c(t) = \frac{1}{2\sinh\omega\tau}\left[\left(xe^{\omega\tau} - x'\right)e^{-\omega t} + \left(x' - xe^{-\omega\tau}\right)e^{\omega t}\right].$$

Define the quantum variables ξ by

$$x(t) = x_c(t) + \xi(t), \quad \xi(0) = 0 = \xi(\tau).$$

The classical solution yields, from Eq. 11.14, the factorization

$$S_0[x_c + \xi] = S_0[x_c] + S_0[\xi].$$

The path integral for the generating functional, up to an irrelevant normalization, is hence given by

$$\int Dxe^S = e^{F_0[x_c;j]}\int D\xi e^{F[\xi;j]},$$

$$F_0[x_c; j] = S_0[x_c] + \int_0^\tau j(t)x_c(t),$$

$$F[\xi; j] = S_0[\xi] + \int_0^\tau dtj(t)\xi(t).$$

The following path integral over the quantum variables $\xi(t)$ needs to be performed:

$$\int D\xi \exp\{F[\xi; j]\}, \quad \xi(0) = 0 = \xi(\tau).$$

The boundary conditions $\xi(0) = 0 = \xi(\tau)$ are satisfied by the Fourier sine expansion for $\xi(t)$,

$$\xi(t) = \sum_{n=1}^{\infty}\sin\left[\frac{n\pi t}{\tau}\right]\xi_n,$$

where, for each n, ξ_n is an independent and real integration variable. Hence, using the orthogonality of the sine functions,

$$\int_0^\tau dt\sin\left[\frac{n\pi t}{\tau}\right]\sin\left[\frac{m\pi t}{\tau}\right] = \frac{\tau}{2}\delta_{n-m}$$

yields

$$F[\xi; j] = -\frac{1}{2}\sum_{n=1}^{\infty}\lambda_n\xi_n^2 + \sum_{n=1}^{\infty} j_n\xi_n,$$

$$\lambda_n = \frac{1}{2}m\tau\left\{\left(\frac{n\pi}{\tau}\right)^2 + w^2\right\}, \quad j_n = \int_0^{\tau} j(t)\sin\left[\frac{n\pi t}{\tau}\right].$$

All the ξ_n variables have decoupled and we obtain, up to a normalization,

$$\int D\xi \exp\{F[\xi; j]\} = \prod_{n=1}^{+\infty}\int_{-\infty}^{+\infty} d\xi_n \exp\left\{-\frac{1}{2}\sum_{n=1}^{\infty}\lambda_n\xi_n^2 + \sum_{n=1}^{\infty} j_n\xi_n\right\}$$
$$= \sqrt{\frac{1}{2\pi\lambda_n}}\exp\left\{\frac{1}{2}\sum_{n=1}^{\infty} j_n\frac{1}{\lambda_n}j_n\right\}.$$

The result further simplifies,

$$\sum_{n=1}^{\infty} j_n\frac{1}{\lambda_n}j_n = \int_0^{\tau} dt\int_0^{\tau} dt' j(t)D(t,t')j(t'),$$

$$D(t,t') = \sum_{n=1}^{\infty}\frac{1}{\lambda_n}\sin\left[\frac{n\pi t}{\tau}\right]\sin\left[\frac{n\pi t'}{\tau}\right].$$

Using the identity

$$\sum_{n=1}^{+\infty}\frac{\cos(\theta)}{a^2+n^2} = \frac{\pi}{2a}\frac{\cosh[a(\pi-|\theta|)]}{\sinh\pi a} - \frac{1}{2a^2} \tag{11.30}$$

yields

$$D\left(t,t'\right) = \frac{1}{mw\sinh w\tau}\sinh w\,(\tau - t)\sinh\left(wt'\right), \quad t > t' \tag{11.31}$$

and hence

$$\frac{1}{2}\int_0^{\tau} dt\int_0^{\tau} dt' j(t)D\left(t,t'\right)j\left(t'\right)$$
$$= \frac{1}{mw\sinh w\tau}\int_0^{\tau} dt\int_0^{t} dt' J(t)\sinh w(\tau - t)\sinh\left(wt'\right)J\left(t'\right). \tag{11.32}$$

Note, from Eq. 11.19, that

$$\int_0^{\tau} j(t)x_c(t) = \frac{x}{\sinh w\tau}\int_0^{\tau} dt J(t)\sinh w(\tau - t)$$
$$+ \frac{x'}{\sinh w\tau}\int_0^{\tau} dt J(t)\sinh(wt). \tag{11.33}$$

The term $S_0[x_c]$ is given in Eq. 11.20; hence, from Eqs. 11.32 and 11.33

$$\int Dxe^S = \exp\{S[x_c; j]\},$$

$$S[x_c; j] = S_0[x_c] + \int_0^\tau j(t)x_c(t) + \frac{1}{2}\int_0^\tau dt \int_0^\tau dt' j(t)D(t,t')j(t'),$$

and, as expected, $S[x_c; j]$ is given by Eq. 11.29.

Furthermore, the generating functional is given by

$$Z[j] = \frac{1}{Z}\int Dxe^S$$
$$= \exp\left\{\frac{1}{2}\int_0^\tau dt \int_0^\tau dt' j(t)D(t,t')j(t') + \int_0^\tau j(t)x_c(t)\right\},$$

with the source-free classical solution $S_0[x_c]$ canceling out due to the normalization by Z.

11.7 Harmonic oscillator's conditional probability

Consider the Lagrangian given by

$$\mathcal{L} = -\frac{1}{2}m\left[(\frac{dx}{dt})^2 + \omega^2 x^2\right].$$

The evolution kernel is the *probability amplitude* that the harmonic oscillator's degree of freedom will reach a final point x_f from initial point x_i in time τ. It is given, up to a normalization constant that cancels out, by

$$K(x_f, x_i; \tau) = \langle x_f|e^{-\tau H}|x_i\rangle$$
$$= \mathcal{N}\exp\left\{-\frac{m\omega}{2\sinh\omega\tau}\left[(x_i^2 + x_f{}^2)\cosh\omega\tau - 2x_i x_f\right]\right\}.$$

The *conditional probability* quantifies the likelihood that the outcome of an experiment will yield the value of the coordinate to be x_i, given that x_i has occurred; it is given by[1]

$$P(x_f|x_i; \tau) = \frac{|K(x_f, x_i; \tau)|^2}{\int dx_f |K(x_f, x_i; \tau)|^2}.$$

The denominator can be simplified as follows:

$$\int dx_f |K(x_f, x_i; \tau)|^2 = \int dx_f e^{-\frac{m\omega}{2\sinh\omega\tau}\left[\left(x_i^2+x_f^2\right)\cosh\omega\tau - 2x_i x_f\right]}$$
$$= \sqrt{\frac{2\pi\sinh\omega\tau}{2m\omega}}\exp\left\{-m\omega x_i^2\tanh\omega\tau\right\},$$

[1] The normalization constant $\mathcal{N}$ cancels out and is set to unity in this section.

yielding the conditional probability given by

$$P(x_f|x_i;\tau) = \sqrt{\frac{m\omega}{\pi \sinh \omega\tau}} \exp\left\{m\omega \tanh \omega\tau x_i^2\right\} |K(x_f, x_i; \tau)|^2.$$

Note that the normalization of the evolution kernel drops out of $P(x_i, x_f; \tau)$.

As expected, one obtains the normalization

$$\int_{-\infty}^{+\infty} dx_f P(x_f|x_i;\tau) = 1.$$

The conditional probability is different for Euclidean time compared to Minkowski time.

11.8 Free particle path integral

A path integral derivation is given of the evolution kernel for a free particle degree of freedom moving in one dimension ($d = 1$), which was obtained earlier in Eq. 5.72 using the eigenfunctions of the free particle Hamiltonian.

Let $\epsilon = t/N$; from Eq. 4.30 the path integral for finite ϵ is given by a multiple integral. Using the infinitesimal form of Eq. 5.72, which also directly follows from the Dirac–Feynman formula, yields (set $\hbar = 1$)

$$\begin{aligned} K(x, x'; \tau) &= \prod_{n=0}^{N-1} K(x_{n+1}, x_n; \epsilon) \\ &= \left(\sqrt{\frac{m}{2\pi\epsilon}}\right) \prod_{n=1}^{N-1} \sqrt{\frac{m}{2\pi\epsilon}} \int_{-\infty}^{+\infty} dx_n e^{-\frac{m}{2\epsilon}\sum_{n=0}^{N-1}(x_{n+1}-x_n)^2} \\ &\equiv \int Dx\, e^{-\frac{m}{2\epsilon}(x-x_{N-1})^2} \cdots e^{-\frac{m}{2\epsilon}(x_2-x_1)^2} e^{-\frac{m}{2\epsilon}(x_1-x')^2}, \end{aligned}$$

$$\text{boundary conditions}: x' = x_N, \quad x = x_0, \tag{11.34}$$

where

$$\int Dx \equiv \left(\sqrt{\frac{m}{2\pi\epsilon}}\right) \prod_{n=1}^{N-1} \sqrt{\frac{m}{2\pi\epsilon}} \int_{-\infty}^{+\infty} dx_n. \tag{11.35}$$

Note the identity

$$\sqrt{\frac{m}{2\pi\epsilon}} \int_{-\infty}^{+\infty} dx_1 e^{-\frac{m}{2\epsilon}(x_2-x_1)^2 - \frac{m}{2\epsilon}(x_1-x')^2} = \sqrt{\frac{1}{2}} e^{-\frac{m}{2}\cdot\frac{1}{2\epsilon}(x_2-x')^2}.$$

One can evaluate the path integral exactly by performing the Dx-integrations recursively, starting from the end with x_1. The successive integrations over the variables $x_1 \to x_2 \to x_3 \cdots \to x_{N-1}$ yield

$$K(x, x'; \tau) = \sqrt{\frac{m}{2\pi N\epsilon}} e^{-\frac{m}{2}\cdot\frac{1}{N\epsilon}(x-x')^2} = \sqrt{\frac{m}{2\pi\tau}} e^{-\frac{m}{2\tau}(x-x')^2}, \tag{11.36}$$

which is the result obtained in Eq. 5.72.

The case for a free particle in arbitrary d-dimensional space follows from Eq. 11.36, since the d-dimensional transition amplitude factorizes into separate one-dimensional components.

To obtain the evolution kernel in Minkowski time, recall from Eq. 5.16 that

$$\tau = it.$$

Hence, from Eq. 11.36, the Minkowski time evolution kernel, denoted by subscript M, and restoring the $\hbar$ in the formula, yields

$$K_M(x, x'; t) = \sqrt{\frac{m}{2\pi\tau}} e^{-\frac{m}{2\tau\hbar}(x-x')^2} = \sqrt{\frac{m}{2\pi i\hbar t}} e^{i\frac{m}{2t\hbar}(x-x')^2}, \tag{11.37}$$

which was obtained in Eq. 5.72 using the free particle Hamiltonian.

11.9 Finite lattice path integral

A rigorous definition of the path integral can be given by approximating it by a finite-dimensional multiple integral. One can then take the continuum limit and obtain the functional integration required for defining a path integral. There is another reason for considering the finite approximation of this integral. For numerical simulation one always approximates the path integral by a finite-dimensional multiple integral.

Hence it is important to have some exact analytical results for a finite system so as to compare the numerical simulations with it. One of the few exact path integral results that one can derive is for the finite approximation of the harmonic oscillator path integral, and in this section the oscillator is analyzed in some detail.

Consider an *open lattice* with boundary values of x_N and x_0; the action is given by

$$S = -\frac{m}{2\epsilon}\sum_{n=0}^{N-1}(x_{n+1} - x_n)^2 - \frac{1}{2}\epsilon\, m\omega^2 \sum_{n=1}^{N-1} x_n^2,$$

$$K(x_N, x_0; \tau) = \langle x_N | e^{-\tau H} | x_0 \rangle = \mathcal{N} \int \prod_{i=1}^{N-1} dx_n e^S.$$

To simplify the analysis, consider time to be periodic with boundary condition $x_N = x_0$. Furthermore, for a symmetric labeling of the lattice sites, consider a periodic lattice of size $2N + 1$ as shown in Figure 11.3.

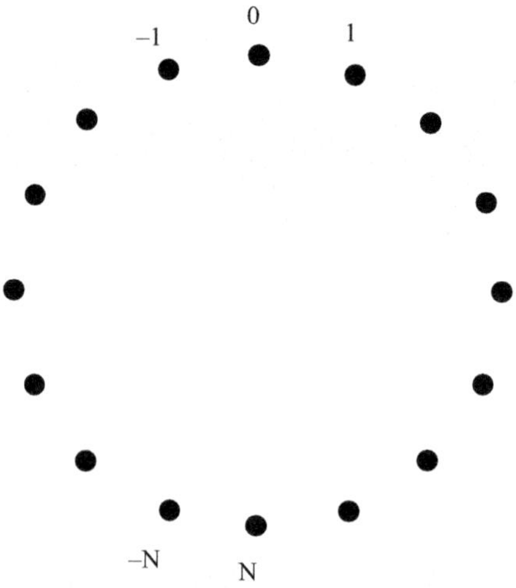

Figure 11.3 Finite periodic lattice, with symmetric numbering of the lattice sites.

The degrees of freedom are periodic in the discrete time index n, where

$$t = n\epsilon, \quad \epsilon = \tau/(2N+1),$$

$$x_n = x_{n+2N+1} = \frac{1}{2N+1} \sum_{p=-N}^{+N} e^{2\pi i p n/(2N+1)} x_p.$$

For simplicity, use the notation $2\pi p/(2N+1) = k$; in this simplified notation one writes

$$x_n = \sum_k e^{ikn} x_k \equiv \frac{1}{2N+1} \sum_{p=-N}^{+N} e^{2\pi i p n/(2N+1)} x_p,$$

and one has the identity

$$\sum_{n=-N}^{N} e^{i(k+k')n} = \sum_{n=-N}^{+N} e^{\frac{2\pi i}{2N+1} n(p+p')} = (2N+1)\delta(p+p') \equiv \delta_{k+k'}.$$

The action is hence given by

$$\begin{aligned} S &= -\frac{m}{2\epsilon} \sum_{n=-N}^{+N} (x_{n+1} - x_n)^2 - \frac{m\omega^2}{2}\epsilon \sum_{n=-N}^{N} x_n^2 \\ &= -\frac{m}{2\epsilon} \sum_{n=-N}^{N} \sum_{k,k'} \left\{ \left(e^{ik} - 1\right) x_k (e^{ik'} - 1) x_{k'} + \epsilon^2 \omega^2 x_k x_{k'} \right\} e^{i(k+k')n} \\ &= -\frac{m}{2\epsilon} \sum_k \left\{ |e^{ik} - 1|^2 + \epsilon^2 \omega^2 \right\} x_k x_{-k}, \end{aligned}$$

where note that, due to the simplified notation being used, one has

$$|1 - e^{ik}|^2 + \epsilon^2\omega^2 = 4\sin^2\left(\frac{\pi p}{2N+1}\right) + \epsilon^2\omega^2.$$

In the momentum basis, both the gradient and quadratic term are diagonalized, and hence the path integral can be performed exactly.

11.9.1 Coordinate and momentum basis

The change of integration variables from real space variables x_n to momentum space variables x_k is analyzed below.

There are $2N + 1$ real integration variables x_n, and these are transformed to $2N + 1$ variables x_k that are complex, and hence there seem to be too many x_k variables. However, note that the x_n coordinates are all real and hence in the Fourier expansion given by $x_n = \sum_k e^{ikn} x_k$, Eq. 11.38 yields

$$x_n^* = x_n \Rightarrow x_k^* = x_{-k}, \quad \text{with } k = 0, \pm 1, \pm 2, \ldots \pm N.$$

Hence, although there are $2N + 1$ complex variables x_k, they are not all independent, and one can choose only the set of complex variables x_k – with $k = 0, 1, 2, \ldots N$ – to be the independent complex variables.

Note the special case of the zero momentum mode, which is real since $x_0 = x_{k=0} = x_{k=0}^*$: real.

In summary, there are $2N + 1$ independent momentum space variables; writing the momentum modes in real and imaginary components yields the $2N + 1$ independent momentum space variables to be

$$x_k = x_k^R + i x_k^I \ : \ k = 1, 2, \ldots N, \quad \tilde{x}_0 \equiv x_{k=0}. \tag{11.38}$$

The change of $2N + 1$ variables from x_n to x_k has a Jacobian equal to one, and yields

$$\prod_{n=-N}^{+N} \int_{-\infty}^{+\infty} dx_n = \int_{-\infty}^{+\infty} d\tilde{x}_0 \prod_{k=1}^{N} \int_{-\infty}^{+\infty} dx_k^R \int_{-\infty}^{+\infty} dx_k^I \equiv \int Dx.$$

11.10 Lattice free energy

To illustrate the workings of the momentum space variables, the partition function is evaluated as follows. Let the lattice simple harmonic Hamiltonian be denoted by H_{Lat}; the partition function for the periodic lattice is given by

$$
\begin{aligned}
Z &= tr(\exp\{-\epsilon(2N+1)H_{\text{Lat}}\} \\
&= \left(\frac{m}{2\pi\epsilon}\right)^{(2N+1)/2} \prod_{n=-N}^{+N} \int dx_n e^S \\
&= \mathcal{N} \int d\tilde{x}_0 \prod_{k=1}^{N} \int dx_k^R dx_k^I e^{-\frac{m}{2\epsilon}\sum_{k=-N}^{N}\{|1-e^{ik}|^2+\epsilon^2\omega^2\}x_{-k}x_k},
\end{aligned}
$$

where H_{Lat} is the lattice Hamiltonian. Note, since $k \equiv 2\pi p/(2N+1)$,

$$
|1-e^{ik}|^2 + \epsilon^2\omega^2 = 4\sin^2\left(\frac{\pi p}{2N+1}\right) + \epsilon^2\omega^2 \equiv d_k.
$$

From Eq. 11.38 one has the simplification

$$
x_{-k}x_k = \left(x_k^R - i x_k^I\right)\left(x_k^R + i x_k^I\right) = \left(x_k^R\right)^2 + \left(x_k^I\right)^2.
$$

Hence the action completely factorizes into *independent* Gaussian integrations for each variable x_k, $k = 0, 1, 2, \ldots N$. This factorization is the reason in the first place why the variables were transformed from real space to momentum space, since in the momentum basis both the quadratic potential and the kinetic energy are diagonal. The partition function is given by

$$
\begin{aligned}
Z &= \left(\frac{m}{2\pi\epsilon}\right)^{\frac{2N+1}{2}} \prod_{k=1}^{N} \int dx_k^R dx_k^I e^{-\frac{m}{2\epsilon}d_k[(x_k^R)^2+(x_k^I)^2]} \int_{-\infty}^{\infty} dx_0 e^{-\frac{m}{2\epsilon}\cdot\epsilon^2\omega^2 x_0^2} \\
&= \left(\frac{m}{2\pi\epsilon}\right)^N \left(\frac{2\pi\epsilon}{m}\right)^N \left[\prod_{k=1}^{N}\left(\frac{1}{d_k}\right)\right] \left(\frac{m}{2\pi\epsilon}\right)^{1/2} \sqrt{\frac{2\pi\epsilon}{m}\cdot\frac{1}{\epsilon^2\omega^2}} \\
&= \sqrt{\frac{1}{d_0}} \prod_{k=1}^{N}\left(\frac{1}{d_k}\right) = \exp\left(-\frac{1}{2}\sum_k \ln(d_k)\right) \\
&= e^{-(2N+1)F},
\end{aligned}
\tag{11.39}
$$

where $d_0 = \epsilon^2\omega^2$ has been used in obtaining Eq. 11.39.

The free energy for the simple harmonic partition function is given by

$$
\begin{aligned}
F &= \text{free energy per lattice site} \\
&= \frac{1}{2}\sum_k \ln(d_k) = \frac{1}{2}\sum_k \ln\left\{4\sin^2\left(\frac{k}{2}\right) + \epsilon^2\omega^2\right\} \\
&= \frac{1}{2}\frac{1}{2N+1}\sum_{p=-N}^{N} \ln\left\{4\sin^2\left(\frac{\pi p}{2N+1}\right) + \epsilon^2\omega^2\right\}.
\end{aligned}
$$

Taking the limit of $N \to \infty$ yields

$$F = \frac{1}{2}\int_{-\pi}^{-\pi} \frac{dk}{2\pi} \ln\left\{4\sin^2\left(\frac{k}{2}\right) + \omega_0^2\right\}, \quad \omega_0^2 = \lim_{\epsilon \to 0} \epsilon^2\omega^2.$$

11.11 Lattice propagator

The propagator (correlator) for the lattice simple harmonic oscillator on a periodic lattice is given by

$$D_n = E[x_{n+m}x_m] = \frac{1}{Z}\int Dx\ x_{n+m}x_m e^S.$$

The propagator does not depend on m due to periodicity of the lattice. Transforming to the momentum space basis yields

$$\begin{aligned}
D_n &= \sum_{k,k'} e^{ik(n+m)} e^{ik'm} \frac{1}{Z}\int DX_k x_p x_k e^S \\
&= \sum_{k,k'} e^{ikn} e^{im(k+k')} \overbrace{x_k x_{k'}} \\
&= \frac{\epsilon}{m}\sum_{k,k'} e^{ikn} e^{im(k+k')} \delta_{k+k'} \frac{1}{|1-e^{ik}|^2 + \epsilon^2\omega^2} \\
&= \frac{\epsilon}{m}\sum_{k} \frac{e^{ikn}}{|1-e^{ik}|^2 + \epsilon^2\omega^2} \\
&\equiv \frac{\epsilon}{8m}\frac{1}{2N+1}\sum_{p=-N}^{+N} \frac{\cos(2\pi pn/(2N+1))}{\sin^2\left(\frac{\pi pn}{2N+1}\right) + \frac{\epsilon^2\omega^2}{4}}.
\end{aligned} \tag{11.40}$$

It can be shown by a direct and tedious calculation that the Fourier sum given in Eq. 11.40 yields

$$D_n = \frac{r^n + r^{2N+1-n}}{2\omega'(1 - r^{2N+1})}, \tag{11.41}$$

where

$$\begin{aligned}
\omega' &= m\omega\left(1 + \frac{\epsilon^2\omega^2}{4}\right)^{1/2}, \\
r &= 1 + \frac{\epsilon^2\omega^2}{2} - \epsilon\omega\left(1 + \frac{\epsilon^2\omega^2}{4}\right)^{1/2} = \left(\frac{\epsilon\omega}{2} - \sqrt{\left(\frac{\epsilon\omega}{2}\right)^2 + 1}\right)^2.
\end{aligned} \tag{11.42}$$

Note the important fact that Eq. 11.41 is convergent for large N because r obeys the bounds

$$0 < r < 1.$$

To define the continuum limit, the lattice parameters are defined in terms of continuum time t and total time τ as

$$n = \frac{t}{\tau}(2N+1), \quad \epsilon = \frac{\tau}{2N+1},$$
$$r = 1 - \frac{\omega\tau}{2N+1} + O\left(\frac{1}{(2N+1)^2}\right),$$
$$\omega' = m\omega + O\left(\frac{1}{(2N+1)^2}\right).$$

For limit $N \to \infty$, the propagator is given by

$$D_n = \frac{r^{n-\frac{2N+1}{2}} + r^{\frac{2N+1}{2}-n}}{2m\omega\left(r^{-\frac{2N+1}{2}} - r^{\frac{2N+1}{2}}\right)},$$
$$r^{2N+1} = \left(1 - \frac{\omega\tau}{2N+1}\right)^{2N+1} \to e^{-\omega\tau},$$

$$r^{n-\frac{2N+1}{2}} = r^{\left(\frac{t}{\tau}-\frac{1}{2}\right)(2N+1)} \to e^{-\omega\tau\left(\frac{t}{\tau}-\frac{1}{2}\right)} = e^{-\omega t+\frac{\omega\tau}{2}},$$

$$\therefore D_n \to \frac{1}{2m\omega}\frac{e^{-\omega t+\frac{\omega\tau}{2}} + e^{\omega t-\frac{\omega\tau}{2}}}{e^{\frac{\omega\tau}{2}} - e^{-\frac{\omega\tau}{2}}}.$$

Hence, one obtains the continuum limit given by

$$D_n = \frac{1}{2m\omega}\frac{\cosh\left(\omega t - \frac{\omega\tau}{2}\right)}{\sinh\left(\frac{\omega\tau}{2}\right)},$$

which agrees with the result obtained earlier in Eq. 11.7.

11.12 Lattice transfer matrix and propagator

The propagator can be found exactly for the lattice simple harmonic oscillator using the exact transfer matrix [Creutz and Freedman (1981)]. The path integral can be expressed as a trace due to the periodic boundary conditions. Hence

$$Z = \int_{(\text{periodic})} DXe^S$$
$$= tr(T^{2N+1}).$$

To obtain the transfer matrix T, consider two adjacent sites; then the integrand of the lattice path integral for Z is given by the product

$$\cdots \langle x_{n+2}|T|x_{n+1}\rangle\langle x_{n+1}|T|x_n\rangle\langle x_n|T|x_{n-1}\rangle \cdots$$

The terms linking variables x_n and x_{n+1} are entirely due to matrix element $\langle x_{n+1}|T|x_n\rangle$, whereas terms like x_n^2 in S can be symmetrically divided between adjacent matrix elements of T. A Hermitian transfer matrix is given by

$$\langle x'|T|x\rangle = e^{-\frac{m\epsilon\omega^2}{4}x'^2}e^{-\frac{m}{2\epsilon}(x-x')^2}e^{-\frac{m\epsilon\omega^2}{4}x^2}.$$

Recall, for $[x, p] = i$, one has

$$\langle x'|e^{-\frac{\epsilon}{2m}\hat{p}^2}|x\rangle = \sqrt{\frac{m}{2\pi\epsilon}}e^{-\frac{m}{2\epsilon}(x'-x)^2},$$

$$\Rightarrow \hat{T} = \sqrt{\frac{2\pi\epsilon}{m}}e^{-\epsilon\frac{m\omega^2}{4}x^2}e^{-\frac{\epsilon}{2m}\hat{p}^2}e^{-\epsilon\frac{m\omega^2}{2}x^2}.$$

If one is working to only lowest order in ϵ, then using the CBH-formula one can easily show that $T \sim e^{-\epsilon\hat{H}}$, where $H = p^2/2m + m\omega^2x^2/2$. For the finite size lattice, we need the operator expression for T that is exact in ϵ, that is, correct to all powers of ϵ.

Note, for $\alpha = \epsilon/m$ and $u^2 = m^2\omega^2$, it can be shown that [Creutz and Freedman (1981)]

$$[x, T] = \frac{\alpha^2u^2}{2}Tx - i\alpha Tp,$$

$$[p, T] = \frac{\alpha^2u^2}{2}Tp + i\alpha u^2\left(1 + \frac{\alpha^2u^2}{4}\right)Tx.$$

Using the commutators given above it can be shown that

$$[T, H] = 0,$$

where

$$H = \frac{1}{2}p^2 + \frac{1}{2}\beta^2x^2,$$

$$\beta = u\sqrt{1 + \frac{\alpha^2u^2}{4}} = m\omega\sqrt{1 + \frac{\epsilon^2\omega^2}{4}} = \omega',$$

where the last relation follows from Eq. 11.42.

The eigenstates of $\hat{H}$ also diagonalize T. Let the creation and annihilation operators be defined as

$$a = \frac{1}{\sqrt{2\omega'}}\left(\omega'x + ip\right),$$

$$a^\dagger = \frac{1}{\sqrt{2\omega'}}\left(\omega'x - ip\right),$$

$$[a, a^\dagger] = 1.$$

One can show that

$$[a, T] = (r - 1)Ta$$

with

$$r = 1 + \frac{\alpha^2 u^2}{2} - \alpha u \left(1 + \frac{\alpha^2 u^2}{4}\right)^{1/2}.$$

Since $|n\rangle$ are eigenstates of T,

$$T|n\rangle = \lambda_n |n\rangle,$$

and the commutator $[a, T]$ yields

$$\begin{aligned} (r-1)Ta|n\rangle &= (aT - Ta)|n\rangle \\ \text{or} \quad (r-1)\lambda_{n-1}|n-1\rangle &= (\lambda_n - \lambda_{n-1})|n-1\rangle, \\ (r-1)\lambda_{n-1} &= \lambda_n - \lambda_{n-1}, \\ \frac{\lambda_n}{\lambda_{n-1}} &= r \;\Rightarrow\; \lambda_n = K r^n \quad (K = \text{constant}). \end{aligned}$$

Hence

$$T = AKr^{(H/W)}.$$

To show that the constant $K = 1$, consider the trace of T

$$\frac{1}{A} trT = K \sum_{n=0}^{\infty} r^{n+1/2} = \frac{Kr^{1/2}}{1 - r}.$$

Note that from Eq. 11.43 it follows that

$$r = \left[\sqrt{1 + \frac{\alpha^2 u^2}{4}} - \frac{\alpha u}{2}\right]^2.$$

Hence

$$\frac{r^{1/2}}{1 - r} = \frac{\sqrt{1 + \frac{\alpha^2 u^2}{4}} - \frac{\alpha u}{2}}{\alpha u \left[\sqrt{1 + \frac{\alpha^2 u^2}{4}} - \frac{\alpha u}{2}\right]} = \frac{1}{\alpha u}.$$

The definition of T yields

$$\frac{1}{A} trT = \int dx \langle x| e^{-\frac{\alpha u^2}{4} x^2} e^{-\frac{\alpha}{2} p^2} e^{-\frac{\alpha u^2}{4} x^2} |x\rangle$$

$$= \int_{-\infty}^{+\infty} dx e^{-\frac{\alpha u^2}{2}x^2} \langle x| e^{-\frac{\alpha}{2}p^2} |x\rangle$$

$$= \sqrt{\frac{2\pi}{\alpha u^2}} \cdot \sqrt{\frac{1}{2\pi\alpha}} = \frac{1}{\alpha u},$$

which yields $K = 1$.

Consequently

$$T = \sqrt{\frac{2\pi\epsilon}{m}} r^{\hat{H}/w}.$$

The path integral is hence given by

$$Z = tr(T^{2N+1})$$

$$= \left(\frac{2\pi\epsilon r}{m}\right)^{\frac{2N+1}{2}} \frac{1}{1 - r^{2N+1}}.$$

Furthermore, the propagator is given by

$$D_n = E\left[x_{n+m}x_m\right] = \frac{1}{Z} tr\left(xT^n x T^{2N+1-n}\right)$$

$$= \frac{1}{2\omega'(1 - r^{2N+1})}\left(r^n + r^{2N+1-n}\right),$$

which is seen to agree with the earlier result given by Eq. 11.41

11.13 Eigenfunctions from evolution kernel

Recall

$$K(x, x'; \tau) = \langle x| e^{-\tau H} |x'\rangle$$

$$= \sum_n e^{-\tau E_n} \langle x|\psi_n\rangle\langle\psi_n|x'\rangle$$

$$= \sum_n e^{-\tau E_n} \psi_n(x)\psi_n^*(x'),$$

where $|\psi_n\rangle$, E_n are the eigenfunctions and eigenvalues of H. Expanding the sum yields

$$K\left(x, x'; \tau\right) = e^{\tau E_0}\psi_0(x)\psi_0^*\left(x'\right) + e^{-\tau E_1}\psi_1(x)\psi_1^*\left(x'\right) + \cdots$$

The evolution kernel can be expanded in a power series of $e^{-\tau E_n}$ term by term, which yields the eigenfunctions.

Consider the harmonic oscillator with

$$K\left(x, x'; \tau\right) = \sqrt{\frac{m\omega}{2\pi \sinh(\omega\tau)}} \exp\left\{-\frac{m\omega}{2\sinh\omega\tau}\left[(x^2 + x'^2)\cosh\omega\tau - 2xx'\right]\right\}.$$

For $\tau \to \infty$, one has to leading order

$$K(x, x'; \tau) \simeq e^{-\frac{\omega\tau}{2}} \sqrt{\frac{m\omega}{2\pi}} e^{-\frac{m\omega}{2}\left(x^2+x'^2\right)},$$

$$\Rightarrow E_0 = \frac{\omega}{2} \quad \psi_0(x) = \left(\frac{m\omega}{2\pi}\right)^{1/4} e^{-\frac{m\omega}{2}x^2}.$$

To next leading order

$$K\left(x, x'; \tau\right) \simeq \psi_0(x)\psi_0(x')\left[e^{-\frac{\omega\tau}{2}} + 2m\omega x x' e^{-\frac{3}{2}\omega\tau} + \left\{\frac{1}{2} + 2m^2\omega^2x^2x'^2 - m\omega(x^2+x'^2)\right\} e^{-\frac{5}{2}\omega\tau} + \cdots\right]$$

and hence by inspection

$$E_1 = \frac{3}{2}\omega \quad , \quad \psi_1(x) = \left(\frac{m\omega}{2\pi}\right)^{1/4} \cdot \sqrt{2m\omega} \cdot x e^{-\frac{m\omega}{2}x^2},$$

$$E_2 = \frac{5}{2}\omega \quad , \quad \psi_2(x) = \left(\frac{m\omega}{2\pi}\right)^{1/4} \frac{1}{\sqrt{2}} \left(2m\omega x^2 - 1\right) e^{-\frac{m\omega}{2}x^2}.$$

11.14 Summary

The simple harmonic oscillator is a leading exemplar in the study of path integrals since many of the key ideas can be fully worked out for the oscillator. Starting from the Hamiltonian, the propagator of the oscillator was derived using state space methods. The oscillator path integral was then defined and the evolution kernel was evaluated, both in the presence and absence of an external current.

The path integral for discrete time and for a finite interval was shown to reduce to an ordinary multiple integral. The propagator for finite and discrete time was exactly evaluated using both the multiple integral formulation and the transfer matrix formulation, showing the consistency of the integral and differential formulations.

The evolution kernel was lastly used to derive the ground state and the first few excited states of the simple harmonic oscillator.

12
Gaussian path integrals

Several path integrals are exactly evaluated here using Gaussian path integration. A few general ideas are illustrated using the advantage of being able to exactly evaluate Gaussian path integrals.

Path integrals defined over a particular collection of allowed indeterminate paths can sometimes be represented by a Fourier expansion of the paths. This leads to two important techniques for performing path integrations:

1. Expanding the action about the classical solution of the Lagrangian;
2. Expanding the degree of freedom in a Fourier expansion of the allowed paths.

Various cases are considered to illustrate the usage of classical solutions and Fourier expansions, and these also provide a set of relatively simple examples to familiarize oneself with the nuts and bolts of the path integral. The Lagrangian of the simple harmonic oscillator is used for all of the following examples; all the computations are carried out explicitly and exactly.

The following different cases are considered:

- Correlators of exponential functions of the degree of freedom are discussed in Section 12.1.
- The generating functional for periodic paths is evaluated in Section 12.2.
- The path integral required for evaluating the normalization constant for the oscillator evolution kernel is discussed in Section 12.3. The path integral entails summing over all paths that start from and return to the same fixed position.
- Section 12.4 discusses the evolution kernel for a particle starting at an initial position x_i and, after time τ, having a final position that is indeterminate.
- The path integral of a free particle in the presence of a fixed external current j is discussed in Section 12.5.
- Section 12.6 discusses the evolution kernel for a system with indeterminate initial and indeterminate final positions.

- Given its importance, the evolution kernel of the oscillator is derived using the Fourier expansion in Section 12.7.

In Section 12.8 the path integral for a free particle in the presence of a magnetic field is exactly evaluated; this example has a new feature in that the magnetic field couples to the *velocity* of the particle. The calculation is carried out in Minkowski time.

12.1 Exponential operators

Gaussian integration has the important property, as seen in Eq. 7.8, that the generating function can be exactly evaluated. This result allows the exact study of exponential of the coordinate operators, namely

$$e^{ix(t)/a},$$

where a is an arbitrary length scale required for making the exponent dimensional. For the case of the simple harmonic oscillator, the degree of freedom $x(t)$ is sometimes called normal since, in the path integral formulation, $x(t)$ can be considered a Gaussian (normal) random variable.

The correlators of the exponential operator can be exactly evaluated. From Eqs. 7.8 and 11.9 the infinite time case yields

$$\begin{aligned} Z[j] &= \frac{1}{Z}\int DXe^{i\int dtj(t)x(t)}e^{S} \\ &= \exp\left\{-\frac{1}{2m\omega}\int dtdt'j(t)e^{-\omega|t-t'|}j(t')\right\}. \end{aligned} \tag{12.1}$$

Consider the correlator of two operators

$$\begin{aligned} G(t,t') &= \frac{1}{Z}\int DXe^{S}e^{ix(t)/a}e^{-ix(t')/a} = \frac{1}{Z}\int DXe^{i\int d\tau j(\tau)x(\tau)}e^{S}, \\ \Rightarrow j(\tau) &= \frac{1}{a}\left[\delta(\tau-t)-\delta(\tau-t')\right]. \end{aligned}$$

Hence

$$\begin{aligned} G(t,t') = \exp\Bigg\{&-\frac{1}{2m\omega a^2}\int d\tau d\tau' e^{-\omega|\tau-\tau'|} \\ &\times\left[\delta(\tau-t)-\delta(\tau-t')\right]\left[\delta\left(\tau'-t\right)-\delta(\tau'-t')\right]\Bigg\} \\ = \exp&\left\{-\frac{1}{m\omega a^2}\left(1-e^{-\omega|t-t'|}\right)\right\}. \end{aligned}$$

Taking the limit of $\omega \to 0$ yields

$$G(t,t') \simeq \exp\left\{-\frac{1}{ma^2}|t-t'|\right\} = e^{-\frac{|t-t'|}{\xi}},$$

where correlation time is given by $\xi = ma^2$.

Note that for $\omega = 0$, namely, the case of a free particle, it can be shown that the correlation of a product of operators

$$e^{ix_1(t)/a_1} e^{ix_2(t)/a_2} \ldots e^{ix_1(t)/a_n}$$

is nonzero only if

$$\frac{1}{a_1} + \frac{1}{a_1} \ldots + \frac{1}{a_n} = 0.$$

12.2 Periodic path integral

The finite time action for a simple harmonic oscillator is

$$S = -\frac{m}{2}\int_0^\tau dt(\dot{x}^2 + \omega^2 x^2). \tag{12.2}$$

The correlator for finite time is given in Eq. 11.6,

$$D\left(t,t',\tau\right) = \frac{1}{Z(\tau)} tr\left(T\left[\hat{x}_t \hat{x}_{t'}\right] e^{-\tau\hat{H}}\right), \quad Z(\tau) = tr(e^{-\tau\hat{H}}).$$

Due to the trace, the propagator is given by the Feynman path integral over all periodic paths, namely for $x(t+\tau) = x(t)$. More precisely, the propagator is given by

$$D\left(t,t',\tau\right) \equiv D_{t-t'} = \frac{1}{Z}\int DXx(t)x(t')e^S,$$
$$x(t+\tau) = x(t).$$

To calculate the propagator, consider the generating function given by

$$Z[j] = \frac{1}{Z}\int DXe^{S+\int j(t)x(t)} = \exp\left\{\frac{1}{2}\int j(t)D_{t-t'}j(t')\right\}. \tag{12.3}$$

All the periodic path $x(t)$ are given by the Fourier expansion

$$x(t) = \sum_{n=-\infty}^{+\infty} e^{2\pi int/\tau} x_n. \tag{12.4}$$

Since

$$\int_0^\tau dte^{2\pi int/\tau} e^{2\pi imt/\tau} = \tau\delta_{n-m}, \tag{12.5}$$

the action is given by

$$S = -\frac{m\tau}{2}\sum_n \left\{\left(\frac{2\pi n}{\tau}\right)^2 + \omega^2\right\} x_n x_{-n}. \tag{12.6}$$

The Fourier modes x_n have all decoupled and one can exactly evaluate the path integral. From Eq. 12.3, using Gaussian integration yields

$$Z[j] = \exp\left\{\frac{1}{2}\int j(t) D_{t-t'} j(t')\right\},$$

where

$$\begin{aligned} D_{t-t'} &= \frac{1}{m\tau}\sum_{n=-\infty}^{+\infty} e^{2\pi i n(t-t')/\tau} \frac{1}{\left(\frac{2\pi n}{\tau}\right)^2 + \omega^2\}} \\ &= \frac{1}{m\tau}\left(\frac{\tau}{2\pi}\right)^2 \sum_{n=-\infty}^{+\infty} \frac{e^{2\pi i n(t-t')/\tau}}{n^2 + (\frac{\omega\tau}{2\pi})^2}. \end{aligned} \tag{12.7}$$

Using the identity given in Eq. 11.30, namely

$$\sum_{n=-\infty}^{+\infty} \frac{e^{in\theta}}{a^2+n^2} = \frac{\pi}{a}\frac{\cosh[(\pi - |\theta|)a]}{\sinh \pi a},$$

yields

$$\begin{aligned} D_{t-t'} &= \frac{1}{m\tau}\left(\frac{\tau}{2\pi}\right)^2 \left(\frac{2\pi}{\omega\tau}\right)\pi \frac{\cosh\left(\frac{\omega\tau}{2} - \omega\tau|t-t'|\right)}{\sinh(\frac{\omega\tau}{2})} \\ &= \frac{1}{2\omega m}\frac{\cosh\left(\frac{\omega\tau}{2} - \omega\tau|t-t'|\right)}{\sinh(\frac{\omega\tau}{2})} \end{aligned}$$

and reproduces the result obtained earlier in Eq. 11.8 using the oscillator algebra.

12.3 Oscillator normalization

The normalization of the simple harmonic oscillator is given by the path integral

$$\int Dz e^{S[z]} = \sqrt{\frac{m\omega}{2\pi \sinh \omega\tau}},$$

$$\text{boundary conditions}: \quad z(0) = 0 = z(\tau).$$

As can be read from the boundary conditions, the path integral corresponds to the degree of freedom starting and ending at the same point after time τ.

A Fourier expansion for $z(t)$ yields

$$z(t) = \sum_{n=1}^{\infty} \sin\left(\frac{n\pi t}{\tau}\right) z_n, \quad \frac{z(t)}{dt} = \sum_{n=1}^{\infty} \left(\frac{n\pi}{\tau}\right) \cos\left(\frac{n\pi t}{\tau}\right) z_n.$$

Note that all the Fourier coefficients $z_n^* = z_n$ are real since $z(t)$ is real.

Using the orthogonality relation,

$$\int_0^\tau dt \sin\left(\frac{n\pi t}{\tau}\right) \sin\left(\frac{m\pi t}{\tau}\right) = \int_0^\tau dt \cos\left(\frac{n\pi t}{\tau}\right) \cos\left(\frac{m\pi t}{\tau}\right) = \frac{\tau}{2}\delta_{n-m}$$

yields the simplification for the action

$$S[z] = -\frac{\tau}{2}\frac{m}{2} \sum_{n=1}^{\infty} \left\{ \left(\frac{n\pi}{\tau}\right)^2 + \omega^2 \right\} z_n^2.$$

The action has completely factorized into a sum over the Fourier modes and the $\int DZ$ path integral factorizes into infinitely many discrete integrations, one for each z_n. We obtain ($\mathcal{N}$ is a normalization constant)

$$\begin{aligned}
\int Dz\, e^{S[z]} &= \mathcal{N}' \prod_{n=1}^{\infty} \int_{-\infty}^{+\infty} dz_n e^{-\frac{\tau}{2}\lambda_n z_n^2} \\
&= \mathcal{N}' \prod_{n=1}^{\infty} \sqrt{\frac{2\pi}{\tau\lambda_n}} = \mathcal{N} \prod_{n=1}^{\infty} \left(\frac{1}{\left(\frac{n\pi}{\tau}\right)^2 + \omega^2} \right)^{1/2} \\
&= \mathcal{N} \prod_{n=1}^{\infty} \frac{1}{\left[1 + \frac{(\omega\tau)^2}{n^2\pi^2}\right]^{1/2}}.
\end{aligned}$$

Since

$$\prod_{n=1}^{\infty} \left[1 + \frac{\theta^2}{n^2\pi^2}\right] = \frac{\sinh\theta}{\theta},$$

the path integral yields

$$\int Dz\, e^{S[z]} = \mathcal{N}\sqrt{\frac{\omega\tau}{\sinh\omega\tau}}.$$

Recall from Eq. 11.16, for the simple harmonic oscillator,

$$K(x', x; \tau) = \tilde{A}\mathcal{N}'\sqrt{\frac{\omega\tau}{\sinh\omega\tau}} e^{S_c}.$$

For $\omega \to 0$, the simple harmonic oscillator reduces to the free particle. The classical action is given by $\lim_{\omega\to 0} S_c(\omega) \to -\frac{m}{2\tau}(x - x')^2$. The evolution kernel for a free particle, from Eq. 11.36, is given by

$$\lim_{\omega \to 0} K(x, x'; \tau) = \sqrt{\frac{m}{2\pi\tau}} e^{-\frac{m}{2\tau}(x-x')^2}.$$

Hence

$$A\mathcal{N}' = \sqrt{\frac{m}{2\pi\tau}},$$

yielding the result

$$K(x', x; \tau) = \sqrt{\frac{m\omega}{2\pi \sinh \omega\tau}} e^{S_c}.$$

12.4 Evolution kernel for indeterminate final position

The simple harmonic oscillator's action is given by

$$S = -\frac{m}{2} \int_0^\tau (\dot{x}^2 + \omega^2 x^2) dt.$$

The transition amplitude of the degree of freedom from the initial to the final position is denoted by $K(x, x'; \tau)$. The transition amplitude $K(x; \tau)$ from x to all possible positions at the final time is

$$K(x; \tau) = \int dx' \langle x | e^{-\tau H} | x' \rangle = \int_{-\infty}^{+\infty} dx' K(x, x'; \tau).$$

There are three ways of obtaining $K(x; \tau)$:

- One can first evaluate the full evolution kernel $K(x, x'; \tau)$ and then integrate over x'.
- Since the boundary condition $[dx(t)/dt]|_{t=\tau} = 0$ is equivalent to integrating over all possible values of the final position x_f, one finds the *classical solution* for $x(t)$ with boundary conditions given by

$$x(0) = x, \quad \left. \frac{dx(t)}{dt} \right|_{t=\tau} = 0. \tag{12.8}$$

  The classical solution provides the classical action that contains the dependence on the final position x.
- Another method for evaluating $K(x_i; \tau)$ is by directly expanding the paths using a Fourier expansion that respects the boundary conditions given in Eq. 12.8, namely

$$x(t) = x + \sum_{n=0}^{\infty} x_n \sin\left[\left(\frac{2n+1}{2}\right) \frac{\pi t}{\tau}\right].$$

Evolution kernel of harmonic oscillator

The simple harmonic oscillator is given by Eq. 11.21. This yields

$$K(x, x'; \tau) = \sqrt{\frac{m\omega}{2\pi \sinh \omega\tau}} \exp\left\{-\frac{m\omega}{2\sinh \omega\tau}\left[\left(x^2 + x'^2\right)\cosh \omega\tau - 2xx'\right]\right\}.$$

Hence, the kernel for reaching an arbitrary point after time τ is given by

$$\begin{aligned} K(x; \tau) &= \int dx' K(x, x'; \tau) \\ &= \mathcal{N} \exp\left\{-\frac{m\omega \cosh \omega\tau}{2\sinh \omega\tau} x^2\right\} \sqrt{\frac{2\pi \sinh \omega\tau}{m\omega \cosh \omega\tau}} \\ &\quad \times \exp\left\{\frac{1}{2}\frac{1}{m\omega}\frac{\sinh \omega\tau}{\cosh \omega\tau}\left(\frac{m\omega}{\sinh \omega\tau}x\right)^2\right\} = \sqrt{\frac{1}{\cosh \omega\tau}} e^F, \end{aligned}$$

where

$$\begin{aligned} F &= -\frac{m\omega}{2\sinh \omega\tau} x^2 \left[\cosh \omega\tau - \frac{1}{\cosh \omega\tau}\right] = -\frac{m\omega x^2}{2\sinh \omega\tau}\frac{\sinh^2 \omega\tau}{\cosh \omega\tau} \\ &= -\frac{m\omega}{2} x^2 \tanh \omega\tau. \end{aligned}$$

Classical solution

The evolution kernel is determined by using the classical solution with appropriate boundary conditions,

$$\frac{\delta S}{\delta x(t)} = 0 = \ddot{x} - \omega^2 x, \quad x_c(t) = Ae^{-\omega t} + Be^{\omega t}.$$

We impose the following boundary conditions on the classical solution:

$$x_c(0) = x; \qquad \frac{dx_c(\tau)}{dt} = 0,$$

which yields

$$x = A + B, \quad \frac{dx(\tau)}{dt} = 0 = -\omega\left(Ae^{-\omega\tau} - Be^{\omega\tau}\right)$$

and hence

$$A = Be^{2\omega\tau} = \frac{x}{1 + e^{2\omega\tau}} e^{2\omega\tau}.$$

The classical solution is given by

$$x_c(t) = \frac{x}{1 + e^{2\omega\tau}} \left[e^{2\omega\tau} e^{-\omega t} + e^{\omega t}\right],$$

with the classical action

$$\begin{aligned} S_c &= -\frac{m}{2}\left[x_c(\tau)\dot{x}_c(\tau) - x_c(0)\dot{x}_c(0)\right] \\ &= \frac{mx}{2}\frac{x\omega}{1+e^{2\omega\tau}}\left[-e^{2\omega\tau}+1\right] \\ &= -\frac{m\omega x^2}{2}\frac{\sinh\omega\tau}{\cosh\omega\tau} = -\frac{m\omega}{2}x^2\tanh\omega\tau. \end{aligned}$$

The evolution kernel is hence

$$K(x;\tau) = \mathcal{N}e^{S_c}.$$

Finding the classical solution does not allow for determination of the normalization $\mathcal{N}$, since one cannot use the composition law to generate the equation for $\mathcal{N}$.

Fourier expansion

The boundary conditions

$$x(0) = x; \qquad \frac{dx(\tau)}{dt} = 0$$

are realized by the Fourier expansion

$$x(t) = x + \sum_{n=0}^{\infty} x_n \sin\left[\left(\frac{2n+1}{2}\right)\frac{\pi t}{\tau}\right].$$

All possible paths obeying the given boundary condition are obtained by varying the real coefficients x_n. The orthogonality equations

$$\begin{aligned} &\int_{t_i}^{t_f} dt \cos\left[n\pi\frac{(t-t_i)}{\tau}\right]\cos\left[m\pi\frac{(t-t_i)}{\tau}\right] \\ &\quad = \int_{t_i}^{t_f} dt \sin\left[n\pi\frac{(t-t_i)}{\tau}\right]\sin\left[m\pi\frac{(t-t_i)}{\tau}\right] = \frac{\tau}{2}\delta_{m-n},\ m,n\geq 1 \end{aligned} \tag{12.9}$$

yield the action

$$\begin{aligned} S = &-\frac{m}{2}\frac{\tau}{2}\sum_{n=0}^{\infty}\left[\frac{(2n+1)^2\pi^2}{4\tau^2}+\omega^2\right]x_n^2 \\ &-\frac{m\omega^2}{2}\int_0^{\tau} dt\left(x^2 + 2x\sum_{n=0}^{\infty}\sin[\left(\frac{2n+1}{2}\right)\frac{\pi t}{\tau}]x_n\right). \end{aligned}$$

Note

$$\int_0^\tau dt \sin\left[\left(\frac{2n+1}{2}\right)\frac{\pi t}{\tau}\right] = -\frac{2\tau}{(2n+1)\pi}\left\{\cos\left[\left(\frac{2n+1}{2}\right)\pi\right]-1\right\}$$
$$= \frac{2\tau}{(2n+1)\pi}.$$

Thus

$$S = -\frac{m}{2}\frac{\pi^2}{8\tau}\sum_{n=0}^{\infty}\left[(2n+1)^2 + \left(\frac{2\omega\tau}{\pi}\right)^2\right]x_n^2 - \frac{m\tau\omega^2}{2}x^2$$
$$-\frac{2\tau m x\omega^2}{\pi}\sum_{n=0}^{\infty}\frac{1}{(2n+1)}x_n.$$

Performing the path integral yields

$$K(x;\tau) = \text{const}\prod_n\left[\frac{2\pi}{(2n+1)^2+(\frac{2\omega\tau}{n})^2}\right]^{1/2} e^F = \mathcal{N}e^F.$$

Using the numerical identity

$$\sum_{n=0}^{\infty}\frac{1}{(2n+1)^4+(\frac{2\omega\tau}{\pi})^2(2n+1)^2} = \frac{\pi^4}{2^6\omega^2\tau^2}\left(1-\frac{\tanh\omega\tau}{\omega\tau}\right)$$

yields

$$F = -\frac{m\tau\omega^2}{2}x^2 + \left(\frac{2\tau m x\omega^2}{\pi}\right)^2 \times \frac{8\tau}{m\pi^2}\times\sum_{n=0}^{\infty}\frac{1}{(2n+1)^2}\left[\frac{1}{(2n+1)^2+(\frac{2\omega\tau}{\pi})^2}\right]$$
$$= -\frac{m\tau\omega^2}{2}x^2 + \frac{32m\tau^3\omega^4}{\pi^4}x^2\frac{\pi^4}{64\omega^2\tau^2}\left(1-\frac{\tanh\omega\tau}{\omega\tau}\right)$$
$$= -\frac{m\tau\omega^2}{2}x^2 + \frac{m\omega^2\tau}{2}\left(1-\frac{\tanh\omega\tau}{\omega\tau}\right)x^2$$
$$= -\frac{m\omega x^2}{2}\tanh\omega\tau,$$

and hence the expected solution

$$K(x;\tau) = \mathcal{N}e^{S_c}.$$

A careful treatment of the normalization of the path integral, as discussed in detail in Section 12.3, can be used to evaluate $\mathcal{N}$.

12.5 Free degree of freedom: constant external source

Consider a free degree of freedom with action with an external current j given by

$$S = -\frac{1}{2}m \int_0^\tau dt \left(\frac{dx}{dt}\right)^2 + j \int_0^\tau dtx$$

and the path integral

$$K(x_f, x_i; \tau) = \int Dxe^S,$$

$$\text{boundary conditions}: \quad x(0) = x_i; \quad x(\tau) = x_f.$$

The general solution for an arbitrary time dependent $j(t)$ given in Eq. 11.29 can be used to perform the path integral. A different approach is to employ a change of variables to simplify the source free quadratic action and then evaluate the path integral.

Consider change of variables,

$$\frac{dx(t)}{dt} = \xi(t) \ \Rightarrow \ x(t) = x_i + \int_0^t dt'\xi(t'),$$

$$x(\tau) = x_i + \int_0^\tau dt'\xi(t').$$

The requirement that $x(\tau) = x_f$ needs to be put in as a constraint in the path integral. Hence

$$\begin{aligned} K &= \int D\xi\delta(x_i + \int_0^\tau dt\xi(t) - x_f)e^S \\ &= \int \frac{d\eta}{2\pi} \int D\xi e^{i\eta(x_i - x_f)} e^{i\eta\int_0^\tau dt\xi(t)} e^S \\ &= \int \frac{d\eta}{2\pi} \int D\xi e^{i\eta(x_i - x_f)} e^{\tilde{S}}. \end{aligned}$$

The action is given by

$$\begin{aligned} \tilde{S} &= i\eta \int_0^\tau dt\xi(t) - \frac{1}{2}m \int_0^\tau dt\xi^2 - j \int_0^\tau dt \int_0^t dt'\xi(t') - jx_i\tau \\ &= -\frac{1}{2}m \int_0^\tau dt\xi^2 + \int_0^\tau dt[i\eta - j(\tau - t)]\xi - jx_i\tau. \end{aligned}$$

Performing the ξ path integral yields[1]

$$K = \int \frac{d\eta}{2\pi} e^{i\eta(x_i - x_f)} e^{\tilde{F} - jx_i\tau}, \tag{12.10}$$

[1] Using the results given in Section 11.8 for the free particle path integral one can obtain the normalization given in Eq. 12.10.

where

$$\tilde{F} = \frac{1}{2m}\int_0^\tau dt[i\eta - j(\tau - t)]^2 = -\frac{\tau}{2m}\eta^2 - \frac{i}{2m}j\eta\tau^2 + \frac{1}{6m}j^2\tau^3.$$

Doing the η-integration yields

$$K(x_f, x_i; \tau) = \mathcal{N}e^F,$$

where

$$\begin{aligned} F &= -\frac{m}{2\tau}(x_i - x_f - \frac{\tau^2}{2m}j)^2 + \frac{1}{6m}j^2\tau^3 - j\tau x_i \\ &= -\frac{m}{2\tau}(x_i - x_f)^2 - \frac{m\tau}{2}(x_i + x_f)j + \frac{1}{24m}j^2\tau^3, \\ \mathcal{N} &= \sqrt{\frac{m}{2\pi\tau}}. \end{aligned}$$

Hence, the evolution kernel is given by

$$K(x_f, x_i, \tau) = \sqrt{\frac{m}{2\pi\tau}}e^F.$$

The normalization is the same as the case of $j = 0$.

12.6 Evolution kernel for indeterminate positions

A particle with paths that have indeterminate initial and final positions can be studied using a Fourier expansion of its possible paths. Let the particle have mass m and spring constant ω, and be subject to an external force j; the particle's Lagrangian and action, from initial time and position t_i, x_i to final time and position t_f, x_f, are given by

$$S = \int_{t_i}^{t_f} dt\mathcal{L}, \quad \mathcal{L} = -\frac{1}{2}m\left(\frac{dx}{dt}\right)^2 - \frac{1}{2}m\omega^2x^2 + jx. \tag{12.11}$$

The transition amplitude is given by

$$K(x; t_i; x_f, t_f) = \langle x_f|e^{-(t_f - t_i)H}|x_i\rangle = \int Dxe^S.$$

For the case when the initial and final positions x_i, x_f are indeterminate the transition amplitude is given by

$$K(t_i, t_f; j) = \int_{-\infty}^{+\infty} dx_i dx_f K(x_i, t_i; x_f, t_f). \tag{12.12}$$

The integration over the initial and final positions that results in $Z(t_i, t_f; j)$ has a simple expression in terms of the boundary conditions imposed on the path integration measure $\int D$. Instead of the initial and final positions being fixed, the paths $x(t)$ now have

$$\frac{dx(t_i)}{dt} = 0 = \frac{dx(t_f)}{dt} \quad : \text{Neumann BCs.} \tag{12.13}$$

The Neumann boundary conditions (BCs) allow one to do an integration by parts of the action given in Eq.(12.11), yielding the action

$$S = -\frac{1}{2}m\int_{t_i}^{t_f} dt x(t)\left[-\frac{d^2}{dt^2} + \omega^2\right]x(t) + \int_{t_i}^{t_f} dt j(t)x(t). \tag{12.14}$$

The generating functional is given by the path integral

$$K(t_i, t_f; j) = \int Dx e^S. \tag{12.15}$$

The path integral $\int Dx e^S$ is performed over all paths (functions) $x(t)$ that satisfy the Neumann boundary conditions given in Eq. (12.13). All such functions can be expanded in a Fourier cosine series as follows:

$$x(t) = a_0 + \sum_{n=1}^{\infty} a_n \cos\left[n\pi\frac{(t-t_i)}{\tau}\right], \quad \tau \equiv t_f - t_i,$$

$$\int DX = \mathcal{N}\prod_{n=0}^{\infty}\int_{-\infty}^{+\infty} da_n \quad : \text{infinite multiple integral,}$$

($\mathcal{N}$ is a normalization constant). The orthogonality equations

$$\begin{aligned}&\int_{t_i}^{t_f} dt \cos\left[n\pi\frac{(t-t_i)}{\tau}\right]\cos\left[m\pi\frac{(t-t_i)}{\tau}\right] \\ &= \int_{t_i}^{t_f} dt \sin\left[n\pi\frac{(t-t_i)}{\tau}\right]\sin\left[m\pi\frac{(t-t_i)}{\tau}\right] = \frac{\tau}{2}\delta_{m-n}, \quad m, n \geq 1\end{aligned} \tag{12.16}$$

yield, for the action given in Eq. 12.14,

$$\begin{aligned}S &= -\frac{1}{2}m\omega^2\tau\left\{a_0^2 + \frac{1}{2}\sum_{n=1}^{\infty}\left[1 + (\frac{n\pi}{\omega\tau})^2\right]a_n^2\right\} \\ &\quad + \int_{t_i}^{t_f} dt j(t)\left\{a_0 + \sum_{i=1}^{\infty} a_n \cos\left[n\pi\frac{(t-t_i)}{\tau}\right]\right\} \\ &= -\frac{1}{2}\sum_{n=0}^{\infty}\kappa_n a_n^2 + \sum_{n=0}^{\infty} j_n a_n,\end{aligned} \tag{12.17}$$

where

$$\kappa_0 = m\omega^2\tau, \quad \kappa_n = \frac{1}{2}m\omega^2\tau\left[1+\left(\frac{n\pi}{\omega\tau}\right)^2\right], \quad n \geq 1,$$

$$j_n = \int_{t_i}^{t_f} dt j(t)\cos\left[n\pi\frac{(t-t_i)}{\tau}\right], \quad n = 0, 1, \ldots \infty.$$

All the Gaussian integrations over the variables a_n have decoupled in the action S given in Eq. 12.17. The path integral has been reduced to an infinite product of single Gaussian integrations, each of which can be performed using Eq. 7.7. Hence, from Eqs. 12.15 and 12.17

$$K(t_i, t_f; j) = \int DXe^S = \mathcal{N}\prod_{n=0}^{\infty}\left[\int_{-\infty}^{+\infty} da_n e^{-\frac{1}{2}\kappa_n a_n^2 + j_n a_n}\right]$$

$$= \exp\left\{\frac{1}{2}\sum_{n=0}^{\infty} j_n \frac{1}{\kappa_n} j_n\right\}, \tag{12.18}$$

yielding

$$K(t_i, t_f; j) = e^{\frac{1}{2}\int_{t_i}^{t_f} dt dt' j(t) D(t,t';t_i,t_f) j(t')}, \tag{12.19}$$

where the function $D(t, t'; t_i, t_f)$ is the propagator for the simple harmonic oscillator. Using Eq. 12.19 to factor out the $j(t)$s from Eq. 12.18 yields

$$D(t, t'; t_i, t_f)$$

$$= \frac{1}{m\omega^2\tau}\left\{1 + 2\sum_{n=1}^{\infty}\cos\left[n\pi\frac{(t-t_i)}{\tau}\right]\left[\frac{1}{1+(\frac{n\pi}{\omega\tau})^2}\right]\cos\left[n\pi\frac{(t'-t_i)}{\tau}\right]\right\}. \tag{12.20}$$

Let $\theta = t - t_i > 0$ and $\theta' = t' - t_i > 0$; then

$$2\sum_{n=1}^{\infty}\frac{\cos(n\pi\theta/\tau)\cos(n\pi\theta'/\tau)}{1+(\frac{n\pi}{\omega\tau})^2}$$

$$= \left(\frac{\omega\tau}{\pi}\right)^2\sum_{n=1}^{\infty}\frac{\cos(n\pi(\theta+\theta')/\tau)+\cos(n\pi(\theta-\theta')/\tau)}{(\frac{\omega\tau}{\pi})^2+n^2}. \tag{12.21}$$

The summation over integer n is performed using the identity[2]

$$\sum_{n=1}^{\infty}\frac{\cos(n\theta)}{a^2+n^2} = \frac{\pi}{2a}\frac{\cosh(\pi-|\theta|)a}{\sinh\pi a} - \frac{1}{2a^2} \tag{12.22}$$

[2] The formula given in Eq. 11.30 is valid for any complex number a, and will be applied in later discussions for a case where a is indeed a complex number. The branch of the square root of a^2 that is taken on the right hand side need not be specified since the rhs is a function of a^2.

and yields the result

$$D(t,t';t_i,t_f)=\frac{\cosh\omega\{\tau-|\theta-\theta'|\}+\cosh\omega\{\tau-(\theta+\theta')\}}{2m\omega\sinh\omega\tau}. \qquad (12.23)$$

Hence, from Eq. 12.23, and since $\tau=t_f-t_i$, the propagator is given by

$$D(t,t';t_i,t_f)$$
$$=\frac{\cosh\omega\{(t_f-t_i)-|t-t'|\}+\cosh\omega\{(t_f-t_i)-(t+t'-2t_i)\}}{2m\omega\sinh\omega(t_f-t_i)}. \qquad (12.24)$$

Note that the propagator (also called a Green's function) satisfies the differential equation

$$m\left[-\frac{d^2}{dt^2}+\omega^2\right]D\left(t,t';t_i,t_f\right)=\delta\left(t-t'\right)\quad \text{: Neumann BCs.}$$

The case of infinite time for Eq. 12.24 is obtained by taking the limit of $t_i\to-\infty$, $t_f\to+\infty$, and yields

$$D(t,t')=\frac{1}{2m\omega}e^{-\omega|t-t'|}, \qquad (12.25)$$

which has been derived earlier using different methods.

12.7 Simple harmonic oscillator: Fourier expansion

Given the importance of the simple harmonic oscillator's evolution kernel, a derivation is given below using the technique of Fourier expansions.

Consider a quantum particle with an initial position x_i and, after time τ, with a final position x_f. The evolution kernel for this case has been obtained earlier in Section 11.6 based on evaluating the solution of the classical equation of motion. The limitation of finding the classical solution is that for many nonlinear Lagrangians, it is usually not possible to obtain the classical trajectory. The Fourier expansion avoids this problem by directly enumerating all the paths with a given initial and final position and hence can be used for a great variety of problems.

Consider the Fourier expansion for the possible paths of the quantum particle,

$$x(t)=x_f+\sum_{n=0}^{\infty}\cos\left[\frac{(2n+1)\pi t}{2\tau}\right]x_n. \qquad (12.26)$$

The boundary conditions are

$$x(0)=x_f+\sum_{n=0}^{\infty}x_n,$$
$$x(\tau)=x_f.$$

At time $t = \tau$, the final condition is automatically satisfied. Note the velocity of the particle at $t = 0$ is zero, since

$$\frac{dx(0)}{dt} = \sum_{n=0}^{\infty} \frac{(2n+1)\pi}{2\tau} \sin[0] x_n = 0.$$

As discussed for the two previous cases, zero velocity at $t = 0$ means that the position $x(0)$ is indeterminate. Hence, to fix the initial position of the particle to be x_i, one needs to put a delta-function constraint into the path integral, namely

$$x_i = x_f + \sum_{n=0}^{\infty} x_n \quad : \text{ constraint.}$$

Hence, the evolution kernel is given by

$$\begin{aligned} K(x_i, x_f; \tau) &= \langle x_f | e^{-\tau H} | x_i \rangle \qquad (12.27) \\ &= \mathcal{N}' \prod_{n=0}^{\infty} \int_{-\infty}^{+\infty} dx_n \, \delta\Big(\sum_{n=0}^{\infty} x_n + x_f - x_i\Big) e^S \\ &= \int Dx \int_{-\infty}^{+\infty} \frac{d\eta}{2\pi} e^{i\eta(\sum_{n=0}^{\infty} x_n + x_f - x_i)} \, e^S, \qquad (12.28) \end{aligned}$$

where $\mathcal{N}'$ is a normalization constant.

Using the orthogonality relation

$$\begin{aligned} &\int_0^\tau dt \sin\left[\frac{(2n+1)\pi t}{2\tau}\right) \sin\left[\frac{(2m+1)\pi t}{2\tau}\right] \\ &= \int_0^\tau dt \cos\left[\frac{(2n+1)\pi t}{2\tau}\right] \cos\left[\frac{(2m+1)\pi t}{2\tau}\right] = \frac{\tau}{2}\delta_{n-m} \end{aligned}$$

yields a simplification for the action,

$$\begin{aligned} S &= -\frac{m}{2} \int_0^\tau dt \left[(\frac{dx(t)}{dt})^2 + \omega^2 x^2(t) \right] \\ &= -\frac{m}{2} \frac{\pi^2}{8\tau} \sum_{n=0}^{\infty} \left[(2n+1)^2 + \left(\frac{2\omega\tau}{\pi}\right)^2 \right] x_n^2 - \frac{m}{2}\omega^2 \tau x_f^2 \\ &\quad + \sum_{n=0}^{\infty} \frac{2m\omega^2\tau}{(2n+1)\pi} (-1)^n x_f x_n, \end{aligned}$$

since

$$\int_0^\tau dt \cos\left[\frac{(2n+1)\pi t}{2\tau}\right] = -\frac{2\tau}{(2n+1)\pi}(-1)^n.$$

All the Fourier modes x_n have completely factorized in the action. Performing the Gaussian path integration (over the Fourier modes x_n) given in Eq. 12.28 yields

$$K(x_i, x_f; \tau) = \mathcal{N} \int_{-\infty}^{+\infty} \frac{d\eta}{2\pi} e^{F(\eta)},$$
$$F(\eta) = \frac{1}{2m} \frac{8\tau}{\pi^2} \sum_{n=0}^{\infty} \frac{1}{(2n+1)^2 + \left(\frac{2\omega\tau}{\pi}\right)^2} \left[i\eta + \frac{2m\omega^2\tau}{(2n+1)\pi}(-1)^n x_f \right]^2.$$

Using the numerical identities

$$\sum_{n=0}^{\infty} \frac{1}{(2n+1)^2 + \left(\frac{2\omega\tau}{\pi}\right)^2} = \frac{\pi^2}{8\omega\tau} \tanh(\omega\tau),$$
$$\sum_{n=0}^{\infty} \frac{(-1)^n}{(2n+1)^3 + \left(\frac{2\omega\tau}{\pi}\right)^2 (2n+1)} = \frac{\pi^3}{8(\omega\tau)^2} \cdot \frac{\sinh^2\left(\frac{\omega\tau}{2}\right)}{\cosh(\omega\tau)},$$
$$\sum_{n=0}^{\infty} \frac{1}{(2n+1)^4 + \left(\frac{2\omega\tau}{\pi}\right)^2 (2n+1)^2} = \frac{\pi^4}{2^5\omega^2\tau^2} \left(1 - \frac{\tanh\omega\tau}{\omega\tau}\right),$$

yields the result

$$F = -\frac{\tanh(\omega\tau)}{2m\omega}\eta^2 - 2ix_f\eta\frac{\sinh^2(\frac{\omega\tau}{2})}{\cosh(\omega\tau)} + \frac{1}{2}x_f^2 m\omega^2\tau\left(1 - \frac{\tanh(\omega\tau)}{\omega\tau}\right)$$
$$-\frac{1}{2}x_f^2 m\omega^2\tau - i\eta(x_i - x_f)$$
$$= -\frac{\tanh(\omega\tau)}{2m\omega}\eta^2 - i\eta\left[x_i - \frac{x_f}{\cosh(\omega\tau)}\right] - \frac{1}{2}x_f^2 m\omega\tanh(\omega\tau).$$

Performing the η integration gives the final result

$$K(x_i, x_f; \tau) = \mathcal{N}\sqrt{\frac{2\pi m\omega}{\tanh(\omega\tau)}} e^G,$$

with

$$G = -\frac{m\omega}{2\tanh(\omega\tau)}\left[x_i - \frac{x_f}{\cosh(\omega\tau)}\right]^2 - \frac{1}{2}x_f^2 m\omega\tanh(\omega\tau)$$
$$= -\frac{m\omega}{2\tanh(\omega\tau)}\left[x_i^2 + x_f^2 - 2\frac{x_i x_f}{\cosh(\omega\tau)}\right] + \xi.$$

The remainder ξ is zero since

$$\xi = -\frac{1}{2}m\omega x_f^2\left[\frac{1}{\tanh(\omega\tau)\cosh^2(\omega\tau)} - \frac{1}{\tanh(\omega\tau)} + \tanh(\omega\tau)\right] = 0.$$

Once G has been determined, the overall normalization of the evolution kernel can be fixed by using the composition law, as in Section 11.5. Hence, the evolution kernel is given by

$$K(x_i, x_f; \tau) = \sqrt{\frac{m\omega}{2\pi \sinh(\omega\tau)}} e^G,$$

which agrees with the result obtained earlier in Eq. 11.21.

12.8 Evolution kernel for a magnetic field

A free charged particle moving in an electromagnetic field couples with the vector potential which, in vector notation, is given by $\mathbf{A}$. The Lagrangian, for $\mathbf{x} = (x, y, z)$, is given in Minkowski time by

$$L = \frac{1}{2}m\dot{\mathbf{x}}^2 + e\mathbf{A} \cdot \dot{\mathbf{x}}, \quad \dot{\mathbf{x}} \equiv \frac{d\mathbf{x}}{dt}. \tag{12.29}$$

The calculation in this section is carried out for Minkowski time to get a flavor for such calculations; in particular, it will be seen that the computations in Minkowski time have a plethora of i factors which are quite unnecessary in obtaining the result for Euclidean time.

Consider a free charged particle interacting with a constant external magnetic field of strength $\mathbf{B}$ acting along the z-direction. The vector potential $\mathbf{A}$ is chosen to be in the symmetric gauge,

$$\mathbf{A} = \frac{1}{2}B(-y, x, 0). \tag{12.30}$$

The Lagrangian is then given by

$$\mathcal{L}(t) = \frac{1}{2}m\left(\dot{x}^2 + \dot{y}^2\right)^2 - m\sigma\left(\dot{x}y - x\dot{y}\right), \quad \sigma = \frac{eB}{2m}, \tag{12.31}$$

where σ is the cyclotron frequency. The particle moves only in the xy-plane and hence the z-axis can be completely dropped from the problem. The evolution kernel is given by the path integral

$$K(x_b, y_b, t_b; x_a, y_a, t_a) = \langle x_b, y_b, |e^{-i(t_b - t_a)H}|x_a, y_a\rangle = \int DxDye^{iS}, \tag{12.32}$$

where H is the Hamiltonian of the system; the action S is given by

$$S = \int_{t_a}^{t_b} dt\mathcal{L}(t).$$

The action is quadratic in the degrees of freedom and hence the evolution kernel can be obtained using the classical solution. The equations of motion for the Lagrangian in Eq. 12.29 yield the classical solution x_c, y_c given by

$$\ddot{x}_c = 2\sigma \dot{y}_c, \quad \ddot{y}_c - 2\sigma \dot{x}_c. \tag{12.33}$$

Shifting the degree of freedom x, y by x_c, y_c

$$x \to x + x_c, \qquad y \to y + y_c$$

and keeping the same notation for the sake of simplicity yields the action

$$S[x + x_c, y + y_c; T] = S_c[x_c, y_c; T] + S_0[x, y; T] + \Delta S, \ T = t_b - t_a, \tag{12.34}$$

$$\Delta S = \int_{t_a}^{t_b} dt \left\{ m(\dot{x}\dot{x}_c + \dot{y}\dot{y}_c) - m\sigma(\dot{x}y_c + y\dot{x}_c - x\dot{y}_c - \dot{y}x_c) \right\}.$$

The shifted quantum degrees of freedom x, y have new boundary conditions given by

$$x(t_a) = 0 = x(t_b), \quad y(t_a) = 0 = y(t_b) \tag{12.35}$$

and lead to the result

$$\Delta S = 0.$$

The decoupling of the classical solution – since $\Delta S = 0$ – from the quantum variables is true for any quadratic Lagrangian.

From Eqs. 12.32, 12.34, and 12.35, the Minkowski time evolution kernel is given by

$$\begin{aligned} K(x_b, y_b; x_a, y_a; T) &= e^{iS_c[x_c, y_c; T]} \int Dx Dy e^{iS_0[x, y; T]} \\ &= \mathcal{N}(T) e^{iS_c[x_c, y_c; T]}. \end{aligned} \tag{12.36}$$

Classical action

The classical action is evaluated using the Lagrangian given in Eq. 12.29,

$$S_c[x_c, y_c; T] = \int_{t_a}^{t_b} dt \left\{ \frac{1}{2} m \left(\dot{x}_c^2 + \dot{y}_c^2 \right) - m\sigma \left(\dot{x}_c y_c - x_c \dot{y}_c \right) \right\}. \tag{12.37}$$

Integrating by parts, Eq. 12.37 becomes[3]

$$S_c[x_c, y_c; T] = \frac{1}{2} m (\dot{x}_b x_b + \dot{y}_b y_b - \dot{x}_a x_a - \dot{y}_a y_a).$$

[3] Since $\int_{t_a}^{t_b} dt (x_c \ddot{x}_c + y_c \ddot{y}_c + 2\sigma \dot{x}_c y_c - 2\sigma x_c \dot{y}_c) = 0$.

Furthermore, integrating Eq. 12.33 gives

$$\dot{x}_c = 2\sigma(y_c + C), \quad \dot{y}_c = -2\sigma(x_c - D), \tag{12.38}$$

where C and D are constants of the integration. Equation 12.38 is the equation for a two-dimensional harmonic oscillator,

$$\ddot{x}_c = -\omega^2(x_c - D), \quad \ddot{y}_c = -\omega^2(y_c + C), \quad \omega = 2\sigma. \tag{12.39}$$

The solutions of Eq. 12.39 are given by

$$x_c = A\cos\omega t + B\sin\omega t + D, \quad y_c = -A\sin\omega t + B\cos\omega t - C.$$

The constants C and D are derived from the boundary conditions,

$$\begin{aligned} x_a &= A\cos\omega t_a + B\sin\omega t_a + D, \quad y_a = -A\sin\omega t_a + B\cos\omega t_a - C, \\ x_b &= A\cos\omega t_b + B\sin\omega t_b + D, \quad y_b = -A\sin\omega t_b + B\cos\omega t_b - C. \end{aligned} \tag{12.40}$$

Solving Eq. 12.40 yields

$$\begin{aligned} B &= \frac{(x_b - x_a)\cos\sigma T' - (y_b - y_a)\sin\sigma T'}{2\sin\sigma T}, \\ A &= -\frac{(x_b - x_a)\sin\sigma T' - (y_b - y_a)\cos\sigma T'}{2\sin\sigma T}, \\ D &= \frac{1}{2}(x_b + x_a) + \frac{(y_b - y_a)\cos\sigma T}{2\sin\sigma T}, \\ C &= \frac{1}{2}(y_b + y_a) + \frac{(x_b - x_a)\cos\sigma T}{2\sin\sigma T}, \end{aligned} \tag{12.41}$$

where $T = t_b - t_a$, $T' = t_a + t_b$. Furthermore, Eq. 12.38 yields

$$x_c\dot{x}_c + y_c\dot{y}_c = 2\sigma(Cx_c + Dy_c).$$

Equation 12.37, together with Eqs. 12.38 and 12.41, yields

$$S_c[x_c, y_c; T] = m\sigma\{C(x_b - x_a) - D(y_b - y_a)\},$$

and we obtain the final result

$$S_c[x_c, y_c; T] = m\sigma\left[\frac{\cos\sigma T\{(x_b - x_a)^2 + (y_b - y_a)^2\}}{2\sin\sigma T} + y_b x_a - x_b y_a\right]. \tag{12.42}$$

The normalization of the evolution kernel $\mathcal{N}(T)$ is given by the consistency equation as discussed in Section 4.8,

$$K(x_b, y_b; x_a, y_a; 2T) = \int d\xi d\eta K(x_b, y_b; \xi, \eta; T) K(\xi, \eta; x_a, y_a; T), \tag{12.43}$$

$$\Rightarrow \mathcal{N}(2T) = \mathcal{N}^2(T)\Lambda(T). \tag{12.44}$$

The evolution kernel is given by Eq. 12.36 and 12.42; performing the Gaussian integration given on the right hand side of Eq. 12.43 yields, from Eq. 12.44, the following recursion equation:

$$\mathcal{N}(2T) = \mathcal{N}^2(T)\frac{i\pi\sin(\sigma T)}{m\sigma\cos(\sigma T)} \Rightarrow \mathcal{N}(T) = \frac{m\sigma}{2\pi i\sin\sigma T}. \tag{12.45}$$

The final result for the Minkowski time path integral, from Eqs. 12.36, 12.42, and 12.45, is thus

$$K(x_b, y_b; x_a, y_a; T) = \mathcal{N}(T)e^{iS_c[x_c, y_c; T]} = \frac{m\sigma}{2\pi i\sin\sigma T}$$
$$\times\exp\Big[\frac{i}{2}\frac{m\sigma\cos\sigma T}{\sin\sigma T}\Big\{(x_b - x_a)^2 + (y_b - y_a)^2\Big\} + im\sigma\Big(y_b x_a - x_b y_a\Big)\Big]. \tag{12.46}$$

The evolution kernel for Euclidean time $\tau > 0$ is given by $\tau = iT$ and

$$K_E(x_b, y_b; x_a, y_a; \tau) = \frac{m\sigma}{2\pi\sinh\sigma\tau}$$
$$\times\exp\Big[-\frac{1}{2}\frac{m\sigma\cosh\sigma\tau}{\sinh\sigma\tau}\Big\{(x_b - x_a)^2 + (y_b - y_a)^2\Big\} + im\sigma\Big(y_b x_a - x_b y_a\Big)\Big].$$

12.9 Summary

The techniques of using classical solutions and Fourier expansions for analyzing path integrals were illustrated using Gaussian path integration. The exact solutions obtained show the flexibility and power of these techniques, which have generalizations to many nonlinear systems. The evolution kernel for various quadratic Lagrangians, in particular of the simple harmonic oscillator, reveal the rich mathematical structure of these relatively simple theories.

The evolution kernel of a charged particle in a magnetic field, evaluated for both Minkowski and Euclidean time, shows the equivalence of these approaches as well as illustrating the relative simplicity of the results in Euclidean time.

Part five

Action with acceleration

Introduction to part five

The most widely used actions in quantum mechanics have a kinetic term that is the velocity squared of the degree of freedom and a potential term that depends on the degree of freedom.[1] The kinetic term fixes the equal time commutation equation of the degree of freedom, as shown in Section 6.7. Furthermore, the velocity term in the Lagrangian entails that all the possible indeterminate paths obey two boundary conditions, which in turn yields a state space that depends on the degree of freedom only. In Section 11.13, this property of the indeterminate paths was used for deriving the state function of the harmonic oscillator.

The action with acceleration has a kinetic term that is given by the *acceleration* squared of the degree of freedom, in addition to the usual velocity and potential terms. It is an example of higher derivative Lagrangians, discussed in Simon (1990). The higher derivative quantum systems have many remarkable properties not present for the usual cases studied so far.

The action with acceleration arises in many diverse fields and has been widely studied; it describes the behaviour of "stiff" polymers, of cell walls, of the formation of microemulsions, the properties of chromoelectric flux lines in quantum chromodynamics, as well as the Big Bang singularity in cosmology. The Euclidean action and path integral were studied by Kleinert (1986), where a list of the applications of the model is given. Bender and Mannheim (2008b) have extensively studied the Lagrangian and Hamiltonian of the model, both in Euclidean and Minkowski time; Mannheim and Davidson (2005) showed that the model in Minkowski time was free of ghost states using the concept of a PT-symmetric Hamiltonian; in Bender and Mannheim (2010) they demonstrated the reality of the model's eigenspectrum. The acceleration action has been applied to

[1] Velocity dependent terms can be included in the potential.

the study of cosmology, and quantum and conformal gravity by Mannheim (2011a) and Mannheim (2011b).

The Euclidean model was studied by Hawking and Hertog (2002) for its role in quantum gravity and in string theory's D-brane dynamics; the Euclidean path integral was used by Fontanini and Trodden (2011) for analyzing the ghost states for Minkowski time.

The acceleration action appears in the study of mathematical finance in describing the time dependence of financial derivative instruments and in the study of interest rates and equity [Baaquie (2010)]. The Euclidean path integral has been studied extensively by Baaquie *et al.* (2012) and Yang (2012) for describing the correlation function of equities. The model has been applied in the study of options by Baaquie and Yang (2014) and to the study of microeconomics by Baaquie (2013b).

The acceleration Lagrangian yields a pseudo-Hermitian Hamiltonian that has been studied in Mostafazadeh (2002) and Wang *et al.* (2010). The analysis of the Euclidean Hamiltonian in this book is a continuation of the study of the Minkowski time Hamiltonian carried out by Bender and Mannheim (2008a), (2008b). The Euclidean case has many new features that are absent for Minkowski time, the most important being a transparent positivity of the Euclidean state space. The path integral for the pseudo-Hermitian Hamiltonian has been studied, starting from the propagator, by Jones and Rivers (2009) and Rivers (2011).

For a special value of the parameters, the acceleration Hamiltonian becomes essentially non-Hermitian. As shown by Mannheim and Davidson (2000), (2005), at the special value of the parameters the Hamiltonian is described by a Jordan block matrix; the Jordan block Hamiltonian was shown to be pseudo-Hermitian by Bender and Mannheim (2011).

The following three chapters analyze the acceleration Lagrangian and Hamiltonian and are largely based on papers by Baaquie (2013c), (2013d). The quantum mechanical system for the acceleration Lagrangian is defined in Euclidean time and using the path integral. Many of the unnecessary complications that appear in Minkowski time formulation – and which obscure key features of the system – are absent in the Euclidean time formulation. Furthermore, applications in biophysics and finance are directly based on the Euclidean formulation and consequently this system is interesting in its own right, regardless of its connection with the theory in Minkowski time.

13

Acceleration Lagrangian

The Lagrangian and path integral for the action with acceleration are defined in Section 13.1 and the quadratic potential is chosen in Section 13.2 so that the path integral can be performed exactly using the classical solution. The propagator is derived from the infinite time action in Section 13.3.

The Hamiltonian is obtained for the Euclidean time Lagrangian using the technique of Dirac brackets in Section 13.4, and it is shown in Section 13.5 that the Hamiltonian yields the expected path integral. The evolution kernel is carefully studied in Section 13.6 to prove that the Lagrangian-based path integral is equivalent to the one derived from the quantum Hamiltonian. It is shown in Section 13.7 how the boundary conditions on the indeterminate paths (integration variables) of the path integral are transformed in going from the Hamiltonian to the action-based formulation of the path integral.

13.1 Lagrangian

Euclidean time is defined by the analytic continuation of Minkowski time $t_M = -it$, where t is Euclidean time. The degree of freedom x does not change in going from Minkowski to Euclidean time but Minkowski velocity v_M picks up an extra factor of i since it is related to Euclidean velocity v, due to Eq. 5.17, by

$$v = -\frac{dx}{dt} = i\frac{dx}{dt_M} = iv_M. \tag{13.1}$$

The minus in the definition of Euclidean velocity leads to $v = iv_M$ and is consistent with $t = it_M$.

The non-Hermitian Euclidean Hamiltonian and path integral are well behaved and a complexification of the degree of freedom is not required.

The Euclidean time Lagrangian with acceleration is given by

$$\mathcal{L} = -\frac{\gamma}{2}\left(\frac{d^2x}{dt^2}\right)^2 - \frac{\alpha}{2}\left(\frac{dx}{dt}\right)^2 - \Phi(x), \tag{13.2}$$

with the acceleration action for finite Euclidean time τ given by

$$\mathcal{S}[x] = \int_0^\tau dt\mathcal{L}. \tag{13.3}$$

The Feynman path integral is given by

$$\mathcal{K}(x_f, \dot{x}_f; x_i, \dot{x}_i) = \int DX\, e^{\mathcal{S}[x]}\Big|_{(x_i, \dot{x}_i; x_f, \dot{x}_f)}, \tag{13.4}$$

$$\int Dx = \tilde{\mathcal{N}} \prod_{t=0}^{\tau} \int_{-\infty}^{+\infty} dx(t), \tag{13.5}$$

where $\tilde{\mathcal{N}}$ is a normalization constant. The paths have four boundary conditions,

$$x(0) = x_i; \quad \frac{dx(0)}{dt} = \dot{x}_i \ \text{ initial position and velocity,} \tag{13.6}$$

$$x(\tau) = x_f; \quad \frac{dx(\tau)}{dt} = \dot{x}_f \ \text{ final position and velocity.} \tag{13.7}$$

To evaluate the path integral exactly using the classical solution, consider the quadratic potential

$$\Phi(x) = \frac{\beta}{2}x^2, \tag{13.8}$$

which yields the Lagrangian

$$\mathcal{L} = -\frac{\gamma}{2}\left(\frac{d^2x}{dt^2}\right)^2 - \frac{\alpha}{2}\left(\frac{dx}{dt}\right)^2 - \frac{\beta}{2}x^2. \tag{13.9}$$

The path integral given Eq. 13.4 is now quadratic and can be evaluated exactly using the classical solution. Let $x_c(t)$ be the classical solution given by

$$\frac{\delta\mathcal{S}[x_c]}{\delta x(t)} = 0 \tag{13.10}$$

that satisfies the boundary conditions,

$$x_c(0) = x_i; \quad \frac{dx_c(0)}{dt} = \dot{x}_i \ \text{ initial position and velocity,}$$

$$x_c(\tau) = x_f; \quad \frac{dx_c(\tau)}{dt} = \dot{x}_f \ \text{ final position and velocity.} \tag{13.11}$$

Consider the change of integration variables, from $x(t)$ to $\xi(t)$,

$$x(t) = x_c(t) + \xi(t), \tag{13.12}$$

with boundary conditions given by

$$\xi(0) = 0; \quad \frac{d\xi(0)}{dt} = 0 \text{ initial position and velocity,}$$
$$\xi(\tau) = 0; \quad \frac{d\xi(\tau)}{dt} = 0 \text{ final position and velocity.} \tag{13.13}$$

The change of variables yields

$$\mathcal{S}[x] = \mathcal{S}[x_c + \xi] = \mathcal{S}[x_c] + \mathcal{S}[\xi],$$
$$\mathcal{K}(x_f, \dot{x}_f; x_i, \dot{x}_i) = \int DX\, e^{\mathcal{S}} = \mathcal{N}(\tau) e^{\mathcal{S}[x_c]}, \tag{13.14}$$
$$\mathcal{N}(\tau) = \int D\xi\, e^{\mathcal{S}[\xi]}.$$

The evolution kernel given in Eq. 13.14 has been evaluated explicitly in Hawking and Hertog (2002) by solving for the classical solution $x_c(t)$ and then obtaining $\mathcal{S}[x_c]$ and $\mathcal{N}(\tau)$.

As can be directly verified from the classical action $\mathcal{S}[x_c]$, the classical solution $x_c(t)$ yields *another* equally valid classical solution $\tilde{x}_c(t)$ given by

$$\tilde{x}_c(t) = x_c(\tau - t), \tag{13.15}$$

with boundary conditions, from Eq. 13.11, given by

$$\tilde{x}_c(0) = x_c(\tau) = x_f; \quad \frac{d\tilde{x}_c(0)}{dt} = -\frac{dx_c(\tau)}{dt} = -\dot{x}_f, \tag{13.16}$$
$$\tilde{x}_c(\tau) = x_c(0) = x_i; \quad \frac{d\tilde{x}_c(\tau)}{dt} = -\frac{dx_c(0)}{dt} = -\dot{x}_i, \tag{13.17}$$
$$\Rightarrow \quad \mathcal{S}[x_c] = \mathcal{S}[\tilde{x}_c],$$
$$\Rightarrow \quad \mathcal{S}_c[x_f, \dot{x}_f; x_i, \dot{x}_i] = \mathcal{S}[x_i, -\dot{x}_i; x_f, -\dot{x}_f]. \tag{13.18}$$

The evolution kernel, from Eqs. 13.14 and 13.18, hence has the symmetry

$$\mathcal{K}(x_f, \dot{x}_f; x_i, \dot{x}_i) = \mathcal{K}(x_i, -\dot{x}_i; x_f, -\dot{x}_f). \tag{13.19}$$

13.2 Quadratic potential: the classical solution

The acceleration Lagrangian given by Eq. 13.9 is re-written with a slight change of notation, in the following manner

$$\mathcal{L} = -\frac{1}{2}\left(a\ddot{x}^2 + 2b(\dot{x})^2 + cx^2\right), \quad S = \int_0^\tau dt \mathcal{L}. \tag{13.20}$$

The parameterization chosen in Eq. 13.20 is more suitable for studying the classical solutions than the one given in Eq. 13.9.

The Euler–Lagrangian equation in Eq. 13.10 yields the equation of motion; the classical solution $x_c(t)$ satisfies the equation of motion

$$a\ddddot{x}_c(t) - 2b\ddot{x}_c(t) + cx_c(t) = 0. \tag{13.21}$$

Let us choose the boundary conditions,

$$\begin{aligned} \text{Initial values}: \ & x(0) = x_f = x_1, \quad \dot{x}(0) = \dot{x}_f = x_2 = -v_f, \\ \text{Final values}: \ & x(\tau) = x_i = x_4, \quad \dot{x}(\tau) = \dot{x}_i = x_3 = -v_i. \end{aligned} \tag{13.22}$$

We define the parameters r and ω by

$$r + i\omega \equiv \sqrt{\frac{b + i\sqrt{ac - b^2}}{a}}. \tag{13.23}$$

From Eq. 13.21 the classical solution of equations of motion is given by

$$x_c(t) = e^{rt}(a_1 \sin \omega t + a_2 \cos \omega t) + e^{-rt}(a_3 \sin \omega t + a_4 \cos \omega t). \tag{13.24}$$

The parameters $a_1, \ldots, a_4$ are obtained from the boundary conditions and are given by Yang (2012) as

$$\begin{aligned} a_1 = \Gamma\Big[& r^2 x_f e^{2r\tau} \sin(2\tau\omega) + \omega v_f e^{2r\tau} - r v_f e^{2r\tau} \sin(2\tau\omega) + r\omega x_f e^{2r\tau} \cos(2\tau\omega) \\ & - r\omega x_f - \omega v_f - 2r^2 x_i e^{r\tau} \sin(\tau\omega) - 2r v_i e^{r\tau} \sin(\tau\omega) \\ & - \omega e^{r\tau}\left(e^{2r\tau} - 1\right) \cos(\tau\omega)\left(v_i + r x_i\right) - \omega^2 x_i e^{r\tau} \sin(\tau\omega) + \omega^2 x_i e^{3r\tau} \sin(\tau\omega)\Big], \end{aligned}$$

$$\begin{aligned} a_2 = \Gamma\Big[& r^2 x_f \left(-e^{2r\tau}\right) + r v_f e^{2r\tau} + r e^{2r\tau} \cos(2\tau\omega)\left(r x_f - v_f\right) - \omega^2 x_f e^{2r\tau} \\ & - r\omega x_f e^{2r\tau} \sin(2\tau\omega) + \omega^2 x_f - \omega v_i e^{r\tau} \sin(\tau\omega) + \omega v_i e^{3r\tau} \sin(\tau\omega) \\ & + \omega^2 x_i e^{r\tau}\left(e^{2r\tau} - 1\right) \cos(\tau\omega) + r\omega x_i e^{r\tau} \sin(\tau\omega) + r\omega x_i e^{3r\tau} \sin(\tau\omega)\Big], \end{aligned}$$

$$\begin{aligned} a_3 = \Gamma e^{r\tau}\Big[& r^2 x_f e^{r\tau} \sin(2\tau\omega) + \omega v_f e^{r\tau} - \omega v_f e^{3r\tau} + r v_f e^{r\tau} \sin(2\tau\omega) \\ & + r\omega x_f e^{3r\tau} - r\omega x_f e^{r\tau} \cos(2\tau\omega) - 2r^2 x_i e^{2r\tau} \sin(\tau\omega) - 2r v_i e^{2r\tau} \sin(\tau\omega) \\ & - \omega\left(e^{2r\tau} - 1\right) \cos(\tau\omega)\left(r x_i - v_i\right) - \omega^2 x_i e^{2r\tau} \sin(\tau\omega) + \omega^2 x_i \sin(\tau\omega)\Big], \end{aligned}$$

$$\begin{aligned} a_4 = \Gamma e^{r\tau}\Big[& r^2 x_f \left(-e^{r\tau}\right) - r v_f e^{r\tau} + r e^{r\tau} \cos(2\tau\omega)\left(r x_f + v_f\right) - \omega^2 x_f e^{r\tau} \\ & + \omega^2 x_f e^{3r\tau} + r\omega x_f e^{r\tau} \sin(2\tau\omega) - \omega v_i e^{2r\tau} \sin(\tau\omega) \\ & - \omega^2 x_i \left(e^{2r\tau} - 1\right) \cos(\tau\omega) - r\omega x_i e^{2r\tau} \sin(\tau\omega) - r\omega x_i \sin(\tau\omega) + \omega v_i \sin(\tau\omega)\Big]. \end{aligned}$$

The term Γ is defined by

$$\Gamma = \frac{1}{\omega^2 + \omega^2 e^{4r\tau} + 2r^2 e^{2r\tau}\cos(2\tau\omega) - 2e^{2r\tau}\left(r^2+\omega^2\right)}. \tag{13.25}$$

The boundary condition given in Eq.13.22 yields the classical action

$$S_c = S_c(x_f, v_f, x_i, v_i) = -\frac{1}{2}\sum_{I,J=1}^{4} x_I M_{IJ} x_J. \tag{13.26}$$

From Eq. 13.18, the classical action has the symmetry

$$\mathcal{S}_c[x_f, v_f; x_i, v_i] = \mathcal{S}[x_i, -v_i; x_f, -v_f] \tag{13.27}$$

and hence the matrices M_{IJ} given in Eq. 13.26 satisfy symmetry,

$$\begin{aligned} M_{11} &= M_{44}, & M_{22} &= M_{33},\\ M_{12} &= -M_{34}, & M_{13} &= -M_{24}. \end{aligned}$$

The action can be simplified to

$$\begin{aligned} S_c(x_f, v_f, v_i, x_i) = &-\frac{1}{2}M_{11}(x_i^2 + x_f^2) - \frac{1}{2}M_{22}(v_i^2 + v_f^2) + M_{12}(x_i v_i - x_f v_f)\\ &+ M_{13}(x_i v_f - x_f v_i) - M_{14}x_i x_f - M_{23}v_i v_f. \end{aligned} \tag{13.28}$$

The solutions for M_{IJ} are given by Yang (2012) and listed below:

$$\begin{aligned} M_{11} &= \Gamma\Big[2ar\omega\left(r^2+\omega^2\right)\left(\omega\left(e^{4r\tau}-1\right) + 2re^{2r\tau}\sin(2\tau\omega)\right)\Big],\\ M_{12} &= \Gamma\Big[\omega^2 e^{4r\tau}\left(2ar^2 - b\right) - 2r^2 e^{2r\tau}\left(2a\omega^2 + b\right)\cos(2\tau\omega)\\ &\quad - \omega^2\left(b - 2ar^2\right) + 2be^{2r\tau}\left(r^2+\omega^2\right)\Big],\\ M_{13} &= \Gamma\Big[4ar\omega e^{r\tau}\left(e^{2r\tau}-1\right)\left(r^2+\omega^2\right)\sin(\tau\omega)\Big],\\ M_{14} &= \Gamma\Big[-4ar\omega e^{r\tau}\left(r^2+\omega^2\right)\left(r\left(e^{2r\tau}+1\right)\sin(\tau\omega) + \omega\left(e^{2r\tau}-1\right)\cos(\tau\omega)\right)\Big],\\ M_{22} &= \Gamma\Big[-2ar\omega\left(\omega\left(-e^{4r\tau}\right) + 2re^{2r\tau}\sin(2\tau\omega) + \omega\right)\Big],\\ M_{23} &= \Gamma\Big[4ar\omega e^{r\tau}\left(r\left(e^{2r\tau}+1\right)\sin(\tau\omega) - \omega\left(e^{2r\tau}-1\right)\cos(\tau\omega)\right)\Big]. \end{aligned}$$

13.3 Propagator: path integral

The quadratic potential given in Eq. 13.8 yields the Lagrangian given in Eq. 13.9, namely

$$\mathcal{L} = -\frac{1}{2}\left[\gamma\ddot{x}^2 + \alpha\dot{x}^2 + \beta x^2\right].$$

A parameterization that is more suitable for studying the Hamiltonian and state space is given by Kleinert (1986),

$$\mathcal{L} = -\frac{\gamma}{2}\left[\ddot{x}^2 + (\omega_1^2 + \omega_2^2)\dot{x}^2 + \omega_1^2\omega_2^2 x^2\right]. \tag{13.29}$$

The Lagrangian is completely symmetric in parameters ω_1 and ω_2. For the case of real ω_1 and ω_2, the entire parameter space is covered by choosing say $\omega_1 > \omega_2$; the roots are chosen accordingly and are

$$\begin{aligned}
\omega_1 &= \frac{1}{2\sqrt{\gamma}}\left(\sqrt{\alpha + 2\sqrt{\gamma\beta}} + \sqrt{\alpha - 2\sqrt{\gamma\beta}}\right), \\
\omega_2 &= \frac{1}{2\sqrt{\gamma}}\left(\sqrt{\alpha + 2\sqrt{\gamma\beta}} - \sqrt{\alpha - 2\sqrt{\gamma\beta}}\right), \\
\omega_1 &> \omega_2 \ \text{ for } \omega_1,\ \omega_2 \text{ real.}
\end{aligned} \tag{13.30}$$

We define the critical value $\alpha_c = 2\sqrt{\beta\gamma}$. The parameters have three branches, namely the real and the complex, plus the critical branch separating them, and these are shown in Figure 13.1:

- Complex branch $\alpha < \alpha_c$.

  Frequencies ω_1, ω_2 are complex,

$$\omega_1 = \omega_2^* = Re^{i\phi} \ : \ R > 0, \ \ \phi \in [-\pi/2, \pi/2]. \tag{13.31}$$

- Real branch $\alpha > \alpha_c$.

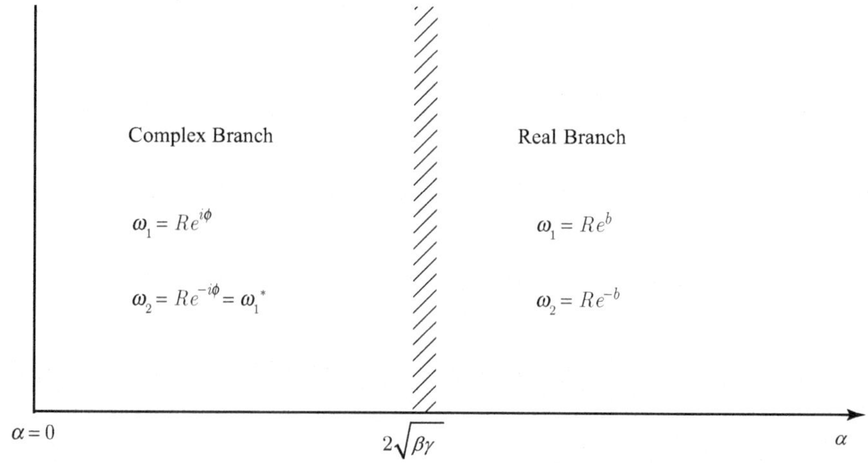

Figure 13.1 Parameter branches for the Euclidean Lagrangian. The critical value of $\alpha_c = 2\sqrt{\beta\gamma}$ is equivalent to $\omega_1 = \omega_2$.

Frequencies ω_1, ω_2 are real and $\omega_1 > \omega_2$ is chosen without any loss of generality,

$$\omega_1 = Re^b, \quad \omega_2 = Re^{-b} \quad : \quad R > 0, \quad b \in [0, +\infty].$$

- Equal frequency $\alpha = \alpha_c$.

  This is a special case that is treated in detail in Chapter 15,

$$\omega_1 = \omega_2, \quad b = 0 = \phi.$$

Note that $\phi \in [-\pi/2, \pi/2]$ for all $\alpha, \beta > 0$, and this is also the range for which the path integral is convergent.

The infinite time path integral is given by

$$Z = \lim_{T \to \infty} tr\left(e^{-TH}\right) = \int Dxe^{\mathcal{S}},$$

$$\mathcal{S} = -\frac{1}{2}\gamma \int_{-\infty}^{+\infty} dt \left[\ddot{x}^2 + (\omega_1^2 + \omega_2^2)\dot{x}^2 + \omega_1^2\omega_2^2 x^2\right].$$

The propagator is given by the path integral

$$G(\tau) = \frac{1}{Z}\int Dxe^{\mathcal{S}}x(t)x(t'). \tag{13.32}$$

The acceleration action is a quadratic functional of the paths $x(t)$ and hence the propagator can be evaluated exactly. We define the Fourier transformed variables that diagonalize the action, namely

$$x(t) = \int_{-\infty}^{+\infty} \frac{dk}{2\pi} e^{ikx} x_k, \tag{13.33}$$

$$\Rightarrow S = -\frac{1}{2}\gamma \int_{-\infty}^{+\infty} \frac{dk}{2\pi}\left[k^4 + (\omega_1^2 + \omega_2^2)k^2 + \omega_1^2\omega_2^2\right] x_{-k}x_k. \tag{13.34}$$

Using Gaussian path integration yields

$$\begin{aligned} G(\tau) &= \frac{1}{\gamma}\int_{-\infty}^{+\infty} \frac{dk}{2\pi} \frac{e^{ik(t-t')}}{(k^2 + \omega_1^2)(k^2 + \omega_2^2)} \\ &= \frac{1}{2\gamma}\frac{1}{\omega_1^2 - \omega_2^2}\left[\frac{e^{-\omega_2\tau}}{\omega_2} - \frac{e^{-\omega_1\tau}}{\omega_1}\right], \quad \tau = |t - t'|, \end{aligned} \tag{13.35}$$

where the last equation has been obtained using counter integration. Figure 13.2a shows a single exponential and Figure 13.2b shows how the acceleration term in the Lagrangian smooths out the "kink" of the exponential at $x = 0$; this smoothing out holds for all branches, including the real and complex branches for ω_1, ω_2.

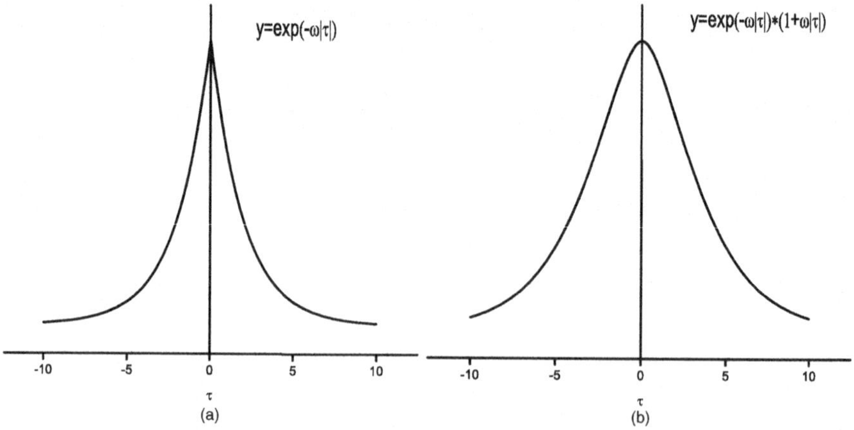

Figure 13.2 (a) Single exponential. (b) Propagator for equal frequency $\exp\{-\omega|\tau|\}[1+\omega|\tau|]$.

13.4 Dirac constraints and acceleration Hamiltonian

The acceleration Lagrangian belongs to the class of higher derivative Lagrangians for which the canonical derivation of the Hamiltonian from the Lagrangian does not hold. The reason is that the term $(d^2x/dt^2)^2$ in the Lagrangian does not yield a pair consisting of a canonical coordinate and momenta, as is required by the canonical framework.

This problem of higher derivative Lagrangians was addressed almost two centuries ago by Ostogradski and discussed by Simon (1990); but instead of following his approach, a more direct route is taken here, based on Dirac's analysis of constrained systems. In this approach, the number of independent degrees of freedom is increased by imposing constraints that re-cast the acceleration Lagrangian into another equivalent form in which the term $(dx^2/dt^2)^2$ is re-written as a first derivative expression, namely as $(dv/dt)^2$, with a constraint that $v = -dx/dt$; the Lagrangian can now be analyzed using the canonical method.

Dirac's analysis shows how to impose a set of consistent *constraints* on the enlarged phase space on which the Hamiltonian is defined so that only the physical phase space is obtained for the higher derivative system.

The Minkowski Hamiltonian for the action acceleration has been obtained by Mannheim and Davidson (2000) using the Dirac bracket approach and this analysis is now carried out for Euclidean space. The canonical transformation that takes the Lagrangian to the Hamiltonian makes no explicit reference to time and hence one needs to write the constraint equation for the Euclidean Lagrangian to obtain the correct Euclidean Hamiltonian.

The Euclidean Lagrangian is given by Eq. 13.29 as follows ($\cdot = d/dt$):

$$\mathcal{L} = -\frac{\gamma}{2}\left[\ddot{x}^2 + (\omega_1^2 + \omega_2^2)\dot{x}^2 + \omega_1^2\omega_2^2 x^2\right]. \tag{13.36}$$

The first step is to re-write the path integral; from Eqs. 5.17 and 13.1, Euclidean velocity v is made equal to $-dx/dt$ by imposing a delta function constraint in the path integral; ignoring the boundary condition for now, one obtains

$$Z = \int Dx \exp\left\{\int dt\mathcal{L}\right\} = \int Dx Dv \exp\left\{\int dt\mathcal{L}\right\}\prod_t \delta\left[v(t) + \dot{x}(t)\right], \tag{13.37}$$

$$= \int Dx Dv D\lambda \exp\left\{\int dt\mathcal{L}_D\right\}, \tag{13.38}$$

with

$$\mathcal{L}_D = -\frac{\gamma}{2}\left[\dot{v}^2 + (\omega_1^2 + \omega_2^2)v^2 + \omega_1^2\omega_2^2 x^2\right] + i\lambda[\dot{x} + v]. \tag{13.39}$$

The equivalent Euclidean Lagrangian $\mathcal{L}_D$ has only first order derivatives and hence can be treated by the canonical method for obtaining the Hamiltonian from the Lagrangian, as discussed in Section 5.1. Compared to the original Lagrangian $\mathcal{L}$, which has one degree of freedom $x(t)$, the equivalent Lagrangian $\mathcal{L}_D$ has three degrees of freedom $x(t)$, $v(t)$, and $\lambda(t)$. Dirac's formalism of constraints removes the redundant degree(s) of freedom.

For simplicity, the notation for Euclidean momenta $p_{x,E}$ and $p_{v,E}$ has been abbreviated to p_x, p_v. Let us define the canonical momenta for Euclidean time as given in Eq. 5.21, which yields

$$p_x = \frac{\partial\mathcal{L}_D}{\partial\dot{x}} = i\lambda, \quad p_v = \frac{\partial\mathcal{L}_D}{\partial\dot{v}} = -\gamma\dot{v}, \quad p_\lambda = \frac{\partial\mathcal{L}_D}{\partial\dot{\lambda}} = 0. \tag{13.40}$$

The canonical Euclidean Hamiltonian is given by expressing all time derivatives in terms of the canonical momenta; hence, using Eqs. 13.39 and 13.40 yields

$$\begin{aligned} H_c &= \dot{x}p_x + \dot{v}p_v + \dot{\lambda}p_\lambda - \mathcal{L}_D \\ &= -\frac{\gamma}{2}p_v^2 + \frac{\gamma}{2}\left[(\omega_1^2 + \omega_2^2)v^2 + \omega_1^2\omega_2^2 x^2\right] - iv\lambda. \end{aligned} \tag{13.41}$$

Note the canonical momenta p_x and p_λ do not appear in the Hamiltonian H_c; hence these two momenta have to be imposed, following the notation and terminology of Dirac, as *constraints* that follow from Eq. 13.40, namely

$$\phi_1 = p_x - i\lambda, \quad \phi_2 = p_\lambda. \tag{13.42}$$

Similar to including a Lagrange multiplier λ to impose a constraint on the Lagrangian, as given in Eq. 13.39, the most general Euclidean Hamiltonian consistent with the constraints is given by

$$\begin{aligned} H_1 &= H_c + u_1\phi_1 + u_2\phi_2 \\ &= H_c + u_1(p_x - i\lambda) + u_2 p_\lambda, \end{aligned} \tag{13.43}$$

where u_1, u_2 are arbitrary functions of the degrees of freedom.

Hamiltonian dynamics yields that the time dependence of any function F of the dynamical variables is given by its Poisson bracket with the Hamiltonian H_1, namely

$$\frac{dF}{dt} = \{F, H_1\}_\mathrm{P} + \frac{\partial F}{\partial t}, \tag{13.44}$$

where the Poisson bracket for arbitrary functions F, G of the dynamical variables x, v, λ is given by

$$\{F, G\}_\mathrm{P} \equiv \frac{\partial F}{\partial x}\frac{\partial G}{\partial p_x} - \frac{\partial G}{\partial x}\frac{\partial F}{\partial p_x} + \frac{\partial F}{\partial v}\frac{\partial G}{\partial p_v} - \frac{\partial G}{\partial v}\frac{\partial F}{\partial p_v} + \frac{\partial F}{\partial \lambda}\frac{\partial G}{\partial p_\lambda} - \frac{\partial G}{\partial \lambda}\frac{\partial F}{\partial p_\lambda}$$

$$\Rightarrow \{F, GH\}_\mathrm{P} = \{F, G\}_\mathrm{P} H + G\{F, H\}_\mathrm{P}. \tag{13.45}$$

In light of the derivation of Mannheim and Davidson (2000), consider the choice of functions $u_1 = -v, u_2 = \gamma\omega_1^2\omega_2$ that, for the Hamiltonian given in Eq. 13.43, yields

$$\begin{aligned} H_1 &= H_c - v(p_x - i\lambda) + \gamma\omega_1^2\omega_2 x p_\lambda \\ &= -\frac{\gamma}{2}p_v^2 + \frac{\gamma}{2}\left[\omega_1^2 + \omega_2^2)v^2 + \omega_1^2\omega_2^2 x^2\right] - v p_x + \gamma\omega_1^2\omega_2 x p_\lambda. \end{aligned} \tag{13.46}$$

The constraints ϕ_1, ϕ_2 must be conserved over time; hence, we need to evaluate the Poisson brackets of the constraints with H_1. A straightforward calculation using Eqs. 13.45 and 13.46 yields

$$\{\phi_1, H_1\}_\mathrm{P} = -\gamma\omega_1^2\omega_2 p_\lambda, \quad \{\phi_2, H_1\}_\mathrm{P} = 0. \tag{13.47}$$

The dynamical variable λ is cyclic since it does not appear in the Hamiltonian H_1. Hence for conjugate variables λ and p_λ

$$\begin{aligned} &\frac{dp_\lambda}{dt} = \{p_\lambda, H_1\}_\mathrm{P} = \frac{\partial p_\lambda}{\partial \lambda}\frac{\partial H_1}{\partial p_\lambda} - \frac{\partial H_1}{\partial \lambda}\frac{\partial p_\lambda}{\partial p_\lambda} = -\frac{\partial H_1}{\partial \lambda} = 0 \\ &\Rightarrow p_\lambda = \text{constant}. \end{aligned}$$

To decouple the degree of freedom λ from the degrees of freedom x and v, choose $p_\lambda = 0$; the resulting physical Hamiltonian, from Eq. 13.46, is

$$
\begin{aligned}
H(v, p_v, x, p_x) &= -\frac{\gamma}{2}p_v^2 + \frac{\alpha}{2}v^2 + \frac{\beta}{2}x^2 - vp_x \\
&= -\frac{\gamma}{2}p_v^2 + \frac{\gamma}{2}\left[\omega_1^2 + \omega_2^2)v^2 + \omega_1^2\omega_2^2 x^2\right] - vp_x. \qquad (13.48)
\end{aligned}
$$

It can be shown that the Poisson brackets amongst x, p_x, v, p_v, and H form a *closed subalgebra*; on this subspace the constraint equations are all conserved. Hence, H is the requisite classical Hamiltonian for the acceleration Lagrangian.

13.5 Phase space path integral and Hamiltonian operator

The classical Hamiltonian yields two pathways for quantization, namely one via the formalism of phase space path integration and the other by imposing commutation equations on the pairs of canonical coordinate and its canonical momentum. These two routes for quantization are discussed below.

The Euclidean Hamiltonian obtained in Eq. 13.48 yields the Lagrangian $\mathcal{L}$ for Euclidean time, which from Eq. 5.31, is given by

$$
\tilde{\mathcal{L}}[v, p_v, x, p_x] = \dot{v}p_v + \dot{x}p_x - H(v, p_v, x, p_x). \qquad (13.49)
$$

As discussed in Eq. 5.22, the Euclidean momentum is pure imaginary and equal to ip, where p is real. Making the change of variables $p_v \to ip_v$, $p_x \to ip_x$ and keeping the same notation for simplicity yields, from Eqs. 13.48 and 13.49, the Euclidean Lagrangian[1]

$$
\begin{aligned}
&\tilde{\mathcal{L}}[v, ip_v, x, ip_x] = i\dot{v}p_v + i\dot{x}p_x - H(v, ip_v, x, ip_x) \equiv \mathcal{L}[v, p_v, x, p_x] \\
&\Rightarrow \mathcal{L}[v, p_v, x, p_x] = i\dot{v}p_v + i\dot{x}p_x - \frac{\gamma}{2}p_v^2 - \frac{\alpha}{2}v^2 - \frac{\beta}{2}x^2 + ivp_x, \qquad (13.50)
\end{aligned}
$$

and the finite time Euclidean action is given by

$$
S[v, p_v, x, p_x] = \int_0^\tau dt\mathcal{L}[v, p_v, x, p_x]. \qquad (13.51)
$$

The Euclidean path integral, from Eq. 5.34, is hence defined by

$$
\begin{aligned}
Z &= \int Dx \exp\{S\} \\
&= \int DxDp_x DvDp_v \exp\{\int_0^\tau dt\mathcal{L}[v, p_v, x, p_x]\}, \qquad (13.52)
\end{aligned}
$$

[1] Using the notation for the Lagrangian given in Eq. 13.9.

with boundary conditions $x' = x(0)$, $v' = v(0)$ and $x = x(\tau)$, $v = v(\tau)$. Equation 13.50 yields

$$Z = \int DxDp_xDvDp_v \exp\left\{\int_0^\tau dt(i\dot{v}p_v + i\dot{x}p_x - \frac{\gamma}{2}p_v^2 - \frac{\alpha}{2}v^2 - \frac{\beta}{2}x^2 + ivp_x)\right\}$$
$$= \int DxDp_xDv \exp\left\{\int_0^\tau dt(i\dot{x}p_x - \frac{1}{2}\dot{v}^2 - \frac{\alpha}{2}v^2 - \frac{\beta}{2}x^2 + ivp_x)\right\}$$
$$= \int DxDv \exp\left\{\int_0^\tau dt(-\frac{1}{2}\dot{v}^2 - \frac{\alpha}{2}v^2 - \frac{\beta}{2}x^2)\right\}\prod_t \delta[v + \dot{x}], \tag{13.53}$$

and we have recovered the path given earlier in Eq. 13.37, thus verifying that the Hamiltonian is indeed correct. Note the sign of the constraint in the equation, namely that $v = -\dot{x}$, yields what is expected, and the fact that Euclidean momentum is pure imaginary consistently yields the required result.

To obtain the quantum Hamiltonian, since the system has constraints, the Dirac brackets are required to obtain the Heisenberg commutators. In particular, the equal time Euclidean commutation equations have to be obtained from the Dirac brackets.

A set of constraints, given by $\phi_\ell, \ell = 1, 2 \ldots M$, defines as per Eq. 5.55, the constraint matrix

$$C_{\ell,\ell'} = \{\phi_\ell, \phi_{\ell'}\}_{\mathrm{P}}. \tag{13.54}$$

The Dirac brackets are given by Eq. 5.56 for arbitrary function $f(q, p)$, $g(q, p)$, by

$$\{f, g\}_{\mathrm{D}} = \{f, g\}_{\mathrm{P}} - \sum_{\ell,\ell'=1}^{M} \{f, \phi_\ell\}_{\mathrm{P}} C^{-1}_{\ell,\ell'} \{\phi_{\ell'}, g\}_{\mathrm{P}}.$$

For degrees of freedom x, p_x, v, p_v, from Eq. 13.42 the constraints are given by

$$\phi_1 = p_x - i\lambda, \quad \phi_2 = p_\lambda, \tag{13.55}$$

and the constraint matrix is given by

$$C_{12} = \{\phi_1, \phi_2\}_{\mathrm{P}} = -i\{\lambda, p_\lambda\}_{\mathrm{P}} = -i = -C_{21}, \tag{13.56}$$
$$\Rightarrow C^{-1}_{12} = i = -C^{-1}_{21}. \tag{13.57}$$

Evaluating the Dirac bracket for only the x, v sector yields

$$\{x, \phi_1\}_{\mathrm{P}} = 1, \quad \{p_x, \phi_1\}_{\mathrm{P}} = \{x, \phi_2\}_{\mathrm{P}} = 0 = \{p_x, \phi_2\}_{\mathrm{P}}, \tag{13.58}$$
$$\{v, \phi_1\}_{\mathrm{P}} = 0 = \{p_v, \phi_1\}_{\mathrm{P}} = \{v, \phi_2\}_{\mathrm{P}} = 0 = \{p_v, \phi_2\}_{\mathrm{P}}. \tag{13.59}$$

From the result above, the Dirac brackets become equal to the Poisson bracket for all the conjugate variables, namely

$$\{x, p_x\}_{\mathrm{D}} = \{x, p_x\}_{\mathrm{P}}, \ldots$$

The canonical quantization for equal Euclidean time is given by

$$[x, p_x] = -\hbar\mathbb{I}\{x, p_x\}_{\rm D} = -\hbar\mathbb{I}, \tag{13.60}$$

$$[v, p_v] = -\hbar\mathbb{I}\{v, p_v\}_{\rm D} = -\hbar\mathbb{I}, \tag{13.61}$$

$$[x, v] = [x, p_v] = [v, p_x] = 0.$$

From the above, the nonzero commutation equations for the degrees of freedom x, v are given by

$$[x, p_x] = -\hbar\mathbb{I} = [v, p_v]. \tag{13.62}$$

Hence, from Eqs. 5.50 and 5.51, the Euclidean momentum operators are given by

$$p_x = \hbar\frac{\partial}{\partial x}, \quad p_v = \hbar\frac{\partial}{\partial v}.$$

In summary, the degrees of freedom for the acceleration Lagrangian are given by (setting $\hbar = 1$)

$$x, \; p_x = \frac{\partial}{\partial x}, \quad v, \; p_v = \frac{\partial}{\partial v}. \tag{13.63}$$

Hence, from Eqs. 13.48 and 13.63, the quantum non-Hermitian Hamiltonian for the acceleration Lagrangian is given by

$$H = -\frac{\gamma}{2}\frac{\partial^2}{\partial v^2} + \frac{\gamma}{2}\left[\left(\omega_1^2 + \omega_2^2\right)v^2 + \omega_1^2\omega_2^2 x^2\right] - v\frac{\partial}{\partial x}, \tag{13.64}$$

and its Hermitian conjugate is

$$H^\dagger = -\frac{\gamma}{2}\frac{\partial^2}{\partial v^2} + \frac{\gamma}{2}\left[\left(\omega_1^2 + \omega_2^2\right)v^2 + \omega_1^2\omega_2^2 x^2\right] + v\frac{\partial}{\partial x} \neq H,$$

since $(v\partial/\partial x)^\dagger = -v\partial/\partial x$.

Noteworthy 13.1 **Important features of the Euclidean Hamiltonian**

The following are some important features in the derivation and form of the Euclidean Hamiltonian H obtained in Eq. 13.64:

- The Euclidean Hamiltonian is not Hermitian due to the $v\partial/\partial x$ term.
- The Euclidean analysis of the Poisson brackets is consistent and one has to explicitly keep track of the signatures of time that differ from the Minkowski derivation.
- The acceleration Lagrangian (incorrectly) seems to have only one degree of freedom, namely $x(t)$. On the other hand, the Hamiltonian H and its state space are each quantum systems with two degrees of freedom, namely x and v; this result can also be seen from the path integral since one needs *two* initial conditions and

two final conditions to define the finite time path integral, reflecting that the initial and final state vectors have two degrees of freedom. The Dirac constraint analysis brings out this feature of the acceleration (higher derivative) Lagrangian.

- The only difference between the initial canonical Hamiltonian H_c given in Eq. 13.41 and the final result given by H in Eq. 13.64 is that the term $iv\lambda$ in H_c has been replaced by $-v\partial/\partial x$. The analysis of the constraints results in the replacement of the nonphysical degree of freedom $i\lambda$ by the physical degree of freedom operator $-\partial/\partial x$.
- It is shown in the path integral analysis of H, analyzed in Section 13.6 below, that the term $-v\partial/\partial x$ has the remarkable property of *constraining*, in the path integral, v to be equal to $-dx/dt$, as required by Eq. 13.1.
- The fact that $-v\partial/\partial x$ is a constraint operator explains why the term is independent of the coupling constants, since the constraint is on the *phase space* itself and not on the forms of interaction allowed in this phase space.

13.6 Acceleration path integral

The path integral derivation of $\mathcal{K}(x_f, \dot{x}_f; x_i, \dot{x}_i)$ given in Eq. 13.14 was done entirely in terms of the coordinate degree of freedom $x(t)$; in particular, the velocity degree of freedom $v(t)$ did not appear in the expression for the path integral.

One would like to interpret the evolution kernel $\mathcal{K}(x_f, \dot{x}_f; x_i, \dot{x}_i)$ as the *probability amplitude* from an initial state of state space to a final state. Such an interpretation of course needs both a state space and a Hamiltonian.

Based on the boundary conditions given in Eq. 13.6, it can be seen that the state space has to have *two independent degrees of freedom*, corresponding to the two initial conditions given by the initial position x and velocity $\dot{x}$. Hence, to have a state space interpretation of the path integral, state space $\mathcal{V}$ is taken to have *two* degrees of freedom, namely (position) coordinate degrees of freedom x and a degree of function reflecting $\dot{x}$ in the state vector. The Hamiltonian has to be chosen in such a manner that an independent velocity degree of freedom v is is introduced and it is *constrained* to be equal to the velocity $-\dot{x}$ of the coordinate degree of freedom x.

Consider two independent degrees of freedom x and v. The completeness equation for the basis states is given by

$$\mathbb{I} = \int_{-\infty}^{+\infty} dxdv|x, v\rangle\langle v, x|, \tag{13.65}$$
$$\langle x, v|x', v'\rangle = \delta(x - x')\delta(v - v').$$

A state space representation of the evolution kernel $\mathcal{K}(x_f, \dot{x}_f; x_i, \dot{x}_i)$ is derived below. It will be shown in this section that the evolution kernel is closely related to the *probability amplitude* of going, in time τ, from the initial state $|x_i, v_i\rangle$ to the final state $\langle x_f, v_f|$ and is given by

$$\mathcal{K}_S(x_f, v_f; x_i, v_i) = \langle x_f, v_f|e^{-\tau H}|x_i, v_i\rangle. \tag{13.66}$$

A similar definition is adopted for the path integral of a pseudo-Hermitian Hamiltonian in Kandirmaz and Sever (2011).

It remains to be seen what is the relation of the probability amplitude $\mathcal{K}_S(x_f, v_f; x_i, v_i)$ defined using the state space and Hamiltonian to the probability amplitude $\mathcal{K}(x_f, \dot{x}_f; x_i, \dot{x}_i)$ defined using the path integral. In particular, as it stands in Eq. 13.66, the initial v_i and final velocity v_f have no relation to the coordinate degree of freedom x. The Hamiltonian has to implement a constraint to set the initial and final state in Eq. 13.66 to have the same boundary conditions given in Eqs. 13.6 and 13.7 for $\mathcal{K}(x_f, \dot{x}_f; x_i, \dot{x}_i)$.

The Hamiltonian, for infinitesimal time $\tau = \epsilon$, is given by the Dirac–Feynman formula as

$$\langle x, v|e^{-\epsilon H}|x', v'\rangle = \mathcal{C}(\epsilon)e^{\epsilon\mathcal{L}(x,x';v,v')}, \tag{13.67}$$

where $\mathcal{C}(\epsilon)$ is a normalization constant that depends only on ϵ. In terms of $v = -\dot{x}$, the discrete time Lagrangian, from Eq. 13.2, is given by

$$\mathcal{L}(x, x'; v, v') = -\frac{\gamma}{2}\left(\frac{\dot{x} - \dot{x}'}{\epsilon}\right)^2 - \frac{\alpha}{2}v^2 - \frac{1}{2}[\Phi(x) + \Phi(x')]. \tag{13.68}$$

A straightforward generalization of the Hamiltonian given in Eq. 13.64 yields a Euclidean Hamiltonian with an arbitrary potential $\Phi(x)$,

$$H = -\frac{1}{2\gamma}\frac{\partial^2}{\partial v^2} - v\frac{\partial}{\partial x} + \frac{1}{2}\alpha v^2 + \Phi(x) = H\left(x, \frac{\partial}{\partial x}, v, \frac{\partial}{\partial v}\right). \tag{13.69}$$

The Hamiltonian and its conjugate both act on a state space $\mathcal{V}$ that has *two* degrees of freedom, namely position coordinate x and velocity degree of freedom v. The state function $|\Psi\rangle$ is given by

$$|\Psi\rangle \in \mathcal{V}, \quad \langle x, v|\Psi\rangle = \Psi(x, v). \tag{13.70}$$

The transition probability amplitude is given by defining $\epsilon = \tau/N$ and inserting $N - 1$ complete sets of states given in Eq. 13.65. Hence, for boundary conditions given by $x_0 = x_i, v_0 = v_i; x_N = x_f, v_N = v_f$, the transition amplitude is given by

$$\begin{aligned}\mathcal{K}(x_f, v_f; x_i, v_i) &= \langle x_f, v_f|e^{-\tau H}|x_i, v_i\rangle \\ &= \prod_{n=1}^{N-1}\int dx_n dv_n \prod_{n=1}^{N}\langle x_n, v_n|e^{-\epsilon H}|x_{n-1}, v_{n-1}\rangle\end{aligned}$$

$$= \prod_{n=1}^{N-1} \int dx_n \langle x_N, v_N | e^{-\epsilon H} | x_{N-1}, v_{N-1} \rangle$$

$$\times \left\{ \prod_{n=1}^{N-1} \int dv_n \langle x_n, v_n | e^{-\epsilon H} | x_{n-1}, v_{n-1} \rangle \right\}. \tag{13.71}$$

The differential operator H given in Eq. 13.69, for $x_n = x, v = v_n;\ x' = x_{n-1}, v' = v_{n-1}$, yields

$$\begin{aligned}
\langle x, v | e^{-\epsilon H} | x', v' \rangle &= e^{-\epsilon H(x, \frac{\partial}{\partial x}; v, \frac{\partial}{\partial v})} \langle x | x' \rangle \langle v | v' \rangle \\
&= e^{-\epsilon H} \int \frac{dp\, dq}{2\pi\, 2\pi} e^{ip(x-x')+iq(v-v')} \\
&= \int \frac{dp\, dq}{2\pi\, 2\pi} e^{-\frac{\epsilon}{2\gamma} q^2 + iq(v-v') - \frac{\epsilon\alpha}{2} v^2} e^{ip(x-x'+\epsilon v) - \epsilon \Phi(x)} \\
&= C\delta(x - x' + \epsilon v) \exp\left\{ -\frac{\gamma}{2\epsilon}(v - v')^2 - \frac{\epsilon\alpha}{2} v^2 - \epsilon \Phi(x) \right\}.
\end{aligned} \tag{13.72}$$

The appearance of the δ-function in Eq. 13.72 yields the following constraint in the path integral:

$$\delta(x - x' + \epsilon v) \quad \Rightarrow v = -\frac{x - x'}{\epsilon}, \tag{13.73}$$

$$\Rightarrow \quad \lim_{\epsilon \to 0} v = -\frac{dx}{d\tau}. \tag{13.74}$$

Equation 13.72 yields the remarkable result that the term $-v\partial/\partial x$ in the Hamiltonian yields a *constraint* on the degree of freedom v so that it is constrained to be $-\dot{x}$ degree of freedom, namely $v = -dx/dt$. It is the delta function constraint on the velocity degree of freedom that leads to its complete elimination in the path integral.

Equation 13.72 yields

$$\langle x_n, v_n | e^{-\epsilon H} | x_{n-1}, v_{n-1} \rangle = C\delta(x_n - x_{n-1} + \epsilon v_n) \exp\{\epsilon \mathcal{L}_n\}, \tag{13.75}$$

where $\mathcal{L}_n$, from Eqs. 13.72 and 13.73, is given by

$$\mathcal{L}_n = -\frac{\gamma}{2\epsilon^2}(v - v')^2 - \frac{\alpha}{2} v^2 - \frac{1}{2}[\Phi(x) + \Phi(x')]. \tag{13.76}$$

The path integral and Lagrangian that appear in Eq. 13.4 make no reference to the integration over the velocity variables. Hence, all the velocity integrations $\int dv_n$ need to be carried out in order to obtain the expression in Eq. 13.4. Remarkably enough, all the $\int dv_n$ integrations can be done *exactly* using the δ-functions that appear in Eq. 13.72.

Equations 13.71 and 13.75 yield

$$\prod_{n=1}^{N-1}\int dv_n\langle x_n, v_n|e^{-\epsilon H}|x_{n-1}, v_{n-1}\rangle = \mathcal{C}^{N-1}\prod_{n=1}^{N-1}\int dv_n\delta(x_n - x_{n-1} + \epsilon v_n)$$
$$\times \exp\{\epsilon\mathcal{L}_n\}. \tag{13.77}$$

The full path integral, from Eqs. 13.71 and 13.77, can be heuristically written, in the continuum notation, as

$$\mathcal{K}(x_f, v_f; x_i, v_i) = \langle x_f, v_f|e^{-\tau H}|x_i, v_i\rangle$$
$$= \int DxDv\prod_{t=0}^{\tau}\delta[\dot{x}(t) + v(t)]\exp\left\{\int_0^\tau dt\mathcal{L}(x, v)\right\}, \tag{13.78}$$
$$\mathcal{L}(x, v) = -\frac{\gamma}{2}\dot{v}^2 - \frac{\alpha}{2}v^2 - \Phi(x),$$

and we recover the expression for the Lagrangian given in Eq. 13.68. To obtain the precise content of the continuum expression given in Eq. 13.78, one needs to go to the lattice and carefully address the issue of the boundary conditions.

13.7 Change of path integral boundary conditions

The initial and final time steps need to be examined carefully to see how the initial and final velocity, v_0 and v_N respectively, can be expressed solely in terms of the coordinate degree of freedom. The four boundary conditions given in Eq. 13.11 are solely in terms of the coordinate degree of freedom $x(t)$, whereas the boundary conditions given in Eq. 13.66, the defining equation for $\mathcal{K}_S(x_f, v_f; x_i, v_i)$, are given in terms of final and initial positions and velocities $x_f, v_f; x_i, v_i$ respectively.

Note that the integrand of Eq. 13.77, for $n = 1$, yields, from Eq. 13.76,

$$\int dv_1\delta(x_1 - x_0 + \epsilon v_1)\exp\{\epsilon\mathcal{L}_1\}$$
$$\mathcal{L}_1 = -\frac{\gamma}{2}\left(\frac{v_1 - v_0}{\epsilon}\right)^2 - \frac{\alpha}{2}v_1^2 - \frac{1}{2}[\Phi(x_1) + \Phi(x_0)]. \tag{13.79}$$

On performing the $\int dv_1$ integration, the delta function constrains $v_1 = -(x_1 - x_0)/\epsilon$; hence, $\mathcal{L}_1$ has the value

$$\exp\{\epsilon\mathcal{L}_1\} = \exp\left\{-\frac{\gamma}{2\epsilon}\left(\frac{x_1 - x_0}{\epsilon} + v_0\right)^2 - \frac{\alpha}{2\epsilon}(x_1 - x_0)^2 + O(\epsilon)\right\}$$
$$= \exp\left\{-\frac{\gamma}{2\epsilon}\left(\frac{x_1 - x_0}{\epsilon} + v_0\right)^2 - \frac{\alpha\epsilon}{2}v_0^2 + O(\epsilon)\right\}$$
$$= \mathcal{C}(\epsilon)\delta(x_1 - x_0 + \epsilon v_0) + O(\epsilon) = \mathcal{C}(\epsilon)\delta(x_1 - x_i + \epsilon v_i), \tag{13.80}$$

where $x_0 = x_i$ and $v_0 = v_i$ are the initial position and velocity.

The final time boundary term for the action, from Eq. 13.75, yields

$$\begin{aligned}\langle x_N, v_N|e^{-\epsilon H}|x_{N-1}, v_{N-1}\rangle &= \mathcal{C}(\epsilon)\delta(x_N - x_{N-1} + \epsilon v_N)\exp\{\epsilon\mathcal{L}_N\}\\ &= \mathcal{C}(\epsilon)\delta(x_f - x_{N-1} + \epsilon v_f)\exp\{\epsilon\mathcal{L}_N\},\end{aligned}$$

since $x_N = x_f$ and $v_N = v_f$.

The path integral over the velocity degrees of freedom yields, in addition to the expected acceleration action, two *extra* delta functions. These delta functions are crucial in changing the boundary conditions for the path integral $\int Dx$ over the coordinate degree of freedom.

Collecting all the results yields the discrete time path integral for the transition probability amplitude expressed solely in terms of the coordinated degrees of freedom, namely

$$\begin{aligned}\mathcal{K}_S(x_f, v_f; x_i, v_i) &= \langle x_f, v_f|e^{-\tau H}|x_i, v_i\rangle\\ &= \tilde{\mathcal{C}}\prod_{n=1}^{N-1}\int dx_n\delta(x_f - x_{N-1} + \epsilon v_f)\delta(x_1 - x_i + \epsilon v_i)\\ &\quad\times\exp\left\{\epsilon\sum_{n=1}^{N}\mathcal{L}_n\right\},\end{aligned} \tag{13.81}$$

where $\tilde{\mathcal{C}}$ is a normalization.

Note that the path integral given in Eq. 13.81, due to the two delta functions in the integrand, has *four* boundary conditions for the coordinate degree of freedom, namely x_i, x_1, x_{N-1}, x_f; the delta functions, in effect, remove two integrations, namely $\int dx_1 dx_{N-1}$ in the path integral given in Eq. 13.81 by fixing the value of $x_1,\ x_{N-1}$. In the derivation of the probability amplitude carried out by Kleinert (1986), the $\dot{x}$ boundary conditions are finally implemented in the discretized path integral by constraining the variables near the end points, namely x_1 and x_{N-1}.

To take the continuum limit we define

$$\dot{x}(t) = \frac{dx(t)}{dt} = \frac{x_n - x_{n-1}}{\epsilon},\quad t = n\epsilon. \tag{13.82}$$

Hence, from Eq. 13.81

$$v_i = \frac{x_1 - x_0}{\epsilon} \to -\dot{x}_i, \tag{13.83}$$

$$v_f = \frac{x_N - x_{N-1}}{\epsilon} \to -\dot{x}_f, \tag{13.84}$$

since $\dot{x}_i = dx(0)/dt$ and $\dot{x}_f = dx(\tau)/dt$.

From Eq. 13.77,

$$\epsilon\mathcal{L}_n = -\frac{\gamma}{2\epsilon}(\dot{x}_n - \dot{x}_{n-1})^2 - \frac{\epsilon\alpha}{2}\dot{x}_n^2 - \epsilon\Phi(x_n). \tag{13.85}$$

Taking the limit of $\epsilon \to 0$ and using $\ddot{x} = [\dot{x}_n - \dot{x}_{n-1}]/\epsilon$ yields

$$\mathcal{L} = -\frac{\gamma}{2}\ddot{x}^2 - \frac{\alpha}{2}\dot{x}^2 - \Phi(x). \tag{13.86}$$

The delta functions for the boundary values of the functional integral $\int Dx$ are constraints that *change the boundary conditions* on the path integral, converting the *two* boundary conditions each for x and v in $\int Dx\,Dv$ to *four* boundary conditions for x in the path integral $\int Dx$. Hence, one obtains the continuum result

$$\mathcal{K}_S(x_f, v_f; x_i, v_i) = \langle x_f, v_f|e^{-\tau H}|x_i, v_i\rangle = \int Dx\delta(v_i + \dot{x}_i)\delta(v_f + \dot{x}_f)e^S\Big|_{(x_i,x_f)}$$

$$= \int Dxe^S\Big|_{(x_i,\dot{x}_i=-v_i;x_f,\dot{x}_f=-v_f)} = \mathcal{K}(x_i, \dot{x}_i = -v_i; x_f, \dot{x}_f = -v_f)$$

$$\Rightarrow\ \mathcal{K}_S(x_f, v_f; x_i, v_i) = \mathcal{K}(x_f, -\dot{x}_f; x_i, -\dot{x}_i).$$

Hence

$$\begin{aligned}\mathcal{K}_S(x_f, v_f; x_i, v_i) &= \langle x_f, v_f|e^{-\tau H}|x_i, v_i\rangle \\ &= \int Dxe^S = \mathcal{K}(x_f, -\dot{x}_f; x_i, -\dot{x}_i).\end{aligned} \tag{13.87}$$

The representation of the path integral in terms of the transition amplitude $\mathcal{K}_S(x_f, v_f; x_i, v_i)$ is necessary to have the amplitude obey the composition law derived in Section 4.8. This is discussed by Hawking and Hertog (2002) and Fontanini and Trodden (2011).

13.8 Evolution kernel

The Hamiltonian in Eq. 13.69 [for the Lagrangian in Eq. 13.86] is given by

$$H = -\frac{1}{2\gamma}\frac{\partial^2}{\partial v^2} - v\frac{\partial}{\partial x} + \frac{\alpha}{2}v^2 + \Phi(x).$$

In terms of the path integrals, from Eqs. 13.4 and 13.78

$$\mathcal{L}(x) = -\frac{\gamma}{2}\ddot{x}^2 - \frac{\alpha}{2}\dot{x}^2 - \Phi(x),\quad \mathcal{L}(x,v) = -\frac{\gamma}{2}\dot{v}^2 - \frac{\alpha}{2}v^2 - \Phi(x),$$

$$\int Dx\exp\{\int_0^\tau dt\mathcal{L}(x)\} = \int Dx\,Dv\prod_{t=0}^{\tau}\delta[\dot{x}(t) + v(t)]\exp\{\int_0^\tau dt\mathcal{L}(x,v)\}.$$

Implicit in the path integrals above are the appropriate boundary conditions.

The symmetry of the evolution kernel given in Eq. 13.19 using the path integral, namely

$$\mathcal{K}(x_f, \dot{x}_f; x_i, \dot{x}_i) = \mathcal{K}(x_i, -\dot{x}_i; x_f, -\dot{x}_f)$$

has a Hamiltonian derivation. Note that $H^\dagger$ can be obtained from Eq. 13.69 by conjugation and is given by

$$H^\dagger = -\frac{1}{2\gamma}\frac{\partial^2}{\partial v^2} + v\frac{\partial}{\partial x} + \frac{\alpha}{2}v^2 + \Phi(x) \neq H. \tag{13.88}$$

From Eq. 13.87

$$\mathcal{K}_S^*(x_f, v_f; x_i, v_i) = \langle x_f, v_f|e^{-\tau H}|x_i, v_i\rangle^* = \langle x_i, v_i|e^{-\tau H^\dagger}|x_f, v_f\rangle. \tag{13.89}$$

For the Hamiltonian given in Eq. 13.69, the only difference in the path integral between H and $H^\dagger$ is to change the sign of $v\partial/\partial x$ to $-v\partial/\partial x$; this in turn yields the constraint $\delta(v - \dot{x})$ for the path integral representation of Eq. 13.89. Hence, on performing the velocity path integral $\int Dv$ for Eq. 13.89, the only change from the earlier case is that the boundary on velocity now has a *minus* sign and yields

$$\langle x_i, v_i|e^{-\tau H^\dagger}|x_f, v_f\rangle = \mathcal{K}_S(x_i, -v_i; x_f, -v_f). \tag{13.90}$$

For the case of K_S being real, which is the case considered in Eq. 13.19, Eqs. 13.89 and 13.90 yield

$$\mathcal{K}_S(x_f, v_f; x_i, v_i) = \mathcal{K}_S(x_i, -v_i; x_f, -v_f), \tag{13.91}$$

from which Eq. 13.19 follows due to Eq. 13.87. The symmetry of the evolution kernel has been shown in Hawking and Hertog (2002) to be correct by a direct evaluation of the classical action $S[x_c]$.

The evolution kernel has an implicit time ordering; explicitly putting in the initial t_i and final t_f time coordinates, the kernel can be written out explicitly in the form

$$\mathcal{K}_S(x_f, v_f, t_f; x_i, v_i, t_i) = \langle x_f, v_f|e^{-|t_f - t_i|H}|x_i, v_i\rangle. \tag{13.92}$$

Using the identity

$$\frac{\partial}{\partial t}\exp\{-\alpha|t|\} = -\alpha\exp\{-\alpha|t|\} + \delta(t)$$

yields, from Eq. 13.92, the following Schrödinger equation with a source:

$$\left[\frac{\partial}{\partial t_f} + H\right]\mathcal{K}_S(x_f, v_f, t_f; x_i, v_i, t_i) = \delta(x_f - x_i)\delta(v_f - v_i)\delta(t_f - t_i). \tag{13.93}$$

This result has been derived for the Euclidean case by Kleinert (1986) and for the Minkowski case by Mannheim (2011b).

13.9 Summary

The Euclidean acceleration Lagrangian leads to a number of new results. The theory has a state space that has two degrees of freedom, namely x and its velocity v. The theory has three branches shown in Figure 13.1. The first branch, given by $\alpha > \alpha_c$, has a pseudo-Hermitian Hamiltonian and a state space with a positive definite norm. The second branch, given by $\alpha < \alpha_c$, has complex parameters in its Hamiltonian and the third critical branch $\alpha = \alpha_c$ – separating the other two branches – consists of an essentially non-Hermitian Hamiltonian consisting of Jordan blocks.

To obtain the Hamiltonian, the acceleration Lagrangian is written as a constrained system and the Dirac constraint method was utilized to obtain a non-Hermitian Hamiltonian. The Hamiltonian in turn was then used for obtaining a path integral representation of the evolution kernel and was shown to reproduce the earlier result, confirming the result obtained for the Hamiltonian. In particular, a derivation was given of the change of boundary conditions in going from the x, v degrees of freedom for the Hamiltonian to the solely x degree of freedom in terms of which the acceleration action is directly defined.

14

Pseudo-Hermitian Euclidean Hamiltonian

There is no conservation of probability in the Euclidean formulation of the acceleration Lagrangian. A narrow definition of a Hamiltonian, and the one that holds in quantum mechanics but is not suitable for our case, is that the Hamiltonian defines a probability conserving unitary evolution. A broader definition of the Hamiltonian is adopted that is consistent with both statistical mechanics and quantum mechanics. Namely, the Hamiltonian – which is equivalent to the transfer matrix in statistical mechanics – is the generator of infinitesimal translations in time for both the Minkowski and Euclidean cases.

The Hamiltonian derived for the acceleration Lagrangian in Chapter 13 is studied in this chapter using the concepts of the state space and operators. The Hamiltonian for unequal and real frequencies ω_1, ω_2 is *pseudo-Hermitian* and is explicitly mapped to a Hermitian Hamiltonian using a nontrivial differential operator. An explicit expression is obtained for the matrix element of the differential operator. The mapping fails for a critical value of the coupling constants.

The Hamiltonian and state space are shown to have real and complex branches that are separated by the critical point. At the critical point, the Hamiltonian is inequivalent to any Hermitian operator, and is shown in Chapter 15 as being equal to an infinite-dimensional block diagonal matrix, with each block being a Jordan matrix.

For the real branch of the theory, the state space of the acceleration action is shown to be a function of two distinct degrees of freedom, namely velocity v and position x, with the two being related by a constraint equation. The state vector for the acceleration Hamiltonian is given by $\psi(x, v)$, and is quite unlike the state vector $\psi(x)$ of a quantum degree of freedom described by a Lagrangian having only the velocity term – namely, without an acceleration term.

The acceleration Lagrangian and its path integral are well defined for all values of ω_1, ω_2, including complex values. The Euclidean Hamiltonian obtained for the acceleration Lagrangian shows that the state space of the acceleration Lagrangian is only well defined for real ω_1, ω_2. For equal frequency $\omega_1 = \omega_2$, the Hamiltonian becomes essentially non-Hermitian, and this is analyzed at length in Chapter 15. For the case of complex ω_1, ω_2 the Hamiltonian is neither Hermitian nor pseudo-Hermitian and the state space is no longer a positive normed space.

In Section 14.1 a similarity transformation $e^{Q/2}$ is discussed that shows that the Hamiltonian H is pseudo-Hermitian. In Section 14.2 the Hamiltonian H is mapped, by a similarity transformation, to a Hermitian Hamiltonian that is the sum of two decoupled oscillators. In Section 14.3 the matrix elements of the similarity transformation are evaluated and are shown in Section 14.4 to yield the expected results. The eigenfunctions of the equivalent Hermitian Hamiltonian are evaluated in Section 14.5 and a few of the excited states of the acceleration Hamiltonian are derived in Section 14.9. The state space for unequal frequencies constitutes a Hilbert space, with all state vectors having a positive norm.

The general features of the state space of the pseudo-Hermitian Hamiltonian are discussed in Section 14.11, and in particular the necessity of introducing a metric on state space that is required for defining the conjugate state vector and scalar product for the state space. The eigenfunctions of the acceleration Hamiltonian are studied in Section 14.6: the propagator for the degree of freedom is evaluated in Section 14.12 using operator methods and in Section 14.13 using the state space analysis. It is shown that the propagator cannot be obtained from any Hermitian Hamiltonian.

14.1 Pseudo-Hermitian Hamiltonian; similarity transformation

The acceleration Hamiltonian, from Eq. 13.64, is given by

$$H = -\frac{1}{2\gamma}\frac{\partial^2}{\partial v^2} - v\frac{\partial}{\partial x} + \frac{\gamma}{2}(\omega_1^2 + \omega_2^2)v^2 + \frac{\gamma}{2}\omega_1^2\omega_2^2 x^2. \tag{14.1}$$

For H to be pseudo-Hermitian, it is sufficient that a similarity transformation $e^{Q/2}$ yields

$$e^{-Q/2} H e^{Q/2} = H_o, \tag{14.2}$$

where H_o is a Hermitian Hamiltonian; it is shown in Section 14.2 that for the real branch H_o consists of a system of two decoupled harmonic oscillators, one each for degrees of freedom x and v.

The Hermitian conjugate Hamiltonian $H^\dagger$, from Eq. 14.2, is given by

$$\begin{aligned} H^\dagger &= e^{-Q/2} H_o e^{Q/2} \\ \Rightarrow H^\dagger &= e^{-Q} H e^{Q} : \text{ pseudo-Hermitian.} \end{aligned} \tag{14.3}$$

Note that

$$\left(e^{-\tau H}\right)^{\dagger} = e^{-\tau H^{\dagger}} = e^{-Q} e^{-\tau H} e^{Q}. \tag{14.4}$$

Equation 14.3 is a definition of a pseudo-Hermitian operator, that differs from its Hermitian conjugate by a similarity transformation. It can be shown that the energy eigenvalues of a pseudo-Hermitian Hamiltonian are either real or appear in complex conjugate pairs, and vice versa. All Hamiltonians that are equivalent to a Hermitian Hamiltonian up to a similarity transformation, as is the case for Eq. 14.2, are automatically pseudo-Hermitian. However, all pseudo-Hermitian Hamiltonians are *not equivalent* to a Hermitian Hamiltonian, and Chapter 15 analyzes such a case.

There is another class of Hamiltonians that have real eigenenergies and the energy eigenstates are complete; these Hamiltonians can be brought to a Hermitian form by a similarity transformation. Such Hamiltonians are referred to by Scholtz *et al.* (1992) as being *quasi-Hermitian.* All quasi-Hermitian Hamiltonians are thus also pseudo-Hermitian, but not all pseudo-Hermitian Hamiltonians are quasi-Hermitian.

A Q is obtained for Euclidean time by analytically continuing the remarkable result obtained by Bender and Mannheim (2008b), and this yields

$$Q = axv - b\frac{\partial^2}{\partial x \partial v}. \tag{14.5}$$

For the real domain where $\omega_1,\ \omega_2$ are real, both coefficients a, b are real and hence

$$Q = Q^{\dagger} : \ \text{Hermitian for } \omega_1,\ \omega_2 \text{ real}. \tag{14.6}$$

The equation for the commutator

$$e^{-Q}\mathcal{O}e^{Q} = \sum_{n=0}^{\infty} \frac{1}{n!} [[[\mathcal{O}, Q], Q] \ldots Q] \tag{14.7}$$

needs to be applied to $\mathcal{O} = x, v, \partial/\partial x, \partial/\partial v$.

To obtain the commutator, note that the n-fold commutator of Q with x, v, $\partial/\partial x$ and $\partial/\partial v$ follows a simple pattern that repeats after two commutations. In particular, note that

$$[x, Q] = b\frac{\partial}{\partial v}, \qquad [[x, Q], Q] = abx, \qquad \ldots$$

$$\left[\frac{\partial}{\partial x}, Q\right] = av, \qquad \left[\left[\frac{\partial}{\partial x}, Q\right], Q\right] = ab\frac{\partial}{\partial x}, \qquad \ldots$$

$$[v, Q] = b\frac{\partial}{\partial x}, \qquad [[v, Q], Q] = abv, \qquad \ldots$$

$$\left[\frac{\partial}{\partial v}, Q\right] = ax, \qquad \left[\left[\frac{\partial}{\partial v}, Q\right], Q\right] = ab\frac{\partial}{\partial v}, \qquad \ldots$$

Carrying out the nested commutators to all orders and summing the result yields, for $a, b > 0$,

$$\begin{aligned}
e^{-\tau Q} x e^{\tau Q} &= \cosh(\tau\sqrt{ab})x + \sqrt{\frac{b}{a}}\sinh(\tau\sqrt{ab})\frac{\partial}{\partial v}, \\
e^{-\tau Q}\frac{\partial}{\partial x}e^{\tau Q} &= \cosh(\tau\sqrt{ab})\frac{\partial}{\partial x} + \sqrt{\frac{a}{b}}\sinh(\tau\sqrt{ab})v, \\
e^{-\tau Q} v e^{\tau Q} &= \cosh(\tau\sqrt{ab})v + \sqrt{\frac{b}{a}}\sinh(\tau\sqrt{ab})\frac{\partial}{\partial x}, \\
e^{-\tau Q}\frac{\partial}{\partial v}e^{\tau Q} &= \cosh(\tau\sqrt{ab})\frac{\partial}{\partial v} + \sqrt{\frac{a}{b}}\sinh(\tau\sqrt{ab})x.
\end{aligned} \tag{14.8}$$

Note that the definitions of the values of a and b are chosen based on $\omega_1 > \omega_2$.

14.2 Equivalent Hermitian Hamiltonian H_O

The fundamental commutation relations given in Eq. 14.8 can be applied to the acceleration Hamiltonian, and the coefficients a, b can be chosen to decouple the x and v degrees of freedom. Consider the equation

$$e^{-Q/2}He^{Q/2} = C_1\frac{\partial^2}{\partial v^2} + C_2 x\frac{\partial}{\partial v} + C_3 v\frac{\partial}{\partial x} + C_4\frac{\partial^2}{\partial x^2} + C_5 x^2 + C_6 v^2. \tag{14.9}$$

To obtain the factorization of the Hamiltonian into two de-coupled oscillators, following Bender and Mannheim (2008b), we choose the following values for a and b:

$$\sqrt{\frac{a}{b}} = \gamma\omega_1\omega_2; \quad \sinh(\sqrt{ab}) = \frac{2\omega_1\omega_2}{\omega_1^2 - \omega_2^2} \quad \Rightarrow \quad \sqrt{ab} = \ln\left(\frac{\omega_1 + \omega_2}{\omega_1 - \omega_2}\right). \tag{14.10}$$

We define

$$\begin{aligned}
A &= \cosh\left(\frac{\sqrt{ab}}{2}\right) = \frac{\omega_1}{\sqrt{\omega_1^2 - \omega_2^2}}, \quad B = \sinh\left(\frac{\sqrt{ab}}{2}\right) = \frac{\omega_2}{\sqrt{\omega_1^2 - \omega_2^2}}, \\
C &= \sqrt{\frac{a}{b}} = \gamma\omega_1\omega_2.
\end{aligned} \tag{14.11}$$

Using the result of Eq. 14.8 and the definitions in Eq. 14.9 yields

$$C_1 = -\frac{1}{2\gamma}A^2 + \frac{\gamma}{2}\left(\frac{B}{C}\right)^2 \omega_1^2\omega_2^2$$
$$= -\frac{1}{2\gamma}\left[\cosh^2\left(\frac{\sqrt{ab}}{2}\right) - \sinh^2\left(\frac{\sqrt{ab}}{2}\right)\right] = -\frac{1}{2\gamma}.$$

Similarly, after some simplifications

$$C_2 = -\frac{CAB}{\gamma} + \gamma\omega_1^2\omega_2^2\left(\frac{AB}{C}\right) = 0,$$
$$C_3 = -(A^2 + B^2) + \gamma(\omega_1^2 + \omega_2^2)\left(\frac{AB}{C}\right) = 0.$$

The constants a and b were chosen so that $C_2 = C_3 = 0$; hence one has

$$C_2 = C_3 = 0 \;\Rightarrow\; \text{determines } a \text{ and } b.$$

The remaining coefficients are given by

$$C_4 = -\frac{AB}{C} + \frac{\gamma}{2}(\omega_1^2 + \omega_2^2)\left(\frac{B}{C}\right)^2 = -\frac{1}{2\gamma\omega_1^2},$$
$$C_5 = -\frac{1}{2\gamma}B^2C^2 + \frac{\gamma}{2}\omega_1^2\omega_2^2A^2 = \frac{\gamma}{2}\omega_1^2\omega_2^2,$$
$$C_6 = -ABC + \frac{\gamma}{2}(\omega_1^2 + \omega_2^2)A^2 = \frac{\gamma}{2}\omega_1^2.$$

Collecting all the results yields

$$e^{-Q/2}He^{Q/2} = H_o$$
$$\Rightarrow H_o = -\frac{1}{2\gamma}\frac{\partial^2}{\partial v^2} - \frac{1}{2\gamma\omega_1^2}\frac{\partial^2}{\partial x^2} + \frac{\gamma}{2}\omega_1^2 v^2 + \frac{\gamma}{2}\omega_1^2\omega_2^2 x^2. \tag{14.12}$$

14.3 The matrix elements of $e^{-\tau Q}$

The Q-operator is given from Eq. 14.5 by

$$Q = axv - b\frac{\partial^2}{\partial x \partial v},$$

where, from Eq. 14.10,

$$\sqrt{\frac{a}{b}} = \gamma\omega_1\omega_2, \quad \sinh(\sqrt{ab}) = \frac{2\omega_1\omega_2}{\omega_1^2 - \omega_2^2}.$$

The finite matrix elements of the Q-operator are required for many calculations involving the state vectors. The exact matrix elements can be obtained by noting that the Hermitian Q-operator exactly factorizes into two decoupled harmonic oscillators by an appropriate change of variables.

Consider the change of variables given by

$$\alpha = \frac{1}{\sqrt{2}}(x+v), \quad \beta = \frac{1}{\sqrt{2}}(x-v), \tag{14.13}$$

$$\frac{\partial}{\partial x} = \frac{1}{\sqrt{2}}\left(\frac{\partial}{\partial \alpha} + \frac{\partial}{\partial \beta}\right), \quad \frac{\partial}{\partial v} = \frac{1}{\sqrt{2}}\left(\frac{\partial}{\partial \alpha} - \frac{\partial}{\partial \beta}\right). \tag{14.14}$$

In these coordinates the α, β sectors completely factorize and yield

$$Q = \frac{a}{2}(\alpha^2 - \beta^2) - \frac{b}{2}\left(\frac{\partial^2}{\partial \alpha^2} - \frac{\partial^2}{\partial \beta^2}\right), \tag{14.15}$$

$$= -\frac{1}{2}b\frac{\partial^2}{\partial \alpha^2} + \frac{1}{2}a\alpha^2 + \frac{1}{2}b\frac{\partial^2}{\partial \beta^2} - \frac{1}{2}a\beta^2. \tag{14.16}$$

Consider the Hamiltonian of a quantum oscillator given by

$$H_{\text{sho}} = -\frac{1}{2m}\frac{\partial^2}{\partial z^2} + \frac{1}{2}m\omega^2 z^2, \tag{14.17}$$

with the transition amplitude given by Eq. 11.21, namely

$$K(z, z'; \tau) = \langle z|e^{-\tau H}|z'\rangle$$
$$= \sqrt{\frac{m\omega}{2\pi \sinh \omega\tau}} \exp\left\{-\frac{m\omega}{2\sinh(\omega\tau)}\left[(z^2 + z'^2)\cosh(\omega\tau) - 2zz'\right]\right\}. \tag{14.18}$$

Comparing the α- and β-sectors of Q with H_{sho} shows that, for a, b real, the α sector is the usual quantum oscillator but the β sector yields a divergent transition amplitude. The result for the β sector is assumed to be given by the analytic continuation of the oscillator transition amplitude; this assumption will later be verified by an independent derivation.

To exploit the quadratic form of the α and β sectors, consider extending the range of m to the real line. This yields

$$\alpha - \text{sector}: \quad b = \frac{1}{m}; \qquad a = m\omega^2 \qquad \Rightarrow \qquad \omega = \sqrt{ab}, \tag{14.19}$$

$$\beta - \text{sector}: \quad b = -\frac{1}{m}; \qquad a = -m\omega^2 \qquad \Rightarrow \qquad \omega = \sqrt{ab}. \tag{14.20}$$

Hence Eqs. 14.18, 14.19, and 14.20 yield

$$\begin{aligned}
&\langle \alpha, \beta | e^{-\tau Q} | \alpha', \beta' \rangle \\
&= N_\alpha N_\beta \exp\left\{ -\frac{1}{2b} \frac{\sqrt{ab}}{\sinh(\tau\sqrt{ab})} \left[(\alpha^2 + \alpha'^2)\cosh(\tau\sqrt{ab}) - 2\alpha\alpha' \right] \right\} \\
&\quad \times \exp\left\{ \frac{1}{2b} \frac{\sqrt{ab}}{\sinh(\tau\sqrt{ab})} \left[(\beta^2 + \beta'^2)\cosh(\tau\sqrt{ab}) - 2\beta\beta' \right] \right\} \\
&= \mathcal{N}(\tau) \exp\left\{ -\frac{1}{2}\mathcal{G}(\tau)(\alpha^2 + \alpha'^2 - \beta^2 - \beta'^2) + \mathcal{H}(\tau)(\alpha\alpha' - \beta\beta') \right\},
\end{aligned}$$

where

$$\mathcal{G}(\tau) = \sqrt{\frac{a}{b}} \coth(\tau\sqrt{ab}), \quad \mathcal{H}(\tau) = \sqrt{\frac{a}{b}} \frac{1}{\sinh(\tau\sqrt{ab})} \tag{14.21}$$

and

$$\begin{aligned}
\mathcal{N}(\tau) &= \sqrt{\frac{\sqrt{ab}}{2\pi b \sinh(\tau\sqrt{ab})}} \cdot \sqrt{\frac{\sqrt{ab}}{-2\pi b \sinh(\tau\sqrt{ab})}} \\
\Rightarrow \mathcal{N}(\tau) &= \frac{i}{2\pi} \sqrt{\frac{a}{b}} \frac{1}{|\sinh(\tau\sqrt{ab})|} = \frac{i}{2\pi} |\mathcal{H}(\tau)|.
\end{aligned} \tag{14.22}$$

Hence a heuristic derivation for Q yields

$$\langle x, v | e^{-\tau Q} | x', v' \rangle = \mathcal{N}(\tau) \exp\left\{ -\mathcal{G}(\tau)(xv + x'v') + \mathcal{H}(\tau)(xv' + vx') \right\}. \tag{14.23}$$

The normalization constant $\mathcal{N}(\tau)$ will be seen to play a crucial role in the normalization of all the state vectors of the Hamiltonian H.

The result obtained by the heuristic method will be verified to be indeed exactly correct, including the normalization constant.

For the real branch, both coefficients $\mathcal{G}(\tau)$ and $\mathcal{H}(\tau)$ are real and hence $Q = Q^\dagger$ is Hermitian.

Note that the form obtained for $e^{-\tau Q}$ in Eq. 14.23 is a major simplification since in general, for Hermitian Q, one would need to evaluate $4 + 12 = 16$ real coefficients. Instead, the form that has been heuristically derived in Eq. 14.23 has reduced the determination of $e^{-\tau Q}$ to that of computing two real coefficient functions $\mathcal{G}(\tau)$ and $\mathcal{H}(\tau)$ and a normalization constant $\mathcal{N}(\tau)$.

The operator Q is unbounded and many of the manipulations are only formally valid. For example, the identity $e^{-\tau Q} e^{\tau Q} = \mathbb{I}$ holds for the matrix elements only in a formal sense. Explicitly evaluating the matrix elements of the product $e^{-\tau Q} e^{\tau Q}$ using Eq. 14.23 yields

$$\langle x, v|e^{-\tau Q}e^{\tau Q}|x', v'\rangle = \int d\xi d\zeta \langle x, v|e^{-\tau Q}|\xi, \zeta\rangle\langle\xi, \zeta|e^{\tau Q}|x', v'\rangle$$
$$= \mathcal{N}^2(\tau)\int d\xi d\zeta e^{-\mathcal{G}(\tau)(xv+\xi\zeta)+\mathcal{H}(\tau)(x\zeta+v\xi)}e^{\mathcal{G}(\tau)(\xi\zeta+x'v')-\mathcal{H}(\tau)(\xi v'+\zeta x')}$$
$$= \mathcal{N}^2(\tau)e^{-\mathcal{G}(\tau)(xv-x'v')}\int d\xi d\zeta \exp\{i\mathcal{H}(\tau)i\xi(v'-v) + i\mathcal{H}(\tau)i\zeta(x'-x)\}$$
$$= \mathcal{N}^2(\tau)\left(\frac{2\pi}{i\mathcal{H}(\tau)}\right)^2 \delta(x-x')\delta(v-v') = \delta(x-x')\delta(v-v')$$
$$\Rightarrow e^{-\tau Q}e^{\tau Q} = \mathbb{I}.$$

To make the above derivation more rigorous one can analytically continue τ back to Minkowski time $t = -i\tau$, do the computation and then analytically continue back to Euclidean time τ. This would give the result above.

14.4 $e^{-\tau Q}$ and similarity transformations

The heuristic derivation for $\mathcal{G}(\tau)$ and $\mathcal{H}(\tau)$ was obtained by an analogy with the oscillator Hamiltonian and cannot be assumed to be correct since the β-sector yields an unstable Hamiltonian. The result needs to be independently verified.

That the matrix elements of $e^{-\tau Q}$ given by Eq. 14.23 are in fact correct is now verified in this section. The fundamental similarity transformations of $e^{-\tau Q}\mathcal{O}e^{\tau Q}$ for operators $\mathcal{O} = x, \partial/\partial x, v$ and $\partial/\partial v$ are directly obtained using the result given in Eq. 14.23 and shown to be identical to the defining equations for Q given in Eq. 14.8.

Recall from Eq. 14.8 that $e^{\tau Q}$ yields the following similarity transformations:

$$I.\ e^{-\tau Q}xe^{\tau Q} = \cosh(\tau\sqrt{ab})x + \sqrt{\frac{b}{a}}\sinh(\tau\sqrt{ab})\frac{\partial}{\partial v},$$
$$II.\ e^{-\tau Q}\frac{\partial}{\partial x}e^{\tau Q} = \cosh(\tau\sqrt{ab})\frac{\partial}{\partial x} + \sqrt{\frac{a}{b}}\sinh(\tau\sqrt{ab})v,$$
$$III.\ e^{-\tau Q}ve^{\tau Q} = \cosh(\tau\sqrt{ab})v + \sqrt{\frac{b}{a}}\sinh(\tau\sqrt{ab})\frac{\partial}{\partial x},$$
$$IV.\ e^{-\tau Q}\frac{\partial}{\partial v}e^{\tau Q} = \cosh(\tau\sqrt{ab})\frac{\partial}{\partial v} + \sqrt{\frac{a}{b}}\sinh(\tau\sqrt{ab})x.$$

Consider the operator equation

$$e^{-\tau Q}xe^{\tau Q} = \cosh(\tau\sqrt{ab})x + \sqrt{\frac{b}{a}}\sinh(\tau\sqrt{ab})\frac{\partial}{\partial v}. \tag{14.24}$$

The matrix element of $e^{-\tau Q}xe^{\tau Q}$, using Eq. 14.24, is given by

$$\langle x, v|e^{-\tau Q}xe^{\tau Q}|x', v'\rangle$$
$$= \left(\cosh(\tau\sqrt{ab})x + \sqrt{\frac{b}{a}}\sinh(\tau\sqrt{ab})\frac{\partial}{\partial v}\right)\delta(x - x')\delta(v - v'). \tag{14.25}$$

Equation 14.23 yields

$$\langle x, v|e^{-\tau Q}|x', v'\rangle = N\exp\{-g(xv + x'v') + h(xv' + vx')\}, \tag{14.26}$$

where

$$g = \mathcal{G}(\tau) = \sqrt{\frac{a}{b}}\frac{\cosh(\tau\sqrt{ab})}{\sinh(\tau\sqrt{ab})},$$
$$h = \mathcal{H}(\tau) = \sqrt{\frac{a}{b}}\frac{1}{\sinh(\tau\sqrt{ab})},$$
$$N = \mathcal{N}(\tau) = \frac{i}{2\pi\mathcal{H}(\tau)}.$$

The left hand side of Eq. 14.25 yields

$$\langle x, v|e^{-\tau Q}xe^{\tau Q}|x, v\rangle = \int d\xi d\zeta \langle x, v|e^{-\tau Q}|\xi, \zeta\rangle\langle\xi, \zeta|\xi e^{\tau Q}|x', v'\rangle$$
$$= N\int d\xi d\zeta\, e^{-g(xv+\xi\zeta)+h(x\zeta+v\xi)}\,\xi\langle\xi, \zeta|e^{\tau Q}|x', v'\rangle.$$

But

$$\xi\exp\{-g(xv + \xi\zeta) + h(x\zeta + v\xi)\}$$
$$= \left[\sqrt{\frac{b}{a}}\sin(\tau\sqrt{ab})\frac{\partial}{\partial v} + \cosh(\tau\sqrt{ab})x\right]e^{-g(xv+\xi\zeta)+h(x\zeta+v\xi)}. \tag{14.27}$$

Hence, we obtain the expected result, given in Eq. 14.25, that

$$\langle x, v|e^{-\tau Q}xe^{\tau Q}|x', v'\rangle = \left[\cosh(\tau\sqrt{ab})x + \sqrt{\frac{b}{a}}\sinh(\tau\sqrt{ab})\frac{\partial}{\partial v}\right]$$
$$\times\langle x, v|e^{-\tau Q}e^{\tau Q}|x', v'\rangle$$
$$\Rightarrow I.\ e^{-\tau Q}xe^{\tau Q} = \cosh(\tau\sqrt{ab})x + \sqrt{\frac{b}{a}}\sinh(\tau\sqrt{ab})\frac{\partial}{\partial v}.$$

Consider

$$\langle x, v|e^{-\tau Q} v e^{\tau Q}|x', v'\rangle = \int d\xi d\zeta \ \langle x, v|e^{-\tau Q}|\xi, \zeta\rangle\langle\xi, \zeta|\zeta e^{\tau Q}|x', v'\rangle$$
$$= \int d\xi d\zeta \ \zeta\langle x, v|e^{-\tau Q}|\xi, \zeta\rangle\langle\xi, \zeta|e^{\tau Q}|x', v'\rangle$$
$$= \left(\sqrt{\frac{b}{a}}\sinh(\tau\sqrt{ab})\frac{\partial}{\partial x} + \cosh(\tau\sqrt{ab})v\right)\delta(x - x')\delta(v - v'), \tag{14.28}$$

yielding the expected result that

$$II. \ e^{-\tau Q} v e^{\tau Q} = \cosh(\tau\sqrt{ab})v + \sqrt{\frac{b}{a}}\sinh(\tau\sqrt{ab})\frac{\partial}{\partial x}.$$

The matrix element has the representation

$$\langle x, v|e^{-\tau Q}\frac{\partial}{\partial x}e^{\tau Q}|x', v'\rangle = \int d\xi d\zeta \langle x, v|e^{-\tau Q}|\xi, \zeta\rangle\langle\xi, \zeta|\frac{\partial}{\partial\xi}e^{\tau Q}|x', v'\rangle$$
$$= N\int d\xi d\zeta \left[\left(-\frac{\partial}{\partial\xi}\right)e^{-g(xv+\xi\zeta)+h(x\zeta+v\xi)}\right]$$
$$\times \langle\xi, \zeta|e^{\tau Q}|x', v'\rangle$$
$$= \int d\xi d\zeta\Big[g\zeta - hv\Big]\langle x, v|e^{-\tau Q}|\xi, \zeta\rangle\langle\xi, \zeta|e^{\tau Q}|x, v\rangle.$$

Using Eq. 14.28 to replace ζ in the above expression yields,

$$e^{-\tau Q}\frac{\partial}{\partial x}e^{\tau Q} = g\left(\sqrt{\frac{b}{a}}\sinh(\tau\sqrt{ab})\frac{\partial}{\partial v} + \cosh(\tau\sqrt{ab})v\right) - hv$$
$$\Rightarrow III. \ e^{-\tau Q}\frac{\partial}{\partial x}e^{\tau Q} = \cosh(\tau\sqrt{ab})\frac{\partial}{\partial x} + \sqrt{\frac{b}{a}}\sinh(\tau\sqrt{ab})v.$$

And lastly, similar to the above derivation

$$\langle x, v|e^{-\tau Q}\frac{\partial}{\partial v}e^{\tau Q}|x', v'\rangle = N\int d\xi d\zeta\left[-\frac{\partial}{\partial\zeta}e^{-g(xv+\xi\zeta)+h(x\zeta+v\xi)}\right]\langle\xi, \zeta|e^{\tau Q}|x', v'\rangle$$
$$= \int d\xi d\zeta\Big[g\xi - hx\Big]\langle x, v|e^{-\tau Q}|\xi, \zeta\rangle\langle\xi, \zeta|e^{\tau Q}|x', v'\rangle.$$

Using Eq. 14.27 to replace ξ in the equation above yields

$$e^{-\tau Q}\frac{\partial}{\partial v}e^{\tau Q} = g\left(\sqrt{\frac{b}{a}}\sinh(\tau\sqrt{ab})\frac{\partial}{\partial v} + \cosh(\tau\sqrt{ab})x\right) - hx$$
$$\Rightarrow IV. \ e^{-\tau Q}\frac{\partial}{\partial v}e^{\tau Q} = \cosh(\tau\sqrt{ab})\frac{\partial}{\partial v} + \sqrt{\frac{b}{a}}\cosh(\tau\sqrt{ab})x.$$

Hence we have verified that the expression for $\langle x, v|e^{-\tau Q}|x', v'\rangle$ given in Eq.14.23 is in fact correct and reproduces all the basic similarity transformations, from I to IV.

14.5 Eigenfunctions of oscillator Hamiltonian H_O

The Euclidean Hamiltonian has two distinct and independent state spaces, namely the state space $\mathcal{V}$ of the non-Hermitian Hamiltonian H and the state space of the Hermitian Hamiltonian H_O, namely $\mathcal{V}_O$.

The oscillator state space $\mathcal{V}_O$ is a Hilbert space, with the norm of two state vectors $|\Psi\rangle$ and $|\chi\rangle$ given by

$$\langle\chi|\Psi\rangle = \int dxdv\chi^*(x, v)\Psi(x, v). \tag{14.29}$$

The oscillator Hamiltonian H_o, from Eq. 14.12, is given by

$$H_o = -\frac{1}{2\gamma}\frac{\partial^2}{\partial v^2} - \frac{1}{2\gamma\omega_1^2}\frac{\partial^2}{\partial x^2} + \frac{\gamma}{2}\omega_1^2 v^2 + \frac{\gamma}{2}\omega_1^2\omega_2^2 x^2.$$

The oscillator structure of the Hamiltonian yields two sets of creation and destruction operators, given by

$$a_v = \sqrt{\frac{\gamma\omega_1}{2}}\left[v + \frac{1}{\gamma\omega_1}\frac{\partial}{\partial v}\right],\quad a_v^\dagger = \sqrt{\frac{\gamma\omega_1}{2}}\left[v - \frac{1}{\gamma\omega_1}\frac{\partial}{\partial v}\right],$$

$$a_x = \sqrt{\frac{\gamma\omega_1^2\omega_2}{2}}\left[x + \frac{1}{\gamma\omega_1^2\omega_2}\frac{\partial}{\partial x}\right],\quad a_x^\dagger = \sqrt{\frac{\gamma\omega_1^2\omega_2}{2}}\left[x - \frac{1}{\gamma\omega_1^2\omega_2}\frac{\partial}{\partial x}\right],$$

$$\Rightarrow [a_i, a_j^\dagger] = \delta_{i-j},\quad i, j = x, v. \tag{14.30}$$

Note that, similar to the case for Minkowski time, the creation operator $a_i^\dagger$ is the Hermitian conjugate of the destruction operator a_i.

From the above

$$v = \sqrt{\frac{1}{2\gamma\omega_1}}(a_v + a_v^\dagger),\quad \frac{\partial}{\partial v} = \sqrt{\frac{\gamma\omega_1}{2}}(a_v - a_v^\dagger),$$

$$x = \sqrt{\frac{1}{2\gamma\omega_1^2\omega_2}}(a_x + a_x^\dagger),\quad \frac{\partial}{\partial x} = \sqrt{\frac{\gamma\omega_1^2\omega_2}{2}}(a_x - a_x^\dagger).$$

The oscillator Hamiltonian is given by

$$H_o = \omega_1 a_v^\dagger a_v + \omega_2 a_x^\dagger a_x + \frac{1}{2}(\omega_1 + \omega_2).$$

The vacuum is defined, as usual, by requiring that there be no excitations, namely

$$a_v|0,0\rangle = a_x|0,0\rangle = 0,$$
$$H_O|0,0\rangle = E_0|0,0\rangle,\quad E_0 = \frac{1}{2}(\omega_1+\omega_2).$$

The coordinate representation of the oscillator vacuum state is given by

$$\langle x,v|0,0\rangle = \left(\frac{\gamma^2\omega_1^3\omega_2}{\pi^2}\right)^{1/4}\exp\left\{-\frac{1}{2}[\gamma\omega_1 v^2 + \gamma\omega_1^2\omega_2 x^2]\right\}.$$

The energy eigenfunctions $|n,m\rangle$ and eigenenergies E_{nm} of H_o are given by

$$H_o|n,m\rangle = E_{nm}|n,m\rangle,\quad \langle n,m|H_o = E_{nm}\langle n,m|, \tag{14.31}$$
$$|n,m\rangle = \frac{a_v^{\dagger n}}{\sqrt{n!}}\frac{a_x^{\dagger m}}{\sqrt{m!}}|0,0\rangle,\quad \langle x,v|m,n\rangle = \langle m,n|x,v\rangle : \text{ real},$$
$$E_{nm} = n\omega_1 + m\omega_2 + E_0 = n\omega_1 + m\omega_2 + \frac{1}{2}(\omega_1+\omega_2). \tag{14.32}$$

The dual eigenfunctions $\langle n,m|$ satisfy the orthonormality equation

$$\langle n',m'|n,m\rangle = \delta_{n-n'}\delta_{m-m'}. \tag{14.33}$$

Hence, the spectral representation of H_o is given by

$$H_o = \sum_{mn} E_{mn}|m,n\rangle\langle m,n|. \tag{14.34}$$

All the state vectors and operators have been defined entirely on state space $\mathcal{V}_O$ with no reference to state space $\mathcal{V}$. The operator $e^{Q/2}$ maps state space $\mathcal{V}_O$ into $\mathcal{V}$.

14.6 Eigenfunctions of H and $H^\dagger$

The Euclidean Hamiltonian is given, from Eq. 14.1, by

$$H = -\frac{1}{2\gamma}\frac{\partial^2}{\partial v^2} - v\frac{\partial}{\partial x} + \frac{\gamma}{2}(\omega_1^2+\omega_2^2)v^2 + \frac{\gamma}{2}\omega_1^2\omega_2^2 x^2,$$

and from Eq. 14.12, the equivalent Hermitian Hamiltonian H_o is given by

$$H = e^{Q/2}H_o\, e^{-Q/2},\quad H^\dagger = e^{-Q/2}H_o\, e^{Q/2}, \tag{14.35}$$
$$H_o = -\frac{1}{2\gamma}\frac{\partial^2}{\partial v^2} - \frac{1}{2\gamma\omega_1^2}\frac{\partial^2}{\partial x^2} + \frac{\gamma}{2}\omega_1^2 v^2 + \frac{\gamma}{2}\omega_1^2\omega_2^2 x^2. \tag{14.36}$$

The left and right eigenfunctions of H are different since H is pseudo-Hermitian; let us denote the right *eigenfunctions* by $|\Psi_{mn}\rangle$ and left *dual eigenfunctions* by $\langle\Psi_{mn}^D|$. The notation $\langle\Psi_{mn}^D|$ is used to denote the dual of the state

$|\Psi_{mn}\rangle$ to differentiate it from the state $\langle\Psi_{mn}|$ that is obtained from $|\Psi_{mn}\rangle$ by transposition and complex conjugation.

The energy eigenfunctions of Hamiltonian H and $H^\dagger$, from Eqs. 14.31 and 14.36, are given by

$$\begin{aligned} &H|\Psi_{mn}\rangle = E_{mn}|\Psi_{mn}\rangle, \quad |\Psi_{mn}\rangle = e^{Q/2}|m,n\rangle, \\ &\langle\Psi^D_{mn}|H = E_{mn}\langle\Psi^D_{mn}|, \quad \langle\Psi^D_{mn}| = \langle m,n|e^{-Q/2} = \langle\Psi_{mn}|e^{-Q}. \end{aligned} \tag{14.37}$$

The eigenfunctions are orthonormal since

$$\langle\Psi^D_{nm}|\Psi_{n'm'}\rangle = \langle\Psi_{nm}|e^{-Q}|\Psi_{n'm'}\rangle = \langle n,m|e^{Q/2}e^{-Q}e^{Q/2}|n',m'\rangle \tag{14.38}$$

$$= \langle n,m|n',m'\rangle = \delta_{n-n'}\delta_{m-m'}. \tag{14.39}$$

Since the eigenfunctions of H_o are complete, the completeness equation for the Hilbert space $\mathcal{V}$ on which the pseudo-Hermitian Hamiltonian H acts is given by

$$\mathbb{I} = \sum_{m,n=1}^{\infty} |\Psi_{mn}\rangle\langle\Psi^D_{mn}|. \tag{14.40}$$

The state space for the Euclidean Hamiltonian H has all positive norm eigenstates.

For many of the computations, it is convenient to separate out the overall normalization constants of the eigenfunctions. We define

$$\begin{aligned} &|\Psi_{mn}\rangle = N_{mn}|\psi_{mn}\rangle, \quad \langle\Psi^D_{mn}| = N_{mn}\langle\psi^D_{mn}|, \\ &\langle\Psi^D_{mn}|\Psi_{mn}\rangle = N^2_{mn}\langle\psi^D_{mn}|\psi_{mn}\rangle = 1, \quad N_{mn} = N^*_{mn} > 0: \text{ real, positive.} \end{aligned} \tag{14.41}$$

One can write the completeness equation given in Eq. 14.40 as

$$\mathbb{I} = \sum_{m,n=1}^{\infty} |\psi_{mn}\rangle N^2_{mn}\langle\psi^D_{mn}|. \tag{14.42}$$

All the states $|m,n\rangle$ are real, as are all the matrix elements of $e^{Q/2}$ given in Eq. 14.52; hence the coordinate representation of $|\Psi_{mn}\rangle$ is real and is given by

$$\Psi_{mn}(x,v) = \langle x,v|\Psi_{mn}\rangle = \Psi^*_{mn}(x,v): \text{ real.}$$

From Eqs. 14.12 and 14.34, the Hamiltonian H has the spectral decomposition

$$\begin{aligned} \langle x,v|H|x',v'\rangle &= \langle x,v|e^{Q/2}H_o\, e^{-Q/2}|x',v'\rangle \\ &= \sum_{mn} E_{mn}\langle x,v|e^{Q/2}|m,n\rangle\langle m,n|e^{-Q/2}|x',v'\rangle \\ &= \sum_{mn} E_{mn}\Psi_{mn}(x,v)\Psi^D_{mn}(x',v'). \end{aligned} \tag{14.43}$$

The evolution kernel has the spectral decomposition given by

$$\langle x, v|e^{-\tau H}|x', v'\rangle = \sum_{mn} e^{-\tau E_{mn}} \Psi_{mn}(x, v)\Psi^D_{mn}(x', v'). \tag{14.44}$$

14.6.1 Dual energy eigenstates

Note that unlike the case for a Hermitian Hamiltonian, the dual eigenfunction $\langle\Psi^D|$ is not the transposition and complex conjugate of $|\Psi\rangle$. The dual eigenstates $\Psi^D_{mn}(x, v)$ are given by

$$\langle\Psi^D_{mn}| = \langle m, n|e^{-Q/2} \quad \Rightarrow \quad \Psi^D_{mn}(x, v) = \langle\Psi^D_{mn}|x, v\rangle = \langle m, n|e^{-Q/2}|x, v\rangle.$$

To obtain the matrix element of a non-Hermitian operator, it is important to note that all operators act *only* on the dual space. This fact is unimportant for Hermitian operators, since acting on the state space or its dual space is equivalent. But this is *not* true for non-Hermitian operators. In particular, for operator $\hat{O}(x, v, \partial_x, \partial_v)$ one has

$$\langle x, v|\hat{O}|\Psi\rangle = \hat{O}(x, v, \partial_x, \partial_v)\Psi(x, v).$$

Since all operators are defined by their action only on the dual space, the matrix element of the operator $\hat{O}(x, v, \partial_x, \partial_v)$ acting on the dual state vector $\langle\Psi|$ is given in terms of the conjugate operator $\hat{O}^\dagger(x, v, \partial_x, \partial_v)$ as

$$\begin{aligned}\langle\Psi|\hat{O}|x, v\rangle &\equiv \left(\langle x, v|\hat{O}^\dagger|\Psi\rangle\right)^* \\ &= \left[O^\dagger(x, v, \partial_x, \partial_v)\Psi(x, v)\right]^*.\end{aligned}$$

To obtain the matrix element $\langle\Psi^D_{mn}|H|x, v\rangle$ consider the eigenfunction equation

$$E_{mn}\Psi^D_{mn}(x, v)^* = \langle\Psi^D_{mn}|H|x, v\rangle^* = \langle x, v|H^\dagger|\Psi^D_{mn}\rangle = H^\dagger\Psi^D_{mn}(x, v).$$

Note that the general form of the Hamiltonian H given in Eq. 14.1 implies that

$$\begin{aligned}H^\dagger(x, v, \partial/\partial v, \partial/\partial) &= H(x, -v, -\partial/\partial v, \partial/\partial x) \\ &= H(-x, v, \partial/\partial v, -\partial/\partial x).\end{aligned}$$

Since all the eigenfunctions are real, there are two possibilities,

$$H^\dagger\Psi^D_{mn}(x, v) = H^\dagger\Psi_{mn}(x, -v) = E_{mn}\Psi_{mn}(x, -v)$$

or

$$H^\dagger\Psi^D_{mn}(x, v) = H^\dagger\Psi_{mn}(-x, v) = E_{mn}\Psi_{mn}(-x, v).$$

Hence, the only difference between $\Psi^D_{mn}(x, v)$ and $\Psi_{mn}(x, v)$ is that $v \to -v$ or $x \to -x$ and one obtains the general result that

$$\Psi^D_{mn}(x, v) = \begin{cases} \Psi_{mn}(x, -v) \\ \text{or} \\ \Psi_{mn}(-x, v) \end{cases} . \tag{14.45}$$

As will be seen for the first two excited states given in Eq. 14.82, the dual eigenvectors are determined precisely in a manner that guarantees a positive norm (Hilbert) state space for the pseudo-Hermitian Hamiltonian H.

This remarkable feature of how the dual state vector is defined can be expected to hold for all the eigenstates. Of the two possibilities stated in Eq. 14.45, it is the Q operator that determines which choice is made for the dual state vector.

The probability amplitude, from Eq. 13.91, is given by

$$\begin{aligned} K_S(x, v; x', v') &= \mathcal{K}_S(x', -v'; x, -v) \\ \Rightarrow \langle x, v|e^{-\tau H}|x', v'\rangle &= \langle x', -v'|e^{-\tau H}|x, -v\rangle. \end{aligned} \tag{14.46}$$

Another proof of Eq. 14.46 is given based on Eq. 14.45. From Eqs. 14.46 and 14.44

$$\begin{aligned} \sum_{mn} e^{-\tau E_{mn}} \Psi_{mn}(x, v)\Psi^D_{mn}(x', v') &= \sum_{mn} e^{-\tau E_{mn}} \Psi_{mn}(x', -v')\Psi^D_{mn}(x, -v) \\ \Rightarrow \Psi_{mn}(x, v)\Psi^D_{mn}(x', v') &= \Psi_{mn}(x', -v')\Psi^D_{mn}(x, -v). \end{aligned} \tag{14.47}$$

To prove that the duality relation given in Eq. 14.45 yields Eq. 14.47 we need to discuss the parity operator $\mathcal{P}$, defined by

$$\begin{aligned} &\mathcal{P} : x \to -x, \quad v \to -v, \quad \mathcal{P}^2 = \mathbb{I} \\ &\Rightarrow \mathcal{P}\Psi_{mn}(x, v) = \Psi_{mn}(-x, -v). \end{aligned} \tag{14.48}$$

It can be readily shown that $\mathcal{P}$ commutes with H,

$$[\mathcal{P}, H] = 0.$$

Hence all the eigenfunctions $\Psi_{mn}(x, v)$ are also eigenfunctions of parity with eigenvalue s; Eq. 14.48 yields

$$\mathcal{P}\Psi_{mn}(x, v) = \Psi_{mn}(-x, -v) = s\Psi_{mn}(x, v), \; s = \pm 1. \tag{14.49}$$

Let us consider the two distinct cases for the dual vector given in Eq. 14.45, and Eq. 14.47 is verified for each case:

- $\Psi^D_{mn}(x, v) = \Psi_{mn}(x, -v)$. Eq. 14.47 yields

$$\begin{aligned} \text{Left hand side}: \; &\Psi_{mn}(x, v)\Psi^D_{mn}(x', v') = \Psi_{mn}(x, v)\Psi_{mn}(x', -v'), \\ \text{Right hand side}: \; &\Psi_{mn}(x', -v')\Psi^D_{mn}(x, -v) = \Psi_{mn}(x', -v')\Psi_{mn}(x, v), \end{aligned}$$

hence, Eq. 14.47 is verified.

- $\Psi^D_{mn}(x, v) = \Psi_{mn}(-x, v)$. Eq. 14.47 yields, using Eq. 14.48,

$$\begin{aligned}
\text{Left hand side}: \quad & \Psi_{mn}(x, v)\Psi^D_{mn}(x', v') = \Psi_{mn}(x, v)\Psi_{mn}(-x', v'), \\
\text{Right hand side}: \quad & \Psi_{mn}(x', -v')\Psi^D_{mn}(x, -v) = \Psi_{mn}(x', -v')\Psi_{mn}(-x, -v), \\
& = s^2\Psi_{mn}(-x', v')\Psi_{mn}(x, v).
\end{aligned}$$

Since $s^2 = 1$, Eq. 14.47 is verified.

It is the nontrivial behavior of the dual state vector, as given in Eq. 14.45, that leads to results that are not allowed for Hermitian Hamiltonians, such as the expression obtained for the propagator and discussed in Section 14.13.

14.7 Vacuum state; $e^{Q/2}$

The vacuum state and the dual vacuum state of the Euclidean Hamiltonian are obtained using the expressions for $e^{Q/2}$ and $e^{-Q/2}$ respectively. The vacuum can be verified to be correct by the direct application of the Hamiltonians H and $H^\dagger$ and hence provides an independent verification of the matrix elements obtained for $e^{\tau Q}$ in Section 14.3.

The vacuum state is given by

$$|\Psi_{00}\rangle = e^{Q/2}|0, 0\rangle, \; H|\Psi_{00}\rangle = E_0|\Psi_{00}\rangle, \; E_0 = \frac{1}{2}[\omega_1 + \omega_2]. \tag{14.50}$$

The coordinate representation of the vacuum state can be directly obtained from the Hamiltonian H by inspection and is given by

$$\begin{aligned}
\Psi_{00}(x, v) &= \langle x, v|\Psi_{00}\rangle = \langle x, v|e^{Q/2}|0, 0\rangle \\
&= N_{00} \exp\left\{-\frac{\gamma}{2}(\omega_1 + \omega_2)\omega_1\omega_2 x^2 - \frac{\gamma}{2}(\omega_1 + \omega_2)v^2 - \gamma\omega_1\omega_2 xv\right\}.
\end{aligned} \tag{14.51}$$

The vacuum state $|\Psi_{00}\rangle$ is real valued and normalizable, with N_{00} the normalization constant. The dual vacuum state obeys

$$\langle\Psi^D_{00}| = \langle 0, 0|e^{-Q/2} \;\Rightarrow\; \Psi^D_{00}(x, v) = \langle\Psi^D_{00}|x, v\rangle = \langle 0, 0|e^{-Q/2}|x, v\rangle.$$

It can be directly verified that the dual ground state, from Eq. 14.51, is

$$\Psi^D_{00}(x, v) = N_{00} \exp\left\{-\frac{\gamma}{2}(\omega_1 + \omega_2)\omega_1\omega_2 x^2 - \frac{\gamma}{2}(\omega_1 + \omega_2)v^2 + \gamma\omega_1\omega_2 xv\right\}.$$

For $\tau = -1/2$, Eq. 14.23 yields the matrix elements

$$\langle x, v|e^{Q/2}|x', v'\rangle$$
$$= \mathcal{N}(\frac{1}{2}) \exp\left\{\sqrt{\frac{a}{b}} \coth\left(\frac{\sqrt{ab}}{2}\right)(xv + x'v') - \sqrt{\frac{a}{b}} \frac{(xv' + vx')}{\sinh\left(\frac{\sqrt{ab}}{2}\right)}\right\}. \quad (14.52)$$

Note that

$$\sqrt{\frac{a}{b}} \coth\left(\frac{\sqrt{ab}}{2}\right) = \gamma\omega_1^2, \quad \sqrt{\frac{a}{b}} \frac{1}{\sinh\left(\frac{\sqrt{ab}}{2}\right)} = \gamma\omega_1\sqrt{\omega_1^2 - \omega_2^2}, \quad (14.53)$$

$$\mathcal{N}(\frac{1}{2}) = \frac{i}{2\pi}\sqrt{\frac{a}{b}} \frac{1}{\sinh\left(\frac{\sqrt{ab}}{2}\right)} = \frac{i}{2\pi}\gamma\omega_1\sqrt{\omega_1^2 - \omega_2^2}. \quad (14.54)$$

Hence, from Eq. 14.52

$$\langle x, v|e^{Q/2}|x', v'\rangle = \mathcal{N}(\frac{1}{2}) \exp\left\{\gamma\omega_1^2(xv + x'v') - \gamma\omega_1\sqrt{\omega_1^2 - \omega_2^2}(xv' + vx')\right\}.$$

From Eq. 14.27, the oscillator vacuum state is given by

$$\langle x, v|0, 0\rangle = \left(\frac{\gamma^2\omega_1^3\omega_2}{\pi^2}\right)^{1/4} \exp\left\{-\frac{\gamma}{2}(\omega_1 v^2 + \omega_1^2\omega_2 x^2)\right\}.$$

Hence

$$\Psi_{00}(x, v) = \langle x, v|e^{Q/2}|0, 0\rangle = \int d\xi d\zeta \langle x, v|e^{Q/2}|\xi, \zeta\rangle\langle\xi, \zeta|0, 0\rangle$$
$$= \mathcal{N}(\frac{1}{2}) \left(\frac{\gamma^2\omega_1^3\omega_2}{\pi}\right)^{1/4} \int d\xi d\zeta e^S,$$

where

$$S = -\frac{\gamma}{2}(\omega_1\zeta^2 + \omega_1^2\omega_2\xi^2 - 2c_1\xi\zeta) - \gamma c_2(\zeta x + v\xi) + \gamma c_1 xv,$$
$$c_1 = \omega_1^2, \quad c_2 = \omega_1\sqrt{\omega_1^2 - \omega_2^2}.$$

Performing the Gaussian integrations over ξ and ζ yields

$$\Psi_{00}(x, v) = N_{00}e^F,$$

where

$$F = \frac{\gamma^2 c_2^2}{2\gamma(\omega_1^3\omega_2 - c_1^2)}[\omega_1 v^2 + \omega_1^2\omega_2 x^2 + 2c_1 xv] + \gamma c_1 xv$$
$$= -\frac{\gamma}{2}[(\omega_1 + \omega_2)\omega_1\omega_2 x^2 + (\omega_1 + \omega_2)v^2] - \gamma\omega_1\omega_2 xv,$$

which is the expected result.

To determine N_{00} note that the Gaussian integration has the matrix

$$M = \gamma \begin{bmatrix} \omega_1^2\omega_2 & -c_1 \\ -c_1 & \omega_1 \end{bmatrix}, \quad \sqrt{\det(M)} = i\gamma\omega_1^{3/2}\sqrt{\omega_1 - \omega_2}.$$

The determinant of matrix M is negative and hence leads to the correct sign for the exponent F of the vacuum state vector.

Using Eq. 14.54, Gaussian integration yields

$$\begin{aligned} N_{00} &= \mathcal{N}(\frac{1}{2})\left(\frac{\gamma^2\omega_1^3\omega_2}{\pi}\right)^{1/4}\frac{2\pi}{\sqrt{\det(M)}} \\ &= \frac{i}{2\pi}\gamma\omega_1\sqrt{\omega_1^2 - \omega_2^2}\left(\frac{\gamma^2\omega_1^3\omega_2}{\pi^2}\right)^{1/4}\frac{2\pi}{i\gamma\omega_1^{3/2}\sqrt{\omega_1 - \omega_2}} \\ \Rightarrow N_{00} &= (\omega_1\omega_2)^{1/4}\sqrt{\frac{\gamma}{\pi}(\omega_1 + \omega_2)}. \end{aligned} \tag{14.55}$$

To derive the dual vacuum state, note from Eq. 14.52

$$\langle x, v|e^{-Q/2}|x', v'\rangle = \mathcal{N}(\frac{1}{2})\exp\left\{-\gamma\omega_1^2(xv + x'v') + \gamma\omega_1\sqrt{\omega_1^2 - \omega_2^2}(xv' + vx')\right\}.$$

A derivation similar to the one for the vacuum state Ψ_{00} yields the dual vacuum state

$$\begin{aligned} \Psi_{00}^D(x, v) &= \langle 0, 0|e^{-Q/2}|x, v\rangle \\ &= N_{00}\exp\{-\frac{\gamma}{2}(\omega_1 + \omega_2)\omega_1\omega_2x^2 - \frac{\gamma}{2}(\omega_1 + \omega_2)v^2 + \gamma\omega_1\omega_2xv\}. \end{aligned} \tag{14.56}$$

The norm of a state is defined by the scalar product of the state with its dual and yields the following norm for the vacuum state

$$\begin{aligned} \langle\Psi_{00}^D|\Psi_{00}\rangle &= \int dxdv\Psi_{00}^D(x, v)\Psi_{00}(x, v) \\ &= N_{00}^2\int dxdv\exp\{-\gamma(\omega_1 + \omega_2)\omega_1\omega_2x^2 - \gamma(\omega_1 + \omega_2)v^2\} = 1. \end{aligned}$$

Recall that Euclidean velocity is related to Minkowski velocity by $v = dx/d\tau = -idx/dt_M = -iv_M$. Hence, the vacuum state $|\Psi_{00}\rangle$ when analytically continued to Minkowski time has a *divergent* norm. Furthermore, the eigenstates generated by applying the creation operators on the vacuum state all have a *divergent* norm. The problem of rendering the Minkowski time state space convergent has been addressed by Bender and Mannheim (2008b).

14.8 Vacuum state and classical action

The definition of the probability amplitude given in Eq. 13.66,

$$\mathcal{K}_S(x_f, v_f; x_i, v_i) = \langle x_f, v_f | e^{-\tau H} | x_i, v_i \rangle,$$

yields, from Eq. 14.44, the infinite τ limit given by

$$\begin{aligned} \lim_{\tau \to +\infty} \mathcal{K}_S(x_f, v_f; x_i, v_i) &\simeq e^{-\tau E_0} \langle x_f, v_f | \Psi_{00} \rangle \langle \Psi_{00} | e^{-Q} | x_i, v_i \rangle + O(e^{-\tau(E_1 - E_0)}) \\ &= e^{-\tau E_0} \Psi_{00}(x_f, v_f) \Psi_{00}^D(x_i, v_i). \end{aligned} \tag{14.57}$$

Hence, from Eq. 14.57, the infinite limit of transition amplitude is the product of the dual vacuum state $\Psi_{00}^D(x_i, v_i)$ and the vacuum state $\Psi_{00}(x_f, v_f)$. On the other hand, the classical solution yields

$$\lim_{\tau \to +\infty} K(x_f, v_f, x_i, v_i) = \lim_{\tau \to +\infty} \mathcal{N}' e^{S_c(x_f, v_f, x_i, v_i)}. \tag{14.58}$$

The infinite τ limits of matrix elements M, given in Section 13.2, are given by

$$\begin{aligned} &\lim_{\tau \to +\infty} M_{11} = 2ra(r^2 + \omega^2), \quad \lim_{\tau \to +\infty} M_{22} = 2ra, \quad \lim_{\tau \to +\infty} M_{12} = 2r^2 a - b, \\ &\lim_{\tau \to +\infty} M_{13} = \lim_{\tau \to +\infty} M_{14} = \lim_{\tau \to +\infty} M_{23} = 0. \end{aligned} \tag{14.59}$$

Therefore, the vacuum state is obtained from Eqs. 13.28 and 14.59 and yields

$$\begin{aligned} \Omega(x, v) &= \lim_{\tau \to +\infty} \mathcal{N} \exp\left(-\frac{1}{2} M_{11} x^2 - \frac{1}{2} M_{22} v^2 - M_{12} x v \right) \\ &= \mathcal{N} \exp\Big(-ra^2(r^2 + \omega^2) x^2 - rav^2 - (2r^2 a - b) x v \Big), \end{aligned} \tag{14.60}$$

where $x_f = x$, $v_f = v$, and $\mathcal{N}$ is fixed by normalizing $\Omega(x, v)$.

To make connection with the earlier parameterization, let us rewrite the Lagrangian in ω_1 and ω_2 parameterization

$$\begin{aligned} \mathcal{L} &= -\frac{1}{2} a \left(\ddot{x}^2 + 2\frac{b}{a} \dot{x}^2 + \frac{c}{a} x^2 \right) \\ &= -\frac{\gamma}{2} \Big(\ddot{x}^2 + (\omega_1^2 + \omega_2^2) \dot{x}^2 + (\omega_1^2 \omega_2^2) x^2 \Big), \end{aligned} \tag{14.61}$$

where ω_1 and ω_2 are[1]

$$\omega_1 = \frac{1}{\sqrt{2\gamma}} \left(\sqrt{b + \sqrt{ac}} + \sqrt{b - \sqrt{ac}} \right),$$

[1] Note that ω_1 and ω_2 in Eq. 13.30 have a different parameterization, given by

$$\omega_1 = \frac{1}{2\sqrt{\gamma}} \left(\sqrt{\alpha + 2\sqrt{\gamma\beta}} + \sqrt{\alpha - 2\sqrt{\gamma\beta}} \right), \quad \omega_2 = \frac{1}{2\sqrt{\gamma}} \left(\sqrt{\alpha + 2\sqrt{\gamma\beta}} - \sqrt{\alpha - 2\sqrt{\gamma\beta}} \right).$$

$$\omega_2 = \frac{1}{\sqrt{2\gamma}}\left(\sqrt{b+\sqrt{ac}} - \sqrt{b-\sqrt{ac}}\right).$$

The definition of variables r and ω in terms of ω_1 and ω_2 is given by

$$\omega_1 = r + i\omega, \quad \omega_2 = r - i\omega.$$

The two parameterizations have the relationship

$$r^2 + \omega^2 = \omega_1^2\omega_2^2, \quad r = \frac{1}{2}(\omega_1 + \omega_2), \quad b = \frac{\gamma}{2}(\omega_1^2\omega_2^2). \tag{14.62}$$

Equations 14.60 and 14.62 hence yield – replacing a by γ to conform to the notation given in Eq. 13.30 – the vacuum state given in Eq. 14.51, namely

$$\Psi_{00}(x,v) = \mathcal{N}\exp\left(-\frac{\gamma}{2}(\omega_1+\omega_2)\omega_1\omega_2 x^2 - \frac{\gamma}{2}(\omega_1+\omega_2)v^2 - \gamma\omega_1\omega_2 xv\right).$$

The definition of the evolution kernel $\mathcal{K}_S(x_f, v_f; x_i, v_i) = \langle x_f, v_f|e^{-\tau H}|x_i, v_i\rangle$ given in Eq. 13.66 is seen to be correct since the evolution kernel obtained from the classical solution also gives the same result for the vacuum state as given by the Hamiltonian in Eq. 14.51.

In particular, there is no need to include the state space metric e^{-Q} in the definition of $\mathcal{K}_S(x_f, v_f; x_i, v_i)$, since the expression $\langle x_f, v_f|e^{-\tau H}|x_i, v_i\rangle$ for the probability amplitude is the matrix elements in a complete basis, and not the probability amplitude between physical states. However, in obtaining the vacuum state the operator e^{-Q} appears, since it defines the dual of the vacuum state $|\Psi_{00}\rangle$ via the expression $\langle\Psi_{00}|e^{-Q}$.

14.9 Excited states of *H*

To illustrate the general features of the eigenfunctions, the first few eigenfunctions are evaluated. Note

$$a_v^\dagger = \sqrt{\frac{\gamma\omega_1}{2}}\left(v - \frac{1}{\gamma\omega_1}\frac{\partial}{\partial v}\right), \tag{14.63}$$

$$a_x^\dagger = \sqrt{\frac{\gamma\omega_1^2\omega_2}{2}}\left(x - \frac{1}{\gamma\omega_1^2\omega_2}\frac{\partial}{\partial x}\right). \tag{14.64}$$

In general, one can find the explicit coordinate representation of any eigenfunction by the following procedure:

$$\begin{aligned}\Psi_{nm}(x,v) &= \langle x, v|e^{Q/2}\left\{\frac{a_v^{\dagger n}}{\sqrt{n!}}\frac{a_x^{\dagger m}}{\sqrt{m!}}\right\}|0,0\rangle \\ &= \langle x, v|e^{Q/2}\frac{a_v^{\dagger n}}{\sqrt{n!}}\frac{a_x^{\dagger m}}{\sqrt{m!}}e^{-Q/2}e^{Q/2}|0,0\rangle\end{aligned}$$

$$= e^{Q/2} \frac{a_v^{\dagger n}}{\sqrt{n!}} \frac{a_x^{\dagger m}}{\sqrt{m!}} e^{-Q/2} \Psi_{00}(x, v), \tag{14.65}$$

where, using Eqs. 14.69 and 14.76 given below, one can explicitly evaluate

$$e^{Q/2} a_v^{\dagger n} a_x^{\dagger m} e^{-Q/2} = \left(e^{Q/2} a_v^{\dagger} e^{-Q/2}\right)^n \left(e^{Q/2} a_x^{\dagger} e^{-Q/2}\right)^m . \tag{14.66}$$

14.9.1 Energy ω_1 eigenstate $\Psi_{10}(x, v)$

The ω_1 single excitation energy eigenstate is

$$\Psi_{10}(x, v) = \langle x, v| e^{Q/2} \left\{ a_v^{\dagger} |0, 0\rangle \right\} = \langle x, v| e^{Q/2} a_v^{\dagger} e^{-Q/2} e^{Q/2} |0, 0\rangle$$

$$\Rightarrow \Psi_{10}(x, v) = e^{Q/2} a_v^{\dagger} e^{-Q/2} \Psi_{00}(x, v). \tag{14.67}$$

The fundamental similarity transformation given in Eq. 14.8 yields

$$e^{Q/2} v e^{-Q/2} = A v - \frac{B}{C} \frac{\partial}{\partial x},$$

$$e^{Q/2} \frac{\partial}{\partial v} e^{-Q/2} = A \frac{\partial}{\partial v} - B C x, \tag{14.68}$$

where, from Eq. 14.11 the coefficient functions are given by

$$A = \frac{\omega_1}{\sqrt{\omega_1^2 - \omega_2^2}}, \quad B = \frac{\omega_2}{\sqrt{\omega_1^2 - \omega_2^2}}, \quad C = \gamma \omega_1 \omega_2.$$

Hence, from Eq. 14.63

$$e^{Q/2} a_v^{\dagger} e^{-Q/2} = \sqrt{\frac{\gamma \omega_1}{2}} e^{Q/2} \left[v - \frac{1}{\gamma \omega_1} \frac{\partial}{\partial v} \right] e^{-Q/2}$$

$$= \sqrt{\frac{\gamma \omega_1}{2}} \left[A v + \frac{BC}{\gamma \omega_1} x - \frac{B}{C} \frac{\partial}{\partial x} - \frac{A}{\gamma \omega_1} \frac{\partial}{\partial v} \right]$$

$$\Rightarrow e^{Q/2} a_v^{\dagger} e^{-Q/2} = \sqrt{\frac{\gamma \omega_1}{2(\omega_1^2 - \omega_2^2)}} \left[\omega_1 v + \omega_2^2 x - \frac{1}{\gamma \omega_1} \frac{\partial}{\partial x} - \frac{1}{\gamma} \frac{\partial}{\partial v} \right]. \tag{14.69}$$

Equations 14.67 and 14.69 yield

$$\Psi_{10}(x, v) = \sqrt{\frac{\gamma \omega_1}{2(\omega_1^2 - \omega_2^2)}} \left[\omega_1 v + \omega_2^2 x - \frac{1}{\gamma \omega_1} \frac{\partial}{\partial x} - \frac{1}{\gamma} \frac{\partial}{\partial v} \right] \Psi_{00}(x, v).$$

Using the explicit representation of the vacuum state $\Psi_{00}(x, v)$ given in Eq. 14.50 yields the final result

$$\Psi_{10}(x, v) = \sqrt{\frac{2\gamma \omega_1}{\omega_1^2 - \omega_2^2}} (\omega_1 + \omega_2) \left[v + \omega_2 x \right] \Psi_{00}(x, v). \tag{14.70}$$

The dual energy eigenstate is defined by

$$\begin{aligned}
\langle\Psi_{10}^D| &= \langle 1,0|e^{-Q/2} = \langle 0,0|a_v e^{-Q/2} \\
\Rightarrow |\Psi_{10}^D\rangle &= e^{-Q/2}a_v^\dagger|0,0\rangle = e^{-Q/2}a_v^\dagger e^{Q/2}e^{-Q/2}|0,0\rangle \\
\Rightarrow \Psi_{10}^D(x,v) &= \langle x,v|\Psi_{10}^D\rangle = e^{-Q/2}a_v^\dagger e^{Q/2}\Psi_{00}^D(x,v).
\end{aligned} \tag{14.71}$$

Similarly to the derivation of Eq. 14.69, Eq. 14.8 yields

$$e^{-Q/2}a_v^\dagger e^{Q/2} = \sqrt{\frac{\gamma\omega_1}{2(\omega_1^2-\omega_2^2)}}\left[\omega_1 v - \omega_2^2 x + \frac{1}{\gamma\omega_1}\frac{\partial}{\partial x} - \frac{1}{\gamma}\frac{\partial}{\partial v}\right], \tag{14.72}$$

and, from Eq. 14.71, we obtain the dual eigenfunction

$$\Psi_{10}^D(x,v) = \sqrt{\frac{\gamma\omega_1}{2(\omega_1^2-\omega_2^2)}}\left[\omega_1 v - \omega_2^2 x + \frac{1}{\gamma\omega_1}\frac{\partial}{\partial x} - \frac{1}{\gamma}\frac{\partial}{\partial v}\right]\Psi_{00}^D(x,v).$$

Using the representation of the vacuum state $\Psi_{00}^D(x,v)$ given in Eq. 14.56 yields the final result

$$\Psi_{10}^D(x,v) = \sqrt{\frac{2\gamma\omega_1}{\omega_1^2-\omega_2^2}}(\omega_1+\omega_2)\,[v-\omega_2 x]\,\Psi_{00}^D(x,v). \tag{14.73}$$

14.9.2 Energy ω_2 eigenstate $\Psi_{01}(x,v)$

The ω_2 single excitation energy eigenstate is given by

$$\begin{aligned}
\Psi_{01}(x,v) &= \langle x,v|e^{Q/2}\left\{a_x^\dagger|0,0\rangle\right\} = \langle x,v|e^{Q/2}a_x^\dagger e^{-Q/2}e^{Q/2}|0,0\rangle \\
\Rightarrow \Psi_{01}(x,v) &= e^{Q/2}a_x^\dagger e^{-Q/2}\Psi_{00}(x,v).
\end{aligned} \tag{14.74}$$

The similarity transformation given in Eq. 14.8 yields

$$\begin{aligned}
e^{Q/2}xe^{-Q/2} &= Ax - \frac{B}{C}\frac{\partial}{\partial v}, \\
e^{Q/2}\frac{\partial}{\partial x}e^{-Q/2} &= A\frac{\partial}{\partial x} - BCv.
\end{aligned} \tag{14.75}$$

Hence, from Eq. 14.64

$$\begin{aligned}
e^{Q/2}a_x^\dagger e^{-Q/2} &= \sqrt{\frac{\gamma\omega_1^2\omega_2}{2}}e^{Q/2}\left[x - \frac{1}{\gamma\omega_1^2\omega_2}\frac{\partial}{\partial x}\right]e^{-Q/2} \\
&= \sqrt{\frac{\gamma\omega_1^2\omega_2}{2}}\left[Ax + \frac{BC}{\gamma\omega_1^2\omega_2}v - \frac{B}{C}\frac{\partial}{\partial v} - \frac{A}{\gamma\omega_1^2\omega_2}\frac{\partial}{\partial x}\right] \\
\Rightarrow e^{Q/2}a_x^\dagger e^{-Q/2} &= \sqrt{\frac{\gamma\omega_2}{2(\omega_1^2-\omega_2^2)}}\left[\omega_2 v + \omega_1^2 x - \frac{1}{\gamma\omega_2}\frac{\partial}{\partial x} - \frac{1}{\gamma}\frac{\partial}{\partial v}\right].
\end{aligned} \tag{14.76}$$

Hence, from Eqs. 14.74 and 14.76

$$\Psi_{01}(x,v) = \sqrt{\frac{\gamma\omega_2}{2(\omega_1^2-\omega_2^2)}}\left[\omega_2 v + \omega_1^2 x - \frac{1}{\gamma\omega_2}\frac{\partial}{\partial x} - \frac{1}{\gamma}\frac{\partial}{\partial v}\right]\Psi_{00}(x,v).$$

Using the representation of the vacuum state $\Psi_{00}(x,v)$ given in Eq. 14.50 yields the final result

$$\Psi_{01}(x,v) = \sqrt{\frac{2\gamma\omega_2}{\omega_1^2-\omega_2^2}}(\omega_1+\omega_2)\,[v+\omega_1 x]\,\Psi_{00}(x,v). \tag{14.77}$$

The dual energy eigenstate, similarly to Eq. 14.78, is defined by

$$\Psi_{01}^D(x,v) = e^{-Q/2}a_v^\dagger e^{Q/2}\Psi_{00}^D(x,v). \tag{14.78}$$

Equations 14.64 and 14.8 yield

$$e^{-Q/2}a_x^\dagger e^{Q/2} = \sqrt{\frac{\gamma\omega_2}{2(\omega_1^2-\omega_2^2)}}\left[-\omega_2 v + \omega_1^2 x - \frac{1}{\gamma\omega_2}\frac{\partial}{\partial x} + \frac{1}{\gamma}\frac{\partial}{\partial v}\right]. \tag{14.79}$$

Using the explicit representation of the vacuum state $\Psi_{00}^D(x,v)$ given in Eq. 14.56 with Eq. 14.78 yields the dual eigenfunction

$$\Psi_{01}^D(x,v) = \sqrt{\frac{2\gamma\omega_2}{\omega_1^2-\omega_2^2}}(\omega_1+\omega_2)\,[-v+\omega_1 x]\,\Psi_{00}^D(x,v). \tag{14.80}$$

We collect the results for the first two single excitation states; the normalization constants are separated out for later convenience; using the value of N_{00} given in Eq. 14.55 yields

$$\begin{aligned}
&\Psi_{10}(x,v) = N_{10}\,[v+\omega_2 x]\,\psi_{00}(x,v),\quad \Psi_{10}^D(x,v) = N_{10}\,[v-\omega_2 x]\,\psi_{00}^D(x,v),\\
&\Psi_{01}(x,v) = N_{01}\,[v+\omega_1 x]\,\psi_{00}(x,v),\quad \Psi_{01}^D(x,v) = N_{01}\,[-v+\omega_1 x]\,\psi_{00}^D(x,v),\\
&N_{10} = \gamma\sqrt{2}\frac{(\omega_1+\omega_2)}{\sqrt{\pi(\omega_1-\omega_2)}}\,\omega_1^{3/4}\omega_2^{1/4},\quad N_{01} = \gamma\sqrt{2}\frac{(\omega_1+\omega_2)}{\sqrt{\pi(\omega_1-\omega_2)}}\,\omega_1^{1/4}\omega_2^{3/4},\\
&E_{10} = \omega_1 + E_{00},\quad E_{01} = \omega_2 + E_{00}.
\end{aligned} \tag{14.81}$$

The eigenstates are orthogonal and normalized, namely

$$\langle\Psi_{10}^D|\Psi_{10}\rangle = 1 = \langle\Psi_{01}^D|\Psi_{01}\rangle,\quad \langle\Psi_{10}^D|\Psi_{01}\rangle = 0.$$

Note the remarkable result that under a duality transformation, the dual eigenstates have a transformation that depends on the eigenstate, as discussed in Eq. 14.45; in particular

$$\Psi_{10}^D(x,v) = \Psi_{10}(-x,v),\quad \Psi_{01}^D(x,v) = \Psi_{01}(x,-v). \tag{14.82}$$

This feature generalizes to all the energy eigenstates and guarantees that the state space, for $\omega_1 > \omega_2$, always has a positive norm.

The first two energy eigenstates above the vacuum state of the pseudo-Hermitian Hamiltonian H are the first excitation of the position degree of freedom x, given by Ψ_{01}, and of the velocity degree of freedom v, given by Ψ_{10}. Note that in the limit of $\omega_1 \to \omega_2$, the eigenstates Ψ_{01} and Ψ_{10} become degenerate. This important property of the energy eigenspectrum will be studied in some detail when the limit of $\omega_1 \to \omega_2$ is taken in Section 15.1.

Eq. 14.81 yields the result

$$\Psi_{10}^D(x,v)\Psi_{10}(x,v) = N_{10}^2\left[v^2 - \omega_2^2 x^2\right]\psi_{00}^D(x,v)\psi_{00}(x,v)$$
$$\Psi_{01}^D(x,v)\Psi_{01}(x,v) = N_{01}^2\left[-v^2 + \omega_1^2 x^2\right]\psi_{00}^D(x,v)\psi_{00}(x,v).$$

For the velocity Lagrangian, the state vector $\psi(x)$ has a unit norm of $\langle\psi|\psi\rangle = 1$ and yields a positive definite density, namely $P(x) = \psi^*(x)\psi(x) \geq 0$; hence, $P(x)$ can be interpreted as a *probability density*. In contrast, although $\langle\Psi^D|\Psi\rangle = 1$, as can be seen from the examples above, $R(x,v) = \Psi^D(x,v)\Psi(x,v)$ can be either positive or negative; hence $-\infty \leq R(x,v) \leq +\infty$ and cannot be interpreted as a probability density function. Consequently, one cannot use the standard quantum mechanical framework to interpret the results of the accelaration Hamiltonian; in particular, it is not clear how to interpret the state functions $\Psi(x,v)$.

14.10 Complex ω_1, ω_2

The application of the acceleration Lagrangian to the study of equities, as shown by Baaquie *et al.* (2012), requires that the parameters ω_1, ω_2 are complex. From Eq. 13.31, consider the parameterization given by

$$\omega_1 = Re^{i\phi},\ \ \omega_2 = Re^{-i\phi}\ \ :\ \ R > 0,\ \ \phi \in [-\pi/2, \pi/2], \tag{14.83}$$
$$\Rightarrow\ \ \omega_1 + \omega_2 = 2R\cos(\phi),\ \ \omega_1\omega_2 = R^2. \tag{14.84}$$

The eigenenergies are given by

$$E_{00} = \frac{1}{2}(\omega_1 + \omega_2) = R\cos\phi,$$
$$E_{mn} = m\omega_1 + n\omega_2 + E_{00} = mRe^{i\phi} + nRe^{-i\phi} + R\cos\phi.$$

As discussed in Section 14.2, the eigenenergies of a pseudo-Hermitian Hamiltonian come in conjugate pairs and are given by

$$E_{nm} = nRe^{i\phi} + mRe^{-i\phi} + R\cos\phi = E_{mn}^*.$$

In other words, energy eigenvalues E_{mn} and E_{nm} are complex conjugates of each other.

From Eq. 14.10, the parameters are given by

$$\sqrt{\frac{a}{b}} = \gamma\omega_1\omega_2 = \gamma R^2, \quad \sinh(\sqrt{ab}) = \frac{2\omega_1\omega_2}{\omega_1^2 - \omega_2^2} = \frac{-i}{\cos(2\phi)}.$$

Since a, b are now complex, the Q-operator is given from Eq. 14.5,

$$Q = axv - b\frac{\partial^2}{\partial x \partial v}$$

is no longer Hermitian.

Since the eigenenergies are complex, there is no sense of the ordering of energies and the concept of a ground state is no longer valid. One can still consider the real part of the eigenenergies and the lowest real energy E_{00} has the following eigenstate given by the continuation of the vacuum state

$$\Psi_{00}(x, v) = N_{00} \exp\left\{-\gamma R^3 \cos(\phi)x^2 - \gamma R \cos(\phi)v^2 - \gamma R^2 xv\right\}. \quad (14.85)$$

Equation 14.85 yields a positive norm state that is well defined for

$$\cos(\phi) > 0 \;\Rightarrow\; -\pi/2 < \phi < \pi/2.$$

Since ω_1, ω_2 are complex, the algebra of the creation and destruction operators given in Eq. 14.30 is no longer valid. Hence, for the complex branch of the parameters, these operators can no longer generate the spectrum of states for H.

14.11 State space $\mathcal{V}$ of Euclidean Hamiltonian

The state space of the non-Hermitian Hamiltonian H, namely $\mathcal{V}$, has a non-trivial metrical structure. This is most clearly seen in the manner in which the dual state vectors are defined as well as the rules of conjugation for operators, called Q-conjugation, that is different from the usual Hermitian conjugation.

To define orthonormality of the eigenfunctions $|\Psi_{nm}\rangle$, one needs to define the scalar product of a state vector $|\Psi\rangle$ with a dual vector $\langle\chi|$, that are elements of $\mathcal{V}$. In particular, one needs to define a dual that is consistent with the similarity transformation given in Eq. 14.12 that maps the non-Hermitian Hamiltonian H to the Hermitian Hamiltonian H_o. Figure 14.1 schematically shows the mapping from the state space of H_o denoted by $\mathcal{V}_0$ to the state space of H denoted by $\mathcal{V}_{\mathcal{H}}$.

A consistent quantum mechanics can be defined by generalizing the scalar product by using a metric W on state space in the following manner [Ballentine (1998)]:

$$\langle\chi|\Psi\rangle_W = \langle\chi|W|\Psi\rangle, \quad W^\dagger = W. \quad (14.86)$$

In analogy with the result for Minkowski time [Mannheim and Davidson (2005)], the metric for the state space $\mathcal{V}$ is given by

$$W^\dagger = W = e^{-Q} \;\Rightarrow\; \langle\chi|\Psi\rangle_Q = \langle\chi|e^{-Q}|\Psi\rangle \;:\; \text{scalar product}. \quad (14.87)$$

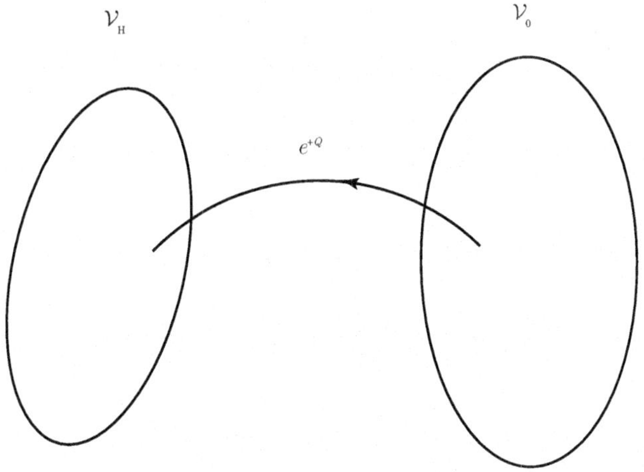

Figure 14.1 A mapping between two state spaces.

Hence two state vectors in $\mathcal{V}$ are orthogonal if

$$\langle\chi|e^{-Q}|\Psi\rangle = 0 \ : \ \text{orthogonal.} \tag{14.88}$$

The norm of a state vector, as exemplified by Eq. 14.38, is given by

$$|\Psi|^2_Q = \langle\Psi|e^{-Q}|\Psi\rangle. \tag{14.89}$$

The definition of the scalar product given in Eq. 14.87 is equivalent to defining the dual vector of $|\Psi\rangle$ by the rule

$$|\Psi\rangle \ \rightarrow \ \text{Dual} \ \rightarrow \ \langle\Psi|e^{-Q} \ : \text{dual vector.} \tag{14.90}$$

In particular, using the definition of the scalar product given in Eqs. 14.87 and 14.90, it follows that all the eigenfunctions $|\Psi_{nm}\rangle$ are orthonormal as shown in Eq. 14.39.

The completeness equation for state space $\mathcal{V}$ is

$$\begin{aligned}\mathbb{I} &= \int_{-\infty}^{+\infty} dxdve^{Q/2}|x,v\rangle\langle x,v|e^{-Q/2} \\ &= \int_{-\infty}^{+\infty} dxdve^{Q/2}|x,v\rangle\Big(\langle x,v|e^{Q/2}\Big)e^{-Q},\end{aligned}$$

with completeness and orthogonality obtained using Eq. 13.65

14.11.1 Operators acting on $\mathcal{V}$

The conjugation of operators in state space $\mathcal{V}$ depends on the state space metric e^{-Q}. Q-conjugation for operator $\mathcal{O}$ is defined by

$$\begin{aligned}\langle\chi|e^{-Q}\mathcal{O}|\Psi\rangle^* &= \langle\Psi|\mathcal{O}^\dagger e^{-Q}|\chi\rangle\\ &= \langle\Psi|e^{-Q}[e^{Q}\mathcal{O}^\dagger e^{-Q}]|\chi\rangle = \langle\Psi|e^{-Q}\mathcal{O}^\dagger_{Q}|\chi\rangle,\end{aligned}$$

where the Q-conjugate of operator $\mathcal{O}$ is defined by

$$\mathcal{O}^\dagger_{Q} = e^{Q}\mathcal{O}^\dagger e^{-Q}$$

and $\mathcal{O}^\dagger$ is the usual Hermitian conjugation.

All operators in state space $\mathcal{V}$ must be Q-conjugated under conjugation. In particular, a Q-self-conjugate operator is given by

$$\begin{aligned}&\mathcal{O}^\dagger_{Q} = e^{Q}\mathcal{O}^\dagger e^{-Q} = \mathcal{O}\\ &\Rightarrow \langle\chi|e^{-Q}\mathcal{O}|\Psi\rangle^* = \langle\Psi|e^{-Q}\mathcal{O}|\chi\rangle: \quad Q\text{-self-conjugate operator.}\end{aligned} \tag{14.91}$$

A Q-self-conjugate operator has real eigenvalues since, from Eq. 14.91, for a normalized eigenfunction $|\Psi\rangle$

$$\begin{aligned}E|\Psi\rangle &= \mathcal{O}|\Psi\rangle, \quad \mathcal{O}^\dagger_{Q} = \mathcal{O}\\ \Rightarrow E\langle\Psi|e^{-Q}|\Psi\rangle &= E = \langle\Psi|e^{-Q}\mathcal{O}|\Psi\rangle,\\ E^* &= \langle\Psi|e^{-Q}\mathcal{O}|\Psi\rangle^* = \langle\Psi|e^{-Q}\mathcal{O}^\dagger_{Q}|\Psi\rangle\\ &= \langle\Psi|e^{-Q}\mathcal{O}|\Psi\rangle = E \quad\Rightarrow E: \ \text{real.}\end{aligned}$$

An obvious but important result is that the state space metric e^{-Q} is Q-conjugate, namely that

$$\begin{aligned}&e^{-Q}_{Q} = e^{Q/2}e^{-Q}e^{-Q/2} = e^{-Q}\\ &\Rightarrow \left(e^{-Q}_{Q}\right)^\dagger = e^{-Q/2}e^{-Q}e^{Q/2} = e^{-Q}: \quad Q\text{-self-conjugate operator.}\end{aligned}$$

The pseudo-Hermitian Hamiltonian H is Q-self-conjugate; from Eq. 14.12

$$H = e^{Q/2}H_o e^{-Q/2}$$

$$\Rightarrow H^\dagger_{Q} = e^{Q}H^\dagger e^{-Q} = e^{Q}e^{-Q/2}H^\dagger_o e^{Q/2}e^{-Q} \tag{14.92}$$

$$= e^{Q/2}H_o e^{-Q/2} = H: \quad Q\text{-self-conjugate operator.} \tag{14.93}$$

All operators acting on state space $\mathcal{V}$ need to be Q-conjugated to obtain the conjugated form of any operator equation. The presence of the operator Q indicates that the operators and state vectors belong to state space $\mathcal{V}$. Examples of operator equations that are only valid on state space $\mathcal{V}$ are given in Eq. 14.8. It can be seen that taking the Hermitian conjugate of any the equations in Eq. 14.8 leads to an inconsistency, whereas taking the Q-conjugate of these same equations leads to a consistent result.

Note that all the properties of the state space $\mathcal{V}$ depend on the operator Q being well behaved. From Eq. 14.10 it can be seen that for $\omega_1 = \omega_2$, the operator Q diverges and hence the Hamiltonian and state space need to be studied from first principles.

To illustrate the role of the Hilbert space metric e^{-Q} consider the time-ordered vacuum expectation value of the Heisenberg operators at two different (Euclidean) times $\tau > 0$. The vacuum state is given by $|\Psi_{00}\rangle$ with $H|\Psi_{00}\rangle = E_{00}|\Psi_{00}\rangle$; $E_{00} \equiv E_0 = (\omega_1 + \omega_2)/2$; using the rule for forming the dual vector of the pseudo-Hermitian Hamiltonian yields

$$\begin{aligned} G(\tau) &= \langle \Psi_{00}^D | x_H(\tau) x_H(0) | \Psi_{00} \rangle = \langle \Psi_{00} | e^{-Q} x_H(\tau) x_H(0) | \Psi_{00} \rangle \\ &= \langle \Psi_{00} | e^{-Q} e^{\tau H} x e^{-\tau H} x | \Psi_{00} \rangle = \langle \Psi_{00} | e^{\tau H^\dagger} e^{-Q} x e^{-\tau H} x | \Psi_{00} \rangle \\ &= \langle \Psi_{00} | e^{-Q} x e^{-\tau (H - E_0)} x | \Psi_{00} \rangle . \end{aligned} \tag{14.94}$$

The term $G(\tau)$ in Eq. 14.94 is the propagator and is analyzed in detail in Section 14.12.

Recall that the probability amplitude is given in Eq. 13.66,

$$\mathcal{K}_S(x, v; x', v') = \langle x, v | e^{-\tau H} | x', v' \rangle ,$$

and the matrix elements of the Hilbert space metric e^{-Q} are given in Eq. 14.23,

$$\langle x, v | e^{-\tau Q} | x', v' \rangle = N \exp \left\{ -\mathcal{G}(\tau)(xv + x'v') + \mathcal{H}(\tau)(xv' + vx') \right\} . \tag{14.95}$$

For both the operators $e^{-\tau H}$ and $e^{-\tau Q}$, there is no need for an extra metric $e^{-\tau Q}$ since Eqs. 13.66 and 14.23 are the matrix elements of the operators in a complete basis $\langle x, v|$ and its dual $|x', v'\rangle$. If the matrix elements of operators $e^{-\tau H}$ and so on are determined for *eigenstates* and *dual eigenstates* of H, then the metric e^{-Q} is required.

One can evaluate the matrix elements of $e^{-\tau H}$ in Hilbert space, using the basis states of $\mathcal{V}$, namely $e^{Q/2}|x, v\rangle$ and the dual basis $\left(\langle x, v| e^{Q/2}\right) e^{-Q} = \langle x, v| e^{-Q/2}$, which yields

$$\langle x, v | e^{-Q/2} e^{-\tau H} e^{Q/2} | x', v' \rangle = \langle x, v | e^{-\tau H_0} | x', v' \rangle . \tag{14.96}$$

The symmetry of the matrix elements of $\langle x, v|e^{-Q/2}e^{-\tau H}e^{Q/2}|x', v'\rangle$ – given above in Eq. 14.96 – is not the symmetry given in Eq. 13.91 for $\langle x, v|e^{-\tau H}|x', v'\rangle$; hence the correct expression for the kernel is given by $\mathcal{K}_S(x, v; x', v') = \langle x, v|e^{-\tau H}|x', v'\rangle$, which in turn yields the correct matrix elements given by the path integral as seen in the results obtained in Section 14.8.

14.11.2 Heisenberg operator equations

In Schrödinger's formulation of quantum mechanics, the time dependence of a quantum system arises solely due to the time evolution of the state vector $|\Psi(t)\rangle$ – with the operators $\mathcal{O}$ being taken to be time-independent. For a pseudo-Hermitian Hamiltonian, the Schrödinger equation yields, for Minkowski time t_M,

$$|\Psi(t)\rangle = \exp\{-it_M H\}|\Psi\rangle, \quad \langle\chi(t)| = \langle\chi|\exp\{it_M H^\dagger\}. \tag{14.97}$$

For Euclidean time $t = it_M$, one has the expressions

$$|\Psi(t)\rangle = e^{-tH}|\Psi\rangle, \quad \langle\chi(t)| = \langle\chi|\, e^{tH^\dagger}.$$

To obtain the Heisenberg operator equations of motion for the pseudo-Hermitian Hamiltonian, it is necessary to define the dual state vectors using Q-conjugation. In particular, for Euclidean time, the time dependent expectation value, using Eq. 14.4, yields the result

$$\begin{aligned} E[\mathcal{O}; t] &= \langle\chi(t)|e^{-Q}\mathcal{O}|\Psi(t)\rangle = \langle\chi|e^{tH^\dagger}e^{-Q}\mathcal{O}e^{-tH}|\Psi\rangle \\ &= \langle\chi|e^{-Q}e^{tH}\mathcal{O}e^{-tH}|\Psi\rangle = \langle\chi|e^{-Q}\mathcal{O}_H(t)|\Psi\rangle. \end{aligned} \tag{14.98}$$

Hence, from Eq. 14.98, due to the choice of the Hilbert space metric e^{-Q}, the Heisenberg time-dependent operator $\mathcal{O}_H(t)$ for the pseudo-Hermitian Hamiltonian is given by the same expression as for the Hermitian Hamiltonian, namely

$$\mathcal{O}_H(t) = e^{tH}\mathcal{O}e^{-tH} \quad \Rightarrow \quad \frac{\partial\mathcal{O}_H(t)}{\partial t} = [H, \mathcal{O}_H(t)].$$

In particular, for the Hamiltonian of the acceleration action given, from Eq. 13.69, by

$$H = -\frac{1}{2\gamma}\frac{\partial^2}{\partial v^2} - v\frac{\partial}{\partial x} + \frac{1}{2}\alpha v^2 + \Phi(x),$$

the Heisenberg equation for the coordinate operator $x_H(t)$, with x being the Schrödinger coordinate operator, yields

$$\begin{aligned} x_H(t) &= e^{tH} x e^{-tH}, \\ \dot{x}_H(t) &= \frac{\partial x_H(t)}{\partial t} = [H, X_H(t)] = -v_H(t). \end{aligned} \tag{14.99}$$

Hence, the identification made in the path integral derivation, namely $\dot{x}(t) = -v(t)$, is seen to hold also as an operator equation for the Heisenberg operators $\dot{x}_H(t)$, $v_H(t)$, as shown by Eq. 14.99.

14.12 Propagator: operators

Constructing the propagator by inserting the complete set of states yields a realization of the propagator in terms of the state space and Hamiltonian. The state space definition of the propagator is given by Eq. 11.6 and yields

$$G(\tau) = \lim_{T \to \infty} \frac{1}{Z} tr\left(e^{-(T-\tau)H} x e^{-\tau H} x\right), \quad \tau = |t - t'|.$$

Note that

$$\lim_{T \to \infty} e^{-TH} \simeq e^{-TE_0} |\Psi_{00}\rangle\langle\Psi_{00}| e^{-Q} = e^{-TE_0} e^{Q/2} |0, 0\rangle\langle 0, 0| e^{-Q/2},$$
$$Z = tr(e^{-TH}) = e^{-TE_0}.$$

Since

$$H = e^{Q/2} H_o e^{-Q/2}, \quad |\Psi_{00}\rangle = e^{Q/2}|0, 0\rangle, \quad \langle\Psi_{00}| = \langle 0, 0| e^{Q/2},$$

the propagator is given by

$$\begin{aligned} G(\tau) &= \lim_{T \to \infty} \frac{1}{Z} tr\left(e^{-(T-\tau)H} x e^{-\tau H} x\right), \quad \tau = |t - t'|, \\ &= \langle\Psi_{00}| e^{-Q} x e^{-\tau(H-E_0)} x |\Psi_{00}\rangle, & (14.100) \\ &= \langle 0, 0| e^{-Q/2} x e^{Q/2} e^{-\tau(H_0-E_0)} e^{-Q/2} x e^{Q/2} |0, 0\rangle. & (14.101) \end{aligned}$$

Note that Eq. 14.100 has been obtained earlier in Eq. 14.94 based on a Hilbert space derivation of the propagator.

From Eq. 14.8,

$$\begin{aligned} e^{-Q/2} x e^{Q/2} &= Ax + BC\frac{\partial}{\partial v} \\ &= A\sqrt{\frac{1}{2\gamma\omega_1^2\omega_2}}(a_x + a_x^\dagger) + BC\sqrt{\frac{\gamma\omega_1}{2}}(a_v - a_v^\dagger). \end{aligned} \tag{14.102}$$

Hence, from Eq. 14.102

$$e^{-Q/2}xe^{Q/2}|0,0\rangle = A\sqrt{\frac{1}{2\gamma\omega_1^2\omega_2}}|0,1\rangle - BC\sqrt{\frac{\gamma\omega_1}{2}}|1,0\rangle,$$

$$\langle 0,0|e^{-Q/2}xe^{Q/2} = A\sqrt{\frac{1}{2\gamma\omega_1^2\omega_2}}\langle 0,1| + BC\sqrt{\frac{\gamma\omega_1}{2}}\langle 1,0|. \tag{14.103}$$

Equation 14.101 yields

$$G(\tau) = A^2\frac{1}{2\gamma\omega_1^2\omega_2}\langle 0,1|e^{-\tau(H_0-E_0)}|0,1\rangle - (BC)^2\frac{\gamma\omega_1}{2}\langle 1,0|e^{-\tau(H_0-E_0)}|1,0\rangle$$
$$= \frac{1}{2\gamma\omega_1^2\omega_2}A^2e^{-\omega_2\tau} - \frac{\gamma\omega_1}{2}(BC)^2e^{-\omega_1\tau}.$$

Note that all the operators and state functions in the equation above are defined solely in state space $\mathcal{V}_0$. However, the coefficients of the various matrix elements, in particular the negative sign on the second matrix element, are a result of the properties of the conjugation operator e^Q and reflect the presence of the underlying state space $\mathcal{V}$; the result obtained could not have been generated by working solely in Hilbert space $\mathcal{V}_0$.

Equation 14.11 yields

$$A^2 = \cosh^2\left(\frac{\sqrt{ab}}{2}\right) = \frac{\omega_1^2}{\omega_1^2-\omega_2^2},$$

$$(BC)^2 = \frac{b}{a}\sinh^2\left(\frac{\sqrt{ab}}{2}\right) = \frac{1}{\gamma^2\omega_1^2\omega_2^2}\cdot\frac{\omega_2^2}{\omega_1^2-\omega_2^2} = \frac{1}{\gamma^2\omega_1^2}\cdot\frac{1}{\omega_1^2-\omega_2^2}.$$

Hence, collecting all the results yields the expected result obtained in Eq. 13.35 that

$$G(\tau) = \frac{1}{2\gamma}\frac{1}{\omega_1^2-\omega_2^2}\left[\frac{e^{-\omega_2\tau}}{\omega_2} - \frac{e^{-\omega_1\tau}}{\omega_1}\right]. \tag{14.104}$$

14.13 Propagator: state space

Recall from Eq. 14.100, that the propagator is given by

$$G(\tau) = \langle\Psi_{00}|e^{-Q}xe^{-\tau(H-E_0)}x|\Psi_{00}\rangle$$
$$= \langle\Psi_{00}^D|xe^{-\tau(H-E_0)}x|\Psi_{00}\rangle. \tag{14.105}$$

The completeness equation for H, from Eq. 14.40, is given by

$$\mathbb{I} = \sum_{m,n=1}^{\infty} |\Psi_{mn}\rangle\langle\Psi^D_{mn}|, \tag{14.106}$$

and yields, from Eq. 14.105,

$$\begin{aligned}
G(\tau) &= \sum_{m,n=1}^{\infty} \langle\Psi^D_{00}|xe^{-\tau(H-E_0)}|\Psi_{mn}\rangle\langle\Psi^D_{mn}|x|\Psi_{00}\rangle \\
&= e^{-\tau\omega_1}\langle\Psi^D_{00}|x|\Psi_{10}\rangle\langle\Psi^D_{10}|x|\Psi_{00}\rangle + e^{-\tau\omega_2}\langle\Psi^D_{00}|x|\Psi_{01}\rangle\langle\Psi^D_{01}|x|\Psi_{00}\rangle, \qquad (14.107) \\
&= e^{-\tau\omega_1}G_1 + e^{-\tau\omega_2}G_2. \qquad (14.108)
\end{aligned}$$

The vacuum state and its normalization, from Eqs. 14.50, 14.55, and 14.56 is

$$\begin{aligned}
&\Psi_{00}(x,v) = N_{00}\psi_{00}(x,v), \quad \Psi^D_{00}(x,v) = \Psi_{00}(x,-v) = \Psi_{00}(-x,v), \\
&\psi_{00}(x,v) = \exp\{-\frac{\gamma}{2}(\omega_1+\omega_2)\omega_1\omega_2 x^2 - \frac{\gamma}{2}(\omega_1+\omega_2)v^2 - \gamma\omega_1\omega_2 xv\}, \\
&N_{00} = (\omega_1\omega_2)^{1/4}\sqrt{\frac{\gamma}{\pi}(\omega_1+\omega_2)}.
\end{aligned}$$

Recall from Eq. 14.81,

$$\begin{aligned}
&\Psi_{10}(x,v) = N_{10}\,[v+\omega_2 x]\,\psi_{00}(x,v), \quad \Psi^D_{10}(x,v) = N_{10}\,[v-\omega_2 x]\,\psi^D_{00}(x,v), \\
&\Psi_{01}(x,v) = N_{01}\,[v+\omega_1 x]\,\psi_{00}(x,v), \quad \Psi^D_{01}(x,v) = N_{01}\,[-v+\omega_1 x]\,\psi^D_{00}(x,v), \\
&N_{10} = \gamma\sqrt{2}\frac{(\omega_1+\omega_2)}{\sqrt{\pi(\omega_1-\omega_2)}}\,\omega_1^{3/4}\omega_2^{1/4}, \quad N_{01} = \gamma\sqrt{2}\frac{(\omega_1+\omega_2)}{\sqrt{\pi(\omega_1-\omega_2)}}\,\omega_1^{1/4}\omega_2^{3/4}.
\end{aligned}$$

Using the coordinate representation for the state vectors yields

$$\begin{aligned}
G_1 &= \langle\Psi^D_{00}|x|\Psi_{10}\rangle\langle\Psi^D_{10}|x|\Psi_{00}\rangle \\
&= N^2_{10}N^2_{00}\int dxdv\, x(v+\omega_2 x)\psi^D_{00}(x,v)\psi_{00}(x,v) \\
&\quad\times\int dxdv\, x(v-\omega_2 x)\psi^D_{00}(x,v)\psi_{00}(x,v) \\
&= -N^2_{10}N^2_{00}\omega_2^2\left[\frac{\pi}{2\gamma(\omega_1+\omega_2)^2(\omega_1\omega_2)^{3/2}}\right]^2 \\
&= -\frac{1}{2\gamma(\omega_1^2-\omega_2^2)^2\omega_1}. \qquad (14.109)
\end{aligned}$$

Similarly

$$
\begin{aligned}
G_2 &= \langle \Psi_{00}^D | x | \Psi_{01} \rangle \langle \Psi_{01}^D | x | \Psi_{00} \rangle \\
&= N_{10}^2 N_{00}^2 \int dx dv\, x(v + \omega_1 x) \psi_{00}^D(x, v) \psi_{00}(x, v) \\
&\quad \times \int dx dv\, x(-v + \omega_1 x) \psi_{00}^D(x, v) \psi_{00}(x, v) \\
&= N_{10}^2 N_{00}^2 \omega_1^2 \left[\frac{\pi}{2\gamma(\omega_1 + \omega_2)^2 (\omega_1 \omega_2)^{3/2}} \right]^2 \\
&= \frac{1}{2\gamma(\omega_1^2 - \omega_2^2)^2 \omega_2}.
\end{aligned} \tag{14.110}
$$

Hence, Eqs. 14.108, 14.109, and 14.110 yield the expected result given in Eq. 14.104, namely that

$$
\begin{aligned}
G(\tau) &= \sum_{m,n=1}^{\infty} \langle \Psi_{00}^D | x e^{-\tau(H-E_0)} | \Psi_{mn} \rangle \langle \Psi_{mn}^D | x | \Psi_{00} \rangle \\
&= e^{-\tau\omega_1} \langle \Psi_{00}^D | x | \Psi_{10} \rangle \langle \Psi_{10}^D | x | \Psi_{00} \rangle + e^{-\tau\omega_2} \langle \Psi_{00}^D | x | \Psi_{01} \rangle \langle \Psi_{01}^D | x | \Psi_{00} \rangle \\
&= e^{-\tau\omega_1} G_1 + e^{-\tau\omega_2} G_2 \\
&= \frac{1}{2\gamma} \frac{1}{\omega_1^2 - \omega_2^2} \left[-\frac{e^{-\omega_1 \tau}}{\omega_1} + \frac{e^{-\omega_2 \tau}}{\omega_2} \right].
\end{aligned} \tag{14.111}
$$

There are a number of remarkable features of the state space derivation. The negative sign that appears in the propagator for the term G_1 is usually taken to be a proof that no unitary theory can yield this result. The reason for this is the following: consider any arbitrary Hermitian Hamiltonian such that $H_A = H_A^\dagger$; the spectral resolution of this Hamiltonian in terms of its eigenstates $|\chi_{mn}\rangle$ is given by

$$
\mathbb{I} = \sum_{mn=1}^{\infty} |\chi_{mn}\rangle \langle \chi_{mn}|. \tag{14.112}
$$

Note that $\langle \chi_{mn}| = \langle \chi_{mn}^D|$, since for a Hermitian Hamiltonian H_A the left and right eigenstates are the dual state vectors of each other. Hence, the propagator for the Hermitian Hamiltonian H_A is given by

$$
\begin{aligned}
G_A(\tau) &= \sum_{m,n=1}^{\infty} \langle \chi_{00}^D | x e^{-\tau(H_A-E_0)} | \chi_{mn} \rangle \langle \chi_{mn}^D | x | \chi_{00} \rangle \\
&= e^{-\tau\omega_1} \langle \chi_{00}^D | x | \chi_{10} \rangle \langle \chi_{10}^D | x | \chi_{00} \rangle + e^{-\tau\omega_2} \langle \chi_{00}^D | x | \chi_{01} \rangle \langle \chi_{01}^D | x | \chi_{00} \rangle \\
&= e^{-\tau\omega_1} \left| \langle \chi_{00} | x | \chi_{10} \rangle \right|^2 + e^{-\tau\omega_2} \left| \langle \chi_{00} | x | \chi_{01} \rangle \right|^2.
\end{aligned}
$$

The result above shows that a Hermitian Hamiltonian defined on a Hilbert space cannot have a propagator such as the one given in Eq. 14.104, except by allowing $|\langle\chi_{00}|x|\chi_{10}\rangle|^2 < 0$, which implies that $|\chi_{10}\rangle$ is a ghost state that has a negative norm.

In contrast, the pseudo-Hermitian Euclidean Hamiltonian H has a positive norm for all the states in its state space; the *duality transformation* in going from $|\Psi_{mn}\rangle$ to $\langle\Psi^D_{mn}|$ provides the negative signs that allow for the propagator given in Eqs. 14.104 and 14.111.

14.14 Many degrees of freedom

Consider the generalization of a Hamiltonian with a quadratic potential given in Eq. 13.69 to many degrees of freedom. For degrees of freedom x_n, $n = 1, 2, \ldots N$ the acceleration action, in matrix notation, is chosen to be

$$H = -\frac{1}{2}\frac{\partial}{\partial v}S^T\frac{1}{\gamma}S\frac{\partial}{\partial v} - v\frac{\partial}{\partial x} + \frac{1}{2}vS^T\alpha S v + \frac{1}{2}xS^T\beta S x, \qquad (14.113)$$

that yields the Lagrangian

$$\mathcal{L} = -\frac{1}{2}\left[\ddot{x}S^T\gamma S\ddot{x} + \dot{x}S^T\alpha S\dot{x} + xS^T\beta S x\right]. \qquad (14.114)$$

A very special choice has been made for H and $\mathcal{L}$ in that all the couplings have the same similarity transformation S connecting the different degrees of freedom. In particular, choosing $S = \mathbb{I}$ would decouple all the different degrees of freedom.

The (real) orthogonal matrix S and the diagonal matrices are given in matrix notation as

$$\begin{aligned} SS^T &= \mathbb{I}, \ \gamma = \text{diag}(\gamma_1, \gamma_2, \ldots, \gamma_N), \\ \alpha &= \text{diag}(\alpha_1, \alpha_2, \ldots, \alpha_N), \ \ \beta = \text{diag}(\beta_1, \beta_2, \ldots, \beta_N). \end{aligned} \qquad (14.115)$$

Let us define new variables, in matrix notation

$$x = S^T z, \ \ v = S^T u. \qquad (14.116)$$

The Hamiltonian in Eq. 14.113 is given by

$$H = -\frac{1}{2}\sum_{n=1}^{N}\frac{1}{\gamma_n}\frac{\partial^2}{\partial u_n^2} - \sum_{n=1}^{N}u_n\frac{\partial}{\partial z_n} + \frac{1}{2}\sum_{n=1}^{N}\alpha_n u_n^2 + \frac{1}{2}\sum_{n=1}^{N}\beta_n z_m^2. \qquad (14.117)$$

The many degrees of freedom Lagrangian, from Eq. 14.114 and similar to Eq. 13.29, is given by

$$\mathcal{L} = -\frac{1}{2}\sum_{n=1}^{N}\gamma_n\left[\ddot{z}_n^2 + (\omega_{1n}^2 + \omega_{2n}^2)\dot{z}_n^2 + \omega_{1n}^2\omega_{2n}^2 z_n^2\right]. \qquad (14.118)$$

The parameterization is a generalization of Eq. 13.30 and is given by

$$\omega_{1n} = \frac{1}{2\sqrt{\gamma_n}}\left(\sqrt{\alpha_n + 2\sqrt{\gamma_n\beta_n}} + \sqrt{\alpha_n - 2\sqrt{\gamma_n\beta_n}}\right), \tag{14.119}$$

$$\omega_{2n} = \frac{1}{2\sqrt{\gamma_n}}\left(\sqrt{\alpha_n + 2\sqrt{\gamma_n\beta_n}} - \sqrt{\alpha_n - 2\sqrt{\gamma_n\beta_n}}\right), \tag{14.120}$$

$$\omega_{1n} > \omega_{2n} \text{ for } \omega_{1n},\ \omega_{2n} \text{ real.}$$

The Hamiltonian H given in Eq. 14.113 is diagonalized by the following generalization of the Q-operator given in Eq. 14.5:

$$Q = \sum_{n=1}^{N} Q_n, \tag{14.121}$$

$$Q_n = a_n z_n u_n - b_n \frac{\partial^2}{\partial z_n \partial u_n}, \tag{14.122}$$

with the following values for a_n and b_n:

$$\sqrt{\frac{a_n}{b_n}} = \gamma_n\omega_{1n}\omega_{2n}, \quad \sqrt{a_n b_n} = \ln\left(\frac{\omega_{1n} + \omega_{2n}}{\omega_{1n} - \omega_{2n}}\right). \tag{14.123}$$

The diagonal Hamiltonian H_o is given similarly to the earlier case. The ground state is given by generalizing Eq. 14.51 and yields

$$\Psi_{00}(x, v) =$$
$$\mathcal{N}\prod_{n=1}^{N}\exp\left\{-\frac{\gamma_n}{2}(\omega_{1n} + \omega_{2n})\omega_{1n}\omega_{2n}z_n^2 - \frac{\gamma_n}{2}(\omega_{1n} + \omega_{2n})u_n^2 - \gamma_n\omega_{1n}\omega_{2n}z_n u_n\right\}.$$

In terms of the original coordinates,

$$\Psi_{00}(x, v) = \mathcal{N}\exp\left\{-\frac{1}{2}\sum_{m,n=1}^{N}\left(P_{mn}x_m x_n + Q_{mn}v_m v_n + 2R_{mn}x_n v_n\right)\right\}, \tag{14.124}$$

where, again in matrix notation

$$P = SpS^T, \quad p_n = \gamma_n(\omega_{1n} + \omega_{2n})\omega_{1n}\omega_{2n},$$
$$Q = SqS^T, \quad q_n = \gamma_n(\omega_{1n} + \omega_{2n}),$$
$$R = SrS^T, \quad r_n = \gamma_n\omega_{1n}\omega_{2n}.$$

14.15 Summary

The Euclidean acceleration Hamiltonian, for the real branch, was shown to be pseudo-Hermitian and was mapped to a Hermitian Hamiltonian, which consists of two decoupled harmonic oscillators. The similarity transformation Q was shown to be an unbounded differential operator and the matrix elements of $e^{\pm\tau Q}$ were exactly evaluated. All the defining commutation equations were evaluated using the matrix elements of $e^{\pm\tau Q}$ to confirm the result obtained.

The state space of the pseudo-Hermitian Euclidean Hamiltonian has a state space metric that is a natural generalization of the state space of quantum mechanics. The Heisenberg operator equations were analyzed to conclude that a state space metric is required for making the theory consistent and leads to a generalized scalar product and to the concept of Q-conjugation discussed by Mostafazadeh (2002).

The state vector and its dual vector were analyzed. It was shown that the symmetry of the evolution kernel obtained using the Hamiltonian, Lagrangian and classical action can be derived using the parity symmetry of the Hamiltonian and duality property of the state functions.

The propagator was evaluated using the algebra of the creation and destruction operators. The first two excited states were computed and the propagator was evaluated a second time using the properties of state space. It was seen that the propagator has a form that is forbidden for a Hermitian Hamiltonian and exists for the acceleration Hamiltonian due to the properties of Q-conjugation required for a positive norm state space.

The state vector $\Psi(x, v)$ does not yield a probability density, as discussed in Section 4.9. Hence the state vector of the acceleration Hamiltonian requires an interpretation quite distinct from the standard quantum case.

As was seen in many of the derivations in this chapter, the results from Minkowski time serve as a useful guide for the derivations; but – given a plethora of i and various $+$ and $-$ signs that differ between the Euclidean and Minkowski results – all the derivations for the Euclidean have to be done from first principles and independently from the Minkowski case.

15

Non-Hermitian Hamiltonian: Jordan blocks

The Euclidean action with acceleration has been analyzed in Chapters 13 and 14 for its path integral and Hamiltonian. In this chapter, the acceleration Hamiltonian is analyzed for the case when it is essentially inequivalent to a Hermitian Hamiltonian: for critical values of the parameters, given by $\omega_1 = \omega_2$ and shown in Figure 13.1, the mapping of the Hamiltonian to an equivalent Hermitian Hamiltonian becomes singular. The Hamiltonian continues to be pseudo-Hermitian but is no longer equivalent to a Hermitian Hamiltonian.

The Hamiltonian for real $\omega_1 > \omega_2$ is pseudo-Hermitian as well as being equivalent to a Hermitian operator $H_0 = H_0^\dagger$, due to the existence of a similarity transformation Q such that

$$H = e^{-Q/2} H_0 e^{Q/2} \Rightarrow H^\dagger = e^{-Q} H e^{Q}, \quad \omega_1 > \omega_2.$$

For the case $\omega_1 = \omega_2$ the coefficients in Q diverge and the Hamiltonian H can no longer be mapped to a Hermitian Hamiltonian. The Hamiltonian is essentially non-Hermitian and is shown in this chapter to be equal to a direct sum of block diagonal matrices, with each block being a Jordan block matrix. The Jordan block itself can be shown to be pseudo-Hermitian, but the transformation cannot be obtained from the diverging Q. The case of the Hamiltonian being a Jordan block is analyzed in detail in this chapter as it forms a quantum mechanical system with its own unique features that is not encountered in Hermitian quantum mechanics.

In Section 15.1 the equal frequency Hamiltonian is obtained; in Sections 15.2 and 15.3 propagator is analyzed and the two lowest lying state vectors for the singular case are evaluated. In Section 15.5 the equal frequency propagator is evaluated using the state vectors of the Jordan block. In Sections 15.6 and 15.7 the Hamiltonian for the Jordan block is derived and the Schrödinger equation for this system is studied.

15.1 Hamiltonian: equal frequency limit

The *critical Hamiltonian* is given for the equal frequency limit. In the equal frequency limit of $\omega_1 = \omega_2$ ($\alpha = 2\sqrt{\beta\gamma}$) the parameters of the Q-operator, namely a and b, given in Eq. 14.10 become divergent and a well defined Q-operator no longer exists. Although the Hamiltonian has special properties for the equal-frequency point, the path integral is well behaved for all (complex and real) values of ω_1, ω_2, including the equal-frequency critical point at $\omega_1 = \omega_2$. Moreover, the non-Hermitian Hamiltonian H remains well defined at the critical point.

The singularity for the Q-operator is due to the fact that the acceleration Hamiltonian H *cannot* be mapped to an equivalent Hermitian Hamiltonian H_0. For the case of $\omega_1 = \omega_2$, non-Hermitian Hamiltonian H is pseudo-Hermitian but is essentially non-Hermitian and has been shown to be expressible as a Jordan-block matrix by Bender and Mannheim (2008a).

The general analysis of the equal frequency Hamiltonian has been carried out for Minkowski time in the pioneering work of Bender and Mannheim (2008a) and the analysis for Euclidean time is similar to their analysis, but with many details that are quite different.

15.2 Propagator and states for equal frequency

To illustrate the general features of the equal frequency limit, the propagator is analyzed from the point of view of the underlying state space. As mentioned at the end of Section 14.9, in the limit of $\omega_1 = \omega_2$ the single excitation eigenstates Ψ_{10}, Ψ_{01} become degenerate, with both eigenstates having energy 2ω.

The purpose of analyzing the propagator is to extract the state vectors that emerge in the limit of $\omega_1 \to \omega_2$.

Since $\omega_1 > \omega_2$, consider the limit of $\epsilon \to 0+$ with

$$\omega_1 = \omega + \epsilon, \quad \omega_2 = \omega_2 - \epsilon, \tag{15.1}$$

which yields, from Eq. 14.32

$$E_{00} = \frac{1}{2}(\omega_1 + \omega_2) \to \omega, \quad E_{10} \to 2\omega + \epsilon, \quad E_{01} \to 2\omega - \epsilon.$$

Consider the limit of $\omega_1 \to \omega_2$ for the state vector expansion of the propagator given by Eq. 14.107,

$$\begin{aligned} G(\tau) &= e^{-\tau\omega_1}\langle\Psi^D_{00}|x|\Psi_{10}\rangle\langle\Psi^D_{10}|x|\Psi_{00}\rangle + e^{-\tau\omega_2}\langle\Psi^D_{00}|x|\Psi_{01}\rangle\langle\Psi^D_{01}|x|\Psi_{00}\rangle \\ &= e^{-\tau\omega}[G_{10} + G_{01}], \end{aligned} \tag{15.2}$$

where, defining $\int dxdvdx'dv' = \int_{x,v,x',v'}$ yields

$$\begin{aligned} G_{10} &= e^{-\epsilon\tau}\langle\Psi^D_{00}|x|\Psi_{10}\rangle\langle\Psi^D_{10}|x|\Psi_{00}\rangle \\ &= e^{-\epsilon\tau}N^2_{00}N^2_{10}\int_{x,v,x',v'} x(v+\omega_2 x)\psi^D_{00}(x,v)\psi_{00}(x,v) \\ &\quad\times x'(v'-\omega_2 x')\psi^D_{00}(x',v')\psi_{00}(x',v') \\ &= e^{-\epsilon\tau}N^2_{00}N^2_{10}\int_{x,v,x',v'} xx'(v+\omega_2 x)(v'-\omega_2 x')P(x,v)P(x',v') \end{aligned} \tag{15.3}$$

and

$$\begin{aligned} G_{01} &= e^{\epsilon\tau}\langle\Psi^D_{00}|x|\Psi_{01}\rangle\langle\Psi^D_{01}|x|\Psi_{00}\rangle \\ &= e^{-\epsilon\tau}N^2_{00}N^2_{01}\int_{x,v,x',v'} x(v+\omega_1 x)\psi^D_{00}(x,v)\psi_{00}(x,v) \\ &\quad\times x'(-v'+\omega_1 x')\psi^D_{00}(x',v')\psi_{00}(x',v') \\ &= e^{-\epsilon\tau}N^2_{00}N^2_{01}\int_{x,v,x',v'} xx'(v+\omega_1 x)\,(-v'+\omega_1 x')P(x,v)P(x',v'). \end{aligned} \tag{15.4}$$

In the limit of $\omega_1 \to \omega_2$ the vacuum state has the following well-defined limit given by

$$\lim_{\epsilon\to 0}\psi_{00}(x,v) = \hat{\psi}_{00}(x,v) = \exp\{-\gamma\omega^3 x^2 - \gamma\omega v^2 - \gamma\omega^2 xv\} \tag{15.5}$$

and yields

$$P(x,v) = \hat{\psi}^D_{00}(x,v)\hat{\psi}_{00}(x,v) = \exp\{-2\gamma\omega^3 x^2 - 2\gamma\omega v^2\}.$$

To leading order in ϵ the normalization constants yield the following:[1]

$$N^2_{10} \to C\frac{\omega_1}{\epsilon},\quad N^2_{10} \to C\frac{\omega_2}{\epsilon},\quad C = \frac{4\gamma^2\omega^3}{\pi}, \tag{15.6}$$

and from Eq. 14.55

$$\hat{N}^2_{00} = \lim_{\epsilon\to 0} N^2_{00} = \frac{2\gamma\omega^2}{\pi}. \tag{15.7}$$

Hence, collecting the above equations,

$$G_{10}+G_{01} = C\hat{N}^2_{00}\int_{x,v,x',v'} xx'\mathcal{F}(x,v;x',v')P(x,v)P(x',v'), \tag{15.8}$$

where

$$\mathcal{F}(x,v;x',v') = e^{-\epsilon\tau}\frac{\omega_1}{\epsilon}(v+\omega_2 x)(v'-\omega_2 x') + e^{\epsilon\tau}\frac{\omega_2}{\epsilon}(v+\omega_1 x)(-v'+\omega_1 x'). \tag{15.9}$$

[1] The definition of the constant C in this chapter is different from the constant with the same notation used in Chapter 14.

Expanding $\mathcal{F}(x, v; x', v')$ to leading order in ϵ yields

$$\begin{aligned}\mathcal{F}(x, v; x', v') &= \frac{1}{\epsilon}\Big[(1-\epsilon\tau)(\omega+\epsilon)\{v+(\omega-\epsilon)x\}\{v'-(\omega-\epsilon)x'\} \\ &\quad + (1+\epsilon\tau)(\omega-\epsilon)\{v+(\omega+\epsilon)x\}\{-v'+(\omega+\epsilon)x'\}\Big] + O(\epsilon) \\ &= 2\Big[(\omega\tau-1)(v+\omega x)(-v'+\omega x')+\omega x(-v'+\omega x') \\ &\quad + \omega(v+\omega x)x'\Big] + O(\epsilon) \\ &= 2\Big[-(v+\omega x)(-v'+\omega x')+\omega\{x+\tau(v+\omega x)\}(-v'+\omega x') \\ &\quad + \omega(v+\omega x)x'\Big]. \end{aligned} \tag{15.10}$$

The expression for $\mathcal{F}(x, v; x', v')$ given in Eq. 15.10 carries information on the state vectors that determine the propagator; to extract this information, the state vectors need to read off from the equation. Recall that the state vectors and their duals are polynomials of x, v multiplied into the vacuum state.

Hence, Eqs. 15.2, 15.8, and 15.10 yield the following [the factor of $e^{-2\omega\tau}$ has been included in the definitions of the state functions for convenience later]:

$$\psi_1(x, v; \tau) = \langle x, v|\psi_1(\tau)\rangle = (v+\omega x)\hat{\psi}_{00}(x, v)e^{-2\omega\tau}, \tag{15.11}$$
$$\psi_1^D(x, v; \tau) = \langle \psi_1^D(\tau)|x, v\rangle = (-v+\omega x)\hat{\psi}_{00}^D(x, v)e^{-2\omega\tau}, \tag{15.12}$$
$$\psi_2(x, v; \tau) = \langle x, v|\psi_2(\tau)\rangle = \omega\{x+\tau(v+\omega x)\}\hat{\psi}_{00}(x, v)e^{-2\omega\tau}, \tag{15.13}$$
$$\psi_2^D(x, v; \tau) = \langle \psi_2^D(\tau)|x, v\rangle = \omega\{x+\tau(-v+\omega x)\}\hat{\psi}_{00}^D(x, v)e^{-2\omega\tau}. \tag{15.14}$$

The dual state vector is defined by $v \to -v$, namely[2]

$$\psi_1^D(x, v; \tau) = \psi_1(x, -v; \tau), \quad \psi_2^D(x, v; \tau) = \psi_2(x, -v; \tau).$$

Note that the subtlety of conjugation for the unequal frequency case – with a different rule for each excited state as given in Eq. 14.82 – has been lost for the equal frequency case since the two excited states have become degenerate.

Collecting the results from Eqs. 15.2, 15.8, 15.9–15.14 yields

$$\begin{aligned} G(\tau) = 2C\hat{N}_{00}^2 e^{\omega\tau}\langle\hat{\psi}_{00}^D|x\Big[&-|\psi_1(\tau)\rangle\langle\psi_1^D(0)| \\ &+ |\psi_2(\tau)\rangle\langle\psi_1^D(0)| + |\psi_1(\tau)\rangle\langle\psi_2^D(0)|\Big]x|\hat{\psi}_{00}\rangle. \end{aligned} \tag{15.15}$$

[2] Bender and Mannheim (2008a), for the case for Minkowski time, define the dual state vector by $x \to -x$, which gives an incorrect result for Euclidean time.

The result in Eq. 15.15 shows that the eigenstates $|\Psi_{10}\rangle$, $|\Psi_{01}\rangle$ that gave the result for the propagator in Eq. 15.2 have been replaced, in the limit of $\omega_1 \to \omega_2$, by new state vectors that are well defined and finite for $\epsilon = 0$.

In state vector notation, Eqs. 15.2 and 15.15 yield

$$\lim_{\omega_1 \to \omega_2} \quad e^{-\tau E_{10}} |\Psi_{10}\rangle\langle\Psi_{10}^D| + e^{-\tau E_{01}} |\Psi_{00}^D\rangle\langle\Psi_{01}^D|$$
$$= 2C\hat{N}_{00}^2 e^{\omega\tau}\Big[- |\psi_1(\tau)\rangle\langle\psi_1^D(0)| + |\psi_2(\tau)\rangle\langle\psi_1^D(0)| + |\psi_1(\tau)\rangle\langle\psi_2^D(0)|\Big]. \tag{15.16}$$

15.3 State vectors for equal frequency

The Hamiltonian for the equal frequency case, from Eq. 14.1, is given by

$$H = -\frac{1}{2\gamma}\frac{\partial^2}{\partial v^2} - v\frac{\partial}{\partial x} + \omega^2 v^2 + \frac{\gamma}{2}\omega^4 x^2. \tag{15.17}$$

The vacuum state is an energy eigenstate with

$$H\hat{\psi}_{00}(x, v) = \omega\hat{\psi}_{00}(x, v), \quad \langle\hat{\psi}_{00}^D|\hat{\psi}_{00}\rangle = \frac{1}{\hat{N}_{00}^2} = \frac{\pi}{2\gamma\omega^2}. \tag{15.18}$$

The state vectors $|\psi_1(\tau)\rangle$, $|\psi_2(\tau)\rangle$ were obtained by analyzing the equal frequency propagator. The state vectors have the interpretation set out below.

15.3.1 State vector $|\psi_1(\tau)\rangle$

The state vector $|\psi_1(\tau)\rangle$ is an energy eigenstate given by the average of the two unequal frequency eigenstates that become degenerate, namely

$$\psi_1(x, v; \tau) = \lim_{\epsilon \to 0}\frac{1}{2}[e^{-\tau E_{10}}\psi_{10}(x, v) + e^{-\tau E_{01}}\psi_{01}(x, v)]$$
$$\Rightarrow \psi_1(x, v; \tau) \equiv \langle x, v|\psi_1(\tau)\rangle = e^{-2\tau\omega}(v + \omega x)\hat{\psi}_{00}(x, v)$$
$$H\psi_1(x, v; \tau) = 2\omega\psi_1(x, v; \tau). \tag{15.19}$$

The first sign of the irreducible non-Hermitian nature of the equal frequency Hamiltonian appears with $|\psi_1(\tau)\rangle$; unlike the norm of all the energy eigenstates, the norm of $|\psi_1(\tau)\rangle$ is zero; namely

$$\langle\psi_1^D(\tau)|\psi_1(\tau)\rangle = e^{-4\omega\tau}\hat{N}_{00}^2\int dxdv(-v + \omega x)(v + \omega x)\hat{\psi}_{00}^D(x, v)\hat{\psi}_{00}(x, v)$$
$$\Rightarrow \langle\psi_1^D(\tau)|\psi_1(\tau)\rangle = 0. \tag{15.20}$$

The norm of the eigenstate being zero is a general feature of a Hamiltonian that is of the form of a Jordan-block and, in particular, is not equivalent to a Hermitian

Hamiltonian (Bender and Mannheim, 2008a). The fact that the eigenstate has zero norm does not prevent the eigenstate being included in the collection of state vectors that, taken together, yields a resolution of the identity operator.

15.3.2 State vector $|\psi_2(\tau)\rangle$

The second state vector $|\psi_2(\tau)\rangle$ that appears for the equal frequency case can be written as the difference of the two unequal frequency eigenstates that become degenerate; for dimensional consistency, the pre-factor of ω is introduced in the $\epsilon \to 0$; hence

$$\begin{aligned}
\psi_2(x, v; \tau) &= \lim_{\epsilon\to 0} \frac{\omega}{2\epsilon}\left[e^{-\tau E_{01}}\psi_{01}(x, v) - e^{-\tau E_{10}}\psi_{10}(x, v)\right] \\
&= \frac{\omega}{2\epsilon} e^{-2\omega\tau}\left[(1+\epsilon\tau)(v+(\omega+\epsilon)x) - (1-\epsilon\tau)(v+(\omega-\epsilon)x)\right] \\
&\quad \times \hat{\psi}_{00}(x, v) \\
\Rightarrow \psi_2(x, v; \tau) &= \langle x, v|\psi_2(\tau)\rangle = e^{-2\tau\omega}\omega\left[x + \tau(v+\omega x)\right]\hat{\psi}_{00}(x, v). \qquad (15.21)
\end{aligned}$$

Time-dependent state vector $|\psi_2(\tau)\rangle$ is not an (energy) eigenstate of H; however, since it results from the superposition of two energy eigenstates, it can be explicitly verified that $|\psi_2(\tau)\rangle$ satisfies the time dependent Schrödinger equation, namely

$$-\frac{\partial\psi_2(x, v; \tau)}{\partial\tau} = H\psi_2(x, v; \tau) \;\Rightarrow\; \psi_2(x, v; \tau) = \exp\{-\tau H\}\psi_2(x, v; 0).$$

$$\text{Initial value} \quad \psi_2(x, v; 0) = \omega x\hat{\psi}_{00}(x, v). \qquad (15.22)$$

Note that $|\psi_2(\tau)\rangle$ has a finite norm and a nonzero overlap with $|\psi_1(\tau)\rangle$; namely, using Eq. 15.6,

$$\langle\psi_2^D(\tau)|\psi_2(\tau)\rangle = \frac{e^{-4\tau\omega}}{2C} = \langle\psi_2^D(\tau)|\psi_1(\tau)\rangle, \;\; C = \frac{4\gamma^2\omega^3}{\pi}. \qquad (15.23)$$

The equal frequency state space continues to have a nonnegative norm; in particular, the norm of the time dependent state $|\psi_2(\tau)\rangle$ is positive definite. Of course, since one is working in Euclidean time, probability is not conserved and one can see from Eq. 15.23 that the norm of the states decays exponentially to zero.

In summary, on taking the equal frequency limit, the two energy eigenstates $|\Psi_{10}\rangle, |\Psi_{01}\rangle$ coalesce to yield a single energy eigenstate $|\psi_1(\tau)\rangle$; a second *time dependent* state $|\psi_2(\tau)\rangle$ appears in this limit, also from the two eigenstates, and takes the place of the loss of one of the eigenstates.

An analysis similar to that carried out for the single excitation level holds for all levels. At each level, all the energy eigenstates collapse into a single energy eigenstate of the equal frequency Hamiltonian with eigenenergy $N\omega$. The eigenstates

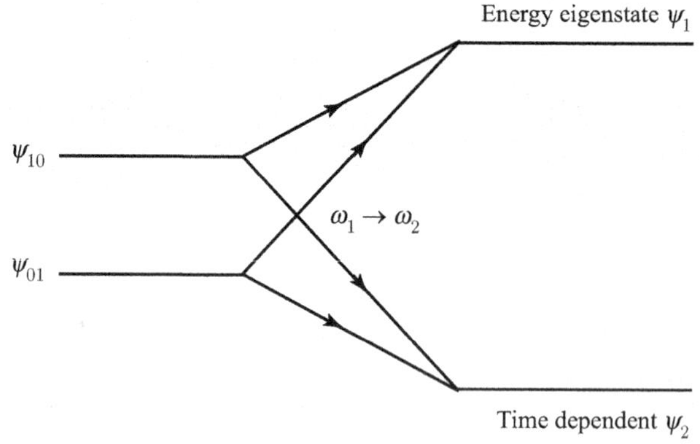

Figure 15.1 The equal frequency limit yields two new states from two energy eigenstates.

that are "lost" are replaced by time-dependent state vectors that are a superposition of the eigenstates of the unequal frequency Hamiltonian. The time-dependent states together with the single eigenstate provide a resolution of the identity. This structure of the equal frequency state space is illustrated in Figure 15.1.

15.4 Completeness equation for 2 x 2 block

We now discuss how the time-dependent state replaces the lost energy eigenstate to provide the complete set of states for the equal frequency case.

The example of the *single excitation* states, created by applying creation operator $a_y^\dagger$ or $a_x^\dagger$ to the harmonic oscillator vacuum state $|0, 0\rangle$, showed that in the limit of $\omega_1 = \omega_2$ the two energy eigenstates $|\Psi_{10}\rangle, |\Psi_{01}\rangle$ were superposed to create new states $|\psi_1(\tau)\rangle, |\psi_2(\tau)\rangle$.

Since the orthogonality of the eigenstates is maintained in the superposition, the mixing of states is only amongst states of a fixed excitation; in other words, states having *two excitations* consisting of applying the creation operator twice, namely $(a_y^\dagger)^2$, $(a_x^\dagger)^2$, or $a_y^\dagger a_x^\dagger$ yield three eigenstates that, under time evolution, only mix with each other in the limit of $\omega_1 = \omega_2$, and so on for all the higher excitation states.

Hence, the resolution of the identity – which is an expression of the completeness of a set of basis states – as shown in Figure 15.2, breaks up into a block-diagonal form, with states of a given excitation mixing, under time evolution, with each other and not with the states of the other blocks.

To illustrate the general result, consider the 2×2 block for the single excitation states. In light of the result obtained in Eq. 15.15, consider the following Hermitian ansatz for the 2×2 block identity operator, with all the state vectors taken at initial time $\tau = 0$. For notational simplicity, let

$$|\psi_1(0)\rangle = |\psi_1\rangle = (v + \omega x)|\hat{\psi}_{00}\rangle, \quad \langle\psi_1^D(0)| = \langle\psi_1^D|, \tag{15.24}$$

$$|\psi_2(0)\rangle = |\psi_2\rangle = x|\hat{\psi}_{00}\rangle, \quad \langle\psi_2^D(0)| = \langle\psi_2^D|. \tag{15.25}$$

Then the identity operator, which is Hermitian, has the following representation for the 2×2 block of Hilbert space:

$$\mathbb{I}_{2\times2} = -P|\psi_1\rangle\langle\psi_1^D| + Q\Big[|\psi_2\rangle\langle\psi_1^D| + |\psi_1\rangle\langle\psi_2^D|\Big] = \mathbb{I}^\dagger_{2\times2}. \tag{15.26}$$

Recall from Eqs. 15.20 and 15.23,

$$\langle\psi_1^D|\psi_1\rangle = 0, \quad \langle\psi_2^D|\psi_2\rangle = \frac{1}{2C} = \langle\psi_2^D|\psi_1\rangle.$$

Hence, from Eq. 15.26 and the above equations

$$\begin{aligned}\mathbb{I}_{2\times2} &= \mathbb{I}^2_{2\times2}\\ &= \frac{1}{2C}\Big\{(-2PQ + Q^2)|\psi_1\rangle\langle\psi_1^D| + Q^2\Big[|\psi_2\rangle\langle\psi_1^D| + |\psi_1\rangle\langle\psi_2^D|\Big]\Big\},\end{aligned}$$

and this yields from Eqs. 15.26 and 15.23

$$\begin{aligned}P &= \frac{1}{2C}\left\{2PQ - Q^2\right\}, \quad Q = \frac{1}{2C}Q^2\\ &\Rightarrow P = Q = 2C = \frac{8\gamma^2\omega^3}{\pi}.\end{aligned}$$

Hence, the completeness equation for the 2×2 block single excitation states is given by

$$\mathbb{I}_{2\times2} = 2C\Big[-|\psi_1(0)\rangle\langle\psi_1^D(0)| + |\psi_2(0)\rangle\langle\psi_1^D(0)| + |\psi_1(0)\rangle\langle\psi_2^D(0)|\Big]. \tag{15.27}$$

The completeness equation above is equal, up to a normalization, to Eq. 15.16.

15.5 Equal frequency propagator

The defining equation for the propagator is, from Eq. 14.105,

$$\begin{aligned}G(\tau) &= \langle\Psi_{00}^D|xe^{-\tau(H-\omega)}x|\Psi_{00}\rangle\\ \Rightarrow \hat{G}(\tau) &= \lim_{\epsilon\to0} G(\tau) = \hat{N}_{00}^2\langle\hat{\psi}_{00}^D|xe^{-\tau(H-E_0)}x|\hat{\psi}_{00}\rangle.\end{aligned} \tag{15.28}$$

The completeness equation can be used to give a derivation of the equal frequency propagator from first principles. Inserting the completeness equation given in Eq. 15.27 into the expression for the equal frequency propagator given in Eq. 15.28 yields

$$\hat{G}(\tau) = 2C\hat{N}_{00}^2 \langle \hat{\psi}_{00}^D | x e^{-\tau(H-\omega)} \times \Big[-|\psi_1(0)\rangle\langle\psi_1^D(0)| + |\psi_2(0)\rangle\langle\psi_1^D(0)| + |\psi_1(0)\rangle\langle\psi_2^D(0)| \Big] x |\hat{\psi}_{00}\rangle. \tag{15.29}$$

Note that Eq. 15.29 is equivalent to the earlier expression given in Eq. 15.15.

It follows from Eqs. 15.19 and 15.21 that

$$\langle x, v| e^{-\tau H} |\psi_1(0)\rangle = \langle x, v|\psi_1(\tau)\rangle = e^{-2\tau\omega}(v + \omega x)\hat{\psi}_{00}(x, v), \tag{15.30}$$

$$\langle x, v| e^{-\tau H} |\psi_2(0)\rangle = \langle x, v|\psi_2(\tau)\rangle = e^{-2\tau\omega}\omega\{x + \tau(v + \omega x)\}\hat{\psi}_{00}(x, v). \tag{15.31}$$

It can be shown that the first and last terms inside the square bracket in Eq. 15.29 cancel. Hence, from Eqs. 15.29, 15.30, and 15.31

$$\begin{aligned}
\hat{G}(\tau) &= e^{-\tau\omega} 2C\hat{N}_{00}^2 \langle \hat{\psi}_{00}^D | x |\psi_2(\tau)\rangle\langle\psi_1^D(0)| x |\hat{\psi}_{00}\rangle \\
&= e^{-\tau\omega} 2C\hat{N}_{00}^2 \int dx dv\, \omega x\{x + \tau(v + \omega x)\} P(x, v) \\
&\quad \times \int dx dv\, x(-v + \omega x) P(x, v) \\
&= e^{-\tau\omega} 2C\omega^2 \hat{N}_{00}^2 [1 + \omega\tau] \times \left[\int dx dv\, x^2 P(x, v) \right]^2.
\end{aligned} \tag{15.32}$$

Performing the Gaussian integrations yields

$$\int dx dv\, x^2 P(x, v) = \frac{1}{2\omega^2 C}. \tag{15.33}$$

Hence

$$\hat{G}(\tau) = e^{-\tau\omega} \frac{\hat{N}_{00}^2}{2\omega^2 C}[1 + \omega\tau] = \frac{1}{4\gamma}\frac{1}{\omega^3} e^{-\omega\tau}[1 + \omega\tau], \tag{15.34}$$

where $C = (4\gamma^2\omega^3)/\pi$ is given in Eq. 15.6 and the normalization constant $\hat{N}_{00}^2 = 2\gamma\omega^2/\pi$ is given in Eq. 15.7.

To verify the equal frequency result obtained for the propagator, consider taking the limit of $\omega_1 \to \omega_2$ in Eq. 14.104, the result being shown in Figure 15.2b. The propagator has the well-defined and finite limit

$$\begin{aligned}
\hat{G}(\tau) &= \frac{1}{4\gamma\epsilon}\frac{1}{\omega_1 + \omega_2} e^{-\omega\tau} \left[\frac{e^{\epsilon\tau}}{\omega - \epsilon} - \frac{e^{-\epsilon\tau}}{\omega + \epsilon} \right] \\
&= \frac{1}{4\gamma}\frac{1}{\omega^3} e^{-\omega\tau}[1 + \omega\tau],
\end{aligned} \tag{15.35}$$

and agrees with the result obtained in Eq. 15.34.

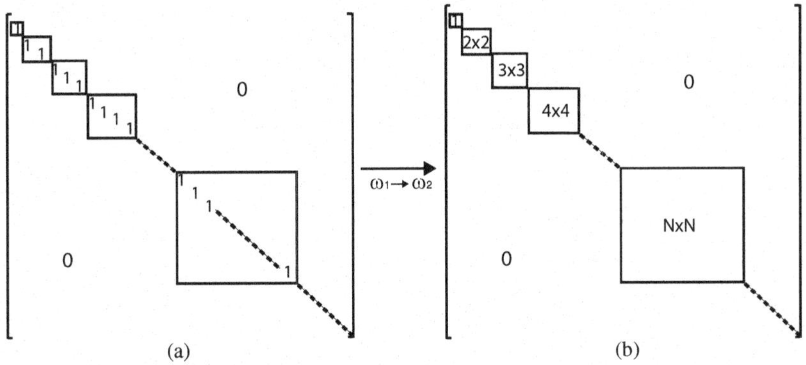

Figure 15.2 (a) Completely diagonal Hamiltonian H for the unequal frequency case. (b) Block-diagonal structure of the Hamiltonian in the equal frequency limit, with each $N \times N$ block being given by an $N \times N$ Jordan block.

The path integral yields a propagator for all values of ω_1, ω_2. The state space approach requires a lot more effort to find the propagator and, in particular, the equal frequency result needs a calculation quite distinct from the unequal case. Furthermore, it is not clear how the state space approach can be used to evaluate the propagator for the case when ω_1, ω_2 are complex. These results show the utility of the path integral which, among other things, allows us to have a deeper understanding of the underlying state space and operator structure of quantum mechanics.

15.6 Hamiltonian: Jordan block structure

In the limit of equal frequencies $\omega_1 = \omega_2$, there is a re-organization of state space into a direct sum of finite-dimensional subspaces, one subspace for each block diagonal component of H, as shown in Figure 15.2. The breakdown of the equivalence to a Hermitian Hamiltonian is due to the fact that, for equal frequencies, H becomes a direct of sum of Jordan blocks.

The total Hilbert space $\mathcal{V}$, for the *equal frequency* case, breaks up into a direct sum of finite-dimensional vector spaces $\mathcal{V}_n$, and is given by

$$\mathcal{V} = \oplus_{n=1}^{\infty} \mathcal{V}_n, \tag{15.36}$$

where $\mathcal{V}_1$ is one dimensional, $\mathcal{V}_2$ is two dimensional and so on.

The coordinate x and velocity v operators are not block diagonal in this representation; the matrix elements of these operators connect the vectors of different subspaces $\mathcal{V}_n$. This feature of the coordinate operator comes to the fore in the calculation of the propagator in the block diagonal basis.

Denoting the finite-dimensional matrix representation of the Hamiltonian by $\mathcal{H}_n$, as shown in Figure 15.2, yields the block diagonal decomposition

$$\mathcal{H} = \oplus_{n=1}^{\infty} \mathcal{H}_n = \oplus_{n=1}^{\infty} a_n \mathcal{J}_{\lambda_n,n}. \tag{15.37}$$

The coefficients a_n are real constants; $\mathcal{J}_{\lambda_n,n}$ is an $n \times n$ Jordan block, specified by its size n and eigenvalue λ_n, and given by[3]

$$\mathcal{J}_{\lambda_n,n} = \begin{bmatrix} \lambda_n & \pm 1 & 0 & 0 & \ddots & \ddots \\ 0 & \lambda_n & \pm 1 & 0 & \ddots & \ddots \\ \ddots & 0 & \lambda_n & \pm 1 & 0 & \ddots \\ \ddots & \ddots & \ddots & \ddots & \ddots & \ddots \\ \ddots & \ddots & \ddots & 0 & \lambda_n & \pm 1 \\ \ddots & \ddots & \ddots & \ddots & 0 & \lambda_n \end{bmatrix}. \tag{15.38}$$

The Hamiltonian is analyzed for the first two blocks; $\mathcal{H}_1$ is one dimensional and $\mathcal{H}_2$ is a 2×2 matrix.

The ground state forms an invariant subspace $\mathcal{V}_1$ with a single element $|e_0\rangle$ proportional to $|\hat{\psi}_{00}\rangle$; for dimensional consistency and to preserve the correct normalization, the mapping is

$$|e_0\rangle = \hat{N}_{00}|\hat{\psi}_{00}\rangle, \quad \langle e_0|e_0\rangle = 1. \tag{15.39}$$

The eigenvalue equation $H|\hat{\psi}_{00}\rangle = \omega|\hat{\psi}_{00}\rangle$ yields the Hamiltonian on $\mathcal{V}_1$ given by

$$\mathcal{H}_1 = \omega, \quad \mathcal{H}_1|e_0\rangle = \omega|e_0\rangle. \tag{15.40}$$

15.7 2×2 Jordan block

A derivation is given of the 2×2 Jordan block structure of the Hamiltonian and state space.

The result given in Eq. 15.40 together with Eq. 15.37 yields

$$\mathcal{H} = \mathcal{H}_1 \oplus \mathcal{H}_2 \oplus \ldots \tag{15.41}$$
$$= \omega \oplus 2\omega \mathcal{J}_2 \oplus \ldots \tag{15.42}$$

It will be shown in this section that

$$\mathcal{J}_2 = \begin{bmatrix} 1 & -1 \\ 0 & 1 \end{bmatrix}. \tag{15.43}$$

[3] The ± 1 terms in the super-diagonal in Eq. 15.38 are allowed, since multiplying $\mathcal{J}_{\lambda_n,n}$ by -1 can switch the sign of the super-diagonal from 1 to -1, and in so doing re-define the eigenvalue to be $-\lambda_n$.

The Jordan block Hamiltonian has been shown by Bender and Mannheim (2008a) to be pseudo-Hermitian; for the case of $\mathcal{J}_2$, it can be seen that

$$\sigma_1 \mathcal{J}_2 \sigma_1 = \begin{bmatrix} 1 & 0 \\ -1 & 1 \end{bmatrix} = \mathcal{J}_2^\dagger, \quad \sigma_1 = \begin{bmatrix} 0 & 1 \\ 1 & 0 \end{bmatrix}.$$

The result above demonstrates that although the Jordan block matrix is pseudo-Hermitian, it is essentially inequivalent to a Hermitian matrix.

Bender and Mannheim (2008a) derive the 2×2 Jordan block for the Minkowski Hamiltonian by defining creation and destruction operators that have a finite limit when $\epsilon \to 0$. In this section, the 2×2 Jordan block for the Euclidean Hamiltonian is directly derived from the state vectors, and the completeness equation is obtained by taking $\epsilon \to 0$ following the procedure in Sections 15.3 and 15.4.

Recall from Eqs. 15.17, 15.24, and 15.25, that the Hamiltonian and state vectors for the equal frequency limit are given by

$$H = -\frac{1}{2\gamma}\frac{\partial^2}{\partial v^2} - v\frac{\partial}{\partial x} + \omega^2 v^2 + \frac{\gamma}{2}\omega^4 x^2,$$
$$|\psi_1\rangle = |\psi_1(0)\rangle = (v + \omega x)|\hat{\psi}_{00}\rangle, \quad \langle \psi_1^D| = \langle \psi_1^D(0)|,$$
$$|\psi_2\rangle = |\psi_2(0)\rangle = \omega x|\hat{\psi}_{00}\rangle, \quad \langle \psi_2^D| = \langle \psi_2^D(0)|.$$

The fact that the state vectors $|\psi_1\rangle$, $|\psi_2\rangle$ form a closed subspace under the action of H points to an invariant 2×2 subspace of the total Hilbert space.

In the 2×2 block space, the Hamiltonian can be represented by a 2×2 Jordan block in a basis fixed by the representation of $|\psi_1\rangle, |\psi_2\rangle$ by two-dimensional column vectors. To obtain this finite-dimensional representation, note that $H|\hat{\psi}_{00}\rangle = \omega|\hat{\psi}_{00}\rangle$ and from Eq. 15.19 $H|\psi_1\rangle = 2\omega|\psi_1\rangle$; hence, the action of H on the state vectors $|\psi_1\rangle, |\psi_2\rangle$ is given by

$$H|\psi_1\rangle = 2\omega|\psi_1\rangle, \tag{15.44}$$
$$H|\psi_2\rangle = -\omega v|\hat{\psi}_{00}\rangle + \omega^2 x|\hat{\psi}_{00}\rangle = -\omega(v + \omega x)|\hat{\psi}_{00}\rangle + 2\omega|\psi_2\rangle$$
$$\Rightarrow H|\psi_2\rangle = -\omega|\psi_1\rangle + 2\omega|\psi_2\rangle. \tag{15.45}$$

Since $|\psi_1\rangle$ is an eigenvector of the Jordan block it is natural to make the identification

$$|\psi_1\rangle \propto \begin{bmatrix} 1 \\ 0 \end{bmatrix}. \tag{15.46}$$

Recall from Eqs. 15.20 and 15.23,

$$\langle \psi_1^D|\psi_1\rangle = 0, \quad \langle \psi_2^D|\psi_2\rangle = \frac{1}{2C} = \langle \psi_2^D|\psi_1\rangle. \tag{15.47}$$

Since $|\psi_1\rangle$ has zero norm, its normalization is fixed by its overlap with $|\psi_2\rangle$. Choosing the normalization consistent with Eq. 15.47 yields

$$\sqrt{C}|\psi_1\rangle = |e_1\rangle = \begin{bmatrix} 1 \\ 0 \end{bmatrix}, \quad \sqrt{C}|\psi_2\rangle = |e_2\rangle = \begin{bmatrix} 1/2 \\ 1/2 \end{bmatrix}, \tag{15.48}$$

with the dual vectors given by

$$\sqrt{C}\langle\psi_1^D| = \langle e_1^D| = \begin{bmatrix} 0, & 1 \end{bmatrix}, \quad \sqrt{C}\langle\psi_2^D| = \langle e_2^D| = \begin{bmatrix} 1/2, & 1/2 \end{bmatrix}. \tag{15.49}$$

Note that $\langle e_1^D|$ is *not* the transpose of $|e_1\rangle$.

The completeness equation for the state space of the 2×2 block has a discrete realization; recall from Eq. 15.27

$$\begin{aligned} \mathbb{I}_{2\times2} &= 2C\Big[- |\psi_1\rangle\langle\psi_1^D| + |\psi_2\rangle\langle\psi_1^D| + |\psi_1\rangle\langle\psi_2^D|\Big] \\ \Rightarrow \mathbb{I}_{2\times2} &= 2\Big[- |e_1\rangle\langle e_1^D| + |e_2\rangle\langle e_1^D| + |e_1\rangle\langle e_2^D|\Big]. \end{aligned} \tag{15.50}$$

The completeness equation for the Jordan block shows that there is a nontrivial metric on the discrete state space $\mathcal{V}_2$.

Using Eqs. 15.48 and 15.49, Eq. 15.50 yields

$$\mathbb{I}_{2\times2} = 2\left\{ - \begin{bmatrix} 0 & 1 \\ 0 & 0 \end{bmatrix} + \frac{1}{2}\begin{bmatrix} 1 & 1 \\ 0 & 0 \end{bmatrix} + \frac{1}{2}\begin{bmatrix} 0 & 1 \\ 0 & 1 \end{bmatrix} \right\} = \begin{bmatrix} 1 & 0 \\ 0 & 1 \end{bmatrix},$$

and we have obtained the expected result.

15.7.1 Hamiltonian

Let $\mathcal{H}_2$ denote the realization of the Hamiltonian as a discrete and dimensionless matrix acting on the two-dimensional state space of the 2×2 Jordan block. Applying Eq. 15.48 to Eqs. 15.44 and 15.45 yields the 2×2 representation

$$\begin{aligned} \mathcal{H}_2|e_1\rangle &= 2\omega|e_1\rangle \;\Rightarrow\; \langle e_1^D|\mathcal{H}_2 = 2\omega\langle e_1^D| \\ \mathcal{H}_2|e_2\rangle &= -\omega|e_1\rangle + 2\omega|e_2\rangle \;\Rightarrow\; \langle e_2^D|\mathcal{H}_2 = -\omega\langle e_1^D| + 2\omega\langle e_2^D|. \end{aligned}$$

The Hamiltonian $\mathcal{H}_2$ – in the $|e_1\rangle$ and $|e_2\rangle$ basis – is proportional to the 2×2 Jordan block matrix and is given by[4]

$$\mathcal{H}_2 = 2\omega\begin{bmatrix} 1 & -1 \\ 0 & 1 \end{bmatrix}. \tag{15.51}$$

The definition of the discrete vectors $|e_1\rangle$ and $|e_1\rangle$ given in Eq. 15.48 requires a rescaling by $\sqrt{C}$ because of dimensional consistency; in contrast, there is no need to rescale $\mathcal{H}_2$ since it has the correct dimension set by ω.

[4] The Euclidean Hamiltonian given in Eq. 15.43 has a -1 for the superdiagonal, unlike the case for the Minkowski Hamiltonian (Bender and Mannheim, 2008a), where it is $+1$.

The Jordan block Hamiltonian given in Eq. 15.43 has only one eigenvalue and this is the reason that the two different eigenstates for the unequal frequencies collapsed into a single eigenstate. The Jordan block limit of $\mathcal{H}_2$ (for equal frequency) shows that $\mathcal{H}_2$ continues to be pseudo-Hermitian but is inequivalent to any Hermitian Hamiltonian since the Jordan block is inequivalent to any Hermitian matrix.

The right eigenvector of $\mathcal{H}_2$ is $|e_1\rangle$ and the left eigenvector of $\mathcal{H}$ is the dual $\langle e_1^D|$; namely

$$\mathcal{H}_2|e_1\rangle = 2\omega|e_1\rangle, \quad \langle e_1^D|\mathcal{H}_2 = 2\omega\langle e_1^D|,$$
$$\Rightarrow \langle e_1^D|e_1\rangle = 0 = \langle \psi_1^D|\psi_1\rangle.$$

Hence, the Jordan block structure shows why the equal frequency eigenstate has a zero norm.

15.7.2 Schrödinger equation for Jordan block

The Schrödinger equation for an arbitrary vector $|e\rangle$ is given by

$$-\frac{\partial}{\partial\tau}|e(\tau)\rangle = \mathcal{H}_2|e(\tau)\rangle.$$

For eigenvector $|e_1\rangle$ the time-dependent solution is

$$-\frac{\partial}{\partial\tau}|e_1(\tau)\rangle = \mathcal{H}_2|e_1(\tau)\rangle = 2\omega|e_1(\tau)\rangle,$$
$$\Rightarrow |e_1(\tau)\rangle = e^{-2\omega\tau}|e_1\rangle, \quad |e_1\rangle = \begin{bmatrix} 1 \\ 0 \end{bmatrix}. \tag{15.52}$$

The time dependence of the state vector $|e_2(\tau)\rangle$ is given by

$$-\frac{\partial}{\partial\tau}|e_2(\tau)\rangle = \mathcal{H}_2|e_2(\tau)\rangle, \quad |e_2(0)\rangle = |e_2\rangle = \begin{bmatrix} 1/2 \\ 1/2 \end{bmatrix}. \tag{15.53}$$

In the 2×2 block representation $|e_2(\tau)\rangle$ is given from the solution obtained in Eq. 15.21, which yields

$$\psi_2(x, v; \tau) = e^{-2\tau\omega}\omega\big[x + \tau(v + \omega x)\big]\hat{\psi}_{00}(x, v)$$
$$\Rightarrow |e_2(\tau)\rangle = e^{-2\tau\omega}\Big[|e_2\rangle + \omega\tau|e_1\rangle\Big] = e^{-2\tau\omega}\begin{bmatrix} 1/2 + \omega\tau \\ 1/2 \end{bmatrix}. \tag{15.54}$$

It can be directly verified using the explicit form for the Hamiltonian given in Eq. 15.43 that the solution for $|e_2(\tau)\rangle$ given in Eq. 15.54 satisfies the Schrödinger equation given in Eq. 15.53.

15.7.3 Time evolution

The Jordan block Hamiltonian is given by Eq. 15.43; a simple calculation yields the evolution operator

$$e^{-\tau\mathcal{H}_2} = e^{-2\omega\tau}\begin{bmatrix}1 & 2\omega\tau\\ 0 & 1\end{bmatrix}. \tag{15.55}$$

The time dependence of the state vectors follows directly from the evolution operator. For eigenvector $|e_1\rangle$ the time-dependent solution is

$$|e_1(\tau)\rangle = e^{-\tau\mathcal{H}_2}|e_1\rangle = e^{-2\omega\tau}|e_1\rangle,$$

which is the expected result as in Eq. 15.52.

The time dependence of the state vector $|e_2(\tau)\rangle$ is given by

$$|e_2(\tau)\rangle = e^{-\tau\mathcal{H}_2}|e_2\rangle = e^{-2\omega\tau}\begin{bmatrix}1/2+\omega\tau\\ 1/2\end{bmatrix}, \tag{15.56}$$

which is the expected result as in Eq. 15.54.

15.8 Jordan block propagator

The equal frequency propagator is given in Eq. 15.28,

$$\hat{G}(\tau) = \hat{N}_{00}^2\langle\hat{\psi}_{00}^D|Xe^{-\tau(H-\omega)}X|\hat{\psi}_{00}\rangle.$$

The position operator X, unlike the Hamiltonian, is not block diagonal for the equal frequency case and connects different subspaces $\mathcal{V}_n$. To determine the propagator, the representation of the position operator X needs to be determined in the 3×3 subspace, which includes the ground state and the 2×2 Jordan block. The operator X has the matrix elements

$$\begin{aligned}\langle\hat{\psi}_{00}^D|X|\hat{\psi}_{00}\rangle &= 0 = \langle\psi_1^D|X|\psi_1\rangle = \langle\psi_2^D|X|\psi_2\rangle = \langle\psi_2^D|X|\psi_1\rangle,\\ \langle\hat{\psi}_{00}^D|X|\psi_1\rangle &= \frac{1}{2\omega C} = \langle\hat{\psi}_{00}^D|X|\psi_2\rangle.\end{aligned} \tag{15.57}$$

Note that the matrix elements of the operator X are zero within a block and are nonzero only for elements that connect vectors from two different blocks.

Since X acts on the $\mathcal{V}_1 \oplus \mathcal{V}_2$ we need to extend the vectors defined on the subspaces $|e_0\rangle \in \mathcal{V}_1$ and $|e_1\rangle, |e_2\rangle \in \mathcal{V}_2$ to the larger space; let us define the following three-dimensional vectors:

$$|e_0\rangle = 1,\quad |E_0\rangle = |e_0\rangle \oplus |0\rangle = \begin{bmatrix}1\\0\\0\end{bmatrix},$$

$$|E_1\rangle = |0\rangle \oplus |e_1\rangle = \begin{bmatrix} 0 \\ 1 \\ 0 \end{bmatrix}, \quad |E_2\rangle = |0\rangle \oplus |e_2\rangle = \begin{bmatrix} 0 \\ 1/2 \\ 1/2 \end{bmatrix}. \tag{15.58}$$

The dual vectors are given by the transpose, except for $\langle E_1^D|$ given by

$$\langle E_1^D| = \big[0, \ 0, \ 1\big]. \tag{15.59}$$

Let the position operator in the block diagonal space be denoted by $\mathcal{X}$; from Eq. 15.57, since all the elements are dimensionless in the Jordan block representation,

$$\begin{aligned} &\langle E_0^D|\mathcal{X}|E_0\rangle = 0 = \langle E_1^D|\mathcal{X}|E_1\rangle = \langle E_2^D|\mathcal{X}|E_2\rangle = \langle E_1^D|\mathcal{X}|E_2\rangle, \\ &\langle E_0^D|\mathcal{X}|E_1\rangle = 1 = \langle E_0^D|\mathcal{X}|E_2\rangle, \end{aligned} \tag{15.60}$$

which yields the representation for the Hermitian matrix $\mathcal{X}$

$$\mathcal{X} = \begin{bmatrix} 0 & 1 & 1 \\ 1 & 0 & 0 \\ 1 & 0 & 0 \end{bmatrix}. \tag{15.61}$$

Since $\mathcal{X}$ is dimensionless, its mapping to the coordinate position operator X needs a dimensional scale; let $\mathcal{X} = \zeta X$. From Eqs. 15.39, 15.49, 15.57, and 15.60,

$$1 = \langle E_0^D|\mathcal{X}|E_1\rangle = \zeta \hat{N}_{00}\sqrt{C}\langle \hat{\psi}_{00}^D|X|\psi_1\rangle = \zeta \frac{\hat{N}_{00}}{2\omega\sqrt{C}}, \tag{15.62}$$

$$\Rightarrow X = \frac{\hat{N}_{00}}{2\omega\sqrt{C}}\mathcal{X}, \quad \zeta = \frac{2\omega\sqrt{C}}{\hat{N}_{00}}. \tag{15.63}$$

Extending the Hamiltonian to the $\mathcal{V}_1 \oplus \mathcal{V}_2$ space yields, from Eq. 15.43,

$$\mathcal{H} = \mathcal{H}_1 \oplus \mathcal{H}_2 = \omega \oplus 2\omega \begin{bmatrix} 1 & -1 \\ 0 & 1 \end{bmatrix} = \omega \begin{bmatrix} 1 & 0 & 0 \\ 0 & 2 & -2 \\ 0 & 0 & 2 \end{bmatrix}.$$

The time evolution operator is given by

$$\exp\{-\tau\mathcal{H}\} = e^{-2\omega\tau} \begin{bmatrix} e^{\omega\tau} & 0 & 0 \\ 0 & 1 & 2\omega\tau \\ 0 & 0 & 1 \end{bmatrix}. \tag{15.64}$$

The completeness equation from Eq. 15.50 has the following extension to $\mathcal{V}_1 \oplus \mathcal{V}_2$:

$$\mathbb{I}_{3\times 3} = |E_0\rangle\langle E_0^D| + 2\Big[-|E_1\rangle\langle E_1^D| + |E_2\rangle\langle E_1^D| + |E_1\rangle\langle E_2^D|\Big]. \tag{15.65}$$

In the block-diagonal basis, the propagator is given by

$$\hat{G}(\tau) = \hat{N}_{00}^2 \langle \hat{\psi}_{00}^D | X e^{-\tau(H-\omega)} X | \hat{\psi}_{00} \rangle$$
$$= \left(\frac{\hat{N}_{00}}{2\omega\sqrt{C}} \right)^2 \langle E_0^D | \mathcal{X} e^{-\tau(\mathcal{H}-\omega)} \mathcal{X} | E_0 \rangle. \tag{15.66}$$

Using the completeness equation given in Eq. 15.65 yields

$$\hat{G}(\tau)$$
$$= \frac{\hat{N}_{00}^2}{4\omega^2 C} \langle E_0^D | \mathcal{X} e^{-\tau(\mathcal{H}-\omega)}$$
$$\times \Big(|E_0\rangle\langle E_0^D| + 2\Big[- |E_1\rangle\langle E_1^D| + |E_2\rangle\langle E_1^D| + |E_1\rangle\langle E_2^D| \Big]\Big) \mathcal{X} | E_0 \rangle$$
$$= \frac{\hat{N}_{00}^2}{2\omega^2 C} \langle E_0^D | \mathcal{X} \Big[- e^{-\omega\tau} |E_1\rangle\langle E_1^D| + e^{\omega\tau} |E_2(\tau)\rangle\langle E_1^D| + e^{-\omega\tau} |E_1\rangle\langle E_2^D| \Big] \mathcal{X} | E_0 \rangle$$
$$= \frac{\hat{N}_{00}^2}{2\omega^2 C} e^{\omega\tau} \langle E_0^D | \mathcal{X} | E_2(\tau)\rangle\langle E_1^D | \mathcal{X} | E_0 \rangle = \frac{\hat{N}_{00}^2}{2\omega^2 C} e^{\omega\tau} \langle E_0^D | \mathcal{X} | E_2(\tau) \rangle, \tag{15.67}$$

since

$$\langle E_1^D | \mathcal{X} | E_0 \rangle = 1 = \langle E_2^D | \mathcal{X} | E_0 \rangle.$$

From Eq. 15.53, the time dependence of $|E_2(\tau)\rangle$ is given by

$$|E_2(\tau)\rangle = e^{-\tau\mathcal{H}} |E_2\rangle = e^{-2\omega\tau} \begin{bmatrix} 0 \\ 1/2 + \omega\tau \\ 1/2 \end{bmatrix} \quad \Rightarrow \quad \langle E_0^D | \mathcal{X} | E_2(\tau) \rangle = 1 + \omega\tau,$$

which yields, from Eq. 15.67, the expected result for the propagator, namely

$$\hat{G}(\tau) = \frac{\hat{N}_{00}^2}{2\omega^2 C} e^{-\omega\tau} (1 + \omega\tau) = \frac{1}{4\gamma\omega^3} e^{-\omega\tau} (1 + \omega\tau).$$

A direct derivation can be given using the matrix representation of the evolution operator given in Eq. 15.64; from Eq. 15.66

$$\hat{G}(\tau) = \left(\frac{\hat{N}_{00}}{2\omega\sqrt{C}} \right)^2 \langle E_0^D | \mathcal{X} e^{-\tau(\mathcal{H}-\omega)} \mathcal{X} | E_0 \rangle.$$

Using

$$\mathcal{X} | E_0 \rangle = \begin{bmatrix} 0 \\ 1 \\ 1 \end{bmatrix}, \quad \langle E_0^D | \mathcal{X} = \big[0, 1, 1 \big]$$

yields, from Eq. 15.64, the expected answer

$$\hat{G}(\tau) = \left(\frac{\hat{N}_{00}}{2\omega\sqrt{C}}\right)^2 e^{-\omega\tau}(2 + 2\omega\tau) = \frac{1}{4\gamma\omega^3} e^{-\omega\tau}(1 + \omega\tau).$$

All the $N \times N$ blocks for the Hamiltonian can be analyzed one by one and it can be shown that they are all equal to a corresponding Jordan block matrix. However, the higher order blocks may not be as simple as $\mathcal{J}_2$ as they can include the direct sum of lower order Jordan blocks.

15.9 Summary

The equal frequency case of the acceleration Lagrangian leads to a Hamiltonian that is essentially inequivalent to any Hermitian Hamiltonian. A carefully chosen limit for the equal frequency leads to a Hamiltonian that is block diagonal, with each block consisting of Jordan block matrices. The state space has zero norm state vectors, even for the Euclidean theory, showing that the fundamental inequivalence of the acceleration Hamiltonian to a Hermitian Hamiltonian holds for both Minkowski and Euclidean time.

The equal frequency propagator was evaluated using various techniques to highlight the different aspects of the Jordan block system that provides a representation of the essentially non-Hermitian sector of the theory. The specific form of the propagator for equal frequency reflects the presence of time-dependent states that are essential in evaluating the propagator. In particular, these time dependent-states appear in the completeness equation and hence are required for spanning out a complete basis for the state space of the equal frequency case.

The quantum mechanics of the 2×2 Jordan block was worked out, with the solution of the Schrödinger equation having only one eigenstate and another time dependent state. A calculation for the propagator was done, block by block, based on the discrete subspaces and finite Jordan block matrices.

Part six

Nonlinear path integrals

16

The quartic potential: instantons

The simple harmonic oscillator is exactly soluble because it has a quadratic potential and yields a linear theory in the sense that the classical equation of motion is linear. Nonlinear path integrals have potentials that typically have a quartic or higher polynomial dependence on the degree of freedom, or the potential can be a transcendental function, such as an exponential. The techniques discussed for quadratic Gaussian path integrals needed to be further developed for nonlinear path integrals. In general, to solve these nonlinear systems, one usually uses either a perturbation expansion or numerical methods.

A perturbation expansion is useful if the theory has a behavior that is smooth about the dominant piece of the action or Hamiltonian; in practice a smooth expansion of the physical quantities yields an analytic series in some expansion parameter, say a coupling constant g, around $g = 0$.

There are, however, cases of physical interest for which nonperturbative effects change the *qualitative* behavior of the theory. Two examples where nonperturbative effects dominate are the following:

- Tunneling through a finite barrier. If one perturbs about the lowest lying eigenstates inside a well, one cannot produce the tunneling amplitude.
- The spontaneous breaking or restoration of a symmetry cannot be produced by perturbing about an incorrect ground state.

Tunneling and symmetry breaking are nonperturbative because these effects depend on g as $\exp\{-1/g\}$, which is nonanalytic about $g = 0$, and hence cannot be obtained by perturbation theory.

The semi-classical expansion is an approximation that is a useful tool for studying nonperturbative effects. The general features of this approximation scheme are discussed for a nonlinear Lagrangian in Section 16.1. Section 16.3 covers nontrivial classical solutions of the double-well potential, called instantons. In Section 16.4 the so called zero mode features of the instantons are discussed, and in Section 16.5

the Faddeev–Popov analysis is applied to the zero mode problem. In Section 16.6 the multi-instanton solutions are obtained, and the transition amplitude, correlators and the dilute gas approximation are then discussed in Sections 16.7, 16.8 and Section 16.9. The double-well potential, in the strong coupling limit, is shown in Section 16.10 to be equivalent to an Ising model, and in Section 16.11 a nonlocal Ising model is shown to produce the double-well potential. Sections 16.12–16.14 discuss spontaneous symmetry breaking and symmetry restoration.

16.1 Semi-classical approximation

Consider the evolution kernel for Euclidean time,

$$K(x, x'; \tau) = \int Dx e^{S[x]/\hbar},$$

where

$$S[x] = \int_0^\tau \mathcal{L} dt \tag{16.1}$$

and $\int Dx$ is the path integration measure. For $\hbar \to 0$, the classical trajectory, for which S is a maximum, that is

$$\frac{\delta S[x_c]}{\delta x(t)} = 0, \tag{16.2}$$

dominates the path integral with the next leading term yielding an expansion in a power series in $\hbar$.

One can expand the paths about the classical path, that is

$$x(t) = x_c(t) + \eta(t), \quad \eta(0) = 0 = \eta(\tau),$$

and expand the action about the classical path. Thus, we have

$$\begin{aligned} S[x] &= S[x_c + \eta] \\ &= S[x_c] + \frac{1}{2} \int dt_1 dt_2 \eta_{t_1} \eta_{t_2} \frac{\delta^2 S[x_c]}{\delta x(t_1) \delta x(t_2)} + 0(\eta^3) \\ &= S_c + S_2, \quad S_c = S[x_c]. \end{aligned}$$

The classical action is a maximum of the Euclidean action, namely

$$\frac{\delta^2 S[x_c]}{\delta x(t_1) \delta x(t_2)} \le 0. \tag{16.3}$$

Hence the quadratic approximation of the action yields a convergent expansion for the path integral,

$$K(x.x';\tau) \simeq \mathcal{N} e^{S_c/\hbar} \int D\eta e^{S_2/\hbar}$$
$$= \mathcal{N} e^{S_c/\hbar} \Bigg/ \sqrt{\det\left[\frac{1}{\hbar}\frac{\delta^2 S[x_c]}{\delta x(t)\delta x(t')}\right]}. \tag{16.4}$$

Note the approximation breaks down if $\det\left[\delta^2 S[x_c]/\delta x(t)\delta x(t')\right] = 0$. Consider a typical Euclidean action

$$S = -\int_0^\tau dt \left(\frac{1}{2}m\dot{x}^2 + V(x)\right). \tag{16.5}$$

Then

$$\frac{\delta^2 S_c}{\delta x(t)\delta x(t')} = -\left(-m\frac{d^2}{dt^2} + V''(x)\right)\delta(t-t'), \qquad (V' = \partial V/\partial x), \tag{16.6}$$

and hence

$$\det\left(\frac{1}{\hbar}\frac{\delta^2 S_c}{\delta x \delta x}\right) \propto \det\left(\frac{1}{\hbar}\left(-m\frac{d^2}{dt^2} + V''(x)\right)\right). \tag{16.7}$$

To evaluate the determinant, the earlier discussion on the simple harmonic oscillator needs to be generalized.

16.2 A one-dimensional integral

The semi-classical expansion is identical to the saddle point method for finite dimension integrals. To see this, consider the integral

$$I(g^2) = \int_{-\infty}^{+\infty} \frac{dx}{\sqrt{2\pi}} \exp\left\{-\frac{1}{2}x^2 - \frac{g^2}{8}x^4\right\}.$$

The integral is well defined for $g^2 > 0$, but can it be analytically continued to $g^2 < 0$? A perturbation expansion around $g^2 > 0$ cannot answer this question and a saddle point expansion will be used to go beyond the perturbation expansion.

The minimum is at $x_0 = 0$, and expanding the integrand about this yields

$$I \simeq \int_{-\infty}^{+\infty} \frac{dx}{\sqrt{2\pi}} e^{-\frac{1}{2}x^2}\left(1 - \frac{g^2x^4}{8} + \frac{g^4x^6}{128}\right)$$
$$= 1 - \frac{3}{8}g^2 + \frac{105}{128}g^4 + 0(g^5) \tag{16.8}$$

: analytic about g = 0.

Let us rewrite I as

$$I = \int \frac{dxdy}{2\pi} e^{-\frac{1}{2}x^2} e^{-\frac{1}{2}y^2} e^{-\frac{i}{2}gyx^2}$$
$$= \int \frac{dy}{\sqrt{2\pi}} e^{-\frac{1}{2}y^2 - \frac{1}{2}\ln(1+igy)}. \tag{16.9}$$

If one expands about $g^2 = 0$ one obtains the same result as Eq. 16.8. However, consider expanding about the minimum of

$$S(y) = -\frac{1}{2}y^2 - \frac{1}{2}\ln(1 + igy). \tag{16.10}$$

The minimum y_0 is given by

$$S'(y_0) = 0 = y_0 + \frac{ig}{2(1 + igy_0)}$$
$$\Rightarrow iy_0(\pm) = \frac{1}{2g}\left(-1 \pm \sqrt{1 + 2g^2}\right).$$

For $g \to 0$, the limiting values are

$$iy_0(-) \to -\frac{1}{g} - \frac{g}{2} \to \infty,$$
$$iy_0(+) \to \frac{g}{2} \to 0.$$

The action about the saddle point $y_0(-)$ yields a negative divergent action, namely

$$\lim_{g\to 0} S(y_0(-)) \to -\frac{1}{2g^2} + \frac{1}{2}\ln\left(-\frac{g^2}{2}\right)$$
$$\to -\infty \qquad \text{divergent.}$$

Hence, the $y_0(+)$ branch is chosen for doing the expansion since it yields the finite action

$$S(y_0(+)) = \frac{1}{8g^2}(\sqrt{1 + 2g^2} - 1)^2 - \frac{1}{2}\ln\left(\frac{1 + \sqrt{1 + 2g^2}}{2}\right), \tag{16.11}$$
$$\equiv S_+^{(0)} \to 0 \text{ as } g \to 0. \tag{16.12}$$

Furthermore, the action has a Taylors expansion about $y_0(+)$ given by

$$S(y) = -\frac{1}{2}y^2 - \frac{1}{2}\ln(1 + igy)$$
$$= \sum_{n=0}^{\infty} \frac{S_+^{(n)}}{n!}(y - y_0(+))^n, \tag{16.13}$$

where the notation $S_+^{(n)} = \frac{\partial^n}{\partial y^n} S(y_0(+))$ is being used:

$$S_+^{(0)} = \frac{1}{8g^2}(\sqrt{1+2g^2} - 1)^2 - \frac{1}{2}\ln\left(\frac{1+\sqrt{1+2g^2}}{2}\right),$$

$$S_+^{(2)} = -\frac{2\sqrt{1+g^2}}{1+\sqrt{1+2g^2}} > 0, \quad S_+^{(3)} = -\frac{8ig^3}{(1+\sqrt{1+2g^2})^3},$$

$$S_+^{(4)} = \frac{48g^4}{(1+\sqrt{1+2g^2})^4}, \quad S_+^{(n)} \sim 0(g^n).$$

Hence, shifting y by $y_0(+)$, Eqs. 16.9 and 16.13 yield

$$I(g^2) \simeq e^{S_+^{(0)}} \int_{-\infty}^{+\infty} \frac{dy}{\sqrt{2\pi}} e^{\frac{1}{2}S_+^{(2)}y^2}\left[1 + \frac{S_+^{(4)}}{4!}y^4 + 0(g^6)\right]$$

$$\cong \frac{e^{\frac{1}{8g^2}(\sqrt{1+2g^2}-1)^2}}{(1+2g^2)^{1/4}}\left[1 + \frac{3g^4}{2(1+2g^2)(1+\sqrt{1+2g^2})^2} + \cdots\right]. \tag{16.14}$$

Note that the second term above is $O(g^4)$, whereas in the expansion given in Eq. 16.8 the second term is only $O(g^2)$.

The nonanalytic structure of the integral about $g^2 = -1/2$, namely that it has a branch cut, is captured in the semi-classical expansion, but is missed in the expansion about $g^2 = 0$ as given in Eq. 16.8. In particular, from Eq. 16.14 it can be seen that the semi-classical expansion provides an analytic continuation of the integral $I(g^2)$ to the range $-1/2 \le g^2 \le +\infty$, whereas the perturbation expansion cannot be extended to $g^2 \le 0$.

The semi-classical expansion is a technique that can capture nonperturbative properties of the path integral and is very useful in quantum mechanics and in quantum field theory.

16.3 Instantons in quantum mechanics

Consider the nonlinear action and Lagrangian for Euclidean time given by

$$S = \int dt\mathcal{L}, \quad \mathcal{L} = -\frac{1}{2}m\dot{x}^2 - V, \tag{16.15}$$

$$V = \frac{g^2}{8}(x^2 - a^2)^2. \tag{16.16}$$

The potential term V is shown in Figure 16.1. The Lagrangian $\mathcal{L}$ has the parity symmetry of being invariant under the transformation $x \to -x$.

Note that if one expands the potential into a polynomial, the quadratic term has a coefficient that is positive; hence any expansion about the quadratic Lagrangian

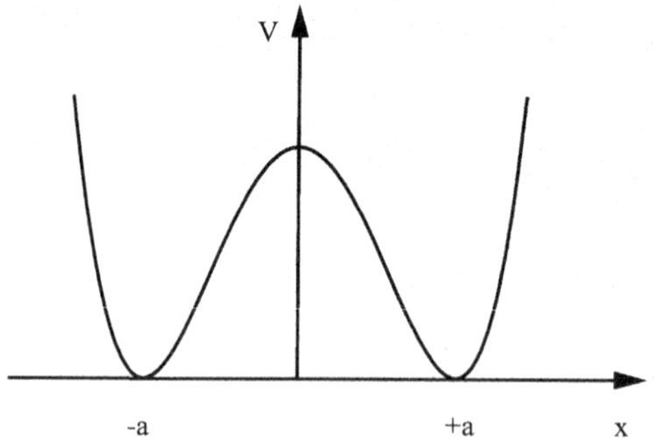

Figure 16.1 The double-well potential for Euclidean time.

will be unstable and lead to a divergent expansion. The reason is that the minima of the Lagrangian are at $\pm a$, as shown in Figure 16.1, and hence the Lagrangian needs to be expanded about one of these minima.

The transition amplitude for Euclidean time τ is given by

$$K(x, x'; \tau) = \int Dx e^S = \langle x | e^{-\tau H} | x' \rangle, \tag{16.17}$$

where H is the Euclidean Hamiltonian. Writing the time integration in a symmetric manner yields the action

$$S = -\int_{-\frac{\tau}{2}}^{+\frac{\tau}{2}} dt \left(\frac{1}{2} m\dot{x}^2 + \frac{g^2}{8}(x^2 - a^2)^2 \right). \tag{16.18}$$

The action is expanded about one of the classical solutions, say $x = a$, similar to the exercise carried out for a single variable in Section 16.2. Consider the shift of variable $x = y - a$; the action is given by

$$\begin{aligned} S &= -\int_{-\frac{\tau}{2}}^{+\frac{\tau}{2}} dt \left(\frac{1}{2} m\dot{y}^2 + \frac{g^2}{8} y^2 (y + 2a)^2 \right) \\ &\simeq -\int_{-\frac{\tau}{2}}^{+\frac{\tau}{2}} dt \left(\frac{1}{2} m\dot{y}^2 + \frac{g^2}{2} a^2 y^2 \right) + 0(y^3). \end{aligned} \tag{16.19}$$

The theory has small oscillations about the minimum at $x = a$, as shown in Figure 16.2; these oscillations spontaneously break the symmetry of $x \to -x$, as is evident from Figure 16.2 and Eq. 16.19.

It is known that a one-dimensional quantum mechanical system cannot spontaneously break any continuous symmetry. However, the system can, in principle, break discrete symmetries. The question that needs to be addressed is whether the perturbative breaking of discrete symmetry of $x \to -x$ by the ground states centered at $x = a$ or $x = -a$, referred to as $x = \pm a$, is only an approximation – with

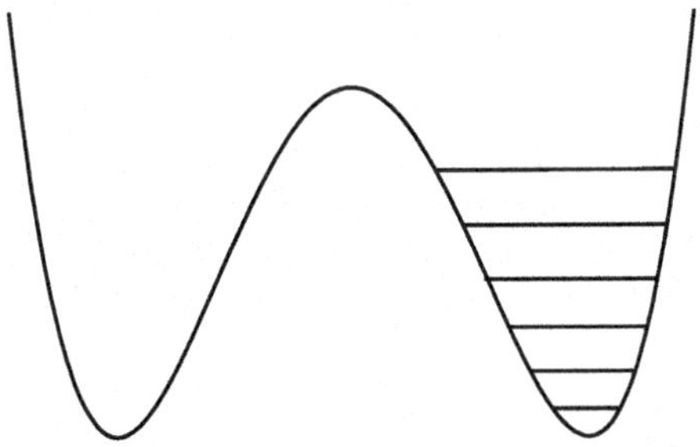

Figure 16.2 The double-well potential for Euclidean time.

nonperturbative effects restoring the parity symmetry? And if symmetry restoration takes place, what is the mechanism by which, in particular for the x^4 potential, the discrete parity symmetry is restored?

The semi-classical expansion for the path integral reveals that there are large quantum fluctuations that are encoded in the neighborhood of an instanton configuration. [A fluctuation is defined as one possible path for the degree of freedom.] It is these large quantum fluctuations – far from the perturbative vacuum defined by $x = \pm a$ – that are responsible for restoring the parity symmetry that is apparently broken by the perturbative vacuum.

Consider the equation of motion [$\dot{x} = dx/dt,\ \ddot{x} = d^2x/dt^2$]

$$0 = \frac{\delta S[x_c]}{\delta x(t)} = m\ddot{x}_c - V'(x_c) \tag{16.20}$$

or

$$m\ddot{x}_c - \frac{g^2}{2}x_c(x_c^2 - a^2) = 0. \tag{16.21}$$

The minimum at $x = \pm a$ are two solutions of the classical equations of motion, and expanding about either of them gives an analytical expansion around $g = 0$.

The equation of motion given in Eq. 16.21 can be re-written as the conservation of energy in the following manner:

$$\frac{d}{dt}\left[\frac{1}{2}m\dot{x}_c^2 - \frac{g^2}{8}(x_c^2 - a^2)^2\right] = 0 = \frac{dE}{dt}$$

$$\Rightarrow\ E = \frac{1}{2}m\dot{x}_c^2 - \frac{g^2}{8}(x_c^2 - a^2)^2. \tag{16.22}$$

Hence, from Eq. 16.22, the Euclidean energy is given by

$$E = \frac{1}{2}m\dot{x}^2 - V(x).$$

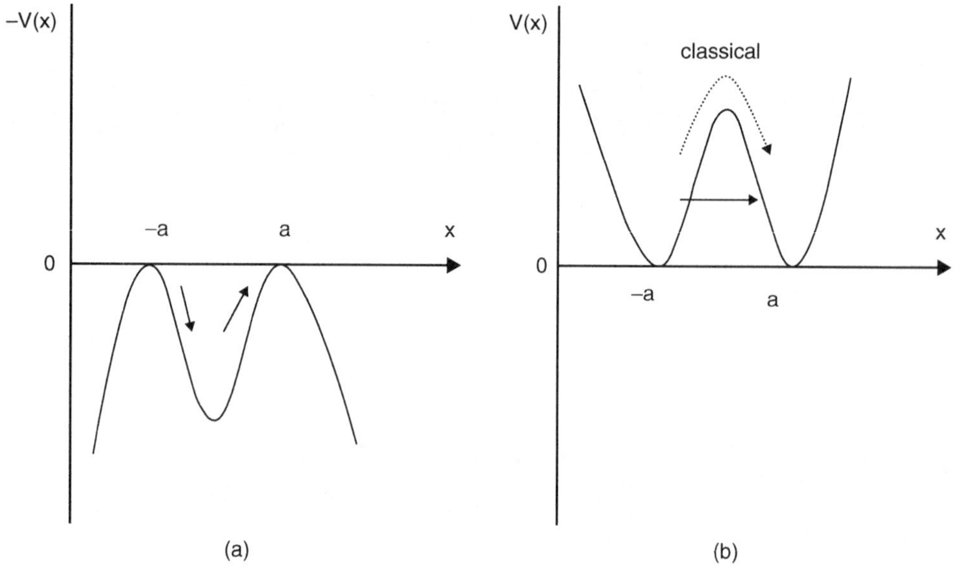

Figure 16.3 The potential for Euclidean time (a), and for Minkowski time (b).

In Minkowski time, the potential is a double well, as shown in Figure 16.3(b). The equations of motion have no classical solutions that go from one well to another when the classical particle does not have an energy E sufficient to cross the potential barrier given by $g^2a^4/8$. In particular, for $E = 0$, which corresponds to the vacuum solutions, there is no trajectory from one well to another.

The Euclidean equations of motion have a nontrivial solution, since it is $-V$ that appears in e^S; hence, the sign of the potential for the Euclidean case has been reversed from the Minkowski case. The potential wells correspond to maxima of two "hills," as shown in Figure 16.3(a), and the classical Euclidean solution corresponds to the particle starting at $x = -a$ and rolling down the hill to reach $x = a$; there is no energy barrier separating the two maxima.

Two zero energy classical solutions are given by the particle being stationary at the two wells $x = \pm a$. The question is, are there other zero energy classical solutions to the equations of motion that contribute to the path integral as well as the $x = \pm a$ solution? The answer is in the affirmative, and we look for these solutions.

For zero energy the classical trajectory, from Eq. 16.22, is given by

$$E = 0 = \frac{1}{2}m\dot{x}^2 - V(x) \;\Rightarrow\; \frac{1}{2}m\dot{x}^2 - \frac{g^2}{8}(x^2 - a^2)^2 = 0. \tag{16.23}$$

The trivial classical solution is given by $x_c = \pm a$ with $\dot{x}_c = 0$ for all time. The nontrivial solution has nonzero velocity and is given by

$$\dot{x} = \pm\frac{g}{2\sqrt{m}}|x_c^2 - a^2| = \pm\frac{g}{2\sqrt{m}}(a^2 - x_c^2), \quad x^2 \le a^2$$
$$\Rightarrow \int_0^{x_\pm} \frac{dy}{a^2 - y^2} = \pm\frac{g}{2\sqrt{m}}\int_{t_c}^{t} dt.$$

Integrating the above equation yields

$$\frac{1}{a}\tanh^{-1}\left(\frac{x_\pm}{a}\right) = \pm\frac{g}{2\sqrt{m}}(t - t_c).$$

Hence the *instanton* and *anti-instanton* classical solutions x_c are given by

$$x_c = x_\pm = \pm a\tanh\left(\frac{\omega(t - t_c)}{2}\right), \qquad \omega^2 = \frac{g^2a^2}{m}. \tag{16.24}$$

The particle starts off with zero velocity at $\pm a$ at $t = -\infty$ and travels very slowly until time t_c when it rapidly crosses over towards $\mp a$ and then slowly rolls to $\mp a$ at $t = +\infty$.

Figure 16.4 shows the shape of the instanton for two different values of ω. As ω becomes large, the particle almost "jumps" in an *instant* from $-a$ to $+a$ at time t_c. Figure 16.5(a) shows a kink (instanton) solution that tunnels from $-a$ to $+a$, and Figure 16.5(b) shows an antikink (anti-instanton) solution that tunnels from a to $-a$.

For this reason, the classical solution is called an instanton or an anti-instanton, depending on whether it tunnels from left to right or from right to left; it also called a kink or an antikink due to the kink-like shape of the classical solution.

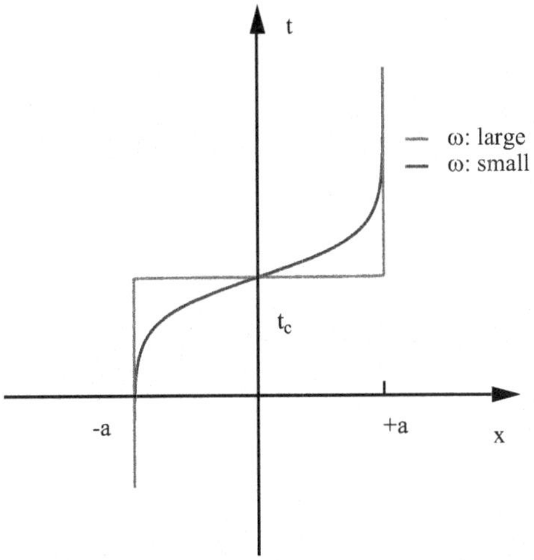

Figure 16.4 The classical instanton solution for different values of $\omega^2 = g^2a^2/m$.

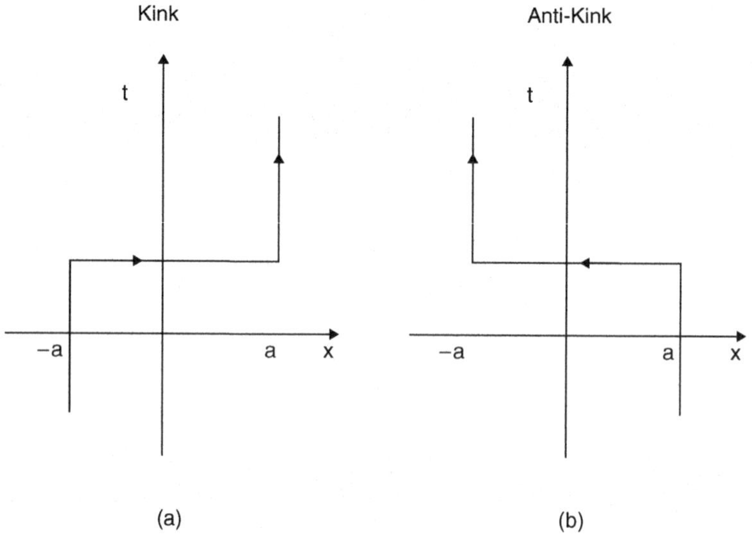

Figure 16.5 Classical solution for large coupling constant g that is (a) a kink or instanton, and (b) an anti-kink or anti-instanton.

The instanton classical action – from the zero energy classical equation given in Eq. 16.23 – is the following:

$$\begin{aligned}S[x_c] &= \int_{\mp\tau/2}^{\pm\tau/2} dt[-\frac{m}{2}\dot{x}_c^2 - V(x_c)] = -m\int_{\mp\tau/2}^{\pm\tau/2} dt\dot{x}_c^2 \\ &== -m\int_{\mp\tau/2}^{\pm\tau/2} dt\dot{x}_c\dot{x}_c = -m\int_{\mp\tau/2}^{\pm\tau/2} dt\frac{dx_c}{dt}\dot{x}_c \\ &= -m\int_{\mp a}^{\pm a} dx_c\dot{x}_c.\end{aligned}$$

Hence, from Eq. 16.23, the classical action for one kink or one anti-kink is given by

$$\begin{aligned}S[x_c] &= -m\int_{\mp a}^{\pm a} dx_c\left(\mp\frac{\omega}{2a}(x_c^2 - a^2)\right) \\ &= \pm\frac{m\omega}{2a}(\frac{1}{3}x_c^3 - x_c a^2)|_{\mp a}^{\pm a} = -\frac{2}{3}m\omega a^2.\end{aligned}$$

Hence, for one kink or anti-kink

$$S_c = -\frac{2m^2\omega^3}{3g^2}.$$

The evolution kernel for the particle to travel from one well to the other is given by

$$K(\pm a, \mp a; \tau) = \langle \pm a | e^{-\tau H} | \mp a \rangle \simeq \mathcal{N} e^{S_c} = \mathcal{N} e^{-\frac{2m^2\omega^3}{3g^2}}$$
$$\equiv \mathcal{N} e^{-\frac{c}{g^2}}, \quad c = \frac{2}{3} m^2 \omega^3$$

: essential singularity around $g = 0$.

A perturbation expansion of the path integral about $g = 0$ to any order cannot produce an essential singularity at $g = 0$.

The classical Lagrangian for a kink or an antikink is given by

$$\mathcal{L}_c = -\frac{1}{2} m \dot{x}_c^2 - V(x_c) = -m \dot{x}_c^2$$
$$= -\frac{ma^2\omega^2}{4} sech^4 \omega \left(\frac{t - t_c}{2} \right).$$

The classical kink Lagrangian, as shown in Figure 16.6, is sharply localized around $t = t_c$ with a width of

$$\Delta t \sim \frac{1}{\omega} = \frac{\sqrt{m}}{ga}.$$

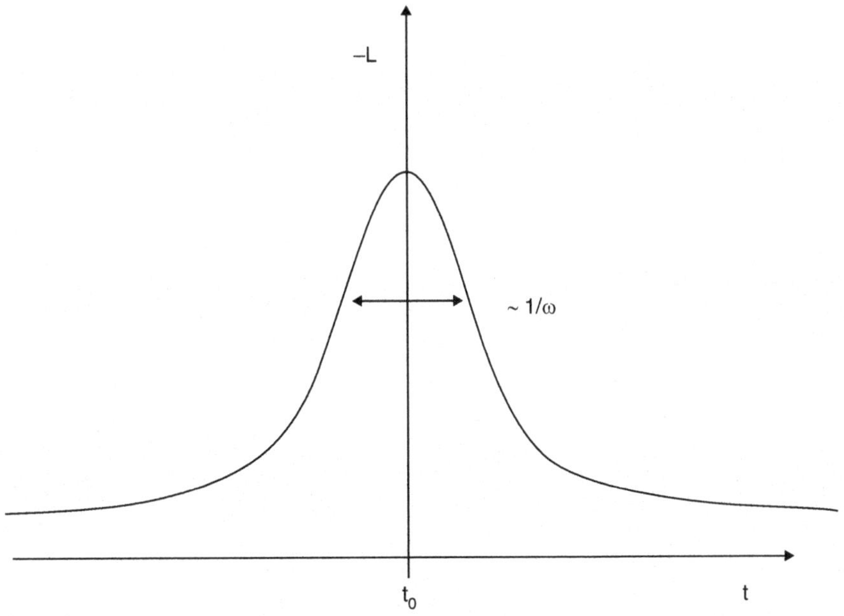

Figure 16.6 Classical Lagrangian for a kink.

The time at which the tunneling occurs, namely t_c, is arbitrary and reflects the fact that the action S, for $\tau \to \infty$, is invariant under translation in time, namely under a shift given by $t \to t+\text{const}$.

At first glance, it seems that, for $g \simeq 0$, $\langle +a|e^{-\tau H}|-a\rangle \sim e^{S_c} = e^{-c/g^2}$ is negligibly small and hence cannot contribute significantly to the transition amplitude. However, this is not true since there is a whole collection of classical solutions, one for each t_c, that are all close to each other, and they all contribute to the transition amplitude. We will show that in fact a more accurate analysis yields

$$\langle +a|e^{-\tau H}|-a\rangle \sim \mathcal{N}e^{-c/g^2}\int_{-\frac{\tau}{2}}^{+\frac{\tau}{2}} dt_c = \mathcal{N}\tau e^{-c/g^2}.$$

Hence, as long as $\tau < e^{-c/g^2}$ the contribution from the classical trajectory is indeed negligible; but for large τ the contribution becomes very large.

There are also multi-kink and multi-antikink solutions with multiple tunneling across the barrier that all contribute in the limit of large time.

There is no exact solution known due to the kink–antikink interaction, but for $g \to 0$, an approximate solution is to compose the multiple kink–antikink solutions from the product of single kink and antikink solutions.

16.4 Instanton zero mode

The double-well action is given by

$$S = -\int dt\left\{\frac{1}{2}m\dot{x}^2 + \frac{g^2}{8}(x^2 - a^2)^2\right\}, \tag{16.25}$$

with the kink, anti-kink classical solution given by ($\omega^2 = a^2g^2/m$)

$$x_c = \pm a\tanh\left(\frac{\omega(t-t_c)}{2}\right).$$

Consider the semi-classical expansion for the transition amplitude. Let the (false) vacua at $\pm a$ be denoted by $\Omega_\pm$. Then

$$\begin{aligned}\langle\Omega_+|e^{-\tau H}|\Omega_-\rangle &= \int dbdb'\langle\Omega_+|b\rangle\langle b|e^{-\tau H}|b'\rangle\langle b'|\Omega\rangle \\ &\cong \Omega_+(a)\langle +a|e^{-\tau H}|-a\rangle\Omega_-(a).\end{aligned} \tag{16.26}$$

The last equation has been obtained using the fact that the false vacua $\Omega_\pm$ are approximately delta-functions (well localized) at points $\pm a$.

Let $x(t) = x_c(t) + \eta(t)$; then, since $\eta(\tau/2) = 0 = \eta(-\tau/2)$, one has from Eq. 16.26

$$\langle +a|e^{-\tau H}|-a\rangle = \int D\eta e^{S[x_c+\eta]}. \tag{16.27}$$

The action has the following semi-classical expansion:[1]

$$\begin{aligned} S[x] &= S[x_c + \eta] \\ &= S[x_c] + \frac{1}{2}\int dt dt' \eta_t \eta_{t'} \frac{\delta^2 S}{\delta x_t \delta x_{t'}} + 0(\eta^3) \\ &= S_c - \frac{1}{2} m \int dt \dot{\eta}^2 - \frac{g^2}{8}\int dt (6x_c^2\eta^2 - 2a^2\eta^2) + 0(\eta^3) \\ &= S_c - \frac{1}{2}\int_{-\tau/2}^{+\tau/2} dt \eta \left(-m\frac{d^2}{dt^2} + \frac{3}{2}g^2 x_c^2 - \frac{g^2a^2}{2}\right)\eta + 0(\eta^3) \\ &= S_c - \frac{1}{2}\int_{-\tau/2}^{+\tau/2} dt dt' \eta(t) M(t,t')\eta(t') + 0(\eta^3), \end{aligned} \tag{16.28}$$

where the matrix elements of the operator M are defined by

$$M(t,t') = \langle t'|M|t\rangle \equiv \left(-m\frac{d^2}{dt^2} + \frac{3}{2}g^2x_c^2 - \frac{g^2a^2}{2}\right)\delta(t-t').$$

To do the $\int D\eta$ path integration one diagonalizes the action using the normal mode expansion of $\eta(t) = \sum_{n=0}^{\infty}\psi_n(t)\eta_n$. The normal modes ψ_n are defined by the eigenfunction equation

$$M|\psi_n\rangle = \lambda_n|\psi_n\rangle. \tag{16.29}$$

The explicit eigenfunction equation, with $\psi_n \equiv \langle t|\psi_n\rangle$, is given by

$$\begin{aligned} \lambda_n\psi_n &= \left(-m\ddot{\psi}_n + \frac{3g^2}{2}x_c^2\psi_n - \frac{g^2a^2}{2}\psi_n\right) \\ &= \left(-m\frac{d^2}{dt^2} + \frac{3g^2a^2}{2}\tanh^2\left(\frac{\omega(t-t_c)}{2}\right) - \frac{g^2a^2}{2}\right)\psi_n. \end{aligned} \tag{16.30}$$

Recall that the classical equation of motion given in Eq. 16.21 yields

$$m\ddot{x}_c - \frac{g^2}{2}x_c^3 + \frac{g^2a^2}{2}x_c = 0.$$

Differentiating the above equation with respect to t, and defining eigenfunction ψ_0 as

$$\psi_0 = \frac{1}{||\dot{x}_c||}\dot{x}_c, \quad ||\dot{x}_c|| = \sqrt{\int_{-\tau/2}^{+\tau/2} dt\dot{x}_c^2}, \tag{16.31}$$

[1] Note that $(x^2 - a^2)^2 = (x_c^2 - a^2)^2 + 6x_c^2\eta^2 - 2a^2\eta^2 + 0(\eta^3)$.

yields

$$m\ddot{\psi}_0 - \frac{3g^2}{2}x_c^2\psi_0 + \frac{g^2a^2}{2}\psi_0 = 0. \tag{16.32}$$

Hence, from Eqs. 16.30 and 16.32, ψ_0 is an eigenfunction such that

$$\lambda_0 = 0 \quad \text{for } \psi_0.$$

The zero eigenvalue eigenfunction, also called the zero mode ψ_0, arises in the action due to the time translation invariance of the action: the instanton can tunnel from $-a$ to $+a$ at any time t_c. This is the reason that the eigenvalue equation for ψ_0 is a *direct result* of the equations of motion, which is also time translation invariant.

The operator M, since $\lambda_0 = 0$, has the spectral representation

$$M = \sum_{n=0}^{\infty} \lambda_n |\psi_n\rangle\langle\psi_n| = \sum_{n=1}^{\infty} \lambda_n |\psi_n\rangle\langle\psi_n|, \tag{16.33}$$

$$\det M = \prod_{n=0}^{\infty} \lambda_n = 0.$$

Note that M is singular, having no inverse, since $\det M = 0$ due to the zero eigenvalue $\lambda_0 = 0$. Hence, we obtain the path integral given in Eq. 16.27, since

$$\langle +a|e^{-\tau H}| - a\rangle = \int D\eta e^{S[x_c+\eta]} \simeq \frac{\mathcal{N}e^{S_c}}{\sqrt{\det M}} = \infty. \tag{16.34}$$

16.5 Instanton zero mode: Faddeev–Popov analysis

The Faddeev–Popov method is of great generality and is indispensable in the quantization of Yang–Mills gauge fields. The elimination of a divergence in the path integral due to a symmetry is analyzed using the Faddeev–Popov approach; the advantage of carrying out this analysis is that it illustrates the main features of the Faddeev–Popov approach in a relatively simple context.

The analysis yields what are called the collective coordinates. The fundamental ingredient in the Faddeev–Popov approach is to explicitly introduce a *constraint* into the path integral that breaks the symmetry, thus removing the singularity in the action; a term has to be introduced to compensate for constraint so as to leave the path integral invariant.

We recall that the zero mode arises due to the invariance of the action under shifting the time coordinate, namely $t \to t$+constant. This symmetry is strictly an invariance of the action only for $\tau \to \infty$, with corrections that are exponentially small and can be ignored in our analysis. In particular, from Eq. 16.18 and for $\tau >> t_*$, the double-well action has the symmetry

$$\begin{aligned} S &= -\int_{-\tau/2}^{+\tau/2} dt \left(\frac{1}{2}m\dot{x}^2(t) + \frac{g^2}{8}\{x^2(t) - a^2\}^2\right) \\ &\simeq -\int_{-\tau/2}^{+\tau/2} dt \left(\frac{1}{2}m\dot{\tilde{x}}^2(t) + \frac{g^2}{8}\{\tilde{x}^2(t) - a^2\}^2\right), \ \tilde{x}(t) = x(t+t_*) \\ &\Rightarrow S[x] = S[\tilde{x}]. \end{aligned} \tag{16.35}$$

For the double well potential, in the Faddeev–Popov approach, one breaks the invariance under $t \to t+t_*$ by directly introducing a fixed value for time, denoted by t_0, directly into the action in the following manner. Following Zinn-Justin (2005), consider the identity

$$1 = \int dp\delta[p - \lambda]. \tag{16.36}$$

To apply Eq. 16.36 to the double-well potential consider the following change of variable

$$p = \int_{-\tau/2}^{+\tau/2} dt\ \dot{x}_c(t)x(t+t_0), \tag{16.37}$$

$$\delta[p - \lambda] = \delta\left[\int_{-\tau/2}^{+\tau/2} dt\ \dot{x}_c(t)x(t+t_0) - \lambda\right], \tag{16.38}$$

which yields

$$dp = dt_0 \int_{-\tau/2}^{+\tau/2} dt\ \dot{x}_c(t)\dot{x}(t+t_0). \tag{16.39}$$

Hence, from Eqs. 16.36 and 16.39

$$1 = \int_{-\tau/2}^{+\tau/2} dt_0 \left[\int_{-\tau/2}^{+\tau/2} dt\dot{x}_c(t)\dot{x}(t+t_0)\right]\delta[\int dt\dot{x}_c(t)x(t+t_0) - \lambda]. \tag{16.40}$$

The path integral given in Eq. 16.27 is re-written – using Eq. (16.40) – in the following manner

$$\begin{aligned} \langle +a|e^{-\tau H}|-a\rangle &= K(a,-a;\tau) = \int Dxe^S \\ &= \int Dxe^S \int_{-\tau/2}^{+\tau/2} dt_0\Big[\int_{-\tau/2}^{+\tau/2} dt\dot{x}_c(t)\dot{x}(t+t_0)\Big]\delta[\int dt\dot{x}_c(t)x(t+t_0) - \lambda]. \end{aligned}$$

The crucial step is to define a change of integration variables as follows[2]

$$x(t) \to \tilde{x}(t) = x(t - t_0). \tag{16.41}$$

Due to Eq. 16.35, the action is invariant under this change of variables and

$$S[x] = S[\tilde{x}]. \tag{16.42}$$

In particular, note that $S[\tilde{x}]$ is independent of t_0. Hence, dropping the tilde on x yields

$$\begin{aligned} K(a, -a; \tau) &= \int Dx e^S \int_{-\tau/2}^{+\tau/2} dt_0 \int_{-\tau/2}^{+\tau/2} dt \dot{x}_c(t) \dot{x}(t) \delta \left[\int dt \dot{x}_c(t) x(t) - \lambda \right] \\ &= \left[\int_{-\tau/2}^{+\tau/2} dt_0 \right] \cdot \int Dx e^S \int_{-\tau/2}^{+\tau/2} dt \dot{x}_c(t) \dot{x}(t) \delta \left[\int dt \dot{x}_c(t) x(t) - \lambda \right]. \end{aligned}$$

The zero mode, namely the variable t_0, has been completely factorized out of the path integral, which no longer has any divergence due to the delta-function constraint.

The zero mode is a physical quantity that reflects that tunneling can take place at any instant in the interval $[-\tau/2, +\tau/2]$. In contrast, for the case of path integrals for gauge fields, the Faddeev–Popov procedure leads to a factorization of the nonphysical gauge degrees of freedom.

It is more convenient to re-write the delta-function constraint as another term in the action. To do this note that the above equation is valid for all λ; hence, multiplying both sides with $\exp\{-\alpha\lambda^2/2)$ and integrating over λ yields

$$K(a, -a; \tau) = \tau \sqrt{\frac{\alpha}{2\pi}} \int Dx \int_{-\tau/2}^{+\tau/2} dt \dot{x}_c(t) \dot{x}(t) e^{S - \frac{\alpha}{2} \left(\int_{-\tau/2}^{+\tau/2} dt \dot{x}_c(t) x(t) \right)^2}. \tag{16.43}$$

Note that since α is an arbitrary parameter, the evolution kernel $K(a, -a; \tau)$ must be independent of it.

The classical field equation for the path integral given in Eq. 16.43 is given by

$$\left[\frac{\delta S}{\delta x(t)} - \alpha \dot{x}_c(t) \int_{-\tau/2}^{+\tau/2} dt' \dot{x}_c(t') x(t') \right] \Bigg|_{x = x_c} = 0. \tag{16.44}$$

From Eq. 16.20,

$$\frac{\delta S[x_c]}{\delta x(t)} = 0 \tag{16.45}$$

[2] Note $x_c(t)$ is unchanged since, unlike $x(t)$, it is not an integration variable.

and

$$\int_{-\tau/2}^{+\tau/2} dt\dot{x}_c(t)x_c(t) = x_c^2(\tau/2) - x_c^2(-\tau/2) = 0.$$

Hence, the classical trajectory $x_c(t)$ defined by the double-well action given by Eq. 16.45 is also a classical solution (maximum) of the modified action given in Eq. 16.44.

Note that, from Eq. 16.31, the modified action can be written as

$$S' = S - \frac{\tilde{\lambda}_0}{2}\langle x|\psi_0\rangle\langle\psi_0|x\rangle, \quad \tilde{\lambda}_0 = \alpha||\dot{x}_c||^2, \tag{16.46}$$

$$\psi_0 = \frac{1}{||\dot{x}_c||}\dot{x}_c, \quad \langle\psi_0|x\rangle \equiv \frac{1}{||\dot{x}_c||}\int_{-\tau/2}^{+\tau/2} dt\dot{x}_c(t)x(t).$$

Expanding the path integral about the classical solution,

$$x(t) = x_c(t) + \eta(t), \tag{16.47}$$

yields, from Eq. 16.28, the action

$$S[x] = S_c - \frac{1}{2}\int_{-\tau/2}^{+\tau/2} dtdt'\eta(t)M(t,t')\eta(t') - \frac{\tilde{\lambda}_0}{2}\langle\eta|\psi_0\rangle\langle\psi_0|\eta\rangle, \tag{16.48}$$

$$= S_c - \frac{1}{2}\int_{-\tau/2}^{+\tau/2} dtdt'\eta(t)\tilde{M}(t,t')\eta(t'), \tag{16.49}$$

where

$$\tilde{M} = \tilde{\lambda}_0|\psi_0\rangle\langle\psi_0| + \sum_{n=1}^{\infty}\lambda_n|\psi_n\rangle\langle\psi_n|.$$

Note from Eq. 16.29 that ψ_0 is a vector that is orthogonal to all the eigenfunctions of operator M and yields the completeness equation

$$\mathcal{I} = |\psi_0\rangle\langle\psi_0| + \sum_{n=1}^{\infty}|\psi_n\rangle\langle\psi_n|.$$

Hence $\tilde{M}$ is a nonsingular operator with determinant given by

$$\det\tilde{M} = \tilde{\lambda}_0\prod_{n=1}^{\infty}\lambda_n \neq 0.$$

Equation 16.43 yields

$$K(a,-a;\tau)=\tau||\dot{x}_c||^2\sqrt{\frac{\alpha}{2\pi}}e^{S_c}\int D\eta\exp\{-\frac{1}{2}\int_{-\tau/2}^{+\tau/2}dtdt'\eta(t)\tilde{M}(t,t')\eta(t')\}$$
$$=\tau||\dot{x}_c||^2\sqrt{\frac{\alpha}{2\pi}}e^{S_c}\times\frac{1}{\sqrt{\det\tilde{M}}}=\tau||\dot{x}_c||^2\sqrt{\frac{\alpha}{2\pi}}e^{S_c}\times\frac{1}{\sqrt{\tilde{\lambda}_0\prod_{n=1}^{\infty}\lambda_n}}$$
$$=\tau||\dot{x}_c||e^{S_c}\times\frac{1}{\sqrt{2\pi\prod_{n=1}^{\infty}\lambda_n}},\quad\text{finite,}\tag{16.50}$$

where Eq. 16.46 has been used for the value of $\tilde{\lambda}_0$.

The evolution kernel obtained above is indeed convergent, unlike the earlier result derived in Eq. 16.34 for which it was divergent. Furthermore, $K(a,-a;\tau)$ given in Eq. 16.50 is indeed independent of α, as expected, since all the eigenvalues λ_n and the constant are independent of α.

16.5.1 Instanton coefficient $\mathcal{N}$

The one instanton transition amplitude is given by

$$\langle -a|e^{-\tau H}|a\rangle=\mathcal{N}e^{S_c}\int_{-\tau/2}^{+\tau/2}dt_c=\mathcal{N}\tau e^{S_c},$$

which is obtained from

$$\langle -a|e^{-\tau H}|a\rangle=\frac{\tau e^{S_c}}{\sqrt{\det(-m\frac{d^2}{dt^2}+V''(x_c))}}.$$

The determinant is apparently zero due to the zero mode giving rise to a zero eigenvalue.

Instead of the Faddeev–Popov approach, another way of factoring out the singularity due to the zero mode is to do an expansion of η that directly takes into account the zero mode of the action. This is accomplished by the expansion

$$x(t)=x_c(t-t_c)+\sum_{n\neq 0}^{\infty}\psi_n(t-t_c)c_n,$$

where t_c is now considered an arbitrary variable. Note that there is no coefficient c_0 multiplying the classical solution $x_c(t-t_c)$; this is because c_0 is replaced by an arbitrary variable t_c, namely the instant at which the particle tunnels from one vacuum state to the other. The change of variables from $x(t)$ to c_n, t_c yields [Das (2006)]

$$Dx = ||\dot{x}_c||dt_c \prod_{n=1}^{\infty} dc_n \quad \Rightarrow \quad \int Dx = ||\dot{x}_c|| \int dt_c \prod_{n=1}^{\infty} \int dc_n. \tag{16.51}$$

By eliminating the zero eigenvalue in the eigenfunction expansion of $\eta(t)$ the zero mode in the determinant has been eliminated and the nonzero and finite determinant is given by

$$\det(-m\frac{d^2}{dt^2} + V''(x_c)) = \prod_{n=1}^{\infty} \lambda_n. \tag{16.52}$$

Hence, the semi-classical expansion in the one instanton approximation yields

$$\begin{aligned} \langle -a|e^{-\tau H}|a\rangle &= \text{const}\, e^{S_c} \left[\prod_{n=1}^{\infty} \int dc_n e^{\frac{-\lambda_n c_n^2}{2}} \right] \int_{-\tau/2}^{+\tau/2} dt_c \\ &= \mathcal{N} \tau e^{S_c}. \end{aligned} \tag{16.53}$$

The coefficient $\mathcal{N}$ that appears in the transition amplitude is evaluated. For the trivial classical solution given by $x_c = \pm a$ in which the particle sits at one of the wells, the classical action is zero. The effective action about one of the wells is given by the simple harmonic oscillator action

$$S_0 = -\int_{-\tau/2}^{+\tau/2} dt \left(\frac{1}{2} m\dot{\eta}^2 + \frac{g^2}{2} a^2 \eta^2 \right),$$

$$\text{boundary conditions: } \eta(-\tau/2) = 0 = \eta(+\tau/2).$$

The zeroth order transition amplitude is given by Eq. 11.21 for the simple harmonic oscillator,[3]

$$\langle 0|e^{-\tau H}|0\rangle_{(0)} = e^{S_c(\pm a)} \int D\eta e^{S_0} = \sqrt{\frac{m\omega}{2\pi\hbar \sinh \omega\tau}}, \tag{16.54}$$

$$\begin{aligned} &\simeq e^{-\omega\tau/2} \left(\frac{m\omega}{\pi\hbar} \right)^{1/2} + O(e^{-\omega\tau}) \\ &\equiv A, \end{aligned} \tag{16.55}$$

where $S_c(\pm a) = 0$ and we recall that $\omega^2 = g^2 a^2/2$. A careful analysis shows that

$$\langle -a|e^{-\tau H}|a\rangle = Ar \int_{-\tau/2}^{+\tau/2} dt_c = Ar\tau,$$

[3] Recall that for the simple harmonic oscillator – restoring the dependence on $\hbar$ – is given by Eq. 11.21 as

$$K(x', x; \tau) = \sqrt{\frac{m\omega}{2\pi\hbar \sinh \omega\tau}} \exp\left\{ -\frac{m\omega}{2\hbar \sinh \omega\tau} \left[(x^2 + x'^2) \cosh \omega\tau - 2xx' \right] \right\}.$$

where A is the coefficient given in Eq. 16.55. The factor r comes from the instanton determinant given in Eq. 16.52 and has been shown by Das (2006) to be

$$r = 2a\sqrt{\frac{2m}{\hbar}}\omega^{3/2}e^{S_c}.$$

16.6 Multi-instantons

Exact multi-instanton classical solutions, characterized by multiple tunnelings, have not yet been found. However, for tunneling at times that are widely separated, and for coupling $g \to \infty$ one can build approximate multiple-instanton solutions from the single kink/antikink solutions. Consider the two kink–antikink ansatz

$$x_c^{(2)}(t) = \pm a^2 \tanh\left(\frac{\omega(t-t_1)}{2}\right)\tanh\left(\frac{\omega(t-t_2)}{2}\right), \quad t_2 \gg t_1 \tag{16.56}$$

as shown in Figure 16.7.

It has been shown by Das (2006) that $x_c^{(2)}$ satisfies the equations of motion up to terms of $O(e^{-\omega(t_2-t_1)})$.

In general a string of widely separated instantons and anti-instantons satisfies the classical equations of motion. In the dilute instanton gas approximation, which is valid for strong coupling $g \gg 1$, an N-instanton classical solution is approximately composed of the product of N instantons and anti-instantons,

$$x_c^N(t) = \prod_{i=1}^{N} x_c(t-t_i)$$

$$\text{with} -\frac{\tau}{2} \leq t_n \leq t_{n-1}\cdots \leq t_1 \leq \frac{\tau}{2}.$$

For $g^2 \to \infty$, one can make further approximations of the N-instanton solution: the cross over from one well to another is instantaneous. Furthermore, the N-instanton configuration, similarly to Eq. 16.56, is constructed by multiplying

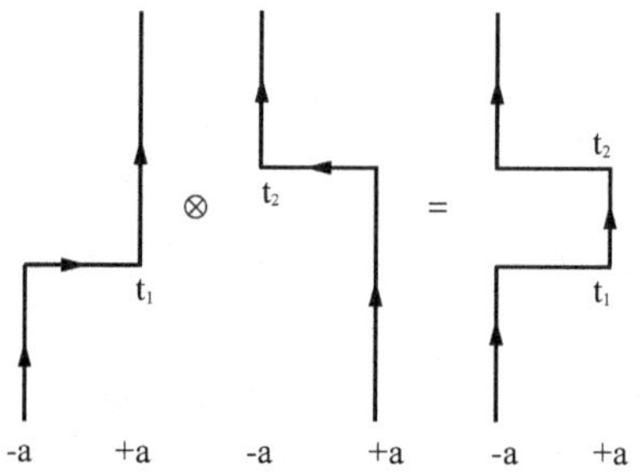

Figure 16.7 A two-kink configuration for the particle's trajectory.

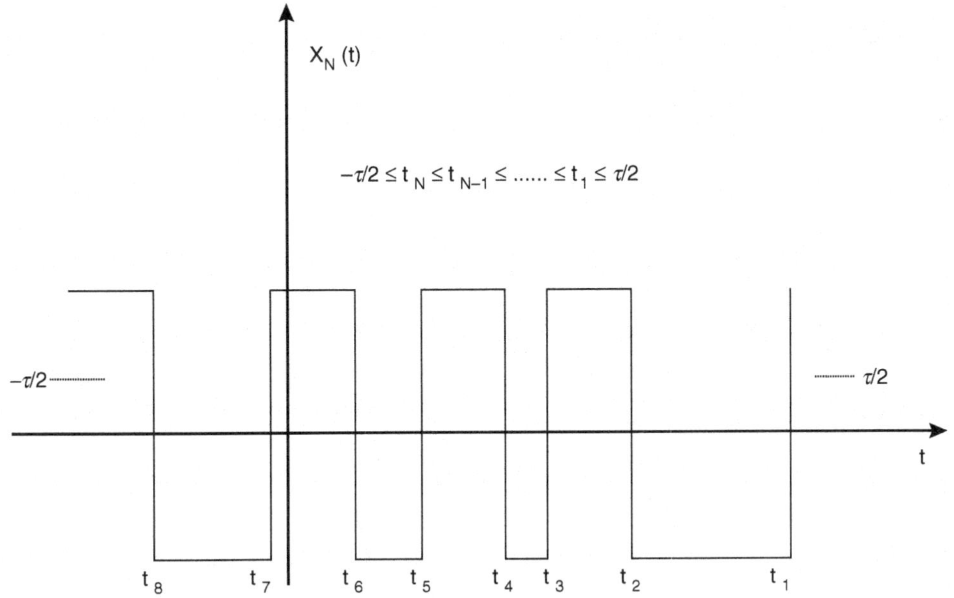

Figure 16.8 A multiple kink–antikink configuration for the particle's classical trajectory.

the configuration of N single instantons, with each instanton tunneling taken to be independent of the other tunnelings. In this approximation, since

$$\lim_{\omega\to\infty} \tanh(\omega\theta) = sgn(\theta), \quad sgn(\theta) = \begin{cases} -1 & \theta < 0 \\ 1 & \theta > 0 \end{cases},$$

the N-instanton classical solution is given by

$$x_c^{(N)}(t) = \pm a^N \prod_{j=1}^{N} sgn(t - t_j). \tag{16.57}$$

A typical N-instanton configuration is shown in Figure 16.8.

16.7 Instanton transition amplitude

In the dilute instanton gas approximation, the transition amplitude is given by

$$\begin{aligned} \langle -a|e^{-\tau H}|a\rangle_N &= Ar^N \int_{-\tau/2}^{\tau/2} dt_1 \int_{-\tau/2}^{t_1} dt_2 \cdots \int_{-\tau/2}^{t_{n-1}} dt_n \\ &= Ar^N \frac{1}{N!}\left(\int_{-\tau/2}^{\tau/2} dt\right)^N \\ &= A\frac{1}{N!} r^N \tau^N, \ N - \text{instanton contribution.} \end{aligned} \tag{16.58}$$

Hence, the full transition matrix element is given by summing over odd N multiple instanton configurations, and yields

$$\langle -a|e^{-\tau H}|a\rangle = \sum_{N=0}^{\infty}\langle -a|e^{-\tau H}|a\rangle_{2N+1} = A\sum_{N=0}^{\infty}\frac{r^{2N+1}}{(2N+1)!}\tau^{2N+1}$$

$$= A\sinh(r\tau) = \frac{1}{2}\sqrt{\frac{m\omega}{\pi\hbar}}\left(e^{-(\frac{\omega}{2}-r)\tau} - e^{-(\frac{\omega}{2}+r)\tau}\right), \quad (16.59)$$

where A is given by Eq. 16.55,

$$A = (\frac{m\omega}{\pi\hbar})^{1/2}e^{-\frac{\omega\tau}{2}}.$$

16.7.1 Lowest energy states

For the double well potential, $\langle a|e^{-\tau H}|a\rangle$ is given by the sum over all even N multiple instanton configurations, starting with $N = 0$. Hence, from Eq. 16.58

$$\langle a|e^{-\tau H}|a\rangle = A\sum_{N=0}^{\infty}\frac{(r\tau)^{2N}}{(2N)!} = A\cosh(r\tau)$$

$$= \frac{1}{2}(\frac{m\omega}{\pi\hbar})^{1/2}[e^{-(\frac{\omega}{2}-r)\tau} + e^{-(\frac{\omega}{2}+r)\tau}].$$

Let the two lowest excited states, as given in Figure 16.9, be denoted by $|-\rangle$ and $|+\rangle$; the completeness equation in this two-state sector is given by

$$\mathbb{I} \simeq |-\rangle\langle -| + |+\rangle\langle +| + O(e^{-4/g^2})^2. \quad (16.60)$$

Hence, the completeness equation given in Eq. 16.60 yields

$$\langle a|e^{-\tau H}|a\rangle \approx \langle a|e^{-\tau H}|+\rangle\langle +|a\rangle + \langle a|e^{-\tau H}|-\rangle\langle -|a\rangle$$

$$= e^{-\tau E_+}|\langle -|a\rangle|^2 + e^{-\tau E_-}|\langle +|a\rangle|^2$$

$$= \frac{1}{2}(\frac{m\omega}{\pi\hbar})^{1/2}[e^{-(\frac{\omega}{2}-r)\tau} + e^{-(\frac{\omega}{2}+r)\tau}].$$

Hence, the two lowest energy eigenvalues are given by

$$E_- \simeq \hbar(\frac{\omega}{2} - r), \quad E_+ \simeq \hbar(\frac{\omega}{2} + r), \quad (16.61)$$

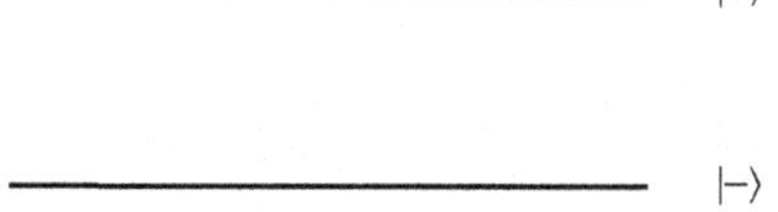

Figure 16.9 Ising configuration and an instanton.

yielding the level splitting given by

$$\Delta E = E_+ - E_- = 2\hbar r = 4\omega^{3/2}\sqrt{2m\hbar}\, e^{S_c}.$$

16.8 Instanton correlation function

The correlation of the double-well potential can be obtained far from the linearized theory by expanding the path integral about the instanton solutions [Polyakov (1987)]. Consider the two point correlator

$$G(T_1, T_2) = \frac{1}{Z}\int Dx\, x(T_1)x(T_2)e^S.$$

A normal mode expansion of the correlation function about $x_c^{(\ell)}$, the ℓ instanton configuration, yields

$$x(t) = \eta_\ell(t - t_c) + x_c^{(\ell)}(t - t_c) = \sum_{n=1}^{\infty} c_n\psi_n(t - t_c) + x_c^{(\ell)}(t - t_c).$$

The fluctuations about the $x_c^{(\ell)}$ are given by η_ℓ. In the semi-classical expansion, the correlator is given by

$$G(T_1, T_2) \simeq \frac{\sum_\ell \int Dx[\eta(T_1 - t_c) + x^{(\ell)}(T_1 - t_c)][\eta(T_2 - t_c) + x^{(\ell)}(T_2 - t_c)]e^{S[\eta + x^{(\ell)}]}}{\sum_\ell \int Dx e^{S[\eta + x^{(\ell)}]}}.$$

Note that $\sum_\ell$ is a sum over the ℓ-instanton and anti-instanton classical solutions, denoted by $x^{(\ell)}$. The expansion of the action yields, to leading order

$$S[x] = S[\eta_\ell + x_c^\ell] = S[x_c^\ell] + S[\eta_\ell] + \ldots$$

The first two classical solutions are $S[x_c^0] = 0,\ S[x_c^1] = S_c$.

Summing over the trivial classical ($\ell = 0$) solution given by $x = \pm a$ and the $\ell = 1$ single instanton and anti-instanton solution yields, to leading order in $\hbar$,

$$G(T_1, T_2) \simeq \frac{a^2 Z_0 + e^{S_c} Z_1 \int dt_c x_c(T_1 - t_c)x_c(T_2 - t_c)}{Z_0 + e^{S_c} Z_1 \int dt_c}$$

$$= \frac{a^2 + Be^{S_c}\int dt_c x_c(T_1 - t_c)x_c(T_2 - t_c)}{1 + Be^{S_c}\int dt_c}, \tag{16.62}$$

$$\Rightarrow G(T_1, T_2) \simeq a^2 + Be^{S_c}\int_{-\infty}^{+\infty} dt_c\Big[x_c(T_1 - t_c)x_c(T_2 - t_c) - a^2\Big]. \tag{16.63}$$

Coefficient Z_0 is given by

$$Z_0 = \int D\eta_0 e^{S[\eta_0]} = \prod_{n=1}^{\infty} \int dc_n e^{-\frac{1}{2}(\lambda_{n,0})^2 c_n^2},$$

where the classical solution $x_c^{(0)}$ is given by $x_c^{(0)} = \pm a$. The effective action $S[\eta_0]$ is given in Eq. 16.19 and $\lambda_{n,0}$ are the eigenvalues for the effective action about the trivial classical solution $\pm a$.

From Eq. 16.51, for the one instanton solution the path integral measure is $D\eta_1 = ||\dot{x}_c|| dt_c \prod_{n=1}^{\infty} dc_n = dt_c D'\eta_1$. The t_c integration couples to the classical solutions and is included in Eq. 16.62. Coefficient Z_1 is given by the remaining integrations,

$$Z_1 = \int D'\eta_1 e^{S[\eta_1]}.$$

The classical solution $x_c^{(1)}$ is the one instanton solution and yields

$$Z_1 = ||\dot{x}_c|| \prod_{n=1}^{\infty} \int dc_n e^{S[\eta_1]} = ||\dot{x}_c|| \prod_{n=1}^{\infty} \int dc_n e^{-\frac{1}{2}(\lambda_n)^2 c_n^2},$$

where the eigenvalues λ_n for the single instanton are given in Eq. 16.30.

Hence, from Eq. 16.62

$$B = \frac{Z_1}{Z_0} = ||\dot{x}_c|| \cdot \frac{\prod_{n=1}^{\infty} \int dc_n e^{-\frac{1}{2}\lambda_n^2 c_n^2}}{\prod_{n=1}^{\infty} \int dc_n e^{-\frac{1}{2}(\lambda_{n,0})^2 c_n^2}}. \tag{16.64}$$

16.9 The dilute gas approximation

The dilute gas approximation, from Eq. 16.57, is based on an approximate description of the double-well potential for large coupling, namely $g \gg 1$ and leads to the simplification that $x_c(t_i - t_c) = \pm a\, sgn(t_i - t_c)$. The dilute gas approximation considers the N-instanton solution to consist of a collection of N noninteracting instanton and anti-instanton solutions. With these assumptions, the double well potential can be solved exactly.

From Eq. 16.63, for $T_2 > T_1$, the integral over the one instanton contribution is given by[4]

$$a^2 B \int_{-\infty}^{+\infty} dt_c\, [sgn(T_1 - t_c) sgn(T_2 - t_c) - 1]$$
$$= a^2 B \left[\int_{-\infty}^{T_1} dt_c - \int_{T_1}^{T_2} dt_c + \int_{T_2}^{+\infty} dt_c - \int_{-\infty}^{+\infty} dt_c \right]$$

[4] The case of $T_1 > T_2$ is obtained in a straightforward manner by exchanging T_1 with T_2.

$$= -2a^2B|T_1 - T_2|, \qquad |T_1 - T_2| \gg \omega^{-1}. \tag{16.65}$$

The integration range is shown in Figure 16.7.

Hence, in the one instanton approximation, the correlator is given by

$$G(T_1, T_2) = a^2(1 - 2B|T_1 - T_2|e^{S_c}).$$

In the dilute instanton gas approximation, the contributions of all the multi-instantons are simply products of the contribution of a single instanton and anti-instanton. The multi-instanton tunneling times $t_{c,n}$ yield

$$x_c^N(t) = \prod_{i=1}^{N} x_c(t - t_{c,i})$$

$$\text{with } T_1 \leq t_N \leq t_{N-1} \cdots \leq t_1 \leq T_2.$$

Note that the instantons are well localized; for large time t, the single instanton has the behavior

$$x_c(t) = \pm a(1 + O(e^{-\omega t}),$$

The action for the superposition of an instanton (I) and an anti-instanton (AI), with separation T_{12}, is given by

$$\begin{aligned} S_2 &\simeq S_c(I) + S_c(AI) + O(e^{-\omega T_{12}}) \\ &= 2S_c, \end{aligned} \tag{16.66}$$

and hence the action for N-instantons is

$$S_N \simeq NS_c. \tag{16.67}$$

Similarly to the derivation of Eq. 16.59, the correlator for the dilute instanton gas is given by

$$\begin{aligned} G(T_1, T_2) &\simeq a^2 \sum_{N=0}^{\infty} (-2B)^N e^{NS_c} \int dt_{c,N} dt_{N-1} \cdots dt_1 \\ &= a^2 \sum_{N=0}^{\infty} (-2B)^N e^{NS_c} \frac{(T_2 - T_1)^N}{N!} \\ &= a^2 e^{-|T_1 - T_2|/\xi}, \end{aligned}$$

where correlation time ξ is given by

$$\xi \simeq \frac{1}{2B} e^{-S_c} = \frac{1}{2B} \exp\left\{\frac{2m^2\omega^3}{3g^2}\right\} \to \infty \text{ as } g \to 0.$$

The nonlinear double well potential gives rise to a nonperturbative correlation length, with an essential singularity at $g = 0$. Any perturbation about the trivial vacua $x = \pm a$ would not yield the correlation length; in contrast, the semi-classical expansion yields a well-defined order by order procedure for evaluating the path integral.

16.10 Ising model and the double well potential

The N-instanton solution, in the dilute gas approximation, can be constructed from N single instanton configurations. Since the overlap of the component one instanton configurations is taken to be negligible, the action is given by

$$S = NS_c = -N\frac{c}{g^2},$$

with tunneling times given by

$$t_1 < t_2 < t_3 \cdots < t_N.$$

From Eq. 16.57 we find that the N-instanton solution, for strong coupling given by $g^2 \gg 1$, reduces to

$$x_c^{(N)}(t) = \pm a \prod_{j=1}^{N} sgn(t - t_j).$$

A single instanton configuration is equivalent to a configuration of the Ising spins, as shown in Figure 16.10. The multi-instanton configuration is now equivalent to the one-dimensional Ising model, as the trajectory of the particle has a value of either $+a$ or $-a$, and the tunneling between these two configurations is equal to a spin flip for the Ising spin.

To make the connection with the Ising model more explicit, consider the limit of $\omega = ag/\sqrt{m} \to \infty$, or equivalently $g^2 \to \infty$; up to irrelevant constants, the potential given in Eq. 16.15 yields

$$\lim_{g\to\infty} e^{-\frac{g^2}{8}(x^2-a^2)^2} \to \delta(x^2 - a^2) = \frac{1}{2a}[\delta(x - a) + \delta(x + a)].$$

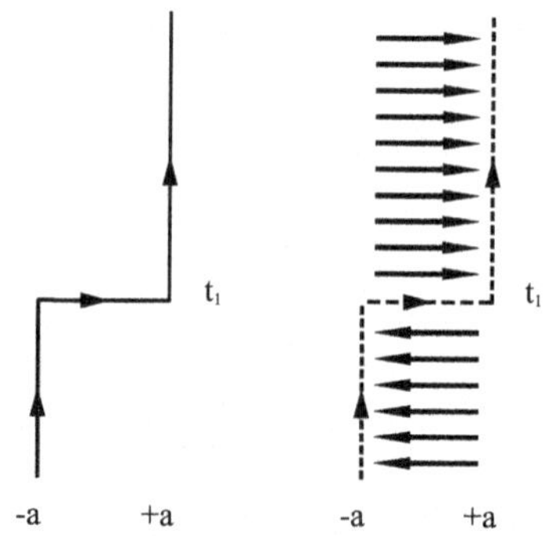

Figure 16.10 Ising configuration and an instanton.

Discretizing time $t = n\epsilon$, the path integral given in Eq. 16.18 yields

$$
\begin{aligned}
Z &= \int Dx e^{-\frac{m}{2\epsilon}\sum_i (x_{i+1}-x_i)^2} \prod_i \delta(x_i^2 - a^2) \\
&= \int Dx e^{\frac{m}{\epsilon}\sum_i (x_i x_{i+1}-x_i^2)} \prod_i \delta(x_i^2 - a^2) \\
&= \sum_{\{\mu_i=\pm 1\}} e^{K\sum_i \mu_i \mu_{i+1}}, \text{ Ising model,}
\end{aligned}
$$

where

$$
\mu_i = x_i/a, \quad K = \frac{ma^2}{\epsilon}.
$$

Recall from Eq. 8.22, the Ising model's transfer matrix is

$$
L = \begin{pmatrix} e^K & e^{-K} \\ e^{-K} & e^K \end{pmatrix}.
$$

The eigenstates and eigenvalues, from Eqs. 8.27 and 8.25, are given by

$$
\begin{aligned}
|\lambda_1\rangle &= \frac{1}{\sqrt{2}}\begin{pmatrix} 1 \\ 1 \end{pmatrix}, \quad \lambda_1 = e^K + e^{-K}, \\
|\lambda_2\rangle &= \frac{1}{\sqrt{2}}\begin{pmatrix} 1 \\ -1 \end{pmatrix}, \quad \lambda_2 = e^K - e^{-K}.
\end{aligned}
$$

Since $\lambda_1 > \lambda_2$, the two states have the following identification:

$$
\begin{aligned}
|\lambda_1\rangle &= \frac{1}{\sqrt{2}}\begin{pmatrix} 1 \\ 1 \end{pmatrix} \equiv |-\rangle, \quad \text{vacuum state,} \\
|\lambda_2\rangle &= \frac{1}{\sqrt{2}}\begin{pmatrix} 1 \\ -1 \end{pmatrix} \equiv |+\rangle. \qquad (16.68)
\end{aligned}
$$

The vacuum state is symmetric under the exchange of $a \to -a$ and hence preserves parity symmetry.

16.11 Nonlocal Ising model

Let us consider the periodic and *non local* Ising model

$$
\begin{aligned}
S &= \frac{K}{2}\sum_{i,j=1}^{N} \mu_i e^{-\alpha|i-j|}\mu_j \\
&= \frac{K}{2}\sum_{ij} \mu_i \mu_j A_{ij}. \qquad (16.69)
\end{aligned}
$$

Using Gaussian integration, the action S has the representation

$$e^S = \prod_{i=1}^{N} \int dx_i \exp[-\frac{1}{2}\sum_{ij} x_i A_{ij}^{-1} x_j + \sum_i \mu_i x_i],$$
$$A_{ij} = K e^{-\alpha|i-j|}.$$

It can be verified that the inverse of A_{ij} is given by

$$A_{ij}^{-1} = \frac{1}{K}[C\delta_{i-j} - B(\delta_{i-j-1} + \delta_{i-j+1})].$$

Then

$$S = -\frac{1}{2K}\sum_{i,j=1}^{N} x_i \Big[C\delta_{i-j} - B(\delta_{i-j-1} + \delta_{i-j+1})\Big] x_j + \sum_{i=1}^{N} \mu_i x_i$$
$$= -\frac{1}{2K}\sum_{i=1}^{N}\{B(x_i - x_{i+1})^2 + (C - 2B)x_i^2\} + \sum_{i=1}^{N} \mu_i x_i.$$

Therefore

$$Z = \sum_{\{\mu\}} e^{S[\mu]} = \prod_{i=1}^{N} \int dx_i e^{S[x]},$$

where

$$S[x] = -\frac{B}{2K}\sum_i (x_i - x_{i+1})^2 - (\frac{C - 2B}{2K})\sum_i x_i^2 + \sum_i \ln(2\cosh x_i)$$
$$= -\frac{B}{2K}\sum_i (x_i - x_{i+1})^2 - V(x),$$

and the potential, shown in Figure 16.11, is given by

$$-V = -(\frac{C - 2B}{2K})x^2 + \ln(2\cosh x) \simeq \begin{cases} -\frac{C-2B}{2K}x^2 + \frac{1}{2}x^2, & x \simeq 0 \\ -\frac{C-2B}{2K}x^2 + |x|. & x \gg 1 \end{cases}$$

For $(C - 2B)/2K > 0$, V is a double well, as shown in Figure 16.12. Hence, the Ising model, for a certain choice of parameters, is equivalent to the double well potential.

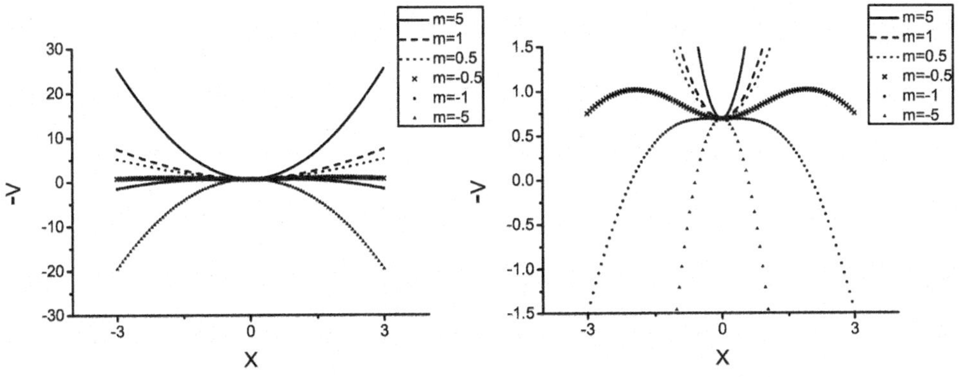

Full range view Small scale of V

Figure 16.11 The potential for different $m = -(C - 2B)/2K$: coupling strengths.

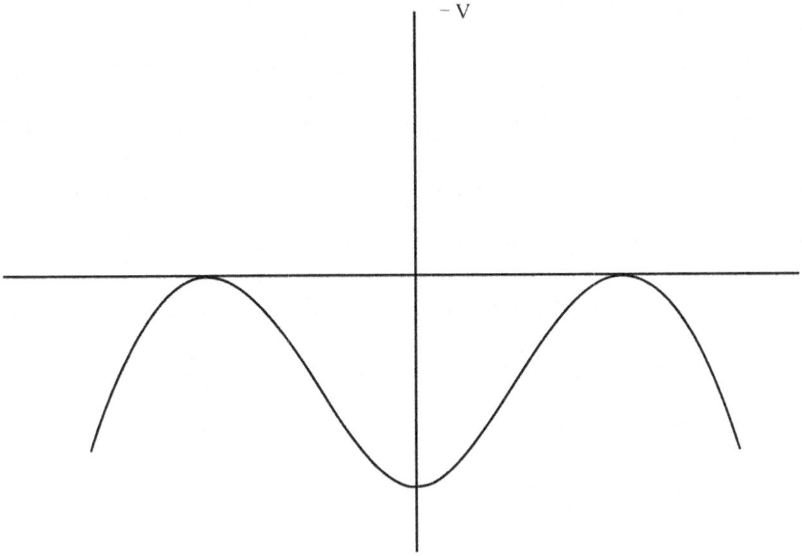

Figure 16.12 Double well potential from an Ising model of a lattice site for $(C - 2B)/2K > 0$.

To determine coefficients B and C, we note that

$$\sum_{l=1}^{N} A_{il} A_{lj}^{-1} = Ce^{-\alpha|i-j|} - B(e^{-\alpha|i-j+1|} + e^{-\alpha|i-j-1|}).$$

For the case of $i = j$,

$$\mathrm{R}HS = C - B(e^{-\alpha} + e^{\alpha}) = 1 \quad \Rightarrow C = 1 + 2Be^{-\alpha}.$$

For the case of $i > j$,

$$\begin{aligned} RHS &= Ce^{-\alpha(i-j)} - B(e^{-\alpha(i-j+1)} + e^{-\alpha(i-j-1)}) \\ &= e^{-\alpha(i-j)}[C - B(e^{-\alpha} + e^{\alpha})] = 0, \end{aligned}$$

and for $i < j$,

$$RHS = e^{-\alpha(i-j)}[C - B(e^{\alpha} + e^{-\alpha})] = 0.$$

Therefore, the case for both $i > j$ and $i < j$ yields

$$B = \frac{C}{2\cosh\alpha} \quad \Rightarrow \quad C = \frac{1}{\tanh\alpha}, \quad B = \frac{1}{2\sinh\alpha}.$$

$(C - 2B)/2K = (\cosh\alpha - 1)/2K\sinh\alpha > 0$ for $\cosh\alpha > 1$ and hence the nonlocal Ising model exhibits a double-well potential for all values of $\alpha > 0$.

16.12 Spontaneous symmetry breaking

Symmetry breaking is quite common in classical physics. For any potential with multiple minima, the particle can be stationary at any of the minima of the potential and consequently satisfy the equations of motion. In the case of the double-well potential given by $V = \frac{g^2}{8}(x^2 - a^2)$, the particle can be classically at rest at either of the two points given by $x = \pm a$, which breaks the $x \to -x$ parity symmetry of the potential.

In quantum mechanics, the Mermin–Wagner theorem states that quantum mechanics cannot break continuous symmetries, but leaves open the possibility of breaking discrete symmetries.

Symmetry in quantum mechanics means the following. Suppose H commutes with some operator R, that is, $[H, R] = 0$. The theory is said to have an unbroken symmetry, or a symmetric state space, if the ground state $|\Omega\rangle$ is invariant under R, namely that it satisfies

$$R|\Omega\rangle = |\Omega\rangle, \text{ symmetric theory.}$$

The theory has spontaneously broken symmetry if

$$R|\Omega\rangle \neq |\Omega\rangle, \text{ spontaneously broken symmetry.} \tag{16.70}$$

In particular, if R is the parity operator defined by $R\hat{x}R = -\hat{x}$, one has

$$\langle\Omega|\hat{x}|\Omega\rangle = -\langle\Omega|R\hat{x}R|\Omega\rangle.$$

It follows from the above equations that

$$\langle\Omega|\hat{x}|\Omega\rangle \begin{cases} = 0, & \text{symmetric,} \\ \neq 0, & \text{nonsymmetric.} \end{cases}$$

16.12.1 Infinite well

Let us consider two infinitely deep potential wells. If the particle is in well I, it has an entire Hilbert space $\mathcal{V}_I$ which is disjoint from the particle in well II; that is

$$\langle \mathcal{V}_I | \mathcal{V}_{II} \rangle = 0,$$

and $\mathcal{V}_I \cap \mathcal{V}_{II} = \phi$, with $|\Omega_I\rangle$, $|\Omega_{II}\rangle$ being the vacuum states.

The discrete symmetry of the potential $V(-x) = V(x)$ is broken by the vacuum since

$$R|\Omega_I\rangle = |\Omega_{II}\rangle \neq |\Omega_I\rangle, \quad \text{nonsymmetric vacuum.}$$

16.12.2 Double well

Let us now consider the double-well potential Hamiltonian

$$H = -\frac{1}{2m}\frac{\partial^2}{\partial x^2} + \frac{g^2}{8}(x^2 - a^2)^2.$$

Let R be the parity operator such that $\langle x|R|\psi_t\rangle = \hat{R}\psi_t(x) = \psi_t(-x)$. Let the "false" vacuum, centered around $x = \pm a$, be denoted by $|\Omega_\pm\rangle$. Then

$$R|\Omega_\pm\rangle = |\Omega_\mp\rangle, \text{ nonsymmetric vacuum.}$$

Hence $|\Omega_\pm\rangle$ spontaneously breaks the R-symmetry. It will be shown that the true vacuum is invariant under R, namely that

$$R|\Omega\rangle = |\Omega\rangle, \quad \text{symmetric vacuum.} \tag{16.71}$$

16.13 Restoration of symmetry

For the finite double-well, it is shown how the symmetry of $x \to -x$ is restored by tunneling. Consider only the ground state sector of the Hilbert space of the double well. To any order in perturbation theory, one has for the lowest energy states of the double well the lowest order Hamiltonian given by

$$H_0 = \begin{pmatrix} E_0 & 0 \\ 0 & E_0 \end{pmatrix}, \quad H_0|\pm a\rangle = E_0|\pm a\rangle,$$

with the degenerate eigenstates, as discussed in Section 16.7, given by

$$|a\rangle \simeq \begin{pmatrix} 1 \\ 0 \end{pmatrix}, \quad |-a\rangle \simeq \begin{pmatrix} 0 \\ 1 \end{pmatrix}.$$

In the limit of $\tau \to \infty$, all the states except the two lowest lying states decouple from the transition amplitude $\langle a|e^{-\tau H}|-a\rangle$. On this two-dimensional subspace of $\mathcal{V}$, the Hamiltonian H is a 2×2 matrix and hence yields the expansion

$$\begin{aligned}\langle a|e^{-\tau H}|-a\rangle &\simeq \langle a|(1-\tau H)|-a\rangle \\ &\simeq -\tau H(a,-a), \quad \langle a|H|-a\rangle \equiv H(a,-a).\end{aligned}$$

Equation 16.53 yields

$$-\tau H(a,-a) = \tau N e^{S_c}.$$

and hence (note the negative sign)

$$H(a,-a) \simeq -N e^{-c/g^2}.$$

The full Hamiltonian is consequently given by

$$H_\Omega = \begin{pmatrix} E_0 & -Ne^{-c/g^2} \\ -Ne^{-c/g^2} & E_0 \end{pmatrix},$$

with eigenstates and eigenvalues

$$\begin{aligned}|\Omega\rangle &= \frac{1}{\sqrt{2}}(|a\rangle + |-a\rangle), \quad E = E_0 - Ne^{-c/g^2}, \quad \text{symmetric}, \\ |1\rangle &= \frac{1}{\sqrt{2}}(|a\rangle - |-a\rangle), \quad E = E_0 + Ne^{-c/g^2}, \quad \text{anti-symmetric}.\end{aligned}$$

The vacuum state $|\Omega\rangle$ was obtained earlier using the Ising approximation in Eq. 16.68. The true vacuum state for the double well is symmetric under the parity operator. The false vacua that to lowest order in perturbation theory start from either of the degenerate perturbative vacua, namely $|\pm a\rangle$, are corrected by the instantons, and symmetry is "restored" in the sense that a calculation for the vacuum state that includes the instanton contributions yields the actual symmetric and unique vacuum state.

Instantons are due to large quantum fluctuations that are represented by tunneling from one false vacuum to the other. The transition amplitudes could not have been obtained by considering only small fluctuations about the false vacua. It is the large fluctuations obtained by perturbing about the N-instanton configurations that restore the symmetry of the false vacuum.

A superposed state is completely nonclassical: the point particle is in an indeterminate state $|\Omega\rangle$, existing simultaneously at two distinct points $\pm a$. This is how quantum mechanics restores parity symmetry.

16.14 Multiple wells

By labeling the two wells as 1 and 2 one can write the Hamiltonian H as

$$H_N = E_0(|1\rangle\langle 1| + |2\rangle\langle 2|) - t(|1\rangle\langle 2| + |2\rangle\langle 1|), \tag{16.72}$$

with $t = Ne(-c/g^2)$. The symmetry operator is then

$$R|1\rangle = |2\rangle, \quad R|2\rangle = |1\rangle.$$

Consider the potential that has minima at sites na, with $n = 1, 2, \ldots N$.

A calculation similar to the double-well case gives to leading order that the effective low energy Hamiltonian is

$$H_N = E_0 \sum_{n=1}^{N} \Big[|n\rangle\langle n| - t\Big(|n\rangle\langle n+1| + |n+1\rangle\langle n| \Big) \Big], \tag{16.73}$$

where for simplicity we assume space is periodic with $|N{+}1\rangle = |1\rangle$. The symmetry of the Hamiltonian is given by the shift operator

$$R|n\rangle = |n+1\rangle, \quad [H_N, R] = 0.$$

One can verify that the eigenstates of H_N are given by

$$|\theta\rangle = \frac{1}{\sqrt{N}} \sum_{n=1}^{N} e^{in\theta}|n\rangle, \quad H_N|\theta\rangle = (E_0 - 2t\cos\theta)|\theta\rangle.$$

with the true ground state given by the symmetric combination

$$|\Omega\rangle = \frac{1}{\sqrt{N}} \sum_{n=1}^{N} |n\rangle, \quad E_\Omega = E_0 - 2t.$$

The θ-parameter is a measurable quantity and occurs in many theories, including the Yang–Mills Lagrangian.

16.15 Summary

Nonlinear path integrals can have qualitative properties that cannot be obtained by perturbing about the linear part of the theory, which is defined by the quadratic piece of the Lagrangian. The double-well potential was chosen to illustrate this aspect of nonlinear path integrals.

Parity is a symmetry of the double-well potential that is broken by the false vacuum of the linearized theory. The path integral was studied to ascertain whether the nonlinearities of the Lagrangian can restore parity symmetry.

The domain of integration over which the path integral is defined is an infinite-dimensional function space. The double-well action has classical multi-instanton solutions that are configurations in function space far from the linear domain. To probe the nonlinearities of the theory, the path integral is expanded about the classical solutions and is evaluated using perturbation theory about the classical solution.

The instantons have a zero mode, which is not constrained by the action, that arises from the instanton being free to tunnel from one false vacuum to the other at any time. The zero modes lead to a *summation* over all instants that the instanton can tunnel and produces a large effect. Examining the lowest energy states, it was seen that tunneling creates an off-diagonal term in the Hamiltonian which cannot be obtained by perturbing about the linearized theory. The off-diagonal term in turn leads to the Hamiltonian having a true vacuum (ground) state that preserves parity.

The multi-instanton expansion can be approximated by a dilute gas of noninteracting instantons. This approximation scheme was used to evaluate the correlation function of the double well; the correlation length obtained shows that it depends on the coupling constants in a nonperturbative manner, further demonstrating the essential distinction between a linear and a nonlinear theory.

17

Compact degrees of freedom

Quantum mechanical degrees of freedom come in many kinds and varieties and are variables that can take values in many different kinds of spaces (manifolds). Compact and noncompact degrees of freedom are two important types. In essence, a compact degree of freedom takes values in a finite range whereas the noncompact take values in an infinite range.

The simplest compact degree of freedom is the Ising spin variable μ that takes two values, namely $\mu = \pm 1$, and which was studied in Chapters 8 and 9. In Chapters 11 to 16, the noncompact degree of freedom x – taking values on the real line, that is $x \in [-\infty, +\infty]$ – was studied for various models.

Compact degrees of freedom have many specific properties not present for the noncompact case, and the focus of this chapter is on these properties. The degrees of freedom that take values in compact spaces for two different cases are discussed, with each case having its own specific features, as follows:

- A degree of freedom taking values on a circle S^1. This case has a nontrivial topological structure that occurs in a wide range of problems, with the simplest case being of a quantum mechanical particle moving on a circle S^1.
- A degree of freedom taking values on a two-dimensional sphere S^2, a compact space. This degree of freedom can represent a quantum particle moving on a sphere and is the simplest case of a quantum particle moving on a curved manifold.

In Section 17.1 the degree of freedom taking values on a circle is introduced, and in 17.2 its multiple classical solutions are derived. The degree of freedom taking values on a sphere S^2 is discussed in Section 17.3 and its Lagrangian is derived in Section 17.4; a divergence arising from the curvature of the sphere is discussed in Section 17.5. Sections 17.6–17.8 apply the S^2 degree of freedom to study the statistical mechanics of the DNA molecule.

17.1 Degree of freedom: a circle

The S^1 degree of freedom takes its values on a circle. The space S^1 is geometrically flat, similarly to the real line, but is a *topologically* nontrivial manifold: S^1 is not simply connected; what this means is that a loop on the circle cannot smoothly be contracted to a single point. One can wind around the circle n times, and all these windings are inequivalent since they cannot be mapped to, say, a single loop. The existence of the winding number is the single most important reflection of the nontrivial topology of S^1.

Consider a particle moving on a circle of radius R; a degree of freedom x then has the property that all points $x + 2\pi nR$, with $n = 0, \pm 1, \pm 2, \ldots \pm \infty$ are equivalent. In particular, the state function $\psi(x)$ has the symmetry[1]

$$\psi(x + 2\pi nR) = \psi(x), \quad n = 0, \pm 1, \pm 2, \ldots \pm \infty.$$

The Hamiltonian is given by

$$H = -\frac{1}{2m}\frac{\partial^2}{\partial x^2}, \qquad x \in [-\pi R, +\pi R].$$

The state function has the normalization

$$1 = \int_{-\pi R}^{+\pi R} dx |\psi(x)|^2.$$

The normalized eigenfunctions of the Hamiltonian are given by

$$\begin{aligned} H\psi_n &= E_n\psi_n, \\ \psi_n &= \frac{e^{inx/R}}{\sqrt{2\pi R}}, \quad E_n = \frac{n^2}{2mR^2}, \qquad n \in \mathbf{Z}. \end{aligned} \tag{17.1}$$

It is convenient to define a new angular variable θ such that

$$\theta = \frac{x}{R} \in [-\pi, +\pi].$$

Then

$$H = -\frac{1}{2mR^2}\frac{\partial^2}{\partial \theta^2},$$

with the orthonormal state functions given by

$$\int_{-\pi R}^{+\pi R} dx \psi_n^*(x)\psi_m(x) = R\int_{-\pi}^{+\pi} d\theta \psi_n^*(\theta)\psi_m(\theta) = \delta_{n-m}.$$

[1] This condition can be relaxed to

$$\psi(x + 2\pi nR) = e^{i\phi}\psi(x), \quad n = 0, \pm 1, \pm 2, \ldots \pm \infty.$$

The evolution kernel, for Euclidean time τ, is given by

$$K(\theta',\theta;\tau) = \langle\theta'|e^{-\tau\hat{H}}|\theta\rangle = \sum_n e^{-\tau E_n}\psi_n(\theta')\psi_n^*(\theta)$$
$$= \frac{1}{2\pi R}\sum_{n=-\infty}^{\infty} e^{-\frac{1}{2}\left(\frac{\tau}{mR^2}\right)n^2} e^{in(\theta'-\theta)}. \tag{17.2}$$

17.1.1 Poisson summation formula

The Poisson summation formula is a useful result in the study of a quantum particle moving on a circle, and is given by the identity

$$\sum_{n=-\infty}^{\infty}\delta(\xi-n) = \sum_{\ell=-\infty}^{\infty} e^{2\pi i\ell\xi}. \tag{17.3}$$

To prove the Poisson summation formula, let

$$f(\xi) = \sum_{n=-\infty}^{\infty}\delta(\xi-n) \;\Rightarrow\; f(\xi+m) = f(\xi), \quad m = \text{integer}.$$

Let $\xi = 0$ and $m = 1$; then $f(1) = f(0)$. Hence $f(\xi)$ is a periodic function on the interval $0 \le \xi < 1$; note that the point $\xi = 1$ has been excluded, since due to periodicity, it is identical to the point $\xi = 0$. Consider the following Fourier expansion of $f(\xi)$:

$$f(\xi) = \sum_{\ell=-\infty}^{\infty} e^{2\pi i\ell\xi} f_\ell,$$
$$\Rightarrow f_\ell = \int_0^1 d\xi e^{-2\pi i\ell\xi} f(\xi) = \sum_{n=-\infty}^{\infty}\int_0^1 d\xi e^{-2\pi i\ell\xi}\delta(\xi-n) = 1,$$
$$\Rightarrow f(\xi) = \sum_{\ell=-\infty}^{\infty} e^{2\pi i\ell\xi},$$

which proves Eq. 17.3. The identity yields

$$\sum_{n=-\infty}^{\infty} f_n = \int_{-\infty}^{\infty} d\xi \sum_{n=-\infty}^{\infty}\delta(\xi-n) f(\xi) = \sum_{\ell=-\infty}^{\infty}\int_{-\infty}^{\infty} d\xi e^{2\pi i\ell\xi} f(\xi), \tag{17.4}$$

where $f(\xi)$ is the extension to a continuous argument ξ of the function f_n defined on a lattice.

17.1.2 The S^1 Lagrangian

The Lagrangian is given by the Dirac–Feynman formula

$$K(\theta',\theta;\tau) = \langle\theta'|e^{-\epsilon H}|\theta\rangle = \mathcal{N}(\epsilon)e^{\epsilon\mathcal{L}(\theta,\theta')}. \tag{17.5}$$

Taking the limit $\epsilon \to 0$ in Eq. 17.2 is hopeless since higher and higher order terms are necessary to evaluate the sum. To take the limit one needs to invert $\tau \to \frac{1}{\tau}$ in Eq. 17.2. The Poisson summation formula given in Eq. 17.4 yields

$$\begin{aligned}\langle\theta'|e^{-\tau H}|\theta\rangle &= \frac{1}{2\pi R}\sum_{n=-\infty}^{\infty} e^{-\frac{\tau}{2mR^2}n^2}e^{in(\theta'-\theta)} \\ &= \frac{1}{2\pi R}\int d\xi \sum_{\ell=-\infty}^{\infty} e^{2\pi i\ell\xi}e^{-\frac{\tau}{2mR^2}\xi^2}e^{i\xi(\theta'-\theta)} \\ &= \sqrt{\frac{m}{2\pi\tau}}\sum_{\ell=-\infty}^{\infty} e^{-\frac{mR^2}{2\tau}(\theta'-\theta-2\pi\ell)^2}.\end{aligned} \tag{17.6}$$

Hence, taking the limit of $\tau = \epsilon$ yields

$$\begin{aligned}\langle\theta'|e^{-\epsilon H}|\theta\rangle &\simeq \sqrt{\frac{m}{2\pi\epsilon}}e^{-\frac{1}{2\epsilon}mR^2(\theta'-\theta)^2} + O\left(e^{-\frac{1}{\epsilon}}\right) \\ &= \mathcal{N}(\epsilon)e^{\epsilon\mathcal{L}},\end{aligned} \tag{17.7}$$

and the Lagrangian is given by

$$\mathcal{L} = -\frac{1}{2}mR^2\left(\frac{\theta'-\theta}{\epsilon}\right)^2 = -\frac{1}{2}mR^2\left(\frac{d\theta}{dt}\right)^2, \quad \theta\in[0,2\pi],$$

$$\text{with } \mathcal{N}(\epsilon) = \sqrt{\frac{m}{2\pi\epsilon}}.$$

17.2 Multiple classical solutions

The evolution kernel is defined by

$$K(\theta',\theta;\tau) = \int D\theta e^{\int_0^\tau \mathcal{L}}. \tag{17.8}$$

The path integral measure, for $\tau_n = n\epsilon$, $N\epsilon = \tau$, is given by

$$\int D\theta = \lim_{N\to\infty}\prod_{n=-N/2}^{+N/2}\int_{\pi}^{+\pi}\sqrt{\frac{m}{2\pi\epsilon}}d\theta_n.$$

The action for the particle on a circle is given by

$$S = -\frac{1}{2} m R^2 \int_0^\tau dt \left(\frac{d\theta(t)}{dt} \right)^2, \quad \theta \in [0, 2\pi].$$

The action is similar to that of a free particle, except that now θ is a *compact variable*, taking values on a circle.

To evaluate the evolution kernel, a semi-classical expansion is carried out about all the classical solutions. The equations of motion are given by

$$\begin{aligned} &\ddot{\theta} = 0 \ \Rightarrow \ \theta_{cl} = (a + bt) \bmod 2\pi, \\ &\text{B.C. } \theta_c(0) = \theta, \ \ \theta_c(\tau) = \theta' + 2\pi\ell, \quad \ell : \text{winding number.} \end{aligned} \tag{17.9}$$

The multiple classical solutions are shown in Figure 17.1 by displaying θ as a noncompact variable. The winding number, shown in Figure 17.2, reflects the fact that $\theta(t)$ is a compact variable that can satisfy the boundary conditions by winding

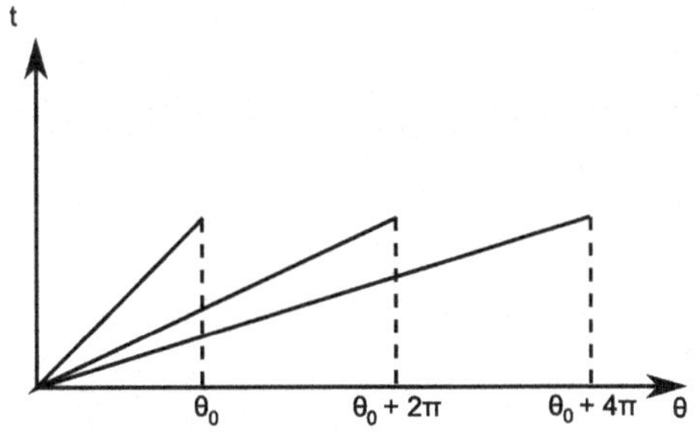

Figure 17.1 Classical solutions satisfy the boundary conditions, shown for zero, one and two windings around the compact direction.

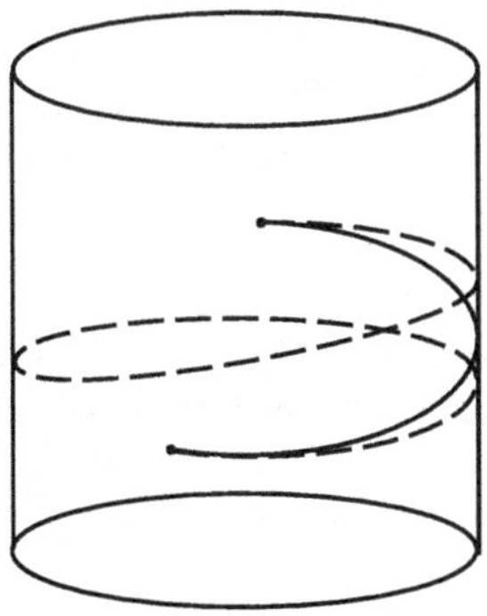

Figure 17.2 Winding number; the unbroken line is the classical solution with no winding and the dashed line is the classical solution with one winding around the compact direction.

around the compact direction. Eq. 17.9 shows that the final position is equal to θ' modulo 2π, since the compact quantum variables $\theta(t)$ are periodic. Hence

$$\theta_c(t) = \theta + \frac{(\theta' - \theta)}{\tau} t + 2\pi \ell \frac{t}{\tau}, \quad \dot{\theta}_{cl} = \frac{1}{\tau}\left(\theta' - \theta + 2\pi l\right).$$

The classical action for the ℓth winding number is given by

$$S_c(\ell) = -\frac{mR^2}{2}\int_0^\tau \dot{\theta}_{cl}^2 dt = -\frac{mR^2}{2}\frac{\left(\theta' - \theta + 2\pi l\right)^2}{\tau}.$$

As expected, the more the classical solution winds, the larger is the velocity it requires; hence, classical solutions with higher and higher winding numbers have more and more negative values for the classical action, as given by $S_c(\ell)$.

A semi-classical expansion of the path integral about all the classical solutions, similar to the one carried in Section 16.6 for the double well potential, yields

$$Z = \int D\theta e^S = N(\tau) \sum_{\ell=-\infty}^{\infty} e^{S_c(\ell)} = N(\tau) \sum_{\ell=-\infty}^{\infty} e^{-\frac{mR^2}{2\tau}(\theta' - \theta + 2\pi\ell)^2}.$$

To obtain the prefactor $\mathcal{N}(\tau)$ one needs to analyze the path integral over the compact variable, namely $\int D\theta e^S$; consider the change of variables

$$\theta(t) = \theta_c(t) + z(t), \quad \text{B.C. } z(0) = z(\tau) = 0. \tag{17.10}$$

This yields, from Eq. 17.9, the action

$$\begin{aligned} S &= -\frac{1}{2}mR^2 \int (\dot{\theta}_{cl} + \dot{z})^2 dt = -\frac{1}{2}mR^2 \int_0^\tau \dot{\theta}_{cl}^2 dt - \frac{1}{2}mR^2 \int \dot{z}^2 dt \\ &= -\frac{1}{2}mR^2 \frac{\left(\theta' - \theta + 2\pi l\right)^2}{\tau} - \frac{1}{2}mR^2 \int \dot{z}^2 dt. \end{aligned} \tag{17.11}$$

The quantum variable $z(t)$ is noncompact, taking values of $-\infty \leq z(t) \leq +\infty$; the fluctuations about the classical solutions are noncompact, making the evaluation of the prefactor an exact Gaussian path integral. The prefactor $\mathcal{N}(\tau)$, in fact, is the same as that for a free particle on the real line and it has been evaluated earlier in Section 5.8.

Since $z(t)$ does not depend on the classical solutions, it completely factorizes from the sum over the classical solutions and yields the final result

$$\begin{aligned} K(\theta', \theta; \tau) &= \left[\int Dz e^{-\frac{1}{2}mR^2 \int \dot{z}^2}\right] \sum_\ell e^{S_{cl}(\ell)} = \mathcal{N}(\tau) \sum_\ell e^{S_{cl}(\ell)} \\ &= \sqrt{\frac{m}{2\pi\tau}} \sum_{\ell=-\infty}^{\infty} e^{-\frac{mR^2}{2\tau}(\theta' - \theta + 2\pi\ell)^2}. \end{aligned} \tag{17.12}$$

The result obtained in Eq. 17.12 was obtained earlier, in Eq. 17.6, using the Poisson summation formula.

The summation n in Eq. 17.2 is a summation over all the discrete momentum states of the particle whereas ℓ in Eq. 17.12 is a summation over all the winding numbers. For a compact variable, the momentum and winding number are "dual" to each other. For noncompact variables, the concept of winding numbers does not exist and hence there is no duality.

17.2.1 Large radius limit

The limit of the radius of S^1, namely $R \to \infty$, takes the circle into the real line. Recall from Eq. 17.2

$$K(\theta', \theta; \tau) = \frac{1}{2\pi R} \sum_{n=-\infty}^{\infty} e^{-\frac{1}{2}\frac{\tau}{m}\frac{n^2}{R^2}} e^{i\frac{n}{R}\cdot R(\theta'-\theta)}. \tag{17.13}$$

Let $p = n/R$; for $R \to \infty$ we have

$$-\infty \leq p \leq +\infty, \qquad \sum_{n=-\infty}^{\infty} = R \int dp.$$

Let us define $R\theta = x$, $R\theta' = x'$. Hence

$$K(x', x; \tau) = \frac{1}{2\pi R} R \int_{-\infty}^{\infty} dz e^{-\frac{\tau}{2m}z^2} e^{iz(x-x')} = \sqrt{\frac{m}{2\pi\tau}} e^{-\frac{m}{2\tau}(x'-x)^2}.$$

Furthermore, from Eq. 17.12

$$\lim_{R\to\infty} K = \sqrt{\frac{m}{2\pi\tau}} \sum_{\ell=-\infty}^{\infty} e^{-\frac{1}{2}\frac{mR^2}{\tau}(\theta'-\theta-2\pi\ell)^2} = \sqrt{\frac{m}{2\pi\tau}} \sum_{\ell=-\infty}^{\infty} e^{-\frac{m}{2\tau}(x'-x-2\pi\ell R)^2}$$

$$\Rightarrow K = \sqrt{\frac{m}{2\pi\tau}} e^{-\frac{m}{2\tau}(x'-x)^2}. \tag{17.14}$$

We have recovered the result for the evolution kernel of a free particle moving on the real line, given in Section 5.8.

17.3 Degree of freedom: a sphere

The main new feature of a degree of freedom that is a circle is the winding number arising from the fact that the space is not simply connected. The degree of freedom taking values on a sphere is exemplified by a quantum particle moving on a two-dimensional sphere S^2. The two-dimensional sphere is simply connected in

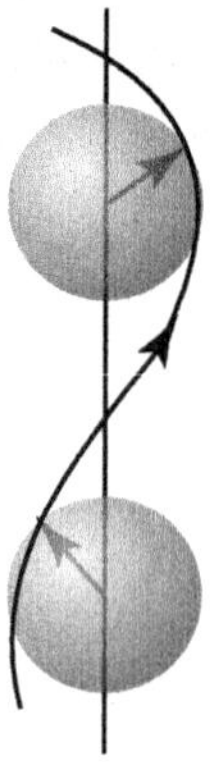

Figure 17.3 Particle moving on a sphere.

that any closed loop on S^2 can be continuously contracted to a point, with every intermediate loop in the mapping being points in S^2. Hence S^2 has a trivial topological structure with no winding number; in particular, unlike the S^1 case, the S^2 degree of freedom does not have multiple classical solutions.

Although S^2 is topologically simple, it has another important geometrical feature, namely that of curvature: the two-dimensional sphere is a Riemannian manifold with constant curvature; for a sphere of radius R the Ricci scalar curvature is constant and given by $2/R^2$.

A particle moving on a sphere is shown in Figure 17.3; the position of the particle on the sphere is specified by two angles, which in spherical coordinates shown in Figure 17.4 are given by the polar and azimuthal angles θ, ϕ respectively.

The three-dimensional Laplacian, in spherical polar coordinates r, θ, ϕ, is given by

$$\vec{\nabla}^2 = \frac{\partial}{\partial x^2} + \frac{\partial}{\partial y^2} + \frac{\partial}{\partial z^2} = \frac{1}{r}\frac{\partial}{\partial r^2}r - \frac{L^2}{r^2},$$

where

$$L^2 = \frac{1}{\sin^2\theta}\frac{\partial}{\partial\varphi^2} + \frac{\partial}{\partial\theta^2} + \cot\theta\frac{\partial}{\partial\theta}, \quad 0 \le \theta \le \pi, \ \ 0 \le \phi \le 2\pi.$$

The position of the particle on the sphere is given by

$$R\,(\cos\theta\sin\varphi, \cos\theta\cos\varphi, \sin\theta)\,.$$

A particle moving on a sphere of fixed radius $r = R$ has $\frac{\partial}{\partial r} = 0$. Two particles at the ends of a rigid rod undergoing rotations form a (linear) rigid rotor. A particle confined to move on the surface of a sphere is equivalent to the motion of a rigid

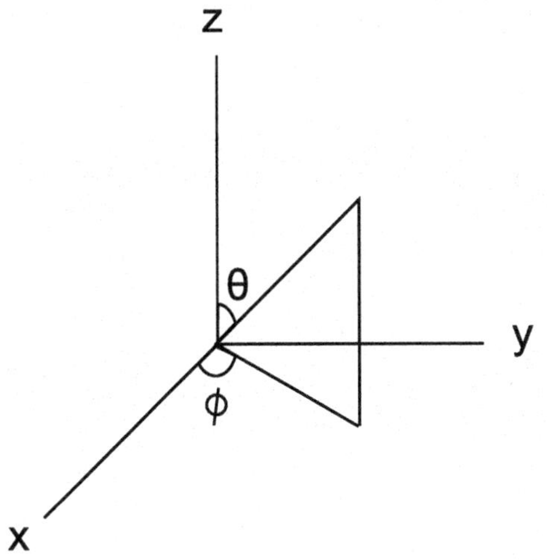

Figure 17.4 Spherical coordinates.

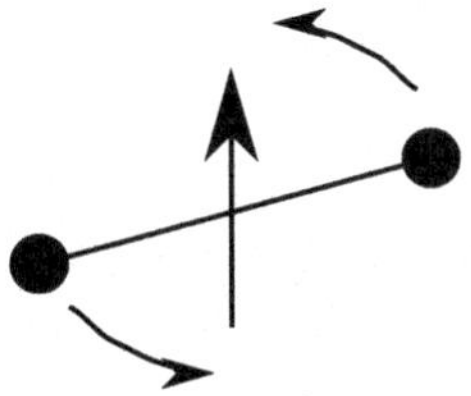

Figure 17.5 A quantum mechanical rigid rotor.

rotor (for example, a diatomic molecule), where I is the moment of inertia of the rigid rotor, which is shown in Figure 17.5.

The Hamiltonian of the rigid rotor is given by

$$H = -\frac{\hbar^2}{2mR^2}L^2 = -\frac{\hbar^2}{2I}L^2, \quad I = mR^2. \tag{17.15}$$

The eigenfunctions are the spherical harmonics $Y_{lm}(\theta, \varphi)$, given by Gottfried and Yan (2003), and satisfy

$$HY_{lm} = \frac{\hbar^2}{2I}l(l+1)Y_{lm}(\theta, \phi), \quad l = 0, 1, 2, \ldots \infty, \quad -l \leq m \leq l.$$

17.4 Lagrangian for the rigid rotor

The rigid rotor is described by two angular degrees of freedom, namely θ, ϕ; the Hilbert space has coordinate basis states $|\theta, \phi\rangle$ and the dual ("momentum") coordinates are $|l, m\rangle$ with

$$\langle \theta, \varphi | l, m \rangle = Y_{lm}(\theta, \varphi).$$

Note that

$$\int_0^{\pi} d\theta \int_{-\pi}^{\pi} d\varphi \sin\theta \, |\theta, \varphi\rangle\langle\theta, \varphi| = \mathbb{I}.$$

The Lagrangian is given by the Dirac–Feynman formula

$$\mathcal{N}(\epsilon)\, e^{\frac{\epsilon}{\hbar}L} = \langle \theta', \varphi' | e^{-\frac{\epsilon}{\hbar}H} | \theta, \varphi \rangle.$$

It is more convenient to use the momentum basis p, q since, for infinitesimal time ϵ, it is equivalent to the spherical harmonic basis $|l, m\rangle$. The completeness equation is given by

$$\int_{-\infty}^{+\infty} \frac{dpdq}{4\pi^2} |p, q\rangle\langle p, q| = \mathbb{I}, \quad \langle \theta, \varphi | p, q \rangle = e^{ip\theta + iq\phi}.$$

The completeness equation yields, from Eq. 17.15

$$\begin{aligned}
\mathcal{N}(\epsilon)\, e^{\frac{\epsilon}{\hbar}L} &= \int \frac{dpdq}{4\pi^2} e^{ip(\theta'-\theta)+iq(\varphi'-\varphi)} e^{-\frac{\epsilon}{2\hbar I}\hbar\left(\frac{q^2}{\sin^2\theta}+p^2-ip\cot\theta\right)} \\
&= \int \frac{dpdq}{4\pi^2} e^{ip\left(\theta'-\theta+\frac{\epsilon\hbar}{2I}\cot\theta\right)+iq(\varphi'-\varphi)} e^{-\frac{\epsilon\hbar}{2I}\left(\frac{q^2}{\sin^2\theta}+p^2\right)} \\
&= \left(\frac{2I}{\epsilon\hbar}\right) \sin\theta\, e^{-\frac{\sin^2\theta}{2\epsilon\hbar}(\varphi'-\varphi)^2} e^{-\frac{I}{2\epsilon\hbar}\left(\theta'-\theta+\frac{\epsilon\hbar}{2I}\cot\theta\right)^2}.
\end{aligned}$$

Hence for $\epsilon \to 0$, the Lagrangian of the rigid rotor is given by

$$L = -\frac{I}{2}\sin^2\theta\dot{\varphi}^2 - \frac{I}{2}\left(\dot{\theta} + \frac{\hbar}{2I}\cot\theta\right)^2. \tag{17.16}$$

The normalization

$$\mathcal{N}(\epsilon) = \left(\frac{2I}{\epsilon\hbar}\right)\sin\theta \tag{17.17}$$

depends on θ and is not a constant. This is a typical case for all path integrals on curved manifolds. The pre-factor $\mathcal{N}(\epsilon)$ cannot be obtained from the Lagrangian and is a result of the state space of the degree of freedom taking values on a curved manifold.

The classical Lagrangian is given by the velocity of the particle on a sphere and is

$$L = -\frac{I}{2}\left(\sin^2\theta\dot{\varphi}^2 + \dot{\theta}^2\right), \quad I = mR^2.$$

Compared to the classical Lagrangian above, the extra term $\frac{\hbar}{2I}\cot\theta$ in Eq. 17.16 is a quantum correction to L.

The path integral is[2]

$$\begin{aligned}
K\left(\theta', \varphi'; \theta, \varphi; t\right) &= \langle\theta', \varphi'|e^{-\tau H}|\theta, \varphi\rangle \\
&= \tilde{\mathcal{N}} \prod_{n=0}^{N-1} \int \left\langle\theta_{n+1}, \varphi_{n+1}\left|e^{-\epsilon H}\right|\theta_n, \varphi_n\right\rangle \prod_{n=1}^{N-1} d\theta_n d\varphi_n \sin\theta_n \\
&= \tilde{\mathcal{N}} \prod_{n=1}^{N-1} \int d\theta_n d\varphi_n \sin\theta_n e^S, \qquad (17.18)\\
S &= \epsilon \sum_{n=0}^{N-1} L\left(\theta_{n+1}, \varphi_{n+1}, \theta_n, \varphi_n, \epsilon\right),
\end{aligned}$$

where $\tilde{\mathcal{N}}$ is a normalization constant independent of θ.

17.5 Cancellation of divergence

The path integral measure, from Eq. 17.18, is given by

$$\prod_{n=1}^{N-1} \sin\theta_n = e^{\sum_{n=1}^{N-1} \ln(\sin\theta_n)}.$$

Note that there is a problem with the measure term as it is apparently divergent since, in the limit $\epsilon \to 0$,

$$\sum_{n=1}^{N-1} \ln\left(\sin\theta_n\right) \to \frac{1}{\epsilon} \cdot \epsilon \sum_n \ln\left(\sin\theta_n\right) = \frac{1}{\epsilon} \int_0^\tau dt\ \ln\left(\sin\theta\left(t\right)\right) \to \infty.$$

The only way to obtain a finite result from the path integral is if all the divergent terms that appear in the path integral exactly cancel. To show the cancellation of the $1/\epsilon$ term, a perturbation expansion is carried out for the path integral in powers of θ. The functional integral $\int D\phi$ is performed while keeping the functional integral $\int D\theta$ fixed. Performing the functional integral $\int D\phi$ generates a term for the θ variable that precisely cancels the divergent term $-\frac{1}{2\epsilon}\int\theta^2$ that comes from the path integral measure.

Note that $0 \le \theta \le \pi$; the minimum value of the action S is around $\theta = \pi/2$. Hence, the angle θ is shifted to $\tilde{\theta}$ as shown in Figure 17.6, so that the shifted angle, in the path integral, is constrained to be near zero due to the action.

[2] Henceforth, unless required, we set $\hbar = 1$.

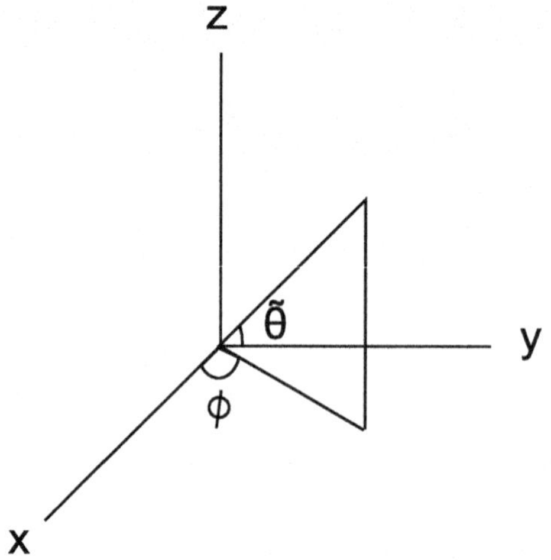

Figure 17.6 Spherical coordinates with angle θ measured from the "forward" y-axis.

Hence, the angle θ is shifted to $\tilde{\theta} = \theta - \pi/2$, with $-\pi/2 \leq \tilde{\theta} \leq \pi/2$; dropping the tilde yields, from Eq. 17.16

$$L = -\frac{I}{2}\cos^2\theta\dot{\phi}^2 - \frac{I}{2}\left(\dot{\theta} + \frac{\hbar}{2I}\tan\theta\right)^2.$$

Let $\tau \to \infty$; the path integral is given by

$$Z = \int D\theta D\phi \left(\prod_t \cos\theta_t\right)\exp\left\{\int_{-\infty}^{\infty} dt L(t)\right\}.$$

Consider the case of the moment of inertia $I \gg 1$ and the action is expanded in powers of $1/I$. Hence

$$L \simeq -\frac{I}{2}\left(1 - \frac{\theta^2}{2} + \cdots\right)^2 \dot{\phi}^2 - \frac{I}{2}\dot{\theta}^2 + O(1).$$

Note that

$$\cos\theta \simeq 1 - \frac{\theta^2}{2} \Rightarrow \prod_t \cos\theta_t \simeq e^{-\frac{1}{2\epsilon}\int_0^\tau dt\theta^2}.$$

Using the notation

$$\int_{-\infty}^{\infty} dt \equiv \int,$$

the partition function is given by

$$Z \simeq \int_{-\infty}^{\infty} D\theta D\phi e^{-\frac{1}{2\epsilon}\int \theta^2} \exp\left\{-\frac{I}{2}\int \dot{\phi}^2 - \frac{I}{2}\int \dot{\theta}^2 + \frac{I}{2}\int \theta^2\dot{\phi}^2 + \cdots\right\}. \tag{17.19}$$

The generating function for a free particle, namely

$$\frac{1}{Z}\int D\phi \exp\left\{-\frac{I}{2}\int \dot{\phi}^2 + \int j\phi\right\} = \exp\left\{\frac{1}{2I}\int j_t D_{t-t'} j_{t'}\right\},$$

has been evaluated in Eq. 12.1.

The correlation function is given by

$$G(t-t') = E[\phi_t \phi_{t'}] = \int \frac{d\omega}{2\pi}\frac{e^{i\omega(t-t')}}{\omega^2}, \tag{17.20}$$

and yields

$$E[\dot{\phi}_t \dot{\phi}_{t'}] = \frac{1}{I}\partial_t \partial_{t'} \langle \phi_t \phi_{t'} \rangle = \frac{1}{I}\int \frac{d\omega}{2\pi} e^{i\omega(t-t')} = \frac{1}{I}\delta\left(t-t'\right).$$

Hence, for discrete time

$$E[\dot{\phi}_t^2] = \frac{1}{I\epsilon}, \tag{17.21}$$

and from Eqs. 17.19 and 17.21

$$\begin{aligned} Z &\simeq \int D\theta e^{-\frac{1}{2\epsilon}\int \theta^2}\left[1 + \frac{I}{2}\int \theta^2 \frac{1}{I\epsilon} - \frac{I}{2}\int \dot{\theta}^2\right] \\ &\simeq \int D\theta \exp\left\{-\frac{I}{2}\int \dot{\theta}^2\right\}, \quad \frac{1}{\epsilon} \text{ terms exactly cancel!} \end{aligned}$$

The divergent piece of the measure term is canceled by a term generated from the kinetic term of the φ variable.

It can be shown that the cancellation holds to all orders in powers of $1/\epsilon$ [Zinn-Justin (1993)]. In summary, all the singular terms in the path integral cancel, yielding perfectly finite results for all computations.

17.6 Conformation of DNA

A polymer is a linear chain of molecules and can be considered to be a string laid out in space. The conformation (shape) of a polymer is statistically determined by the likelihood of it taking the various allowed shapes. The statistical mechanics of the system – in equilibrium at temperature T – is described by the competition between the entropy of the chain of length L and the energy required for bending it.

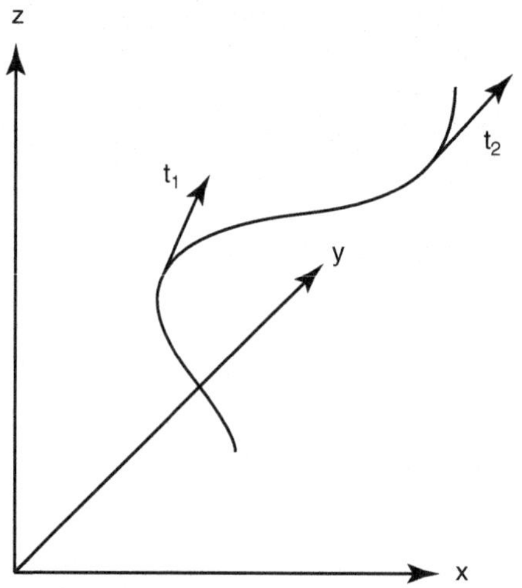

Figure 17.7 The trajectory of a particle specified by its tangent vectors.

A polymer in space is shown in Figure 17.7. If one does not need to know the location of the polymer in three-dimensional space, then as shown in Figure 17.7, the polymer's configuration is completely determined by specifying its three-dimensional tangent vector along its length.

Let us consider a DNA molecule, with total length L, that is free to move in a solute, discussed by Phillips *et al.* (2008). The curve of the DNA in three-dimensional space is parameterized by a parameter s; the vector $\mathbf{t}(s)$ that is tangential to the shape of the DNA specifies the shape of the DNA, as shown in Figure 17.8. The parameterization is chosen so that, for every s, the tangential vector has unit length, namely $\mathbf{t}(s) \cdot \mathbf{t}(s) = 1$, and takes values on a two-dimensional sphere S^2.

In terms of the spherical polar angles given in Figure 17.4, the tangent vector has the coordinates

$$\mathbf{t} = (\sin\theta\cos\phi, \sin\theta\sin\phi, \cos\theta)\,.$$

The degree of freedom for the DNA is taken to be the tangent vector $\mathbf{t}$, which takes all possible allowed values along the curve occupied by the polymer (DNA); the random configurations of the DNA can be modeled by assigning a probability distribution for the different configurations of $\mathbf{t}(s)$.

The energy of bending is given by

$$H = -k_B T S = k_B T \times \frac{\xi}{2}\int_0^L ds \left(\frac{d\mathbf{t}}{ds}\right)^2,$$

where 2ξ is the correlation length of DNA and the partition function is

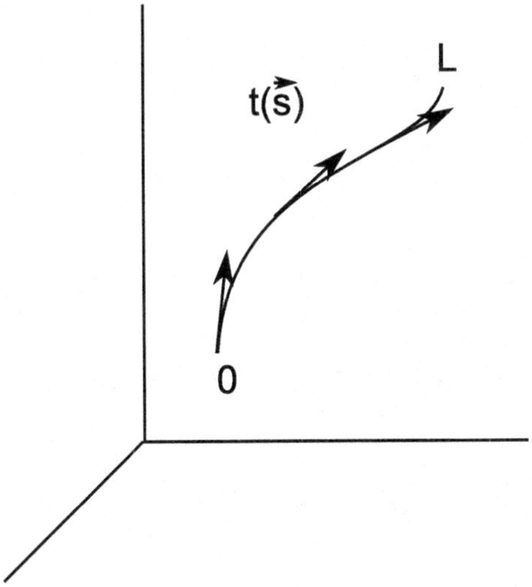

Figure 17.8 DNA polymer in three-dimensional space.

$$Z = \int D\mathbf{t} \exp\left\{-\frac{H}{k_B T}\right\} = \int D\mathbf{t} e^S; \quad S = \frac{1}{k_{BT}} H.$$

The statistical mechanics of the DNA molecule is modeled by the following "action" and Lagrangian which, to leading order in ξ, are given by

$$S = \int_0^L ds \mathcal{L}, \quad \mathcal{L} = -\frac{\xi}{2}\left(\frac{d\mathbf{t}}{ds}\right)^2 = -\frac{\xi}{2}\sin^2\theta\dot{\phi}^2 - \frac{\xi}{2}\dot{\theta}^2.$$

The role of Euclidean time is played by the parameter s that runs along the curve.

Since the action is a nonlinear functional of the degrees of freedom, the partition function needs to be evaluated perturbatively by expanding the action about $\theta = \phi = 0$.

The correlation function of the z-component of the tangent vectors is given by

$$G\left(s, s'\right) = E[t_z(s)t_z(s')].$$

In order to calculate the force required to *stretch* the DNA we need to include a force term applied along $\hat{z}$ in the action S that is given by $f\int_0^L t_z ds$. Hence

$$\mathcal{L} = -\frac{\xi}{2}\left(\frac{d\vec{t}}{ds}\right)^2 + f t_z = -\frac{\xi}{2}\left(\sin^2\theta\dot{\phi}^2 + \dot{\theta}^2\right) + f\cos\theta.$$

17.7 DNA extension

The chain's extension is given by $\mathcal{P} = \int_0^L \cos\theta(s) ds$. Note that if all the tangent vectors are parallel, then $\theta = 0$ and $z = L$ leading to the full extension of the DNA. The average extension is given by

$$\bar{\mathcal{P}} = E[\mathcal{P}] = \frac{1}{Z}\int \mathrm{D}t \int_0^T \cos\theta ds e^S = \frac{\partial \ln Z}{\partial f}.$$

The small force limit $f \simeq 0$, which yields the expansion

$$Z(f) = \int \mathrm{D}t e^S \left[1 + f\int_0^L \cos\theta(s) ds + \frac{f^2}{2}\int_0^L dsds' \cos\theta(s)\cos(s') + \cdots\right]$$
$$= 1 + G_1 + G_2 + O(f^3).$$

For $\xi \gg 1$ the Lagrangian is approximately given by

$$\mathcal{L} \simeq -\frac{\xi}{2}\dot{\phi}^2\left(1 - \frac{1}{2}\theta^2\right)^2 - \frac{\xi}{2}\dot{\theta}^2 \simeq -\frac{\xi}{2}\dot{\phi}^2 - \frac{\xi}{2}\dot{\theta}^2 + O(\theta^4).$$

For making the computation well defined we introduce a regulator ω and consider

$$\mathcal{L} = -\frac{\xi}{2}\dot{\phi}^2 - \frac{\xi}{2}\left(\dot{\theta}^2 + \omega^2\theta^2\right).$$

The limit $\omega \to 0$ will be taken at the end of the calculation. The propagator, from Eq. 11.9, is approximately given by

$$E[\theta(s)\theta(s')] \simeq \frac{e^{-\omega|s-s'|}}{2\omega\xi}.$$

For the first term note that

$$G_1 = f\int_0^L E[\cos\theta(s)]ds,$$
$$E[\cos\theta(s)] = \frac{1}{2Z}\int D\theta\left(e^{i\theta(s)} + e^{-i\theta(s)}\right)e^{S_0} = \frac{1}{Z}\int D\theta e^{i\theta(s)}e^{S_0}.$$

Hence, for $j_l = \delta(s - l)$, we have from Eqs. 12.1 and 17.20

$$\langle\cos\theta\rangle = e^{-\frac{1}{2}\int j_l D_{ll'} j_{l'}} = e^{-\frac{1}{4\omega\xi}} \Rightarrow \lim_{\omega\to 0}\langle\cos\theta\rangle = e^{-\frac{1}{4\omega\xi}} \to 0$$
$$\Rightarrow G_1 = 0.$$

The second term is given by

$$G_2 = \frac{f^2}{2}\int_0^L dsds' G\left(s, s'\right), \quad G\left(s, s'\right) = E[\cos\theta(s)\cos(s')].$$

The path integral is approximated by

$$G\left(s, s'\right) \simeq \frac{1}{Z}\int D\theta D\phi e^S \cos\theta(s)\cos\theta(s')$$

$$\simeq \frac{1}{Z}\int D\theta e^{-\int_0^L \frac{\xi}{2}(\dot{\theta}^2+\omega^2\theta^2)ds}\cos\theta(s)\cos\theta(s')$$
$$= \frac{1}{2}E\left[e^{i\{\theta(s)+\theta(s')\}} + e^{i\{\theta(s)-\theta(s')\}}\right].$$

Let us define the external current for the two terms, respectively, by

$$\int dl j_i(l)\theta(l), \quad i = 1, 2$$
$$\Rightarrow j_1(l) = \delta(l-s) + \delta\left(l-s'\right), \qquad j_2(l) = \delta(l-s) - \delta\left(l-s'\right).$$

The first term for $G\left(s, s'\right)$ is zero since

$$\lim_{\omega\to 0} exp\left(-\frac{1}{4\omega\xi}\int dldl'\left(\delta_{l-s}+\delta_{l-s'}\right)e^{-\omega|l-l'|}\left(\delta_{l'-s}+\delta_{l'-s'}\right)\right)$$
$$= \lim_{\omega\to 0}\exp\left(-\frac{1}{2\omega\xi}\left(1+e^{-\omega|s-s'|}\right)\right) \to 0.$$

Hence, the correlator is given by

$$G\left(s, s'\right) = \lim_{\omega\to 0}\frac{1}{2}\exp\left(-\frac{1}{4\omega\xi}\int dldl'\left(\delta_{l-s}-\delta_{l-s'}\right)e^{-\omega|l-l'|}\left(\delta_{l'-s}-\delta_{l'-s'}\right)\right)$$
$$= \lim_{\omega\to 0}\frac{1}{2}\exp\left(-\frac{1}{2\omega\xi}\left(1-e^{-\omega|s-s'|}\right)\right) = \frac{1}{2}\exp\left(-\frac{|s-s'|}{2\xi}\right).$$

The correlation length is 2ξ, where ξ is called the persistence length.

17.8 DNA persistence length

DNA can be thought of as a jointed chain having rigid links of persistence length ξ, as shown in Figure 17.9. Typically, $\xi \simeq 50$ nm. The distance between base pairs is approximately 0.33 nm, hence ξ is about 150 base pairs. Total length of the DNA is $L \simeq 16\,\mu$m and $f \simeq 0.1$ pN [Phillips *et al.* (2008)].

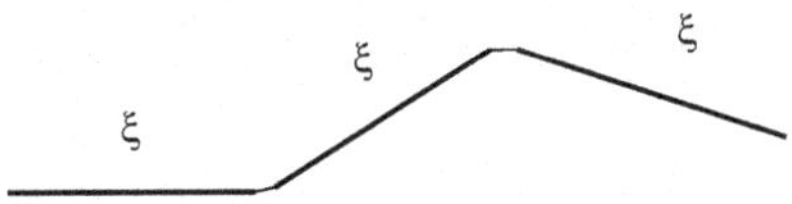

Figure 17.9 A DNA polymer with finite correlation length is equivalent to a freely jointed rigid chain.

For $f \simeq 0$, the correlator yields the partition function

$$Z(f) = 1 + \frac{f^2}{2}\int_0^L dsds'G\left(s, s'\right) = 1 + \frac{f^2}{4}\int_0^L dsds'e^{-\frac{|s-s'|}{2\xi}}$$
$$\cong 1 + \frac{1}{2}f^2 L\xi \text{ for } L \gg \xi.$$

Hence, from the equation above

$$f \simeq 0 \ \Rightarrow \ \frac{E[\mathcal{P}]}{L} = \frac{1}{L}\frac{\partial \ln Z(f)}{\partial f} = f\xi, \quad \text{DNA extension per unit length.}$$

The other limit is that of a large force $f \gg 1$, in which case the action is approximated by

$$L \simeq -\frac{\xi}{2}\dot{\phi}^2 - \frac{\xi}{2}\dot{\theta}^2 + f\left(1 - \frac{1}{2}\theta^2 + \cdots\right) = -\frac{\xi}{2}\dot{\phi}^2 - \frac{\xi}{2}\left\{\dot{\theta}^2 + \frac{f}{\xi}\theta^2\right\},$$
$$\mathcal{P} = \int_0^L ds\, \cos\theta \simeq L - \frac{1}{2}\int_0^L ds\, \theta^2.$$

In this approximation, using the result of the simple harmonic oscillator given in Eq. 11.10 (with $m = \xi$ and $\omega^2 = f/\xi$),

$$E[\theta(s)\theta(s')] = \frac{1}{2\xi\sqrt{f}}e^{-(\sqrt{f}/\sqrt{\xi})|s-s'|}$$

yields the result

$$E[\mathcal{P}] = \frac{1}{Z}\int D\theta D\phi \int_0^L ds \cos(\theta(s))e^S \simeq L - \frac{1}{2}\int_0^L ds\, \theta^2(s)e^S$$
$$= L - \frac{1}{2}\int_0^L ds \frac{1}{2\xi\sqrt{f}}$$
$$\Rightarrow \quad \frac{E[\mathcal{P}]}{L} = 1 - \frac{1}{4\xi\sqrt{f}}.$$

For intermediate f, the following equation interpolates between small f and large f

$$f\xi \simeq \frac{E[\mathcal{P}]}{L} + \frac{1}{4\left(1 - E[\mathcal{P}]/L\right)^2} - \frac{1}{4}.$$

The graph of force versus extension for the DNA is shown in Figure 17.10.

The path integral for the DNA's conformation can answer more complicated questions such as what is the likelihood of the DNA looping and interacting. Intersections are sometimes crucial for obtaining the full information encoded in the base pairs. When the DNA loops, there are special proteins sitting on the DNA that lock the intersections and hence bringing otherwise distant base pairs into close proximity [Phillips *et al.* (2008)].

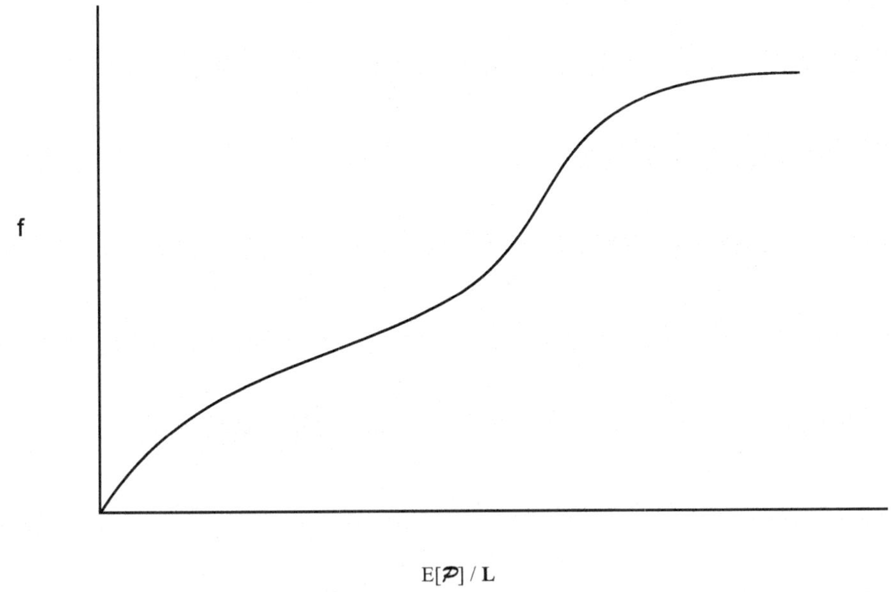

Figure 17.10 Extension of the DNA.

17.9 Summary

The two cases of compact degrees of freedom that were studied, namely S^1 and S^2, have qualitative features that occur widely in nonlinear theories.

The S^1 degree of freedom is periodic and hence is topologically nontrivial, having multiple classical solutions that are classified by the number of times the classical path winds around S^1. Furthermore, the momentum is discretized due to the periodicity of the degree of freedom.

The path integral was defined using the Hamiltonian. The S^1 theory has the simplifying feature that the semi-classical expansion yields the exact path integral. The Poisson summation formula allowed the semi-classical expansion to be exactly re-summed and leads to the Lagrangian description of the degree of freedom. In effect, the Poisson summation formula interchanges the sum over the winding number of the classical solutions with the discrete momentum of the periodic degree of freedom.

The conformational properties of the DNA molecule – in equilibrium at finite temperature – were modeled by an S^2 degree of freedom. It was seen that the mathematics for describing the statistical mechanics of a system by a path integral is identical to that used for describing a quantum system, with the difference lying only in the interpretation of the results.

The key new feature of the S^2 degree of freedom is that S^2 is a manifold with constant nonzero curvature. This leads to a nontrivial measure for the path integral, which, in turn, gives rise to apparent divergences in the path integral. The divergence arising from the nontrivial measure is a generic feature of all path integrals for degrees of freedom that take values in a curved manifold.

An expansion in powers of the inverse of the moment of inertia I showed that, to lowest order, all the divergences exactly cancel; it can be shown that this cancellation takes place to all orders. The path integral yields finite and well-defined results for all physical quantities. Various experimentally observable properties of the DNA were derived to illustrate the flexibility and utility of modeling the system using a path integral.

18
Conclusions

All the chapters of the book have examined and studied various aspects of path integrals and Hamiltonians, which in turn exemplified different aspects of quantum mathematics. The principles of quantum mathematics, stated in Chapter 2, can be summarized as follows:

- The fundamental degrees of freedom $\mathcal{F}$ form the bedrock of the quantum system.
- A linear vector state space $\mathcal{V}$ based on the degrees of freedom $\mathcal{F}$ provides an exhaustive description of the quantum system.
- Operators $\mathcal{O}$, which includes the Hamiltonian H, represent the physical properties of the degree of freedom and act on the state space $\mathcal{V}$. Observable quantities are the matrix elements of the physical operators.
- A spacetime description of quantum indeterminacy is encoded in the Lagrangian $\mathcal{L}$ and the Dirac–Feynman formula relates it to the Hamiltonian.
- The path integral provides a representation of all the physical properties of a quantum system. In particular, the path integral yields the correlation functions of the degrees of freedom as well as the probability amplitudes for quantum transitions.
- The interconnection of the path integral with the underlying Hamiltonian and state space is a specific feature of quantum mathematics that distinguishes path integration from functional integration in general.

A point not emphasized in this book, but noteworthy, is that for many problems the path integral approach is more powerful and useful than Heisenberg's operator formulation or Schrödinger's state vector approach.

To illustrate this fact, consider the case of the acceleration Lagrangian, discussed in Chapter 13; there are three branches of its parameters, namely the real and complex frequency branches as well as the equal frequency branch – and which are schematically represented in Figure 13.1. It was shown in Chapters 14 and 15 how the real and equal frequency branches are connected to their respective Hamiltonians and state spaces; but, as of now, no such explanation has been possible for the complex branch. In contrast, the path integral provides a rigorous and exhaustive description of all the branches of the acceleration Lagrangian, including the complex branch.

The advantage of the path integral over the other formulations becomes more apparent in the quantization of nonabelian gauge fields.

The fundamental principles of quantum mathematics extend in a straightforward manner to quantum field theory, which describes a quantum system having infinitely many degrees of freedom. Many new features of quantum mathematics arise in quantum field theory, in particular that of renormalization, but the underlying principles remain unchanged.

This book aims at presenting a pedagogical introduction to quantum mathematics in the context of the path integral and the Hamiltonian; all the chapters have been carefully chosen to address different aspects of its underlying principles using various models and examples. The emphasis is on the mathematical and theoretical analysis that are required for understanding these models – without biasing the analysis by tying the derivations to any specific quantum system. Furthermore, the focus has been on the methods that help to unravel the inner workings of the path integral; in particular, the methods have been exemplified by carrying out, explicitly and in great detail, the steps required for elucidating the mathematical properties of these models.

Path integrals are widely used in theoretical physics and, in particular, the Standard Model of particle physics is founded on the path integral. Given its mathematical depth and power, and in particular that it is based on the calculus of infinitely many variables, there is an expectation that path integrals can be applied to non-quantum systems as well. The question naturally arises as to why should this expectation be realized and, furthermore, if so then how should the applicability of path integrals, and of quantum mathematics in general, extend beyond quantum systems?

The 'why' has no real answer but one can only hope that *what is sauce for the goose is sauce for the gander*; in other words, since quantum mathematics has worked flawlessly for quantum systems, one can hope that it may work equally well for other non-quantum systems as well. This indeed has been the case for calculus; recall that calculus was invented by Newton for explaining and describing mechanics; but it was realized, over time, that calculus can be applied to a vast variety of systems, from the physical to the social sciences. As we will see below, the case of quantum mathematics is turning out to be similar to calculus.

There is no fixed formula or prescription of 'how' to make the connection of quantum mathematics to other non-quantum disciplines. A few examples were treated, such as in Chapter 7 on Stochastic systems and in Chapter 7 on DNA in thermodynamic equilibrium, where the path integral was applied to classical random systems. Quantum mathematics can be applied to non-quantum systems only by using analogical and metaphorical reasoning, since one needs to go beyond disciplinary boundaries and 'see' what is similar, what is alike, between two different domains of knowledge.

To illustrate how quantum mathematics can be employed to analyze non-quantum systems, the examples quoted in Section 2.12 – on the application of quantum mathematics to phase transitions, to finance and to pure mathematics – are examined below.

Systems undergoing *phase transitions* are said to be critical and have fluctuations of all scales, from the very large to the very small. A fluctuation is defined to be a specific configuration of the underlying physical constituents of a system. For example, when iron is cooled, at a critical temperature T_c it undergoes a continuous (second order) phase transition from its non-magnetic to its magnetic phase. For analyzing its phase transition, one can think of iron as being composed of an Ising spin variable, pointing up or down, and located at every atom. As the temperature approaches T_c, domains form in iron for which all the spins are aligned in the same direction; each domain can be considered to be one fluctuation. At the critical temperature T_c, domains of *all* scales (sizes) become equally important, from the very small to the very large, and the system is said to be scale invariant.

Wilson's study of phase transitions led to his visionary insight that quantum field theory describes a system containing infinitely many length scales, and which greatly clarified the otherwise enigmatic procedure of renormalization. Wilson then went on to describe critical systems using the mathematics of quantum field theory: the quantum degrees of freedom are analogically related to the fluctuations of a critical system. Predictions based on this analysis have been confirmed by numerous experiments.

The reverse analogy also holds, namely that all renormalizable quantum field theories correspond to classical systems that undergo a continuous phase transition.

The procedure of renormalization, based on separately studying each scale of the problem, and its connection to phase transitions are briefly discussed in Section 9.6.

In *finance*, the price of an option is analogous to the state function that, unlike the case of quantum mechanics, can be directly observed. The time evolution of the option price is driven by a differential operator, which is analogous to the Hamiltonian operator. Interest rates consist of a curve in future time that evolves randomly; interest rates are mathematically described by a two dimensional nonlinear quantum field theory that is quite different from what appears in physics.

The modeling of different financial instruments using quantum mathematics has been tested against market data and found to be consistently accurate.

The results in *mathematics* found by Donaldson and Jones can both be obtained in the mathematical framework quantum field theory. The properties of four dimensional manifolds found by Donaldson have been shown by Witten to be equivalent to the correlation functions of a topological supersymmetric Yang–Mills quantum field theory in four dimensions. The Jones polynomial, which classifies knots and links, has been shown by Witten to be equal to the expectation value of Wilson loops evaluated for a Chern–Simon quantum field theory in three dimensions.

The connection of quantum field theory with diverse branches of mathematics has led to many new and unforeseen results in geometry, topology and other subjects in pure mathematics.

In all the three examples quoted above, the path integral formulation of quantum field theory was employed, and that too for Euclidean time. Furthermore, there is no $\hbar$ and there is no quantum phenomenon; instead, the key ideas of these non-quantum disciplines have been interpreted in the symbols and structures of quantum mathematics. Critical systems are described by equilibrium statistical mechanics for which there is no element of time and hence there is no probability and no state vector; results in mathematics make no reference to any physical process; in finance, quantum mathematics is applied to a superstructure that has been created by society.

In many cases apparently different and distinct disciplines are connected via a labyrinth of methodological tunnels and hidden links – and form a body of knowledge that is transdisciplinary, interdisciplinary, multidisciplinary, and crossdisciplinary. One needs to have a point of view that reflects this integrated landscape of knowledge, being equally familiar with quantum systems as well as other disciplines. One can then try and find the right analogy, the correct mapping – if indeed such a mapping exists at all – to apply the theoretical ideas and quantitative structures of quantum mathematics to these disciplines. In turn, as was the case for phase transitions, the applications may throw new light on the mathematical structures of quantum mathematics that could be of relevance to physics.

In conclusion, path integrals and Hamiltonians provide a collection of flexible and powerful mathematical tools for addressing a broad spectrum of problems. The formalism of quantum mathematics, by encompassing wide-ranging subjects such as physics, mathematics, engineering, finance to the social sciences, provides a unified and integrated view of myriad domains of knowledge; and in doing so it also opens up new vistas for a quantitative and mathematical description of existing, as well as of potentially new, disciplines and phenomena.

References

Baaquie, Belal E. (2004). *Quantum Finance*. Cambridge, UK: Cambridge University Press.

Baaquie, B. E. (2009). Interest rates in quantum finance: The Wilson expansion and Hamiltonian. *Phys. Rev. E*, **80**(Oct), 046119.

Baaquie, Belal E. (2010). *Interest Rates and Coupon Bonds in Quantum Finance*. Cambridge, UK: Cambridge University Press.

Baaquie, B. E. (2013a). Financial modeling and quantum mathematics. *Computers and Mathematics with Applications*, **65**, 1665–1673.

Baaquie, B. E. (2013b). Statistical microeconomics. *Physica A*, **19**(1), 4400–4416.

Baaquie, Belal E. (2013c). Action with acceleration I: Euclidean Hamiltonian and path integral. *Int. J. Mod. Phys. A*, **28**, 1350137.

Baaquie, Belal E. (2013d). Action with acceleration II: Euclidean Hamiltonian and Jordan blocks. *Int. J. Mod. Phys. A*, **28**, 1350138.

Baaquie, Belal E. (2013e). *The Theoretical Foundations of Quantum Mechanics*. New York, USA: Springer.

Baaquie, B. E. and Martin, F. (2005). Quantum psyche: Quantum field theory of the human psyche. *NeuroQuantology*, **3**(5), 7–42.

Baaquie, B. E. and Yang, Cao. (2014). Option volatility and the acceleration Lagrangian. *Physica A*, **393**, 337–363.

Baaquie, Belal E., Yang, Cao, Lau, Ada, and Tang, Pan (2012). Path integral for equities: Dynamic correlation and empirical analysis. *Physica A*, **391**(4), 1408–1427.

Ballentine, L. E. (1998). *Quantum Mechanics: A Modern Development*. Singapore: World Scientific.

Bell, J. (2004). *Speakable and Unspeakable in Quantum Mechanics*. 2nd edn. Cambridge, UK: Cambridge University Press.

Bender, Carl M. and Mannheim, Philip D. (2008a). Exactly solvable $\mathcal{PT}$-symmetric Hamiltonian having no Hermitian counterpart. *Phys. Rev. D*, **78**(Jul), 025022.

Bender, Carl M. and Mannheim, Philip D. (2008b). No-ghost theorem for the fourth-order derivative Pais–Uhlenbeck oscillator model. *Phys. Rev. Lett.*, **100**(Mar), 110402.

Bender, C. M. and Mannheim, P. D. (2010). PT symmetry and necessary and sufficient conditions for the reality of energy eigenvalues. *Phy. Lett. A*, 1616.

Bender, Carl M. and Mannheim, Philip D. (2011). PT-symmetry in relativistic quantum mechanics. *Phys. Rev. D*, **84**(Nov), 105038.

Creutz, M. and Freedman, B. (1981). A statistical approach to quantum mechanics. *Annals of Physics*, **132**, 427–462.

Das, A. (2006). *Field Theory: A Path Integral Approach.* Singapore: World Scientific.

Dirac, P.A.M. (1964). *Lectures on Quantum Mechanics.* New York: Belfer Graduate School of Science Monographs Series 2.

Dirac, P. A. M. (1999). *The Principles of Quantum Mechanics.* 4th edn. Oxford, UK: Oxford University Press.

Feynman, R.P. (1948). Space-time approach to non-relativistic quantum mechanics. *Rev. Mod. Phys.*, **20**(Apr), 367–387.

Feynman, R.P. and Hibbs, A.R. (1965). *Quantum Mechanics and Path Integrals.* New York, USA: McGraw-Hill.

Fontanini, Michele and Trodden, Mark. (2011). Tackling higher derivative ghosts with the Euclidean path integral. *Phys. Rev. D*, **83**(May), 103518.

Gottfried, K. and Yan, T-.M. (2003). *Quantum Mechanics.* 2nd edn. Heidelberg, Germany: Springer.

Haven, E. and Khrennikov, A. (2013). *Quantum Social Science.* Cambridge, UK: Cambridge University Press.

Hawking, S. W. and Hertog, Thomas (2002). Living with ghosts. *Phys. Rev. D*, **65**(May), 103515.

Jones, H.F. and Rivers, R.J. (2009). Which Green functions does the path integral for quasi-Hermitian Hamiltonians represent? *Physics Letters A*, **373**.

Kandirmaz, N. and Sever, R. (2011). Path integral solution of PT-/non-PT-symmetric and non-Hermitian Hulthen potential. *Acta Polytechnica*, **0**(0), 38–42.

Kardar, M. (2007). *Statistical Theory of Fields.* Cambridge, UK: Cambridge University Press.

Kleinert, H. (1986). Path integral for second derivative Lagrangian. *J. Math. Phy.*, **27**(Dec), 3003–3013.

Kleinert, H. (1990). *Path Integrals in Quantum Mechanics, Statistics, and Polymer Physics, and Financial Markets.* Singapore: World Scientific.

Landau, L. D. and Lifshitz, L. M. (2003). *Quantum Mechanics: Non-Relativistic Theory.* Third edn. Vol. 3. Amsterdam, Holland: Elsevier Science.

Mannheim, P.D. (2011a). Comprehensive solution to the cosmological constant, zero-point energy, and quantum gravity problems. *Gen. Rel. Gravitation*, **43**, 703–750.

Mannheim, P.D. (2011b). Making the case for conformal gravity. *Foundations of Physics*, **37**, 532.

Mannheim, Philip D. and Davidson, Aharon (2000). Fourth order theories without ghosts. *arXiv preprint hep-th/0001115.*

Mannheim, P. D. and Davidson, A. (2005). Dirac quantization of the Pais–Uhlenbeck fourth order oscillator. *Phys. Rev. A*, **71**(Apr), 042110.

Mostafazadeh, A. (2002). Pseudo-Hermiticity versus PT-symmetry III: Equivalence of pseudo-Hermiticity and the presence of antilinear symmetries. *Journal of Mathematical Physics*, **43**(8), 3944–3951.

Papon, P., Leblond, J., and Meije, P. H.E. (2002). *The Physics of Phase Transitions.* New York, USA: Springers.

Phillips, R., Kondev, J., Theriot, J., and Garcia, H. (2008). *Physical Biology of the Cell.* New York, USA: Garland Science.

Polyakov, A. M. (1987). *Gauge Fields and Strings.* New York, USA: Harwood Academic Publishers.

Risken, M. (1988). *The Fokker–Planck Equation.* Heidelberg, Germany: Springer.

Rivers, R.J. (2011). Path integrals for Pseudo-Hermitian Hamiltonians. *International Journal of Theoretical Physics*, **50**, 1081–1096.

Scholtz, F. G., Geyer, H. B., and Hahne, F. J. W. (1992). Quasi-Hermitian operators in quantum mechanics and the variational principle. *Annals of Physics*, **213**(*1*), 74101.

Simon, J.Z. (1990). Higher-derivative Lagrangians, nonlocality, problems, and solutions. *Phys. Rev. D*, **41**(Jun), 3720–3733.

Stapp, H. P. (1963). The Copenhagen interpretation. *Am. J. Phys.*, **40**(*8*), 1098.

Vanderbilt, D. (1990). Soft self-consistent pseudopotentials in a generalized eigenvalue formalism. *Phys. Rev. B*, **41**(Apr), 7892–7895.

Wang, Q., Chia, S., and Zhang, J. (2010). PT symmetry as a generalization of Hermiticity". *Journal of Physics, A*, **43**, 295301.

Weinberg, S. (2013). *Lectures on Quantum Mechanics*. Cambridge, UK: Cambridge University Press.

Wilson, K. G. (1969). Non-Lagrangian models in current algebra. *Physical Review*, **179**, 1499–1512.

Wilson, Kenneth G. (1983). The renormalization group and critical phenomena. *Rev. Mod. Phys.*, **55**(Jul), 583–600.

Wilson, K. G. and Zimmermann, W. (1972). Operator product expansions and composite field operators in the general framework of quantum field theory. *Communications in Mathematical Physics*, **24**(2), 87–106.

Witten, E. (1989). Quantum field theory and Jones polynomial. *Communications in Mathematical Physics*, **121**(3), 351–399.

Wu, T.T. and Yu, M.L. (2002). Theory and application of Fermi pseudo-potential in one dimension. *Journal of Mathematical Physics*, **43**(12), 5949.

Yang, Cao (2012). *Higher Derivative Models and Libor Market Model in Quantum Finance*. Ph.D. thesis, National University of Singapore, Department of Physics, 2 Science Drive 3, Singapore 117551.

Zinn-Justin, J. (1993). *Quantum Field Theory and Critical Phenomenon*. Oxford, UK: Oxford University Press.

Zinn-Justin, J. (2005). *Path Integrals in Quantum Mechanics*. Oxford, UK: Oxford University Press.

Index